MURRAY GELL-MANN

Selected Papers

World Scientific Series in 20th Century Physics

Published

Vol. 22 A Quest for Symmetry — Selected Works of Bunji Sakita
edited by K. Kikkawa, M. Virasoro and S. R. Wadia

Vol. 23 Selected Papers of Kun Huang (with Commentary)
edited by B.-F. Zhu

Vol. 24 Subnuclear Physics — The First 50 Years: Highlights from Erice to ELN
by A. Zichichi
edited by O. Barnabei, P. Pupillo and F. Roversi Monaco

Vol. 25 The Creation of Quantum Chromodynamics and the Effective Energy
by V. N. Gribov, G. 't Hooft, G. Veneziano and V. F. Weisskopf
edited by L. N. Lipatov

Vol. 26 A Quantum Legacy — Seminal Papers of Julian Schwinger
edited by K. A. Milton

Vol. 27 Selected Papers of Richard Feynman (with Commentary)
edited by L. M. Brown

Vol. 28 The Legacy of Léon Van Hove
edited by A. Giovannini

Vol. 29 Selected Works of Emil Wolf (with Commentary)
edited by E. Wolf

Vol. 30 Selected Papers of J. Robert Schrieffer — In Celebration of His 70th Birthday
edited by N. E. Bonesteel and L. P. Gor'kov

Vol. 31 From the Preshower to the New Technologies for Supercolliders — In Honour of Antonino Zichichi
edited by B. H. Wiik, A. Wagner and H. Wenninger

Vol. 32 In Conclusion — A Collection of Summary Talks in High Energy Physics
edited by J. D. Bjorken

Vol. 33 Formation and Evolution of Black Holes in the Galaxy
— Selected Papers with Commentary
edited by H. A. Bethe, G. E. Brown and C.-H. Lee

Vol. 35 A Career in Theoretical Physics, 2nd Edition
by P. W. Anderson

Vol. 36 Selected Papers (1945–1980) with Commentary
by Chen Ning Yang

Vol. 37 Adventures in Theoretical Physics
— Selected Papers with Commentaries
by Stephen L. Adler

Vol. 38 Matter Particled — Patterns, Structure and Dynamics
— Selected Research Papers of Yuval Ne'eman
edited by R. Ruffini and Y. Verbin

Vol. 39 Searching for the Superworld — A Volume in Honour of Antonino Zichichi on the Occasion of the Sixth Centenary Celebrations of the University of Turin, Italy
edited by S. Ferrara and R. M. Mössbauer

Vol. 40 Murray Gell-Mann — Selected Papers
edited by H. Fritzsch

Vol. 41 Many-Body Theory of Molecules, Clusters, and Condensed Phases
edited by N. H. March and G. G. N. Angilella

For information on Vols. 1–21, please visit http://www.worldscibooks.com/series/wsscp_series.shtml

World Scientific Series in 20th Century Physics Vol. 40

MURRAY GELL-MANN

Selected Papers

editor

Harald Fritzsch

University of Munich, Germany

World Scientific

NEW JERSEY • LONDON • SINGAPORE • BEIJING • SHANGHAI • HONG KONG • TAIPEI • CHENNAI

Published by

World Scientific Publishing Co. Pte. Ltd.

5 Toh Tuck Link, Singapore 596224

USA office: 27 Warren Street, Suite 401-402, Hackensack, NJ 07601

UK office: 57 Shelton Street, Covent Garden, London WC2H 9HE

Library of Congress Cataloging-in-Publication Data

Gell-Mann, Murray.
[selections. 2009]
Murray Gell-Mann : selected papers / edited by Harald Fritzsch.
p. cm. -- (World Scientific series in 20th century physics ; v. 40)
Includes bibliographical references.
ISBN-13: 978-981-283-684-7 (hardcover : alk. paper)
ISBN-10: 981-283-684-5 (hardcover : alk. paper)
ISBN-13: 978-981-4261-62-3 (pbk. : alk. paper)
ISBN-10: 981-4261-62-9 (pbk. : alk. paper)
1. Particles (Nuclear physics) 2. Mathematical physics. I. Fritzsch, Harald, 1943– II. Title.
QC793.2.G452 2009
530.092--dc22

2009005667

British Library Cataloguing-in-Publication Data

A catalogue record for this book is available from the British Library.

The editor and publisher would like to thank the following publishers of the various journals and books for their assistance and permissions to include the selected reprints found in this volume:

American Institute of Physics
American Physical Society
Cambridge University Press
EDP Sciences
Elsevier
John Wiley and Sons, Inc.
National Academy of Sciences (USA)
North-Holland Publishing Co.
Oxford University Press
Springer
University of California Press
W. A. Benjamin

While every effort has been made to contact the publishers of reprinted papers prior to publication, we have not been successful in some cases. Where we could not contact the publishers, we have acknowledged the source of the material. Proper credit will be accorded to these publications in future editions of this work after permission is granted.

H. Fritzsch (left) and M. Gell-Mann

Murray Gell-Mann

The American physicist Murray Gell-Mann has contributed more than anybody else to the theoretical understanding of the physics of elementary particles.

Gell-Mann, the son of Austrian immigrants, was born in New York City on September 15, 1929. His father, Arthur Gell-Mann, came from the city of Czernowitz in the Bukowina district, which is today part of the Ukraine. He studied at the University of Vienna and planned to become a high school teacher. He cut short his studies and moved to the United States so as to help his parents, who had emigrated to New York and were not doing well. He created a language school in Manhattan, which was successful for a while. When it failed, because of the new immigration laws and the Great Depression, he got a job in a bank.

Arthur Gell-Mann was very interested in science, especially in astronomy, physics and mathematics. Gell-Mann's mother, Pauline, was a housewife who had ambitions for her son Murray. She encouraged him to play the piano. She also tried to get him a scholarship to a private school. Finally it was his music teacher who succeeded in doing that. Murray Gell-Mann's brother Benedict was nine years older than Murray and taught him a great deal about plants, animals (especially birds and butterflies), and the preservation of natural ecological systems. The two brothers spent a large part of their time outdoors, studying Nature, and in museums, such as the American Museum of Natural History and the Metropolitan Museum of Art.

At the age of 14, Murray Gell-Mann graduated from high school and obtained a scholarship at Yale University, where his major subject was physics. He finished his studies in June 1948 as Bachelor of Science, when he was 18. He went to the Massachusetts Institute of Technology for his graduate studies. His Ph.D.-advisor there was Victor Weisskopf, and he obtained his doctorate in January 1951, whereupon he became a member of the Institute for Advanced Study in Princeton. The director of the Institute, Prof. Robert Oppenheimer, encouraged Gell-Mann in his work on problems in elementary particle physics. In 1952, Gell-Mann joined the Physics Department of the University of Chicago. There he worked together with Marvin Goldberger and had close contact with Enrico Fermi.

In 1955, Gell-Mann married J. Margaret Dow, who came from England and was working at the Institute in Princeton as the assistant to an archaeology professor. They had two children, a daughter Lisa, born in 1956, and a son Nicholas, born in 1963.

In the same year, Gell-Mann left Chicago and accepted a professorship at the California Institute of Technology in Pasadena, California. In 1967, Gell-Mann obtained the prestigious Robert Andrews Millikan professorship at Caltech. In 1993, he retired and went to the Santa Fe Institute in Santa Fe, New Mexico, which he had helped to found and which he regarded as one of his most satisfying achievements.

Gell-Mann obtained the Nobel Prize in Physics in 1969, mainly for his work on the classification of the hadrons, based on the approximate symmetry SU(3), known as the SU(3) of "flavor."

Some of Gell-Mann's early papers were on the newly discovered V-particles, found in cosmic ray experiments. They were copiously produced in nucleon collisions, but decayed slowly. Gell-Mann found a quantum number, strangeness, which was conserved by the strong and electromagnetic interactions. The new hadrons had nonzero strangeness, unlike the familiar ones, and so could not decay except through the weak interaction.

In 1953, Gell-Mann worked with Francis Low on problems of quantum field theory. They introduced the concept of the renormalization group and studied the behavior of the coupling constants as functions of momentum transfer. The methods of Gell-Mann and Low were very useful later in the theory of quantum chromodynamics. Also, Gell-Mann's student Kenneth Wilson applied those methods to condensed matter physics, where the results were so important that he was awarded the Nobel Prize.

In 1956, the violation of parity symmetry in the weak interactions was discovered. Richard Feynman and Murray Gell-Mann proposed in 1957 that the weak interactions are described by left-handed currents with a V–A structure (V: vector current, A: axial vector current). The same idea was published by Robert Marshak and George Sudarshan. The V–A theory of the weak interactions was used later in the gauge theory of the weak interaction, developed by Sheldon Glashow, Abdus Salam, and Steven Weinberg.

In 1961, Gell-Mann succeeded in describing the new hadrons, found in cosmic ray and accelerator experiments. He used an appproximate symmetry scheme based on the group SU(3). This group is an extension of the isospin group SU(2) and includes what amounts to the strangeness quantum number. Yuval Ne'eman in Israel also proposed this symmetry group.

The baryons were described in the SU(3) scheme as an octet, consisting of the two nucleons, the three sigma baryons, the λ baryon and the two ξ baryons. Likewise the mesons formed an octet, consisting of the three π mesons, the four K mesons and the η meson. The excited baryons formed a decimet, consisting of the four δ resonances, the three σ^* resonances, the two ξ^* resonances and the ω resonance.

The group SU(3) of "flavor" gives a symmetry which, in contrast to the isospin symmetry, is strongly violated. Gell-Mann and the Japanese theorist Okubo succeeded in describing the symmetry breaking in a simple way that led to formulae relating the masses of particles within each octet or decimet.

The negatively charged omega particle had not been seen in experiments when Gell-Mann predicted the existence of the particle and its mass and the search for it started. In 1964, it was found at the Brookhaven National Laboratory (BNL) in the United States. The mass of 1672 MeV corresponded accurately to the value predicted by Gell-Mann. The discovery of the ω particle was a triumph for the SU(3) symmetry and was a factor in the award of the Nobel Prize to Gell-Mann in 1969.

Gell-Mann described the electromagnetic and weak interactions of the hadrons by the electromagnetic and weak currents, which could be classified in terms of the SU(3) symmetry. He proposed that the commutators of these currents at equal times had simple properties, given by a current algebra. The predictions that could be derived from the current algebra agreed with experimental results.

The SU(3) symmetry had a strange feature. The observed particles could be classified into octets or decimets, but the simplest representation, the triplet, was not observed. Gell-Mann and also his student George Zweig, who worked in 1964 at CERN in Geneva, noticed that the baryon octets and decimet could be constructed by multiplying three triplet representations. Likewise the meson octet could be constructed by multiplying a triplet by an antitriplet. The electric charges of these triplet objects, derivable from the electric charges of the observed particles, came out to be $2/3$ and $-1/3$. Thus the triplet objects could not be known, directly observable particles, which have integral charges. In this sense the triplet particles were not "real." They could be described as confined inside known particles, with three of them, roughly speaking, constituting a neutron or proton.

Gell-Mann published a short letter in *Physics Letters* with the title "*A Schematic Model of Baryons and Mesons.*" He called the triplets quarks. The three objects were named u (for up), d (for down) and s (for strange). They were assumed to be fermions with spin $1/2$. The proton had the structure (uud), the neutron (udd). The strange particles contained one, two or three s-quarks corresponding to strangeness -1, -2, and -3 respectively. The λ baryon had the structure (uds). The ω particle consisted of three s-quarks: (sss). In his paper in *Physics Letters* Gell-Mann also mentioned a quark–gluon field theory, analogous to quantum electrodynamics, as a model for a possible theory of the strong interactions.

In 1968, the scaling behavior of the cross section for electron–proton scattering was discovered. Richard Feynman speculated that inside the proton were pointlike objects, which he called partons. Gell-Mann and his German collaborator Harald Fritzsch instead worked out a generalization of the current algebra to lightlike distances, the light-cone current algebra. They assumed that the commutators of the currents could be abstracted approximately from the free quark model at large momentum transfers. In that case they could derive the observed scaling behavior. The partons were, of course, quarks, antiquarks, and gluons behaving more or less like free particles at large momentum transfers.

The hypothesis of the quarks had a problem, which can be seen, for example, by considering the omega particle with the structure (sss). The spin wave function of the state is completely symmetric in the three s-quarks. Since these s-quarks are in the ground state, the space wave function must also be symmetric. This violates the Pauli principle, which states that under the exchange of two fermions the wave function must be antisymmetric.

In 1971, Gell-Mann proposed, together with William Bardeen and Harald Fritzsch, that the quarks have a new, exactly conserved quantum number, which they called color. Each quark has an index, running from 1 to 3. The transformations among these indices lead to a new symmetry group, the color group SU(3). The observed hadrons have to be singlets under this color symmetry. This implies that the wave function of a baryon is totally antisymmetric with respect to the color indices of the quarks, and the Pauli principle is not violated. Bardeen, Fritzsch, and Gell-Mann showed that the electromagnetic decay rate of the neutral π meson is correctly described by the color scheme, but that if one were to leave out the color index in the quark picture, the theoretical decay rate would disagree with the experimental result by a factor of 9. The quarks are color triplets. The assumption that the observable hadrons are color singlets requires that the quarks be permanently confined inside the hadrons.

In 1972, Fritzsch and Gell-Mann (as well as Gross and Wilczek) constructed a new field theory of the quarks. They postulated that the color group SU(3) is a gauge group, which yields a Yang–Mills theory of the quarks, interacting through massless vector gluons. Fritzsch, Gell-Mann, and Heinrich Leutwyler discussed in 1973 several interesting properties of such a theory in a paper in *Physics Letters* with the title: *Advantages of the Color Octet Gluon Picture.* Fritzsch and Gell-Mann later named this theory "Quantum Chromodynamics" (QCD).

The QCD theory has the property that it is asymptotically free, as emphasized by Gross and Wilczek and by Politzer. This means that the effective coupling constant describing the interaction of the quarks and gluons gets smaller at small distances. Thus the quarks behave as nearly pointlike particles in the lepton–hadron scattering at very high momentum transfers, and the observed scaling behavior follows from that. At small momentum transfers or large distances the interaction becomes strong, and the confinement of the quarks and gluons is a consequence.

Gell-Mann strongly supported the emerging string theory in particle physics. He thought that superstring theory might lead to a theory of all particles and forces.

Together with James Hartle, Gell-Mann has been working for some years on the interpretation of quantum mechanics, especially in the light of quantum cosmology. They have nearly completed their project, which emphasizes decoherent alternative coarse-grained histories of the universe, and the result is a view of quantum mechanics in which there is little "weirdness" and a greatly reduced role for a human observer. The universe "measures" itself.

Gell-Mann is interested in many areas outside physics, e.g. in problems of historical linguistics and in policy studies of sustainability. He is also an active bird watcher. For many years he worked as a director of the J. D. and C. T. MacArthur Foundation, where he supported initiatives to improve the preservation of biological diversity, to foster peace and international cooperation, and to promote scientific work relevant to mental health and human development.

The physics research of Murray Gell-Mann contributed greatly to the successful Standard Model of particle physics. The development of the Standard Model was the main result of fundamental physics in the second part of the 20th century.

CONTENTS

Commentary Notes

1. The Garden of Live Flowers (p. 25)

This article is a contribution to a book of essays celebrating the 90th birthday of Victor Weisskopf, who was Gell-Mann's thesis supervisor at MIT. Here Gell-Mann describes how Weisskopf urged his students to resist the temptation to tackle the biggest issues right away, working at first on more modest problems that might yield useful results rather quickly, and to temper a love of formalism with a "pedestrian" approach based on physical insight and closely connected with experiment. Gell-Mann goes on to recount how that advice led him to follow certain research paths that did not aim directly at the big questions but nevertheless turned out to lead there. He compares this situation to the "garden of live flowers" in *Through the Looking-glass* (the companion volume to *Alice in Wonderland*), where Alice had to head in what seemed to be the wrong direction in order to get where she wanted to go.

Gell-Mann discusses some of his research on symmetry and conservation laws, starting with his work on strangeness in 1952 and 1953. His first two papers on that subject are reproduced here, one published and the other unpublished but widely circulated and influential (see paper #2). He goes on to discuss his contributions to the theory of the weak interaction, leading to the coupling of a charged current with its hermitian conjugate, an interaction that could be mediated by a spin-one intermediate boson, with the current having the $V - A$ (vector minus axial vector) form and some special symmetry properties in the case of the hadrons. Then he recounts some of the research on the "eightfold way" approximate symmetry of the hadrons based on the Lie group SU(3), followed by the quark picture, which explained many of these features.

He remarks on how knowledge of quarks and gluons and of intermediate bosons brought the semi-empirical research closer to the desired dynamical theory that became the standard model. But he emphasizes even more the magic of gauge theory, which allows the theorist to go immediately from discovering symmetry to constructing the dynamical theory based on that symmetry. Thus understanding the perfectly conserved but hidden color variable for hadrons led directly to quantum chromodynamics, the quantum field theory of quarks and gluons.

Gell-Mann describes another line of research, besides the work on symmetries, that led to important advances at the fundamental level. That was extracting from field theory results that held to all orders of perturbation theory and thus could be considered to be exact. One example of this is the so-called "renormalization group" discovered by Francis Low and Gell-Mann in 1953. When applied to quantum chromodynamics many years later it led theorists to understand the "asymptotic freedom" of quarks, for example those in the proton, at high momentum transfer.

Another example of extracting results from field theory was the work (much of it with Marvin Goldberger) on the crossing relation in scattering amplitudes, on dispersion relations (or, more generally, the analytic properties of the amplitudes), and on generalized unitarity relations. These three features of scattering amplitudes on the mass shell permitted Gell-Mann to propose the reformulation of field theory using only mass-shell amplitudes. That is what Geoffrey Chew later called "S-matrix theory," claiming it was somehow different from field theory. Superstring theory, which may help in the search for a unified theory of all the particles and forces, has been formulated only with the use of the mass-shell method. It is not known whether it can be reformulated as a field theory. Even though the conventional and the mass-shell approaches to Lagrangian field theory are presumably equivalent, it is conceivable that superstring theory might be different.

In all of these cases Gell-Mann shows that studying symmetries and abstracting exact results from field theory could lead to explicit dynamics, and he compares that with Alice's experience in the garden of live flowers. He credits Weisskopf's advice with helping to point the way toward that kind of research.

2. **Strangeness** (p. 38)

In this essay Gell-Mann writes about an early episode in his scientific career. After doing graduate work with Victor Weisskopf at MIT, he became a postdoctoral fellow at the Institute for Advanced Study in Princeton, where he collaborated with Francis Low. In 1952, Gell-Mann moved to the University of Chicago and joined the group led by Enrico Fermi. He collaborated with Marvin Goldberger on pion–nucleon field theory and on abstracting from such theories general principles like crossing symmetry and analyticity (dispersion relations).

In order to understand the peculiar properties of the new hadrons found in the cosmic rays, Gell-Mann introduced a new quantum number, which he called strangeness. The nucleons were assigned strangeness 0. The newly discovered Λ hyperon was assigned the strangeness -1, likewise the three Σ hyperons. The Ξ-hyperons had strangeness -2. The negatively charged K-meson had strangeness -1, while the K-meson with positive charge had strangeness $+1$. The strangeness quantum number was shown to be conserved by the strong and electromagnetic interactions, but not by the weak interaction. Thus hadronic decays with strangeness change would have to take place via the weak interaction.

The strange particles are isotopic spin multiplets that are displaced in electric charge from the familiar nucleon and pion multiplets. Strangeness is twice the displacement of the center of charge; its conservation by the strong and electromagnetic interactions follows from the conservation by those interactions of baryon number and the z-component of isotopic spin.

The idea of strangeness explained in a simple way why the new particles were produced copiously in collisions via the strong interactions, but decayed very slowly due to the weak interactions. In a collision, a new particle with strangeness -1 could be produced by the strong interaction together with a particle with strangeness $+1$, with the total strangeness change being zero. For example, a negatively charged Σ hyperon could be produced together with a positively charged K-meson. However, a positively charged Σ hyperon could not be produced together with a negatively charged K-meson, since both particles have strangeness -1. Likewise, two neutrons could not turn into two Λ hyperons.

Today we describe the strange particles using the quark model. The strangeness of a particle is minus the number of strange quarks in the particle. Thus the Λ hyperon has the quark structure (uds) and it has strangeness -1. It might have been better to reverse the sign of strangeness, in which case the strangeness would count the number of the strange quarks in a hadron, not minus the number of strange quarks.

In the paper "Isotopic Spin and New Unstable Particles," written in 1952, and the preprint "On the Classification of Particles," written in 1953, the idea of strangeness was introduced and discussed, but the expression "strangeness" was not yet used because the *Physical Review* objected to the use of words like "curious" or "strange."

3. **Quantum Electrodynamics at Small Distances** (with *F. E. Low*) (p. 52)

In 1954, Murray Gell-Mann and Francis Low worked on some important consequences of the renormalization program of quantum electrodynamics (QED). Somewhat similar research was carried out by Petermann and Stueckelberg. The method employed is usually called the "renormalization group" method. Years later, it was very successfully applied by Kenneth Wilson (who had been a student of Gell-Mann) to problems in condensed matter physics, especially ones involving phase transitions. He was awarded the Nobel Prize in 1982 for that work.

Gell-Mann and Low introduced a cutoff to make the renormalization constants finite in perturbation theory. They noticed that the amplitudes in QED were then finite in perturbation theory as the mass of the electron was decreased to zero. That enabled them to relate the renormalization program to the behavior of certain features of the theory at high energies or large momentum transfers.

In particular, Gell-Mann and Low exhibited some important constraints on the functional form of the momentum transfer (or energy) dependence of the electrical interaction of a test charge, the so-called vacuum polarization. (The lowest order correction had been calculated in the 1930's.) In quantum electrodynamics the effective coupling constant increases with the energy. With the LEP accelerator at CERN one has found that the fine structure constant at 200 GeV is about $1/127$, while at very low energies it is close to $1/137$. The observed increase agreed perfectly with the old theoretical prediction of QED.

The method of Gell-Mann and Low played a crucial role around 1972 and 1973 in the development of the theory of quantum chromodynamics (QCD). It was found that the renormalized coupling in QCD tends to zero at high energies (or small distances). This feature is called asymptotic freedom. It explains why the quarks in deep inelastic scattering behave almost like pointlike objects.

Conversely, the coupling constant increases at large distances, and perturbation theory cannot be used in that case. But the increase suggests that the quarks and gluons, the basic quanta of QCD, cannot be observed as free particles — they are permanently confined inside the hadrons. A rigorous proof of the confinement property is still missing.

4. Behavior of Neutral Particles under Charge Conjugation
(with *A. Pais*) (p. 65)

In this article Gell-Mann and Pais discuss neutral mesons that are different from their antiparticles, unlike the photon or the neutral pion. A conservation law, in this case the conservation of strangeness, forbids transitions between particle and antiparticle, but the conservation law is not valid for the weak interaction. These mesons are described by a complex field. Under charge conjugation the field is transformed into its Hermitian conjugate. They introduce the sum and the difference of the field and its conjugate. The corresponding states must show a difference in their weak decays. As a result, the two states have different lifetimes and their masses are slightly different.

The authors showed that if charge conjugation were exactly conserved, then the nonleptonic decay would yield two pions for one state and three pions for the other. The same would be true if parity times charge conjugation were conserved, but we now know that even that is not exactly correct.

The predictions of Gell-Mann and Pais were mostly right. The two different mesons are the long-lived neutral kaon K_L and the short-lived kaon K_S. Their lifetimes differ by a factor of about 556. The nonleptonic decay of K_L gives mostly three pions. The meson K_S decays mainly into two pions.

5. Sixth Annual Rochester Conference (Rochester, 1956)
Field Theory on the Mass Shell (p. 68)
The Nature of the Weak Interaction (p. 72)

An important new consequence of Gell-Mann's work on dispersion relations is presented in this contribution at the Sixth Rochester Conference in 1956. Some of that work had been done in collaboration with Marvin Goldberger. They had used very general principles like microscopic causality to express the real part of the photon–nucleon scattering amplitude in the forward direction in terms of the imaginary part, which is proportional to the total cross section for photons on nucleons producing anything. Now Gell-Mann used much more general dispersion relations relating the real part of an amplitude to an integral over its imaginary part. That imaginary part could be related to bilinear forms in the same and other amplitudes using a generalization of unitarity. Finally, use could be made of the famous "crossing relation." In this way the entire S-Matrix (all scattering amplitudes on the mass shell) for a given field theory could be obtained from the basic interactions involved or from some boundary conditions. This was the program that was later discussed and called "S-Matrix theory" by Geoffrey Chew. Both Gell-Mann and Chew referred to Heisenberg's hope that the S-Matrix could somehow be obtained from fundamental principles. In his talk Gell-Mann showed as an illustration how the program would work in perturbation theory in a field theory such as quantum electrodynamics.

An important feature of field theory on the mass shell (or "S-Matrix theory") is that one deals with the real charges and masses and it is not necessary to worry about the unrenormalized quantities and sweep infinities under the rug. What was a logarithmic infinity in charge renormalization, for example, appears only as logarithmic behavior at high momentum transfers of a term in the perturbation expansion of a physical quantity, the vacuum polarization.

The second contribution of Gell-Mann to that conference concerns the weak interaction. He starts with the Puppi triangle, describing the universality of the weak interactions for electron, muon, and nucleons. He adds the strangeness-changing weak interaction that yields the weak decays of the strange particles. Consequences for the lifetimes of the hyperons and of the K-mesons are discussed.

6. **Theory of the Fermi Interaction** (with *R. P. Feynman*) (p. 75)

In this paper Feynman and Gell-Mann describe the weak interaction correctly as a universal Fermi interaction with a $V-A$ (vector minus axial vector) current. (A similar proposal was made by Sudarshan and Marshak.) The lepton portion of the current couples each charged lepton to a neutrino, which can be described using a two-component spinor field since only the left-handed neutrino and right-handed antineutrino interact. The electrons emitted in β-decay are then expected to be polarized along their direction of motion; the amount of polarization is the velocity divided by the velocity of light. They apply their formalism to muon decay and calculate the muon lifetime. It agrees very well with the observed lifetime. The Fermi constant for muon decay is equal to the Fermi constant for β-decay through the vector current. (The strangeness-preserving vector current is taken to be a component of the isotopic spin current and therefore conserved by the strong interaction. The small correction required by the existence of the strangeness-changing vector current is not treated here even though that current is discussed.) Feynman and Gell-Mann find an interaction which is universal, which involves two-component neutrinos, and which preserves invariance under CP and T although both C and P conservation are both maximally violated. That interaction is capable of being transmitted by a charged intermediate vector boson. A non-leptonic parity-violating weak coupling of hadrons is predicted.

The $V-A$ theory was in disagreement with more than a half dozen experimental results on β-decay, but it was so beautiful that the authors proposed it anyway, suggesting that all those results were wrong.

7. The Eightfold Way: A Theory of Strong Interaction Symmetry (p. 81)

In this preprint Gell-Mann discusses the approximate SU(3) symmetry that he calls the "eightfold way." Similar work at about the same time was done by Y. Ne'eman. The SU(3) symmetry extends the isotopic spin symmetry, described by the group SU(2). While the isotopic spin symmetry is almost exact, broken only by the electromagnetic and weak interactions, the SU(3) symmetry is substantially violated. Electric charge and baryon number remain conserved.

The ground state baryons are described by the octet representation of SU(3), while the light vector and pseudoscalar mesons belong to the octet and singlet representations. In this preprint, Gell-Mann introduces the λ matrices, which were afterwards called the Gell-Mann matrices. He calculates the structure constants of the group SU(3). The physical meanings of the eight charges of the SU(3) group are described. The charges are obtained by integrating the time components of the corresponding current densities. Gell-Mann proposes the current algebra of these charges, given by the commutation relations of the charge densities.

The SU(3) symmetry is broken by the eighth component of the unitary spin. Gell-Mann derives a first-order mass formula where a linear combination of the masses of the λ and σ baryon equals the average of the masses of the nucleon and the ξ baryon. This formula, later called the Gell-Mann–Okubo mass formula, agrees very well with the experimental data.

Gell-Mann predicts a ninth pseudoscalar meson, an SU(3) singlet, which is today called the η' meson. He predicts, that this meson should decay, like the η meson, into two photons. This meson was found in experiments and has a mass of about 958 MeV.

8. Symmetries of Baryons and Mesons (p. 128)

In this article in *Phys. Rev.* (1962) Gell-Mann again discusses SU(3) symmetry and also current algebra, the equal time commutation relations of the time components of the eight vector and eight axial vector currents. Integrated over space they generate the group SU(3) $\times$ SU(3). He derives generalized Goldberger–Treiman relations for the eight baryons. As in the previous paper the baryons and mesons are classified as members of octets and singlets.

The baryons, related by the SU(3) symmetry to the nucleons, and the mesons form octets of the group SU(3). Gell-Mann suggests that the symmetry breaking is provided by a term that transforms as an octet under SU(3). From this assumption he obtains the first-order relation.

This relation agrees very well with the masses found in the experiments. It was also derived by Susumi Okubo and is called the Gell-Mann–Okubo relation. The baryons of spin-3/2, including the delta resonances, transform as a decimet under SU(3). In this case the Gell-Mann–Okubo relation becomes very simple, namely equal mass splitting between the states with strangeness 0, 1, 2 and 3, in excellent agreement with experiment. In fact, in a discussion at the 1962 International Conference on High Energy Physics, printed here as #9, Gell-Mann was able to predict the existence and the mass of the ω^- particle with charge -1 and strangeness -3. The approximately predicted mass was 1685 MeV. Two examples were found in 1964 at the Brookhaven National Laboratory after two years of work. The particle was produced in collisions of negatively charged K-mesons and protons. Its mass is 1672 MeV, agreeing well with the prediction of Gell-Mann.

9. **Prediction of the Ω^- Particle,** from 1962 Int. Conf. on High-Energy Physics (p. 146)

Gell-Mann discusses first the mass formula (Gell-Mann–Okubo relation) for the baryons and for the mesons. In the case of the vector mesons the mass formula does not seem to work. Now we understand that this is due to the mixing between the singlet vector meson and one of the octet vector mesons. The η' meson is nearly a state consisting of a strange quark and its antiquark, which is a mixture of singlet and octet in the SU(3) framework.

The new point is the application to the δ resonances. Gell-Mann derives the equal spacing mass relation for the decimet and predicts the existence of a new baryon state with strangeness -3 and isospin 0, the "Ω^-." He predicts the mass to be approximately 1685 MeV. Two years after the CERN conference the Ω^- was found in Brookhaven, with a mass of 1672 MeV. There were two examples, with different predicted decay modes.

10. **Elementary Particles of Conventional Field Theory as Regge Poles** (with *M. L. Goldberger*) (p. 147)

Chew and Frautschi made the ingenious suggestion that the hadrons could be composed of one another in a self-consistent "bootstrap" scheme. They suggested that the hadron states would then behave differently from the elementary particles of conventional Lagrangian field theory. In particular, they would correspond to poles in the angular momentum plane that move on Regge trajectories, while the elementary particles of ordinary field theory would correspond to fixed poles at integral or half-integral spins. In this paper, Gell-Mann and Goldberger showed that when vector bosons are involved (as in gauge theories) perturbation-theory calculations suggest that spin-1/2 fermions fall on moving trajectories when interactions are included. The distinction between fixed and moving poles in the J plane could therefore not be used to tell bootstrapped hadrons from "elementary" ones.

11. **A Schematic Model of Baryons and Mesons** (p. 151)

This is one of the most important articles by Gell-Mann, and the shortest one – only two pages. He sent it to *Physics Letters* because he was afraid that the referees at *Phys. Rev.*

Letters would reject it because the nucleon was treated not as elementary but as composed of new entities, with fractional charges, which were probably confined inside observable objects like the nucleon. He called those entities quarks, referring (for the spelling) to the book "Finnegans Wake" by James Joyce. In this book the word "quarks" is introduced on p. 383 with the cry "Three quarks for Muster Mark!" Both the numbers 3 and 8 play a big role in SU(3) symmetry and in the quark model. More importantly, the nucleon is composed, roughly speaking, of three quarks.

Gell-Mann starts with a short discussion of the Eightfold Way classification of the baryons and mesons. He constructs the baryons and mesons as systems composed of quarks, which are triplets of SU(3) and must have the electric charges 2/3 and $-1/3$, and antiquarks. Baryons consist (roughly) of three quarks, while mesons are bound states of quark and antiquark. A three-quark system can be a singlet, an octet, or a decimet of SU(3). Thus the observed baryons can easily be described.

Mesons are either singlets or octets, in agreement with observation. Gell-Mann uses the names u, d and s for the quarks. The proton has the structure (uud) and the neutron (udd). The λ hyperon contains one s-quark: (uds). The u-quark has electric charge 2/3, while the d and s quarks have electric charge $-1/3$, in units of the proton charge e. Gell-Mann can also describe the SU(3) violation in a simple way by introducing formal quark masses. The strange quark must have a larger mass than the u and d quarks, which are both very light but differ in mass.

Gell-Mann discusses briefly a field theory model in which the quarks interact with a neutral vector meson field, which he later called a gluon. He uses this model for the abstraction of the current algebra of the electromagnetic and weak currents.

Gell-Mann also discusses the possibility that free quarks might exist in nature. One of the quarks would have to be stable, since the electric charge is exactly conserved. If free quarks existed, they would probably be useful in industry, for example in catalyzing thermonuclear reactions. Gell-Mann refers to such quarks as "real." Here Gell-Mann emphasizes the possibility that the quarks are not "real" but confined and never free. As a mechanism, he suggests letting their masses go to infinity while the masses of the bound states are finite. In 1966, in the subsequent paper #12, he suggests instead that they are bound by an infinite potential well, which is what we believe today.

12. Current Topics in Particle Physics,

from Proceedings of the XIII Int. Conf. on High-Energy Physics (p. 153)

Gell-Mann was asked to give an introductory lecture for the "Rochester conference" in Berkeley. At the beginning he spoke about the various features of the Regge trajectories. Then he introduced the quark model for the description of the approximate symmetries of the baryons and mesons. Roughly speaking, the baryons are three-quark bound states and the mesons are antiquark–quark bound states. He assumed the wave function of the baryons to be totally symmetric, apparently violating the Pauli principle. (Years later, with the discovery of the color variable, it became clear that there was no violation.) He discussed the SU(6) symmetry for the spins of the three flavors of quarks — the lowest baryons are described by a 56-representation.

Gell-Mann described the quarks as mathematical entities, connected to representations of current algebra. He treated the confinement of quarks much as we do today, with an infinite potential well for quark–quark interactions. As in earlier work, he described confined quarks as not "real," meaning that they could not be handled singly or used in industry.

He also mentioned that resonances with exotic quantum numbers might exist, which in the quark model could consist of three quarks and a quark–antiquark pair (baryons) or two quarks and two antiquarks (mesons).

He introduced the algebra of SU(3) $\times$ SU(3) for the eight vector and eight axial vector charges and also the local current algebra of the charge densities, where in the commutator of two densities a density appears multiplied by a δ function in space. The local current algebra does not contradict the principles of relativistic quantum mechanics, since in the relativistic quark model it is realized. The charged weak current is described in terms of a superposition of the d and the s quark fields, with coefficients proportional to the cosine and sine respectively of the famous angle. Using the commutator of two axial-vector charge densities, he derives the Adler–Weisberger relation, which connects the ratio of the axial-vector and vector coupling constants of the nucleon to the neutrino cross section. Furthermore, he uses PCAC to get a sum rule that could be verified by experiment. It connects the difference of the charged pion–nucleon total cross sections to the ratio of the coupling constants. This sum rule is obeyed by the experimental data.

Gell-Mann concludes his lecture by speaking about CP-violation, which he connects to the superweak model. Today we know that the CP violation can be described very well by the Standard Model, involving 6 flavors of quarks and three kinds of neutrino along with three kinds of charged lepton. The fact that there are three rather than two lepton families allows a complex phase factor to enter, leading to CP violation, as described by Kobayashi and Maskawa.

13. Behavior of Current Divergences under $SU_3 \times SU_3$
(with *R. J. Oakes* and *B. Renner*) (p. 160)

In 1968 Gell-Mann wrote, together with Robert Oakes and Bruno Renner, the article "Behavior of Current Divergences under SU(3) $\times$ SU(3)". In particular they studied the transformation properties of the divergences of the axial-vector currents and the strangeness-changing vector current. They concluded that the breaking of the SU(2) $\times$ SU(2) symmetry is less strong than the breaking of the SU(3) symmetry. In the quark model this is easily understood. The masses of the u and d quarks are both close to zero compared to the mass of the strange quark, even though the ratio of the u and d quark masses is not close to one. Also, the hadronic mass scale is large compared to the two light quark masses.

14. **Light Cone Current Algebra** (with *H. Fritzsch*) (p. 165)

In 1971 Gell-Mann wrote, together with Harald Fritzsch, the article: "Light Cone Current Algebra", which appeared in the Proceedings of the Conference on Duality in Tel Aviv (1971). This article was based on the lectures of Gell-Mann and Fritzsch at that conference. A similar article appeared before as a preprint in 1971, based on a lecture given by Gell-Mann at the conference in Coral Gables in January 1971. It was published in the proceedings.

In this article the experimental results on deep inelastic scattering of electrons on nuclear targets are related to the behavior of current commutators near the light cone. It is postulated that the current commutators near the light cone approach the commutators abstracted from a free quark model. The commutators are given in terms of bilocal operators, which in the quark model are products of quark and antiquark fields, separated by a light-like distance.

Fritzsch and Gell-Mann use the algebraic system obtained from the quark model to derive various sum rules. They obtain the same results as those obtained by Feynman in his "parton" model, if the electrically charged "partons" are identified with the quarks and antiquarks. They also speculate about new neutral objects, which could not be seen directly in deep inelastic scattering, but would appear in the energy–momentum tensor, which enters into current algebra. They identify these objects with the gluons, which provide the interaction among the quarks. In this article the possibility is mentioned that the u, d, and s quarks each come in three forms, later called "colors." In this way one can understand how three quarks with a symmetrical ground state wave function can form a baryon without violating the Pauli principle. The antisymmetry of the wave function, required by the Pauli principle, is provided by the color index, while the wave function in the other variables is symmetric. This gives for electron–positron annihilation into hadrons at high energy a cross section that is three times larger than naively expected. At high energy but below the charm threshold, the predicted ratio of the hadronic cross section to the cross section for producing muon pairs is 2 instead of 2/3, which would be obtained in the quark model without color.

15. **Light-Cone Current Algebra, π^0 Decay, and e^+e^- Annihilation** (with *W. A. Bardeen* and *H. Fritzsch*) (p. 199)

In 1972 Gell-Mann wrote, together with William Bardeen and Harald Fritzsch, a contribution to the Proceedings of the Conference on Conformal Invariance (1972). They point out that the cross-section for electron–positron annihilation can be directly related to the amplitude for the electromagnetic decay of the neutral pion. They realize that the pion decay is correctly described if the quarks have the color quantum number, but that if color is absent the decay rate is a factor of 1/9 too small. They discuss in detail the advantages of exactly conserved color with color confinement and compare this idea with other schemes, in particular with the proposal of paraquarks by O. Greenberg and the proposal of M. Han and Y. Nambu for a broken color SU(3) symmetry, which would allow the possibility of integral charges for the nine quarks, but would permit colored objects to emerge and be detected individually.

16. Quarks

Lecture given at the XI Internationale Universitätswochen für Kernphysik, Schladming (1972) (p. 212)

Gell-Mann distinguishes between constituent quarks and current quarks. In the constituent quark model the low-lying bound and resonant states of the baryons act like three-quark states and those of the mesons like quark–antiquark states. Other configurations, e.g. a baryon consisting of four quarks and one antiquark, are called exotic states, but they should exist in the hadronic spectrum, at least in the continuum.

The low-lying meson and baryon states are classified in terms of a symmetry group U(6) $\times$ U(6) $\times$ O(3), where one U(6) group is for the quarks (three quark states and two spin states), the other one for the antiquarks, and O(3) describes the angular momentum L. For $L = 0$ one obtains nine pseudoscalar and nine vector mesons. For $L = 1$ one finds tensor mesons, axial vector mesons, and scalar mesons. For the baryons one has for $L = 0$ the 56-representation of the symmetry group SU(6), i.e. three quarks in a totally symmetric state with respect to space, spin, and "flavor" SU(3). In this way one obtains the baryon octet with spin-1/2 and the decimet with spin-3/2 with positive parity. The states with $L = 1$ are described by a 70-representation of SU(6). These are the baryons with negative parity.

Gell-Mann introduces the color quantum number to avoid the problem with the Pauli principle for the quarks. He calls this quark statistics. The baryons are color singlets with a totally antisymmetric color wave function: RGB − GRB + BRG − RBG + GBR − BGR (R: red, B: blue, G: green). The hadrons act as if they are made up of quarks, but the quarks do not have to be "real" particles, capable of emerging singly from baryons and mesons and being used in industry.

The weak and electromagnetic currents of the hadrons are written in terms of quark bilinears and are color singlets. Gell-Mann introduces the light cone current algebra of these currents. The commutator of two currents near the light cone is given in terms of bilocal currents, depending on two space-time variables x and y. If x and y are equal, one has a normal current. The scaling functions, measured in deep inelastic scattering, are given by the nucleon matrix elements of the bilocal currents.

Finally, Gell-Mann discusses the disconnected part of the current commutator, which is connected to the cross-section for electron–positron annihilation at high energies. Using the color quantum number and the three quark flavors u, d, and s, Gell-Mann predicts the value 2 for the ratio of the cross section for electron–positron annihilation into hadrons to the cross section for producing muon pairs. (Of course that value must be increased as more flavors need to be included at higher energies.) In the same way he calculates the rate for the electromagnetic decay of the neutral pion. Both agree very well with the results of experiments.

17. Current Algebra: Quarks and What Else? (with *H. Fritzsch*)
Proceedings of the XVI Int. Conf. on High Energy Physics (p. 241)

For the proceedings of the XVI International Conference on High Energy Physics in Chicago, Gell-Mann wrote, together with Harald Fritzsch, the article "Current Algebra: Quarks and What Else?", based on a lecture by Gell-Mann at the conference. They discuss again the light cone algebra of the currents and of the bilocal operators.

The most important part of the paper is the first section, dealing with color octet gluons. The authors start by describing the color variable and the requirement that "real" hadrons (i.e. those capable of being observed in isolation) be color singlets. They review some of the evidence for color, for example from the neutral pion decay rate and the cross-section for the transformation of electron and positron into hadrons. Of course the main virtue of color is that it explains quark statistics, as they point out. They go on to mention the quark–gluon theory, assuming that the gluons are color octets and are described by a Yang–Mills theory. This article is the first one in which the Yang–Mills theory of color octet gluons coupled to colored quarks is mentioned. That is, of course, the theory which Fritzsch and Gell-Mann later called quantum chromodynamics (QCD).

18. Advantages of the Color Octet Gluon Picture
(with *H. Fritzsch* and *H. Leutwyler*) (p. 262)

In 1973 Gell-Mann wrote, together with Harald Fritzsch and Heinrich Leutwyler, the letter called "Advantages of the Color Octet Gluon Picture". This is an early article on quantum chromodynamics as the field theory of the strong interactions. They assume that color is an unbroken symmetry. Only color singlets are allowed as physical states. Thus quarks and gluons are permanently confined. Several advantages of the color octet theory are discussed. In particular, the SU(3) singlet axial vector current has an anomalous divergence, which is given by a term bilinear in the gluonic field strengths. This explains why there are only eight massless Goldstone bosons if the three quark masses are set to zero. Naively one would expect nine such bosons.

More importantly, if we had a color singlet gluon instead of an octet there would be an approximate SU(9) symmetry in the approximation of neglecting the u, d, and s quark masses, as pointed out also by Okun. Since there is no sign of such an approximate symmetry, one must reject the singlet gluon model. There is no corresponding difficulty with the color octet gluon theory.

19. Complex Spinors and Unified Theories (with *P. Ramond* and *R. Slansky*) (p. 266)

Gell-Mann, Ramond, and Slansky discuss the theories of so-called grand unification of the strong, electromagnetic, and weak interactions, based on the groups SU(5) and SO(10). In the SU(5) theory the fermions of one family are described by the sum of the representations $\bar{5} + 10$. The representation $\bar{5}$ contains the anti-d-quark and the electron and its neutrino, and the 10-representation contains the d-quark, the u-quark, the anti-u-quark and the positron. The sum $\bar{5} + 10$ is anomaly-free. In the SO(10) theory the 16-dimensional spinor representation breaks up into $1 + \bar{5} + 10$ of SU(5).

The fermion masses can be generated by a Higgs mechanism. The square of the spinor representation is given by the sum $10 + 126$. The decomposition of the 126-representation under SU(5) gives: $126 \Rightarrow 1 + 45 + 10 + \bar{15} + \bar{5} + 50$. An operator transforming like the singlet of SU(5) would break the SO(10) symmetry down to SU(5) and would give a Majorana mass term to the unobserved neutrino, which has to be very large. The 10-representation of SO(10) would give rise to Dirac masses for the charged fermions and the neutrinos. The Dirac mass for the neutrino leads directly to a small effective mass of the neutrinos, given roughly by the square of the Dirac mass divided by the Majorana mass. That situation might explain the small neutrino masses. It is now called the "see-saw" mechanism.

20. Particle Theory: From *S*-Matrix to Quarks (p. 273)

In this paper, Gell-Mann describes his many hesitations and uncertainties as he tried to understand the elementary particles and their interactions. Often he would be faced with two or more possibilities with arguments supporting each of them. He felt forced to choose rather than merely describe the possibilities and their consequences. In many cases, that was painful. He lists a number of difficulties and confusions, most of which were finally cleared up.

The first example has to do with what has been called the "renormalization group," studied in quantum electrodynamics by Gell-Mann and Low and also by Stueckelberg and Petermann. (See #3.) Gell-Mann and Low introduced the function that they called ψ, related to one that was introduced later and called β. That function has physical significance in the theory. Depending on how it behaves at large values of its argument, one can have three possible situations: a finite unrenormalized coupling constant, an infinite unrenormalized coupling constant, or a self-contradictory theory for any nonzero value of the renormalized (observed) coupling constant. Landau and his associates guessed the last, based on arguments from perturbation theory, but Gell-Mann pointed out to them that their perturbation-theoretic methods could not reveal the behavior of ψ as its argument went to infinity. Gell-Mann was not sure which possibility was right, although he later suspected that Landau and Company made the right choice for the wrong reasons and that only asymptotically free theories are consistent. The controversy seems to be unresolved to this day.

Some of the issues had to do with Yang–Mills theories and their possible renormalizability, as well as the related question of what a "soft" mechanism for generating particle masses would look like. Another had to do with whether dispersion theory (what Geoffrey Chew called "S-matrix" theory) was really different from field theory or just a different way of describing field theory. Gell-Mann has always believed the latter. In superstring theory, however, it is not known whether a field theory formulation is possible.

In a number of cases confusion was caused by experimental error. The β-decay interaction was thought to be of the scalar and tensor forms instead of vector and axial vector. The decay of the charged pion into electron and neutrino or positron and antineutrino was missed for a time. The relative parity of λ and σ hyperons was believed to be negative at one time, in contradiction to the eightfold way scheme.

There was theoretical confusion over whether the electron and muon were coupled to the same neutrino or different ones. Correspondingly, was there a fourth kind of quark (with charm)? What about the angle relating the strangeness-changing and strangeness-preserving weak currents for the hadrons? Was it 45 degrees or more like 15?

Gell-Mann describes how he wanted the electric charge operator to be a generator of a semi-simple group, leading to a zero sum for the charges in a representation. That led him to doubt, for a while, the existence of charm, which spoiled that zero sum for the three flavors of quark. If one thinks more broadly, however, in terms of a unified Yang–Mills theory, one should put in the leptons as well as the quarks and then each family (or "generation") would have a zero sum for the charges: $3(2/3\ -1/3) + 1(0 - 1)$. Gell-Mann draws a lesson from this and other cases. Some ideas are useful right away; others need to be put off to a later stage in the work, when it will become clear how they apply.

That same lesson applies to Gell-Mann's hesitation in presenting what was later called QCD in the written version of his lecture at the 1972 "Rochester" meeting in Chicago. He was worried that the correct theory might be one based on colored strings. That is why the written version (#17) mentions QCD in a low-key way instead of forthrightly as in the actual lecture. The string notion may have great importance for the ultimate unified theory, including gravitation, but it is a distraction at a less advanced stage of description.

The paper refers to a number of additional confusions and difficulties. A particularly important one has to do with applying the Yang–Mills idea to the hadrons. Before the advent of color, it was hard to see how to do that. If the strong interaction was of that type, based on the SU(3) of flavor, then it collided with the needed Yang–Mills theory for the weak and electromagnetic interactions. Without color, there just was not enough room for the two theories. That was very frustrating for Gell-Mann and Glashow, who showed how to generalize the idea of Yang and Mills to other Lie groups than SU(2) and then tried to use that idea for both the strong and weak interactions.

Gell-Mann mentions his efforts to construct a field theory of the weak interactions with Feynman. They introduced the charged intermediate bosons and showed that the muon could decay quickly into an electron and a photon, unless there were two neutrinos, one for the electron and one for the muon. The neutral boson would lead to strangeness-changing terms that would cause a decay of the kaon into two muons, which was not observed. They did not suggest cancelling the strangeness-changing terms by inventing a fourth quark.

Subsequently Gell-Mann describes how he arrived at the idea of colored quarks. In 1963 and 1964 he played with Greenberg's para-Fermi statistics as a possible way of obtaining the baryons from three quarks in a symmetrical state of space, spin, and flavor variables. In 1971 he and H. Fritzsch introduced the exactly conserved color quantum number and speculated about the perfect confinement of color. They showed that color with detectable particles as color singlets gave the same results for statistics as parastatistics with detectable particles as fermions and bosons. In 1972, Gell-Mann and Fritzsch discussed the theory that they later called quantum chromodynamics.

21. Remarks Given at the Celebration of Victor Weisskopf's 80th Birthday (p. 297)

Gell-Mann wrote this essay on the occasion of the 80th birthday of Victor (Viki) Weisskopf in September 1988. Gell-Mann started graduate work when he went to MIT 40 years earlier. He shared an office with Marvin Goldberger and David Jackson, among others. Victor Weisskopf was his thesis advisor.

In this paper he describes the wide-ranging discussions between Viki and his students and post-docs over dinner and how the memory of those discussions helped inspire Murray to be a co-founder of the highly interdisciplinary Santa Fe Institute, where he has worked since his retirement from Caltech. As an example of a topic of discussion he mentions Zipf's Law, approximately valid in a wide variety of subjects and still somewhat mysterious in origin.

22. Quantum Mechanics in the Light of Quantum Cosmology (with *J. B. Hartle*) (p. 303)

In this paper Gell-Mann and Hartle present their interpretation of quantum mechanics, applying it to the universe as a whole. The usual interpretation involves dealing with a smaller system and with an observer outside it, typically a physicist with a piece of apparatus making measurements of it. This is rather strange as a fundamental description of quantum mechanics, since for billions of years there were no physicists. Does that mean there was no quantum mechanics? Of course not. Clearly the usual approach is a special one applicable to situations in which measurements are made. In the more general interpretation one can discuss the quantum mechanics of the universe (quantum cosmology) and note that "measurement situations" can arise without "observers," as when fission tracks are produced in mica containing atoms with spontaneously fissionable nuclei. The tracks are there whether a physicist (or a chinchilla or a cockroach) looks at them or not.

Instead of restricting attention to subsystems in special states, the general approach considers an immense set of alternative histories of the universe. In order for probabilities to be assigned to the alternative histories, it is necessary that they be decoherent with respect to one another. Otherwise, for exclusive alternative histories A and B, the probability of A or B would not be equal to the sum of the probabilities of A and of B, as the laws of probability require. Now perfectly fine-grained histories of the universe do not decohere, so one must be dealing with coarse-grained histories. The character of the coarse graining involved is of the greatest importance.

In the usual approach it is supposed that there is a classical world (Landau and Lifshitz actually postulate that there are classical laws in addition to quantum mechanics). It is taught that when a measurement takes place the classical world takes over, although it doesn't matter much at which point in a chain of connections the classical world becomes involved. In the approach of Gell-Mann and Hartle it is the coarse graining that causes the histories to follow "quasi-classical" behavior. They obey classical laws approximately but with many small fluctuations and with occasional major branchings of histories, as in a "measurement situation." In the Stern–Gerlach experiment, the quantum variable (the z-component of the elecron's spin) becomes very strongly correlated with its spatial position and then with the development of a photographic grain in one place rather than another and then with a chain of effectively classical results. The coarse graining that permits this kind of correlation is thought to be the one that follows integrals over small volumes of densities of conserved or nearly conserved quantities at discrete times with a small interval separating them. The volumes are big enough so that they contain enough inertia to resist quantum fluctuations and small enough for local equilibrium to be reached. The time interval is just long enough to allow for decoherence. Clearly the details of the coarse graining are history-dependent as well as time-dependent. For example, the size of one of the "small volumes" depends on the density of matter that is present there at a particular time on a particular branch of history.

The interpretation of Gell-Mann and Hartle does not give different results from the "Copenhagen interpretation" when a physicist is making a measurement, but it does lead to a different view of situations like the EPRB experiment, where a neutral meson at rest decays into two photons with equal and opposite momenta. If the linear polarization of one of them is registered, then the linear polarization of the other is determined and likewise for the circular polarization. But it is on different branches of history that these different polarizations are determined. There is no reason to refer to this kind of situation as reflecting "nonlocality." (Unfortunately it is often called "nonlocal" by definition.) There is no signal and no causation travelling instantaneously or faster than the speed of light from the site of one registration to the site of the other. It would be correct to say that if the EPRB experiment were to be interpreted classically then one would need nonlocality or negative probabilities or both to understand the results. But the situation is not classical.

In the interpretation of Gell-Mann and Hartle there are three kinds of information: the action function of elementary particle theory, the initial state vector or density matrix of the universe, and the specification of features of the particular history that occurs.

A particular simple law for the initial condition of the universe may lead to the thermodynamic arrow of time, to the existence of classical space-time, to the large scale homogeneity and isotropy of the universe, and to its approximate spatial flatness.

Some of the observable characteristics of the elementary particles might be quantum-probabilistic, with a probability distribution that depends on the initial conditions. Even the fundamental constants, like the fine structure constant or Newton's constant of gravity for electrons, might conceivably be cosmic accidents.

In classical physics probabilities result from our ignorance. In quantum theory they can be fundamental as well. In most presentations of quantum mechanics the external classical observer plays an important role. A measurement is carried out through contact with this classical domain. Such an interpretation is inadequate for cosmology. In a theory of the whole universe there can be no fundamental division into observer and the observed system.

Gell-Mann and Hartle point out that quantum mechanics is best and most fundamentally understood in the context of quantum cosmology. The theory yields probabilities for alternative coarse-grained histories of the universe obeying decoherence, in fact what they call "medium decoherence," which is much stronger than mere "consistency."

23. Dick Feynman — The Guy in the Office Down the Hall (p. 326)

On February 15, 1988 Richard Feynman died. Gell-Mann and Feynman were colleagues at Caltech for 33 years. Gell-Mann wrote about him in an issue of Physics Today containing many tributes to Feynman. After some personal anecdotes he turned to Feynman's work on quantum mechanics.

In Feynman's Ph.D. dissertation at Princeton University, he considered (building on work of Dirac) a new approach to quantum mechanics, which is called the path integral method. Later on, using that method, Feynman introduced the famous Feynman diagrams. In the article Gell-Mann describes the advantages of the path integral method for many problems. Feynman's method is in particular very useful in quantum cosmology. Gell-Mann describes how he and James Hartle used Feynman's method in an interpretation of quantum mechanics suitable for quantum cosmology. See #22.

In any theory that quantizes Einsteinian gravitation, the whole fabric of spacetime is subject to quantum fluctuations and thus conventional methods of describing quantum mechanics, involving a Hamiltonian and a simple time variable, tend to fail. However, Hartle has shown how the sum-over-histories method can yield a slight generalization of usual quantum mechanics that is perfectly consistent with general relativity or an extension thereof. Thus Feynman can be credited with this real advance in physical theory. Much of his work, however brilliant, involved greatly improved mathematical methods of dealing with known theories, but here he took an important step toward the future unified theory that will reconcile quantum mechanics and general relativity.

Earlier, Feynman had shown how to quantize general-relativistic gravitation in perturbation theory, after practicing (at Gell-Mann's suggestion) on Yang–Mills theory.

24. Time Symmetry and Asymmetry in Quantum Mechanics and Quantum Cosmology (with *J. B. Hartle*) (p. 331)

Gell-Mann and Hartle discuss the exact or approximate time reversal symmetry of the fundamental laws of physics and the time asymmetries of the observed universe. Our universe shows time reversal asymmetries on various levels:

1. The thermodynamic arrow of time.
2. The psychological arrow of time.
3. The arrow of time of retarded electromagnetic radiation.
4. The arrow of time given by the CP non-invariance of the weak interactions.
5. The arrow of time of the expansion of our universe.
6. The arrow of time supplied by the growth of inhomogeneity in the expanding universe.

These time asymmetries could arise from time-symmetric dynamical laws, solved with time-asymmetric boundary conditions. Quantum cosmology is concerned with the theory of the boundary conditions of our universe. It is the general context in which one should investigate the origin of the observed time asymmetries.

Gell-Mann and Hartle discuss first the arrow of time in quantum mechanics. There is an asymmetry between future and past, exhibited in the formula for probabilities in quantum mechanics. That formula contains a nontrivial density matrix at the beginning of time but only a unit matrix at the final time. The authors discuss possible generalizations of quantum mechanics in which there are density matrices at both ends. They even consider the possibility of a situation symmetrical between past and future, with a particular time picked out as invariant. Such theories would not yield conventional causality. Rather, such causality, dependent on the past condition, would become more and more approximate as time goes on, gradually yielding to reverse causality, dependent on the future condition. Gell-Mann and Hartle are not advocating this possibility, just exploring it for the light it throws on the issue of time-reversal symmetry and asymmetry.

The weak interactions exhibit a violation of CP-invariance. Since field theory is CPT invariant, there must also be a violation of T-invariance. This is a very small effect in laboratory experiments, but it is of central importance in the evolution of the matter content of our universe. Together with the non-conservation of baryon number it is thought to be responsible for the emergence of a matter-dominated universe from a postulated initial equality of matter and antimatter.

The violation of T-symmetry could in principle be either the result of a T violation in the fundamental Hamiltonian of the weak interactions, or due to asymmetries in the cosmological boundary conditions of the universe, or an asymmetry characteristic of our particular epoch and spatial location. These three possibilities are discussed in general terms. Of course most physicists, including Gell-Mann and Hartle, connect the violation with the specific mechanism put forward by Kobayashi and Maskawa.

25. Progress in Elementary Particle Theory, 1950–1964

Int. Symp. on the History of Particle Physics, Fermilab, 1985 (p. 360)

In this contribution Gell-Mann describes some features of the history of particle physics, mostly as he experienced it, up to and including the time when the quark model was introduced. This talk differs from #20 not only in the time interval covered but also in the approach. Though in #20 he emphasized the hesitations and confusions that accompanied the various theoretical advances, in this talk he concentrates on the advances themselves, as if there had been no (or very few) doubts or blind alleys. Of course he mentions important experimental evidence relevant to theory, but he deliberately glosses over wrong results just as he does over wrong ideas.

He goes methodically through the various insights, one after another, bearing on quantum field theory (including the mass-shell formulation), on hadrons and the strong interaction, on leptons, on the electromagnetic interaction, and on the weak interaction.

He ends by going beyond his assigned period of time and mentioning the widespread hope that a finite unified theory of the particles and interactions will be found. He suggests that perhaps superstring theory will help in the search for that theory.

26. Nature Conformable to Herself (p. 378)

The title of this article is taken from Isaac Newton, who wrote in "Opticks":

"For Nature is very consonant and conformable to herself..." He was referring, among many other things, to an insight that he had when the University of Cambridge was closed on account of the plague and he retired to his mother's farm in Lincolnshire. There he had an idea that some biographers attribute to his seeing an apple fall to the ground in his mother's orchard. It occurred to him that the apple could be obeying the same law of gravity that governed the moon and the planets even though the scales involved are so different.

In this article Gell-Mann offers an answer to the frequently asked question, "Why should beauty or elegance be a successful criterion in choosing a theory in fundamental physics?" His answer is connected with Newton's observation.

Gell-Mann starts from some simple assumptions:

1) The fundamental laws of Nature are not constructs of the human mind; they are really there and physicists keep getting closer to them as they pursue their research.
2) There is a unified quantum theory of the elementary particles and the forces of Nature; physicists approach it by exploring higher and higher energies and shorter and shorter distances, "peeling the skins of the onion."

3) That theory has the property of very approximate self-similarity; its manifestations at different scales of energy or length tend to resemble one another.
4) As a result, each new theoretical discovery is expressed in mathematical language that is not so different from the language used for previous discoveries, as in these examples:
   a) Coulomb's law and Newton's law of gravitation,
   b) Yang–Mills theory and electromagnetism,
   c) the weak interaction and the electromagnetic interaction,
   d) the exact SU(3) symmetry of color and the approximate symmetry of three flavors of quark.

The new theoretical advance looks simple and therefore beautiful or elegant because it can be expressed easily in terms of mathematics that is not so different from what is familiar.

This rough self-similarity is a property of the fundamental law of the particles and forces. It does not depend on human beings. An alien intelligence would find the same situation, although the mathematics would probably have a different form.

Gell-Mann ends by discussing the concept of emergence and the slogan that he associates with it: "You do not need to put in something more in order to get something more."

27. Quarks, Color, and QCD (p. 382)

In this article Gell-Mann reminisces in considerable detail about his experiences with the quark model and QCD. He recalls how he and others made their way through a thicket of issues involving current algebra, the existence of quarks, quark statistics, quark confinement, color, neutral pion decay, hadron production by e^+ and e^-, Yang–Mills theory, asymptotic freedom, "partons," light-cone current algebra, deep inelastic scattering, and anomalies, all culminating in the formulation of QCD and its contribution to the standard model.

Among the papers to which Gell-Mann alludes here are those in this volume numbered 11 through 18. Particular emphasis is given to the idea of perfectly conserved color, with confinement of color non-singlets, as suggested by Fritzsch and Gell-Mann, giving the right rate of neutral pion decay and the right cross-section for hadron production in electron–positron collisions below and above the charm threshold. They went on to discuss the idea of color-octet gluons in a perfect Yang–Mills theory based on SU(3). The discovery by 't Hooft and (with more awareness of the implications) by Gross and Wilczek and by Politzer, that such a theory had decreasing coupling strength at high momentum transfers was very important. It led to the speculation that the coupling strength might go rapidly to infinity at low momentum transfers, thus explaining confinement of color. That connected up with ideas about asymptotic freedom, Feynman's "partons," light-cone current algebra, and so forth.

QCD was soon accepted as the correct theory of hadrons and the strong interaction. The name was invented by Gell-Mann in the summer of 1974 and used by him and Fritzsch, by H. Pagels, and then by everyone else.

28. Effective Complexity (with *S. Lloyd*) (p. 391)

The authors discuss what is meant by complexity or its opposite, simplicity. They start with a rough definition of the "effective complexity" of an entity as the length of a highly compressed description of its regularities. The distinction between what is treated as regular and what is treated as incidental depends on a kind of coarse graining, the source of some of the context dependence of effective complexity. For the minimum description length the authors use the algorithmic information content (AIC) of Chaitin, Kolmogorov, and Solomonoff, applied to the description encoded into a string of bits. The effective complexity of a string is low, for a given string length, when the string is very regular and also when it has few regularities. It is only in the intermediate region between order and disorder that the complexity can be high. The AIC of a string, for a given universal computer, is the length of the shortest program that causes that computer to print out the bit string and then halt. Here the authors are breaking up the AIC of their string into two terms, one (the effective complexity) representing the AIC of the regularities and the other the AIC of the remaining features of the string.

By analogy with statistical mechanics, the authors describe the bit string by embedding it in an ensemble, a set of strings with probabilities. The ensemble represents the regularities.

Information and ignorance are really the same thing; if one receives a letter and reads it, the amount of information gained is the same as the amount of ignorance before the letter arrives. The usual Shannon formula for information is employed. In statistical mechanics, the same formula yields the usual entropy after multiplication by $k \ln 2$.

The authors discuss the role of the quantity Y, which is the AIC of the ensemble, and the measure I of ignorance, which is close to the average contingent AIC of the members, given the ensemble. (Here "close" and "approximately" mean "within a few bits.") For the ensemble to represent the string describing the entity under consideration, the first step is to bring the sum $I + Y$ close to its minimum value, which is approximately K, the AIC of the string. The second step, as in the familiar Jaynes prescription, is to maximize I subject to constraints that set ensemble averages of certain quantities equal to the values of those quantities for the string describing the entity. As in statistical mechanics, the resulting probability distribution for the ensemble contains those averages as parameters. (In the simplest case in statistical mechanics, one has the Maxwell–Boltzmann distribution over the energy, with the temperature as the lone parameter.)

When these two steps are completed, one can identify Y with the effective complexity. The approximate formula $K = I + Y$ exhibits the break-up of the AIC of the string into that of its regularities and that of its "random" or incidental features. If the number of parameters is small, that complexity is small. If the number is large, one can have a great deal of effective complexity. Kolmogorov, in defining his "minimum sufficient statistic," did not introduce the constraints under discussion here and consequently every entity would come out simple if one used his quantity as a measure of complexity.

The authors then take up the issue of logical depth, defined by Charles Bennett roughly as follows. It is the number of steps or the length of time that the universal computer needs, at a minimum, to go from a program to the string under consideration, averaged over programs in a way that favors shorter ones.

Consider the set of energy levels of a heavy nucleus. Sixty years ago that seemed to require a very complex description, but today we believe that an accurate calculation of the positions of all levels is possible, using a simple theory: the quantum chromodynamics of the quarks and gluons, combined with the quantum electrodynamics of the quarks and photons. Thus in principle the levels are simple. But the computation time required is too long for the calculation to be performed using existing hardware and software. The system has little effective complexity but considerable logical depth. It is "pseudocomplex." The authors treat this kind of situation by introducing a more general *AIC* that is a function of the maximum computation time allowed for generating the bit string. As that time goes to infinity, one gets the usual *AIC*. In the case of pseudocomplexity, one can have a much higher value for finite but still fairly large times. The time scale over which most of the change takes place is similar to the logical depth.

At the end of their article, Gell-Mann and Lloyd pose the question of what happens to this discussion when the information or ignorance or entropy is given by an unorthodox formula, as in Tsallis's q-statistics, which reduces to normal statistics for $q = 1$.

29. Asymptotically Scale-Invariant Occupancy of Phase Space Makes the Entropy S_q Extensive (with *C. Tsallis* and *Y. Sato*) (p. 403)

The q-statistics of Tsallis depends on a single real parameter q. It reduces to Boltzmann–Gibbs statistics for $q = 1$. For other values of q the entropy does not have the additive property for a system composed of two independent subsystems. For that reason it is usually called "non-extensive." Of course it should not be applied to such a system. The authors of this paper show that for a sequence of systems of larger and larger size and with strong internal correlations one can have a situation approaching additivity at large sizes for some particular value of q not equal to 1. Thus q-statistics can actually yield extensivity in a case where B-G statistics would not do so.

30. Quasiclassical Coarse Graining and Thermodynamic Entropy (with *J. B. Hartle*) (p. 409)

In this article Gell-Mann and Hartle first review their approach to quantum mechanics as proposed in paper #22. They go on to discuss the coarse graining that yields a quasiclassical realm of the usual type. A realm is a set of exclusive and exhaustive projection operators onto ranges of values of certain quantities at a set of discrete times. (Those times correspond to a coarse graining of the continuous time in the history of the universe.) A realm yields a branching tree of alternative coarse-grained decoherent histories of the universe. The quasiclassical realms under discussion divide space into small cells and deal with integrals over those cells of densities of conserved or nearly conserved quantities. The set of projection operators at a given time depends on the branch of history up to that time. For example, the size of a cell depends on the density of matter in the vicinity and that is, of course, branch-dependent.

The authors emphasize that the coarse graining involved is closely related to the coarse graining used in classical statistical mechanics, resulting in a "hydrodynamic" description of the universe, with local equilibrium and with approximate closed sets of equations of motion. Near the classical limit, the quantum-mechanical system tends to follow the classical equations with many small fluctuations and occasional large branchings, as in "measurement" situations. The entropy associated with the coarse graining involved in a quasiclassical realm is thus connected with the usual thermodynamic entropy.

31. Progress in Elementary Particle Theory, 1946–1973 (p. 425)

This chart, issued as a Caltech preprint, extends the one in #25 to a much longer period of time, as indicated in the title. The spirit is the same, however. Comments such as the ones that accompanied the chart in #25 are lacking here but are replaced in a number of cases by brief remarks incorporated into the chart itself.

The history is divided into several intervals of time: 1946–1952, 1952–1962, 1963–1967, and 1968–1973. The whole article gives a picture of how ideas about the particles and interactions evolved during those periods, leading up to the standard model and to speculations about how to go beyond that model.

The Garden of Live Flowers

Murray Gell-Mann

Murray Gell-Mann, La Jolla, California, January 1991 (Photo courtesy of the Institute for Advanced Physics Studies, La Jolla, California)

Murray Gell-Mann (1929–), who became regarded by some as Albert Einstein's scientific successor, was born in New York City, the youngest son of Austrian immigrants. He entered Yale University just after his fifteenth birthday, earned his PhD degree at MIT when he was 21, and after teaching at the University of Chicago he was appointed full professor at Caltech when he was 26. There he became R. A. Millikan Professor of theoretical physics in 1967.

Gell-Mann's domination of the complex and popular subject of elementary particle physics is illustrated by the following story. For the 1966 International Conference on High-Energy Physics at Berkeley, the organizers planned to have five or six experts give progress reports on special areas of this rapidly growing field. After some wrangling over whom should be asked to present these papers, someone made the inspired suggestion that Gell-Mann do the whole thing. In a 90-minute talk he covered the whole field authoritatively, having worked on almost all aspects of it.

In about 1961 Gell-Mann and, independently, the Israeli physicist Yuval Ne'eman progressed to a new theory that Gell-Mann named the eight-fold way after the eight ways of right living recommended to Buddhists. Isotope multiplets were grouped into supermultiplets according to a mathematical structure called the group SU(3). This led to prediction of an eight meson called η^0, which was subsequently discovered.

The real world is not perfectly symmetrical under SU(3) theory; by making assumptions about how the symmetry is broken in nature, one can predict relations among masses of the particles. In 1962 Gell-Mann predicted in this way the existence of a particle called Ω^-. The 1964 discovery of this particle underlined the importance of the theory. It was the SU(3) theory that led Gell-Mann to speculate on the existence of fundamental entities out of which all strongly interacting particles could be built. The nucleon, for example, is composed of three of these entities which he named quarks (James Joyce, in Finnegans' Wake: "Three quarks for Muster Mark!"). Gell-Mann was awarded the 1969 Nobel Prize in Physics for his contributions and discoveries concerning the classification of elementary particles and their interactions.

Graduate students in theoretical physics who plan to study the fundamental laws of nature are very often impressed with "formalism"—the formal apparatus of their subject. Learning the beautiful equations of quantum field theory and of Einstein's general-relativistic theory of gravitation, some of them dream of inventing something equally important and mathematically elegant. Not being able to do so right away, they may become discouraged or else waste their time in the futile pursuit of grand theories for which the ground is not yet prepared. They do not always appreciate the difficult and tortuous processes of thought that had to be gone through before those splendid structures could be erected, nor the comparatively simple steps that had to be taken, one by one, before the final result could be obtained.

I suffered, at least as much as other students, from an infatuation with beautiful formalism. Working with Viki Weisskopf was a most effective remedy against the excesses of such an infatuation. He never ceased to harp on the importance of "pedestrian" work in theoretical physics and on understanding, by means of simple arguments, the physical meaning of a theory and its implications. He also stressed the need to learn about the relevant experimental evidence and to use theory to interpret that evidence and predict the results of new experimental work.

In this essay I point out that much of my research in elementary particle theory can be regarded as flowing from a struggle between a natural predilection for formal theory and an awareness of Viki's advice. That struggle led me to try a number of compromises between the "pedestrian" approach and the attempt to construct grandiose theories. In a number of cases those compromises turned out to be of just the right sort to help lay out the paths toward the standard model and toward superstring theory. The situation might be compared to that in the garden of live flowers in *Through the Looking-Glass*, where an attempt to walk straight toward a beautiful flower bed was quite futile, but striking out in a different direction made it possible to reach the objective.

I completed my Ph.D. dissertation, after half a year of dawdling, in January, 1951 and proceeded to the Institute for Advanced Study in Princeton, where a postdoctoral appointment had been waiting for me since September, 1950. There I collaborated with Francis Low on a useful exercise in formalism, which clarified the theoretical underpinnings of the bound-state equation for two bodies in quantum field theory, the simplest approximation to which had been found by Hans Bethe and Edward E. Salpeter.

After joining the faculty of the University of Chicago in January, 1952, I collaborated with Marvin L. "Murph" Goldberger on another project in quantum field theory. At that time, of course, the nucleon and pion were generally taken to be elementary particles, and many theorists believed that they obeyed the equations of pseudoscalar meson theory. We started from that theory and made the approximation of ignoring the "closed loops" that give rise to meson-meson interactions. Using no other approximation, we succeeded in writing down a new kind of relativistic equation for the nucleon–pion system that would hold for all coupling

Murray Gell-Mann. (Photo by NARA, courtesy of the AIP Emilio Segrè Visual Archives.)

strengths. In perturbation theory, it reproduced the results of all the Feynman diagrams except those involving closed loops.

Again I was participating in a formal exercise, even though Murph and I planned to use it for making an interesting calculation. We knew that the pion–nucleon resonance that Enrico Fermi and his collaborators were discovering downstairs (although Enrico hated the idea that it was a resonance!) was predicted by pseudoscalar meson theory with static nucleons in the strong coupling limit and even for intermediate coupling. In the corresponding relativistic meson theory (even if closed loops were omitted so that it looked like a relativistic generalization of the static theory), only the weak coupling approximation of perturbation theory was available. Our equation, we believed, would allow a strong coupling approximation to be developed, which might yield the same sequence of resonances as in the nonrelativistic theory; an interpolation suitable for intermediate coupling could then make contact with the experimental results. However, we never found such a strong coupling method. In the long run, that didn't matter, because it later became clear that nucleons and pions are composed of quarks (along with some contribution from gluons and quark-antiquark pairs), and so nobody cares any more what happens in relativistic pseudoscalar meson theory. Moreover, since that theory is not asymptotically free, it may not even be consistent.

Meanwhile, I was spending some time thinking about another problem entirely, one where the puzzle was of a rather primitive character, so that formalism was not very much involved. For several years, cosmic ray experimentalists had been collecting evidence for particle states (which I later named strange particles) that were copiously produced but decayed slowly, with lifetimes of the order of a ten-billionth of a second. After arriving in Chicago, I began to speculate about how this result could come about. I concluded that the strange particles must be strongly interacting objects (what we now call hadrons) but prohibited by a strong interaction conservation law from decaying into any hadron combination permitted by conservation of energy. The weak interaction, violating the selection rule, would then induce a comparatively slow decay. (By the way, thinking about the forces of nature, other than gravitation, in terms of these three interactions was not at all standard at that time.)

Searching for an appropriate conservation law, I naturally turned to the one then being confirmed by the pion-nucleon scattering experiments going on in the basement, namely the conservation of isotopic spin, respected by the strong interaction but not the weak or electromagnetic ones. I imagined, for example, that the particle we now think of as the sigma baryon might have $I=5/2$, with six charge states ranging from $+3$ to -2. Its decay into pion plus nucleon, with $I=1/2$ or $3/2$, would be forbidden by the strong interaction but might be permitted by the weak one. The difficultly with this idea soon became apparent, especially after conversations with Murph Goldberger and Ed Adams. The electromagnetic interaction could change $I=5/2$ into $I=3/2$ without exacting the huge penalty in the decay rate characteristic of the weak interaction. Only while lecturing on this puzzle at the Institute for Advanced Study in May, 1952 did I suddenly discover, by means of a slip of the tongue, that the difficulty would be resolved if the sigma baryon had $I=1$. In general, an isotopic multiplet with its center of charge displaced from the usual position (1/2 for a baryon, 0 for a meson) could be a multiplet of strange particles. Twice the displacement is what I later called the ''strangeness'' of the multiplet; total strangeness would be conserved by the strong and electromagnetic interactions but not the weak one. I even hypothesized correctly that the weak interaction would violate the conservation of strangeness by one unit at a time.

For complicated reasons I delayed almost a year and a half before publishing this idea, and even then I published only one of the three preprints that I circulated on the subject. (It is reproduced here, along with one of the unpublished ones.) I have told the story of my adventures with strangeness in much more detail elsewhere— that story has been touched on here only in order to illustrate the point about the influence of Viki's teaching on my work. By thinking about symmetry principles, which yield conservation laws and selection rules, I had found, like others before me, one way to reconcile elegance with qualitative insights relevant to experiment.

A number of us theorists spent the next twenty years trying, at least part of the time, to understand the symmetry rules governing the elementary particles and their interactions. For example, I worked, along with others, on the $J_\beta{}^+J_\beta$ form of the weak interaction (which suggested it was mediated by an intermediate boson with spin one); on the structure of that weak current (and of the related electromagnetic current) for hadrons; and on the approximate ''eightfold way'' symmetry of the hadrons, which led to the proposal of quarks as their fundamental constituents. We were evidently accomplishing a useful task, but it was not always obvious to everyone how useful it really was. After all, we were still putting off the task of constructing real dynamical theories of the weak and strong interactions. Eventually, we would have to realize the ambitious graduate student's dream, to discover the beautiful sets of equations describing those interactions and the particles participating in them. Weren't we still wandering around in the foothills of a tall mountain range that had to be scaled before those equations could be found?

Not really. By the time we had succeeded in identifying the patterns and the associated symmetries of the elementary particle system, the elegant theories were only a few steps beyond. One simple reason is that we were finding, by theoretical methods, the underlying particles, such as the intermediate bosons and the quarks. But a deeper reason is that, at least at the level of the standard model, the theories we needed were all gauge theories, with dynamics based on symmetry. The model was the Yang–Mills theory, originally based on the Lie group SU_2, which Shelly Glashow and I generalized to any simple compact Lie group or any product of such groups and of U_1 factors. (Even Einstein's general-relativistic theory of gravitation can be regarded as a gauge theory, albeit of a different kind, in which the relevant symmetries belong to the inhomogeneous Lorentz group.)

Once we had recognized the quark pattern underlying the hadrons and then, a few years later, understood that the quarks come in three colors, with the color SU_3 group exactly conserved, the step from symmetry to hadron dynamics was not very difficult to take. The strong interaction is described by a Yang–Mills gauge theory of quarks and gluons based on that color symmetry group—in other words quantum chromodynamics (QCD).

Similarly, the weak interaction is described, along with electromagnetism, by a Yang–Mills gauge theory with broken symmetry, one based on the Lie group $SU_2 \times U_1$ generated by the weak and electromagnetic charges. Thus the standard model consists in great part of a gauge theory built on the exact and approximate symmetries of the elementary particle system.

In most of science, the identification of pattern precedes the discovery of mechanism. For instance, Hans Wegener perceived the pattern of continental drift about half a century before plate tectonics was discovered. But in fundamental physical theory, the connection between pattern and mechanism is unusually close.

Discovering symmetries of the elementary particle system—and thus laying the groundwork for the construction of detailed dynamical theories of the interactions—was one way to implement the lesson I had learned from Viki in graduate school about making progress toward understanding dynamics without attempting directly the creation of a beautiful formal theory. But I made use of another way as well. That second method consisted of abstracting general results from quantum field theory—results that could be found by looking at the set of all Feynman diagrams but did not require the validity of a weak coupling approximation to be correct.

One case of that kind of abstraction is the discovery, in renormalizable quantum field theories, of what came to be known as the renormalization group. Francis Low and I found it by noticing that it was implied by the renormalizability itself, and we observed its operation in the perturbation expansion of quantum electrodynamics, where it applied to all orders.

Almost twenty years later, our work on quantum electrodynamics was extended (by Politzer, by Gross and Wilczek, and by 't Hooft) to non-Abelian gauge theories, that is Yang-Mills theories, and it was found that there the crucial quantity can have the opposite sign, yielding asymptotic freedom. This observation gave support to the idea of a color SU_3 Yang–Mills theory of quarks and gluons (which I subsequently named quantum chromodynamics), because an asymptotically free theory might predict the confinement of colored particles and might also have the required properties to agree with the findings of deep inelastic electron scattering experiments. As is well known, QCD is now believed to possess both of those features.

A different example of abstraction of results from quantum field theory is provided by the work during the 1960s on current algebra, in which commutation rules for charge densities were taken from a field theory of quarks coupled to a single neutral gluon. That theory was known to be wrong, but might still supply current algebra formulae that would hold in the correct field theory. I compared the process of abstracting results in this way to a recipe for cooking pheasant meat between two slices of veal, which are then thrown away. The veal was the theory with a single gluon.

Another instance of the abstraction of general results was the research that Murph Goldberger and I, among others, carried out during the years 1953 to 1955. Looking for results that held to all orders of the perturbation expansion (and therefore presumably held exactly) in quantum field theory, we found dispersion relations—some already known and some new—and also the crossing principle. We applied them to some specific problems, such as the scattering of photons from nucleons, where the underlying dynamical theory was unknown. (In fact that theory involves quarks and gluons, about which no one had yet dreamed.) We noticed, too, that in this work we could restrict ourselves to scattering amplitudes on the "mass shell," that is to say, amplitudes in which particles always obeyed the free-particle relation between energy and momentum. We did not have to deal with virtual amplitudes, which violate that relation. (Instead, we did have to work with on-shell amplitudes possessing complex values of energy or momentum or both.) Dispersion relations were connected with the analytic properties of the amplitudes expressed as functions of complex relativistically invariant kinematic variables on the mass shell. The unitary condition could be generalized to these complex values.

It soon became clear that generalized unitarity, crossing symmetry, and dispersion relations, applied to amplitudes on the mass shell but with complex variables, permitted an alternative formulation of quantum field theory. In 1956, I described that "dispersion theory" approach at the annual Rochester Conference. In my talk, I referred in passing, almost as a joke, to Heisenberg's ideas about the S-matrix.

Five years later, what we called dispersion theory was embraced and relabeled "S-matrix theory" by Geoffrey Chew, who tried to distinguish it from quantum field theory. I have always maintained, however, that it is merely the on-shell formulation of quantum field theory and not a distinct subject. In any case, Chew and Frautschi used the on-shell method to propose, for the hadrons, the "bootstrap principle," according to which the particles are all made up out of one another in a consistent manner, so that the same list of particles gives the constituents, the particles exchanged to yield the forces among the constituents, and bound states resulting from those forces. At first, the bootstrap idea was employed in approximations involving very few particle states. I suggested, in discussions with my colleagues at Caltech, that it would be much better to treat an infinite number of particle states crudely rather than to beat to death a very restricted problem such as that of the rho meson alone. In the initial approximation, all those states would have zero width, but in a higher approximation the excited states would all acquire non-zero widths as a result of their ability to decay. Our Caltech postdoctoral fellows Dolen, Horn, and Schmid made use of this notion in their seminal paper on "duality." The Veneziano model provided a beautiful example of duality with an infinite number of states (even though it was marred by including only bosons and by having, as one of those bosons, a spin zero particle with negative mass squared). It was soon pointed out that the Veneziano model could be regarded as a string theory.

In 1971, Neveu and Schwarz (making use of some results of Pierre Ramond) proposed superstring theory, which contains both bosons and fermions and has (as shown later) no problems with negative masses squared or negative probabilities. The apparent difficulty arising from the prediction of a zero mass particle of spin two was resolved by Scherk and Schwarz when they changed the scale of the theory by a factor of around 10^{19}, applied it to all elementary particles (not just the hadrons), and identified the spin two particle with the graviton. For the first time in history, a serious candidate had appeared for the role of unified quantum field theory of all the particles and interactions. Superstring theory predicts Einstein's general-relativistic theory of gravitation in a suitable approximation. Moreover, it provides the only known way to incorporate general relativity into quantum field theory without encountering preposterous infinite corrections to low order results.

Many of us theorists hope that superstring theory will soon be reformulated in a way that includes off-mass-shell amplitudes, in other words as a quantum field theory with a well-defined action functional. That will permit the utilization of the sum-over-paths method, with its many applications, including cosmological ones. But since superstring theory arose as an expression of the bootstrap principle in the mass-shell formulation, we can see that the practice of abstracting principles from the Feynman diagrams of quantum field theory ended up once again bearing fruit in the form of a dynamical theory, this time one that may be universal in its applicability.

By searching for patterns or structure, whether by looking for exact and broken symmetries or by abstracting general results from quantum field theory while the correct specific theory was still unknown, it was possible to make progress toward those beautiful equations, not by hastening toward them but by walking in other directions, as in the garden of live flowers.

The following paper is reproduced in its original form.

Isotopic Spin and New Unstable Particles*

M. Gell-Mann
Department of Physics and Institute for Nuclear Studies, University of Chicago, Chicago, Illinois
(Received August 21, 1953)

Peaslee[1] has considered the interesting possibility that the principle of charge independence, now believed to hold for nucleons and pions, may extend to the new unstable particles as well. In order to discuss this suggestion, let us suppose that both "ordinary particles" (nucleons and pions) and "new unstable particles" (V_1, V_4, τ, etc.) have interactions of three kinds:

(i) Interactions that rigorously conserve isotopic spin. (We assume these to be strong.)

(ii) Electromagnetic interactions. (Let us include mass-difference effects in this category.)

(iii) Other charge-dependent interactions, which we take to be very weak.

Peaslee inquires whether the quasi stability of the V_1^0 may be accounted for in this way if we assume it has isotopic spin 5/2. With respect to (i) the decay into pion and nucleon is absolutely forbidden. Interactions of type (iii) are supposed to be weak enough to account for the long observed lifetime of $\sim 3\times 10^{-10}$ second. However, he concludes that effects of type (ii) will cause transitions in a very much shorter time than this, since, for example, each electromagnetic interaction can change the isotopic spin of the system by one unit.

Recently Pais[2] has made the ingenious proposal that the new unstable particles differ from the ordinary ones in possessing one unit of "orbital isotopic angular momentum" and a negative "isotopic parity." If we then re-interpret (i) as referring to conservation of total isotopic angular momentum and isotopic parity, we see that as far as (i) is concerned, the decay of new unstable particles into ordinary is forbidden. Also, these particles will always be produced in even numbers, as Pais had suggested earlier.[3] Moreover, effects of type (ii) conserve isotopic parity, as Pais has introduced it, and so do not contribute to instability of the new particles.

In connection with the work of Peaslee and of Pais, the author would like to put forward an alternative hypothesis that he has considered for some time, and which, like that of Pais, overcomes the difficulty posed by electromagnetic interactions. Let us suppose that the new unstable particles are fermions with integral isotopic spin and bosons with half-integral isotopic spin. For example, the V_1 particles may form an isotopic triplet, consisting of V_1^+, V_1^0, and V_1^-. The τ^+ and V_4^0 may form an isotopic doublet, which we may call τ^+ and τ^0. To each of these particles there would presumably correspond an antiparticle,[4] which we shall denote by means of square brackets.

*Reprinted from THE PHYSICAL REVIEW, Vol. 92, No. 3, 833–834. November 1, 1953.

[1] D. C. Peaslee, Phys. Rev. **86**, 127 (1952).

[2] A. Pais (unpublished). The author is indebted to Professor Pais for the communication of his results prior to publication.

[3] A. Pais, Phys. Rev. **86**, 663 (1952).

In this scheme, (ii) is ineffective in causing decay because it can change isotopic spin only by integers, whereas in $V_1^0 \to \pi^- + p$, for example, the isotopic spin is 1 on the left and 1/2 or 3/2 on the right. Only interactions of type (iii), which do not respect isotopic spin at all, can lead to decay. Moreover, the new unstable particles again are produced only in even numbers.

There is no difficulty associated with stating a generalized Pauli principle for each kind of new unstable particle.[4] For example, let us postulate that the wave function of a collection of V_1's must be totally antisymmetric in space, spin, and isotopic spin. If the wave function of two V_1's is antisymmetric in space and spin, as it would be for particles of identical charge, then the total isotopic spin must be 0 or 2, which includes $V_1^+ V_1^+$, $V_1^- V_1^-$, and $V_1^0 V_1^0$. If the total isotopic spin is 1, the wave function is to be symmetric in space and spin, which is all right since the charges are then not identical. Similarly, the postulate that the wave function of a collection of τ's must be totally symmetric in space, spin, and isotopic spin leads to no contradictions.

It should be noted that according to this scheme the conservation of the z component of isotopic spin is more stringent than conservation of charge.[5] To see this, let us remark that the τ^+ and τ^0 have z components equal to $+1/2$ and $-1/2$, respectively, like the proton and neutron. Correspondingly the antiparticles $[\tau^+]$ and $[\tau^0]$ have z components equal to $-1/2$ and $+1/2$, respectively, like the antiproton and antineutron. Thus we see that the reactions $\pi^- + p \to V_1^0 + \tau^0$ and $\pi^- + p \to V_1^- + \tau^+$ are allowed, while the reactions $\pi^- + p \to V_1^0 + [\tau^0]$ and $\pi^- + p \to V_1^+ + [\tau^+]$ are forbidden, although all four are allowed by conservation of charge. In order to produce anti-τ's it would be necessary to resort to a reaction like $\pi^- + p \to n + \tau^+ + [\tau^+]$ or $\pi^- + p \to n + \tau^0 + [\tau^0]$.

In a similar fashion, all reactions of the form nucleon+nucleon$\to V_1 + V_1$ and all reactions of the form τ+nucleon$\to V_1 + \pi$ are forbidden, while reactions such as nucleon+nucleon$\to V_1 + \tau$+nucleon or $[\tau]$+nucleon$\to V_1 + \pi$ are allowed.

---

[4]We postulate the principle of invariance under the operation of charge conjugation, which carries every particle into its antiparticle. In the case of charged particles, such as the electron and the π^+, it is obvious that the antiparticles are the positron and the π^-, respectively. A neutral particle, however, may or may not be identical with its antiparticle. Among neutral fermions, it is necessary that the neutron and the antineutron be distinct, while the question of whether the neutrino and antineutrino are distinct is one that must be settled by experiment. Among neutral bosons, the γ ray and π^0 are apparently identical with their respective antiparticles, but there is no reason to believe that this is a general rule. We suppose here that the τ^0 is a neutral boson which is not identical with its antiparticle. A model for such a situation is provided by picturing the τ particle as a complex of a nucleon and an anti-V_1, while the $[\tau]$ is pictured as the corresponding complex of antinucleon and V_1.

[5]Of course the conservation of charge is absolute, while the conservation of the z component of isotopic spin can be violated by interactions of type (iii). Such violations should, however, play no important role in production phenomena.

On the Classification of Particles

M. Gell-Mann

Department of Physics and Institute for Nuclear Studies, University of Chicago, Chicago, Illinois

In a previous communication [to be referred to as (*I*)] the suggestion [M. Gell-Mann, Phys. Rev. (in press)*] was advanced that the "curious particles" (such as V_1, V_1, and τ) may be fermions with integral isotopic spin and bosons with half-integral isotopic spin. It has been assumed that strong interactions [type (i)] conserve isotopic spin, and it is clear that electromagnetic and mass-difference effects [type (ii)] conserve the z-component of isotopic spin. Then only very weak interactions [type (iii)] affect the z-component and lead to processes such as the decay of the curious particles into ordinary ones (nucleons and pions).

It is the purpose of this note to point out that

a) "Cascade V-particles" fit into the scheme in a natural way [C. D. Anderson *et al.*, Bull. Am. Phys. Soc. **28**, No. 5 (1953), papers H3 and H4].

b) The scheme suggests a classification of particles and their production and decay mechanisms.

c) The classification may have value even if the original idea should turn out to be wrong.

The language and notation of (*I*) will be freely employed.

Let us first remark that according to the isotopic spin assignments of (*I*) we have the following relations between the charge Q (in units of the proton charge) and the z-component of isotopic spin:

pion $(\pi^{(+)}, \pi^{(0)}, \pi^{(-)})$:	$Q = I_z$
nucleon $(n^{(+)}, n^{(0)})$:	$Q \approx I_z + \frac{1}{2}$
anti-nucleon $(\overline{n^{(+)}}, \overline{n^{(0)}})$:	$Q \approx I_z - \frac{1}{2}$
$(V_1{}^{(+)}, V_1{}^{(0)}, V_1{}^{(-)})$:	$Q \approx I_z$
$(\overline{V_1{}^{(+)}}, \overline{V_1{}^{(0)}}, \overline{V_1{}^{(-)}})$:	$Q \approx I_z$
$(\tau^{(+)}, \tau^{(0)})$:	$Q \approx I_z + \frac{1}{2}$
$(\overline{\tau^{(+)}}, \overline{\tau^{(0)}})$:	$Q \approx I_z - \frac{1}{2}$

Only those reactions can take place which conserve both Q and I_z, unless the weak interactions have time to intervene, as they do in the decay of the curious particles, when I_z changes by $\pm 1/2$.

*The reprint of the paper is attached.

The relation $Q=I_z$ characterizes both a V_1-particle and a system consisting of $n+\bar{\tau}$. In fact, whenever the production of a V_1 is allowed by our rules, so is that of $n+\bar{\tau}$, and vice versa. In this sense we may say that V_1 is *equivalent* to $n+\tau$ and write $V_1 \sim n+\tau$. (In the same way $n \sim n+\pi$, $\pi \sim n+n$, $\tau \sim n+\bar{V}_1$, etc.)

Now suppose there is a particle Λ that is equivalent to $n+2\bar{\tau}$. Then Λ is characterized by $Q=I_z-1/2$ and consequently cannot decay rapidly into either n + pions or V_1 + pions. In fact, if the decay mechanism always changes I_z by $\pm 1/2$, then Λ will decay by cascade

$$\Lambda \to V_1+\pi \text{ (slow)}, \quad V_1 \to n+\pi \text{ (slow)}$$

provided its mass is sufficiently large. We may thus identify Λ with the "cascade V-particle" of Anderson and co-workers. Since $\Lambda \sim n+2\bar{\tau}$, we must assign to this particle a half-integral isotopic spin; if $I=1/2$, then there are $\Lambda(0)$ and $\Lambda(-)$ particles.

Corresponding to the fermion Λ there might exist a boson $x \sim 2\tau$ along with its anti-particle $\bar{x} \sim 2\bar{\tau}$. For x we would have the relation $Q=I_z+1$ and for $\bar{x}$ the relation $Q=I_z-1$. If x has a suitable mass it too would decay by cascade by the process

$$x \to \tau+\pi \text{ (slow)}, \quad \tau \to \text{pions (slow)}$$

and $\bar{x}$ would behave similarly. Both x and $\bar{x}$ would have integral isotopic spin. If $I=0$, then x is positive and $\bar{x}$ negative.

Let us now represent each particle with which we are concerned as equivalent to a number of nucleons or anti-nucleons plus a number of τ's or $\bar{\tau}$'s. If we treat an anti-nucleon as minus one nucleon and a $\bar{\tau}$ as minus one τ, then we may write relations of the form

$$\text{particle} \sim xn+y\tau$$

and we may construct a table of particles as follows:

$\bar{\Lambda}$ $(-1)n+(2)\tau$ $Q=I_z+\frac{1}{2}$	$x(\xi)$ $(0)n+(2)\tau$ $Q=I_z+1$	$(1)n+(2)\tau$ $Q=I_z+3/2$
$\bar{V}_1$ $(-1)n+(1)\tau$ $Q=I_z$	τ $(0)n+(1)\tau$ $Q=I_z+\frac{1}{2}$	$(1)n+(1)\tau$ $Q=I_z+1$
$\bar{n}$ $(-1)n+(0)\tau$ $Q=I_z-\frac{1}{2}$	π $(0)n+(0)\tau$ $Q=I_z$	n $(1)n+(0)\tau$ $Q=I_z+\frac{1}{2}$

$(-1)n+(-1)\tau$ $Q=I_z-1$	$\bar{\tau}$ $(0)n+(-1)\tau$ $Q=I_z-\frac{1}{2}$	V_1 $(1)n+(-1)\tau$ $Q=I_z$
$(-1)n+(-2)\tau$ $Q=I_z-3/2$	$\bar{x}(\xi)$ $(0)n+(-2)\tau$ $Q=I_z-1$	Λ $(1)n+(2)\tau$ $Q=I_z-\frac{1}{2}$

JOURNAL DE PHYSIQUE
Colloque C8, supplément au n° 12, Tome 43, décembre 1982 *page* C8-395

STRANGENESS

M. Gell-Mann

Lauritsen Laboratory of Physics, California Institute of Technology, Pasadena, CA 91125, U.S.A.

Résumé - L'exposé ne prétend pas être une mise au point historique de la naissance du concept d'étrangeté. C'est simplement un récit de souvenirs personnels sur les idées et l'atmosphère de la période 1951 - 1953. Les raisons qui ont guidé l'auteur pour introduire "l'étrangeté" sont développées et il dit comment il a eu à convaincre et surmonter les oppositions.

Abstract - This paper is not a history of the discovery of the strangeness, but rather a contribution to such a history, consisting of personal reminiscences. The atmosphere and ideas of the period 1951 - 1953 are described. The author explains the reasons that led him to introduce the concept of the strangeness and how he had to convince people and to overcome oppositions.

I have not prepared a history of the discovery of strangeness, but rather a contribution to such a history, consisting entirely of personal reminiscences. I will not be able to discuss how Nishijima and his colleagues arrived at similar conclusions. Some of them are here, including Nishijima himself, and I hope that they can comment on it. Also I have not carefully studied the published material, not even my own published material, which is itself very sparse, and so I can't claim in any way to be giving a presentation that belongs in the realm of historical research. Rather it resembles a story told by an old farmer near a peat fire recollecting his youth, or something of that sort. Such accounts are often recorded these days.

Let me try first to recall briefly, especially to the younger people here , if there are any, what it was like at that time, 1951 to 1953.

Strange particles had been discovered experimentally, as you heard from many of those who took part in the work. Such particles were not considered respectable, especially among theorists. I am told (Dick Dalitz, who is here, can perhaps confirm it) that when he wrote his excellent paper on the decay of the tau particle into three pions Dalitz was warned that it might adversely affect his career, because he would be known as the sort of person who worked on that kind of thing. Second, speculation by theorists in the physics journals was not considered particularly respectable. In fact theoretical physics itself had not been respectable during the decade prior to 1948, when the muon didn't have the properties of the meson, and, even worse, theorists dealt with field theory, which, as soon as you tried to correct the lowest order, gave infinity. Apparent defects in theory had led to a situation in which theorists hung their heads in shame all the time and were not taken very seriously. Well, those defects had just been remedied at the time of which I am speaking, but theorists were still not encouraged to speculate. The journals did welcome innumerable articles on perturbation calculations in field theory, even when the coupling was strong and the theory, for example the pseudoscalar meson theory, was not very useful.

Not everyone thought at that time in terms of strong, weak, and electromagnetic interactions, but I was one of those who did. The weak interactions had been unified by Puppi, by Klein, by Lee, Rosenbluth, and Yang, by Tiomno and Wheeler, and so forth around 1949. Another concept that had just recently been clarified was that of baryon conservation, discussed by Wigner in 1949, and no one understood why it should be exact in the absence of a long range vectorial force to accompany the baryon charge. These days one says : "Well, probably it isn't exact". Very simple !

Now the problem of strange particles was in the air in 1951 and 1952, the puzzle of why they were produced copiously, at a reasonable fraction of the rate of production of pions, but decayed very slowly.

I arrived at Chicago as an instructor in January 1952 and I worked on a variety of subjects. One of them was, of course, the ($g\ \bar{N}\ \tau\gamma_5 N\ \pi$) field theory, which was so popular at that time, but at least Goldberger and I were studying a way to do non-perturbative calculations in that theory. I also collaborated with an experimental physicist called Telegdi on isotopic spin physics in the nuclear domain. These were my main occupations. As a sideline I began to look into the strange particles in the the winter and spring of 1952.

Isotopic spin was again much in vogue. Although the charge independence of nuclear forces should have settled the usefulness of isotopic spin many years earlier, as indicated in the beautiful talk by Professor Kemmer, in fact it had been falling out of favor for reasons that were obscure to me ; but there was a great revival of interest in isotopic spin as a result of the work at Chicago, where Fermi and his collaborators found that pion-nucleon scattering was indeed charge independent, so that the pion had isotopic spin one, the nucleon had isotopic spin one half, and the vector sum was conserved. I had always been interested in isotopic spin conservation, and early in the winter of 1952 I began to wonder whether isotopic spin could explain the behavior of strange particles. I tried assigning $I = 5/2$ to a strange baryon, assuming that there would be many as yet undiscovered charged states. Isotopic spin conservation would then prevent the V_1° particle, as Λ° was called, from decaying into nucleon plus pion, and conservation of energy would prevent its going into nucleon plus two pions. The Q value was quite well known.

I soon realized though, in thinking about it and also in discussing it with Ed Adams and Murph Goldberger, that the electromagnetic interaction would ruin the scheme, by changing isotopic spin by one unit. I dropped the idea. Many years later I heard that Okun, when he had an idea that sounded good but to which there seemed to be fatal objections, was given the advice that he should publish the idea with the objections. It never entered my mind to do that, but it was done by another physicist, Dave Peaslee, whom I had never met but who had been my predecessor as graduate student and assistant to Viki Weisskopf at MIT, and who was at Columbia. Apparently he had the same idea, found the same objection, and published the idea with the objection. It appeared as a letter to the Physical Review on April 1, 1952. (No connection intended with the "poisson d'avril"). I didn't read the article at the time, I only glanced at it for a few seconds, but a couple of days ago I tried to read it and found it difficult to follow.

A few weeks after that, probably in May of 1952, I paid a visit to the Institute for Advanced Study, where I had spent the previous year. While I was there someone asked me whether I had read Peaslee's letter. I described the situation, and I was then asked to get up and talk for a few minutes in the seminar room on the idea and why it wouldn't work. I don't recall exactly who was there but I think Francis Low, T.D. Lee Abraham Pais, and various others. In my explanation, as I got to the $I = 5/2$ proposal I made a mistake, a slip of the tongue, and said $I = 1$. I paused and didn't go on with the talk for a minute or two because I was thinking to myself "$\Delta I = 1$ or 0, $\Delta I_z = 0$ are the rules for electromagnetism ; if we need $\Delta I = 1/2$ and $\Delta I_z = \pm 1/2$ for decay electromagnetism will have trouble doing that, and the problem is solved. "I went on, but at the end I said : "by the way, a few minutes ago I got what I think is the right idea. If this V particle belongs to a triplet, plus, zero and minus, with $I = 1$, electromagnetism will have great difficulty causing decay ; we don't know of any kind of electromagnetic interaction that will change isotopic spin by a half unit, or the z component by a half unit, and so the decay can be weak "I might have gotten very excited about it at that time but in fact the audience was not very enthusiastic.

Let me say a word now about getting ideas in that way. Years later in Aspen, Colorado, we had a discussion at the Aspen Center for Physics about how one gets ideas in physics, in poetry, in painting, and in other subjects. There were two painters, one poet, and a couple of other people. I spoke about this incident involving a slip of the tongue. The others spoke about some of their problems. It was agreed that in all these quite different domains one sometimes tries to achieve something that is not permitted by the traditional framework. It is necessary to go outside the usual framework in some way in order to accomplish the objective. In theoretical physics this frustration usually appears as a paradox. But a paradox is after all just one way of having your path blocked ; in art the blocking is manifested differently. Having filled your mind with the problem and the difficulty you may then find that in an odd moment while driving or shaving or while asleep and dreaming (as in the case of Kekulé and the benzene ring) or through a slip of the tongue as in this case one may suddenly find the path unblocked. Perhaps the solution comes, in the language of the psychoanalyst (a language that is not very popular in scientific circles today), from the preconscious mind, the portion of our mind that is just out of awareness.

To return to May, 1952, the audience, as I said, was not very enthusiastic. Abraham Pais came up and started to tell me that he had just written a long paper on associated production of strange particles with an even-odd rule. The strong interaction allowed even plus odd going to even plus odd or even plus even going into odd plus odd, but only a weak interaction would allow odd into even. My idea as I had described it (and I had mentioned that it would obey this kind of rule) was, he said, just a subcase of his idea and therefore not very important. What I should have done was to point out quickly that after all there were some differences. Isotopic spin was already familiar and not a new ad hoc symmetry. I would lead to an additive conservation law for I_z with experimental consequences. For example, neutron plus neutron going to what we would now call Λ plus Λ would be forbidden whereas that was allowed according to his scheme and was supposed to be one of the principal tests of the idea of associated production. Also the charge multiplets would be seen to conform to the new idea singlet and triplet for the baryons, doublet for the mesons, and this could be verified by observation of the states. But I did'nt like to stress the importance of my own work and I didn't say much.

In my subsequent papers I have often started, in explaining the work on strangeness, from associated production and from the elegant paper of Abraham Pais. In fact, though I was unacquainted with his work and did not proceed from associated production. I learned about associated production just as I invented the scheme. But logically, for purposes of explanation, it was better to discuss associated production first and then the special idea of the connection with displaced isotopic multiplets, (for example, in the Scientific American article that I wrote later with Ted Rosenbaum). Historically, it was inaccurate.

Now associated production as it turned out had been treated earlier, particularly in Japan in 1951 in the Progress of theoretical Physics by three sets of authors : by Nambu, Nishijima and Yamaguchi, by Oneda, and by Miyazawa. I have just tried to read, in the last few days, the papers by these authors and although I have never referred in my subsequent work to Miyazawa, I noticed that his paper was actually very good. He used the bound state approach but that didn't make much difference ; the effect of it was to predict more or less the correct situation of associated production. Oneda's work was somewhat less perspicuous but he certainly had the notion. Nambu, Nishijima, and Yamaguchi wrote up in an encyclopedic manner all possible explanations, but laid particular emphasis on associated production as being an interesting possibility although apparently contradicted by experiment. They even included the idea of high angular momentum as one possible explanation in their series of letters. Theirs is a very nice piece of work that is not usually mentioned. Feynman told me later that he had thought of the idea of associated production in 1951 and immediately began to talk with the Caltech experimentalists who were doing some very good work on strange particles. I don't believe they are represented here, but theirs was one of the important laboratories at that time. They told Feynman that associated production did not seem to be correct. Cosmic rays were apparently not ideally suited for finding associated production and the experimentalists discouraged Feynman from continuing in that direction. He therefore took up the idea of high angular momentum as the way in which a particle could be restrained for a long time from decaying while being produced copiously. Fermi, on a visit to Caltech, discussed the same thing and the two of them collaborated a little bit at long distance on the idea of high angular momentum as an alternative explanation.

From Princeton, I proceeded to make my first visit to Europe in June 1952. Here in Paris, Bernard d'Espagnat, whom I knew from Chicago, kindly introduced me to the research group from the Ecole Polytechnique and Louis Leprince-Ringuet generously invited me, a complete stranger, to the meeting of his research group at his country home in Courcelles-Frémoy in Burgundy. Peyrou was away, unfortunately, and it was some years before I met him. But among those present were Bernard Gregory, whom I knew slightly from our graduate days at M.I.T., Louis Michel, Jacques Prentki, André Lagarrigue, Francis Muller, Agnès Lecourtois, and many many others, some of whom are here today.

It was a wonderful experience to meet them and in many cases we have been friends now for thirty years.

At Courcelles-Frémoy I gave the first talk on strangeness after the slip of the tongue in Princeton, but I went very easy on isotopic spin, because at that time it was considered extremely difficult to explain to experimentalists.

On returning to Chicago in the Fall of 1952 I related my idea in detail to the weekly seminar of the Institute for Nuclear Studies (now named after Fermi). It was a kind of Quaker meeting where one could get up and say anything one wanted. Fermi was unfortunately absent, but Dick Garwin was there and at the end of my little talk he was very negative, saying he couldn't see what use my idea could possibly be. Again, if I had been less averse to promoting my ideas, I would have explained that it had all sorts of experimental consequences such as the distribution of charge states, the prohibition of $n + n$ going to $\Lambda + \Lambda$, and so forth. He was at that moment engaged in the experiment on $n + n$ giving $\Lambda + \Lambda$, which gave a negative result. (I am told that Pontecorvo, by then called ПОНТЕКОРВО, did this experiment independently in the Soviet Union, but Garwin was doing it just then in Chicago). If only we had conversed in more detail the world would have become aware of strangeness much sooner, I should mention that Garwin later apologized handsomely for his skepticism.

I became discouraged again and put away strange particles for a while. I worked with Goldberger on the crossing theorem, dispersion relations, and other exact general results extracted from field theory. Early in the summer of 1953 I went to Urbana, Illinois, where Francis Low and I did our work on the renormalization group and on the spectral formulae for propagators, published more than a year later. It seems that I could not publish anything without leaving an interval of at least a year or a year and a half.

I mentioned the strangeness idea in its complete form to Francis Low and T. D. Lee, who had both been present at the slip of the tongue a year earlier. They were somewhat impressed, but I think not very much, probably because I did not explain things very forcefully. At that time I disliked giving a clear presentation in the didactic style, probably in reaction to my father who was a private teacher of languages and had a very didactic style.

I returned to Chicago late in July 1953, when it was terribly hot. I found a draft induction notice. The secretary of the Institute Director had failed to send in the yearly notification to the draft board of my being engaged in research at the University of Chicago, and the draft notice was the result. I imagined that I would immediately be drafted and sent to Korea. The fighting was over but guard duty in Korea would not be ideal for working on theoretical physics and I decided to write up strangeness immediately, after fifteen months'delay, on the grounds that it would be amusing to have this in print while I was over there in the Army.

I never did go into the Army but I did write up the paper. Valentine Telegdi kindly lent me a desk in his air-conditioned lab on which to do the writing. Only equipment was thought to require air-conditioning at that time ; there was no air-conditioning for brains. I often wanted to have a wax pencil that would melt at the same temperature at which I became incapable of thinking, so that I could say I needed the air-conditioning for my equipment, but it never worked. However, on this occasion air-conditioning was available and, at the desk of Valentine 's Manchurian student, I started writing up the idea of strangeness.

Meanwhile, Pais had come in the summer of 1953 to the idea that the even-odd rule he had proposed the year before would come from an orbital isotopic spin, which would be added to an intrinsic isotopic spin to make the total, and that it was the parity of this orbital isotopic spin that would give the even-odd rule. I heard in the summerof 53 that he was invited to the world conference in Kyoto and that he was making a big splash there with these ideas about orbital angular momentum in isotopic space. Jealousy was another reason why I decided I would put forward the strangeness scheme. I thought that it was probably correct and I resented the publicity being given to the scheme of Pais, which I was convinced was wrong !

In my letter to the Physical Review, I placed great emphasis on the conservation of I_z, which is equivalent to the conservation of strangeness, and I showed how $n + n \to \Lambda + \Lambda$ is forbidden. (I still referred to Λ as V_1° but it had just been named Λ at Bagnères de Bigorre). I pointed out that $\pi^- + p \to \Lambda^\circ + K^\circ$ or $\Sigma^- + K^+$ is allowed but $\Sigma^+ + K^-$ is forbidden (here I use the names we invented later).

At the same time I wrote a companion piece, which is exhibited in the next room. In August 1953, they were distributed together as preprints to laboratories all over the world. The companion piece was called : "On the Classification of Particles" and it went much further than the other preprint. It went into great detail on the multiplet structure and on the existence of doubly strange cascade particles, of which two had been seen, one by Armenteros et al. and one by the group at Caltech. Herb Anderson , who was most enthusiastic about strangeness right from the first day he heard about it, had brought me a copy of the preprint from Caltech and had challenged me to explain the cascade particle and to include the explanation in my written work. I predicted what we now call the Ξ° particle to accompany the Ξ^- and I also suggested that to explain the decay in two steps via Λ we should postulate that weak non-leptonic strange decays obey $|\Delta I_z| = 1/2$ (or strangeness changing by one unit).

Although I didn't use the word strangeness yet, I did have the quantity,which I called y , and in effect gave the formula

$$Q = I_z + \frac{N}{2} + \frac{y}{2}$$

In fact, I described each particle as equivalent to N nucleons and $y\tau$ mesons, and I explained that "equivalent to" meant having the same difference between I_z and charge. Evidently N is the baryon number.

I wrote a third paper around the same time (about September 1953), which I didn't even distribute widely as a preprint, in which I suggested that not only was $|\Delta I_z|$ equal to a half in weak nonleptonic decays but also that ΔI was approximately equal to a half, thereby explaining Dalitz's work on the isotopic spin of the pions in τ decay. You start with isotopic spin one half for the τ particle , you add a half unit, and you get either one or zero. But in the charged state of pions you can't have I=0, so you have only I = 1, and therefore the three pions in the final state have a pure isotopic spin of one to the extent that $\Delta I = 1/2$ is correct. Why I didn't publish the second and third preprints right then I don't know. It seems that I just had to let things ripen for a year or two.

Much of my work was included the next summer in the article in the Proceedings of the Glasgow Conference, July 1954, which I wrote together with Abraham Pais. We included three models in that paper, after discussing the high angular momentum hypothesis , which we said we didn't believe. First we gave his orbital isotopic angular momentum scheme, then my strangeness scheme, and then a third one, which generalized strangeness to restore the symmetry around charge one-half in the baryon system. We said that the last model was very speculative and also that it didn't appear to be right, because the experiments didn't seem to find the extra states. Looking back on it in the last few days I realized that what the third scheme amounts to, if we supply a couple of missing states, is assuming a charmed quark with a mass similar to that of the strange quark.

Anyway I then didn't bother to send the 1953 preprint "On the Classification of Particles" for publication since its content was mostly included in a section of the joint paper for the Glasgow meeting in 1954.

Now let me return to the paper that I did send off in August 1953. It is also on display in the next room : Isotopic Spin and New Unstable Particles. That was not my title, which was : Isotopic Spin and Curious Particles. Physical Review rejected "Curious Particles". I tried "Strange Particles", and they rejected that too. They insisted on : "New Unstable Particles". That was the only phrase sufficiently pompous for the editors of the Physical Review. I should say now that I have always hated the Physical Review Letters and almost twenty years ago I decided never again to publish in that journal, but in 1953 I was scarcely in a position to shop around.

They also objected to the neutral boson being different from the neutral anti-boson ; that was a very sore point. Their referees couldn't understand how K^{o} could be different from $\bar{K}^{o}$. I didn't know what to do to convince them. I tried saying merely "It's all right, they can be like that, "but failed to change their minds. Then a thought occurred to me. I decided, in order to learn about neutral mesons, to look up the paper by Nick Kemmer in which he had proposed the isotopic triplet. I had met him that previous summer of '52 in Cambridge, where he was very kind to me and I was very impressed with him. I discovered that a large portion of his paper was devoted to showing that a neutral boson does not have to be different from its anti-particle ! What he did was to take the Pauli-Weisskopf theory of the charged scalar particle and take away the charge, which left him with a neutral particle *different from its anti-particle. Then he argued at great length that it was* not absolutely necessary to have it that way. He wrote the complex field as a real field plus i times another real field and pointed out that it was possible to use just one of the real fields and omit the other. In that way he was able to get what we now call the π^{o} and adjoin it to π^{+} and π^{-} to make the isotopic spin triplet. When I recounted this story to the Physical Review they finally agreed that it was O.K. to have a neutral boson different from its anti-particle.

In the meantime, though, I was reminded that you could take the two real parts and consider them as real fields if you wanted to. That was to be useful later.

Another thing I had to do for the Physical Review was to explain that the generalized Pauli principle was applicable to fermions with integral isotopic spin and to bosons with half integral isotopic spin. It was widely believed that there was a mathematical demonstration that fermions had to be isofermions and bosons had to be isobosons because that was the only way the Pauli principle could be generalized to include isotopic spin. It simply wasn't true, and I succeeded in pointing that out.

Around the same time, August or September 1953, the first accelerator results were being obtained on strange particles. One or two associated production events were observed. I called Brookhaven to find out in the case of the charged associated production whether they had seen what we would now call $\Sigma^{+} + K^{-}$, which would make me unhappy, or $\Sigma^{-} + K^{+}$, which would be good. (At that time $\Sigma^{\pm}$ was called $V_1^{\pm}$). I phoned Brookhaven and got hold of Courtenay Wright, an experimentalist from Chicago who was visiting Brookhaven. He asked : "What possible difference does it make ? who cares ?" I said merely "I care ; please find out". He asked me to hold the phone, he was gone quite a while, and when he came back he said : "I checked and they are sure that it is V_1^{-}". I let out a cheer over the telephone, which mystified Courtenay Wright, but which meant that the one event that had been seen was compatible with strangeness.

On a visit back to Urbana I saw Geoff Chew, who had been away during the summer. Chew was much taken with the cosmic ray result that K^{+} production predominated over K^{-} production and he said mine was the first theory he had ever heard of that would explain it. He gave a colloquium a couple of days later in which he presented the strangeness theory from that point of view, as an explanation of the predominance of K^+ over K^-. T.D. Lee wrote me to say that he had just invented a new scheme, but that it occured to him that maybe it was the same one that I had told him about (It was in fact the same). That was a very nice thing for him to do.

Fermi then returned to Chicago and I went to see him. That was an important moment. He sounded very skeptical when I told him about explaining the strange particles by means of displaced isotopic spin multiplets. He said he was convinced more than ever that high angular momentum was the right explanation. I was a great admirer of Fermi; I also liked him very much and enjoyed his company. I was unhappy when he rejected my scheme. A day or two later, though, I did something that no gentleman is supposed to do, I read someone else's mail. I was in the office late in the evening, and out of boredom I started to look at what our secretary (I think her name was Vivian) was typing. It was a reply from Enrico to Giuseppe Cocconi, who had written him that he was looking at the consequences of Fermi's and Feynman's proposal of high angular momentum for the new particles and that he had gotten some nice mathematical results that he wanted to communicate. Enrico wrote him as follows, more or less, (I para - phrase because I don't remember the exact words or indeed the language) : "Dear Cocconi, I was pleased to receive your results. However, I should tell you that here at Chicago Gell Mann is speculating about a new scheme involving displaced isotopic spin multiplets and perhaps that is the explanation of the curious particles rather than high angular momentum". I stopped instantly being depressed, but for a while I was somewhat annoyed at Enrico.

On a visit to New York and Princeton in September 1953, I gave the name strangeness to this quantity $\mathcal{Y}$, and after talking with Serber and Lee at Columbia I decided that it was necessary to postulate a triplet and a singlet, that the mass difference was just too great between V_1^o and the charged V_1's for them to form a triplet and that there must be a Σ^o which decayed by γ emission to Λ . I was predicting three new neutral particles : Σ^o, Ξ^o, and, with K^o being different from $\bar{K}^o$, a second neutral K particle, the properties of which I was then thinking about. All of those neutral objects were the objects of experimental searches during the next year or so.

In June of 1953, at Bagnères de Bigorre, it had been recommended that baryons be assigned a capital Greek letter and mesons a small Greek letter. I decided to use Σ for the triplet, Ξ for the new doublet. For the bosons it was very complicated. We had θ for the decay into two π's and τ for the decay into three π's, and people were very confused about the relation between the two. Some wanted to use K for θ and τ together although it is not a small Greek letter, and was intended as a generic term for strange bosons. I wanted to use κ, but κ was assigned to a leptonic decay mode discovered by O'Ceallaigh et al., and so we were stuck with the Latin capital letter K.

Back in Chicago I gave a colloquium and then or later Fermi attacked at least one of my ideas, namely my statement that the electromagnetic interaction would have $\Delta I = 1$, $\Delta I_z = 0$. Fermi said it was not necessarily true, that in fact it could also have $|\Delta I_z| = 1/2$. $|\Delta S| = 1$ and he wrote down an electromagnetic interaction that would have that character, namely

$$F_{\mu\nu}(\bar{n}q_{\mu\nu}\Lambda + herm.conj.)$$

a gauge-invariant coupling through which n and Λ would be interconverted directly by the emission of a photon. I replied to Fermi that he was violating what I considered to be a fundamental principle of electromagnetism, namely that electromagnetism doesn't do dirty little jobs for people, but has a coupling that flows directly from the properties of matter. In fact, as I explained it in the new few weeks, thinking about Fermi's objection, we have the physicist's equivalent of the biblical "Fiat lux" "Let there be light" which looks like this : Take the Lagrangian without light and then let p_μ go into $p_\mu - eA_\mu$. This is what I called pompously the principle of minimal electromagnetic interaction . As Valentine Telegdi kindly pointed out to me, it was merely a generalization of Ampère's law.

Fermi made another objection at a course that I gave during the Fall of 1953 (or perhaps the winter of 1954) in Chicago. Whenever Enrico came to a seminar, a lecture, a colloquium, or a course, if he didn't like anything he interrupted. The interruption was not a minor matter ; it continued until Enrico felt happy about what the speaker was saying, which often took essentially forever, that is to say the seminar ended, Enrico was still not happy, and the speaker never finished what he was going to say. If it was a course, as in this case, the course could be blocked for a week or two, while at each class he came in and started objecting where he had left off at the end of the previous class. At my course his principal objection was to the idea that one could have a neutral boson different from its anti-particle. I thought, "Here we go again, just like the Physical Review. I only hope he isn't the referee with whom I had all the trouble". Finally he came up with a clinching argument. He said, "I can write K^o = A + iB, where A and B are both real fields with definite charge conjugation, and you have in each case a neutral particle that is its own charge conjugate". Well, I had already been through this and I was able to answer : "Yes , that's true, but in the production of strange particles, because of strangeness conservation it is the K^o and $\bar{K}^o$ that matter ; in the decay, if it is into pions or photons or both, then it will be your A and B that matter and that have different lifetimes." I don't remember whether my reply was delivered in class or privately afterwards. I think it must have been the latter because I hadn't explained the strangeness theory in detail to the class. Anyway, Fermi's objections gradually subsided. This was the origin of the work on K_1^o and K_2^o because A is K_1^o and B is K_2^o.

I didn't write it up though for another year and when I did it was with Abraham Pais, who gave me much encouragement in publishing the idea. Again it required that peculiar neurotic gestation period of a year or a year and a half before I could manage to publish. It is very strange !

Of course charge conjugation, which was so important in this argument, later had to be amended to CP, but then the argument went through exactly the same with CP as it had previously with C. Finally CP was found to be violated too and even K_1^o and K_2^o got slightly mixed, but that is of course a much later story, dating from 1964.

I think it was early in 1955 at Rochester that I discussed the weak leptonic decays of strange particles, with the rules $\Delta I = 1/2$ and $\Delta S/\Delta Q = +1$, but I am not sure whether the discussion appeared in the proceedings.

In July, 1955, in Pisa, I finally gave a straighforward full and didactic presentation of all these ideas in public, and published it. Nishijima also waited until about then to give a full presentation of the whole scheme, with the classification of particles, the selection rules, and everything. He also must have thought of all these things earlier and perhaps he can explain the delay in his case ! Meanwhile the experimental labs had been sent copies of the preprints, even the ones that weren't published. They all knew of the predictions of Σ^o, Ξ^o and the second kind of neutral K particle, K_2^o (K_1^o was the θ^o , which was very well known) and various experimentalists very kindly sent me beautiful signed photographs of the events in which these predicted particles were unambiguously found. Jack Steinberger was particularly nice about sending me such photos and in his published comments about the usefulness of my predictions. Later on, in another connection, he sent Feynman and me a photograph inscribed "You may stuff this and hang it".

During those years I was concerned with renormalization group, with dispersion relations, crossing relations, and combining these with unitarity to make a theory of the S-matrix, and so forth. Strange particles were only a part of my work. But they helped me to get those souvenirs from my experimental friends that I treasure to this day.

Thank you.

DISCUSSION

N. KEMMER.- May I add a brief observation on Prof Gell Mann's account of finding in my 1938 paper support for his proposal to introduce a neutral particle with a distinct antiparticle ? In fact, when I wrote that paper, that kind of particle seemed the more natural thing to have and my lengthy discussion on how to introduce a neutral "pion" that had a real state vector was supposed to be a justification of this strange step. When Pauli and Weisskopf first showed how to quantize a charged boson field it was easy to see the link between their field, based on what we then always called the relativistic Schrödinger equation which later somehow get to be called the Klein-Gordon equation) and the non-relativistic Schrödinger equation. Schrödinger's expression for probability density stood in a very simple relationship to the Pauli-Weisskopf charge density : confined to wave packets with only $E > 0$ or only $E < 0$ components they were essentially the same thing. Whether the particles described were charged or not, this seemed the natural way of interpreting the relativistic Schrödinger equation. The relation between the charged and uncharged bosons on this view was the same as between electron and Dirac neutrinos. The equivalent step to passing from Dirac to Majorana neutrinos for the boson case was just to make the Pauli-Weisskopf ψ real. This seemed to present a problem : there was no longer an easy way of linking anything in the Pauli-Weisskopf formalism to the non relativistic probability density $|\psi|^2$. I think I was quite clear in my mind that this point could be settled and in 1946 I asked my research student at Cambridge K. J. Le Couteur to do this (Proc. Camb. Phil. Soc. 44(1947) 229). I think that the "great detail" of my discussion of this point in my paper is explained by my awareness of this problem.

Professor Nishijima was then invited to give the contribution he had been asked to prepare on the subject of strangeness :

K. NISHIJIMA.- The history of the discovery of strangeness has been told in detail by Professor Gell-Mann, and I would like to add to it personal reflections from another corner of the world.

The experimental observations of V particles by the Pasadena and Manchester groups [1] gave us a strong stimulus to start working on this problem. The natural question to be asked was that of how to reconcile their abundance with their longevity. Years before we had a similar problem and its solution was given by the recognition of the existence of two kinds of mesons, π and μ . Such an idea could work only once, however, and could not be extended to cover V particles.

Many groups in Japan started to work on this challenge, and each group reached its own solution. In order to compare and exchange ideas among them, a symposium was held on July 7, 1951 in Tokyo. There were reports by Nambu, Yamaguchi and myself [2] from Osaka, by Miyazawa [3] from Tokyo, and by Oneda [4] from Tohoku.

Although various models of V particles had been presented by different groups, everybody had recognized that one thing was almost in common. That was the pair production of V particles. It struck all of us that this could be the only way to prevent V particles from decaying rapidly through strong interactions. At that time, however, cosmic ray experiments did not seem to support this idea. We had to wait for the Cosmotron experiment [5] in 1953 to confirm the pair production of V particles.

Meanwhile various theoretical ideas had been reorganized and reformulated. For instance, the formulation of the pair production of V particles in terms of the so-called even-odd rule by Pais [6], and the reformulation of various selection rules originally discovered by Fukuda and Miyamoto [7].

In the even-odd rule one assigns an integral quantum number to each hadron. What is relevant is whether that quantum number is even or odd, and one assigns a sort of parity to each hadron. Let us call it V parity to distinguish it from space parity. One assigns even V parity to nucleons and pions and odd V parity to V particles. Then this multiplicative quantum number is postulated to be conserved in strong interactions or in production processes, but it is then postulated to be violated in weak interactions or in decay processes.

I called this multiplicative quantum number the V parity tentatively, but it was really a forerunner of the space parity in the sense that both are conserved in strong interactions but are violated in weak interactions. It is interesting to recall that the violation of space parity in weak interactions shocked the world whereas the corresponding aspect of V parity slipped in without calling any resistance.

Introduction of the multiplicative quantum number was not sufficient, however, in interpreting the experimental data that had been accumulated by then. First of all, the cascade particle Ξ^- decaying into $\Lambda^\circ+\pi^-$ had already been known. In order to forbid Ξ^- from decaying into this channel through strong interactions one has to assign even V parity to Ξ^-, but then one cannot forbid the decay $\Xi^- \to n+\pi^-$ through strong interactions.

Another difficult problem was the interpretation of the positive excess of the heavy mesons then known experimentally [8]. The identified charges of the most of the observed heavy mesons were positive, and it was one of the key problems to explain this property since the multiplicative quantum number was of limited capability.

From a theoretical point of view we did not have a basic principle which enabled us to assign V parity to each hadron.

A great leap forward was made when the cosmotron at Brookhaven started to operate in 1953. The experiment by Fowler, Shutt, Thorndike and Whittemore [5] clearly revealed the pair production of V particles that could not be confirmed by cosmic ray experiments. The abundance of V particles also assured us of the fact that they are produced by strong interactions.
Since strong pion-nucleon interactions are charged independent and observed deviations from it are rather small [9], strong interactions of V particles must also respect charge independence in order not to disturb the charge independence in pion-nucleon interactions.

Once charge independence is assumed for V particles the next step is the isospin assignment to V particles, through which the concept of strangeness emgerged.

Therefore, I think that the key issue in the introduction of strangeness consists in the charge independence hypothesis. Once this postulate is made, everything follows automatically. Charge independence is respected by strong interactions but is violated by electromagnetic interactions and small mass differences among members of an isospin multiplet. I_3, the third component of the isospin is respected by both of them. It is violated only by weak interactions such as the beta-decay. These properties remind us of V parity, and it seemed to be convenient to described the properties of V particles in terms of I_3.

At that time the only established hadrons were nucleons and pions :

$$p,\ n \text{ and } \pi^+,\ \pi^\circ,\ \pi^-$$

The relationship between the charge and I_3 may be most simply given by

$$\Delta Q = e\Delta I_3$$

The increase of I_3 by one results in the increase of the charge also by one unit. Strangeness or hyper charge is introduced as a constant of integration of this difference equation.

$$Q = e(I_3 + \frac{Y}{2}), \quad Y = B + S \qquad (2)$$

From the cosmotron experiment it was natural to assign $I = 0$ to Λ° and $I = 1$ to Σ^+, (Σ°), Σ^-, where Σ° was not directly seen and was assumed to decay into $\Lambda^\circ + \gamma$ in a short time. Then we had to assign $I = 1/2$ to the heavy mesons or the K mesons.

C8-405

If one considers a system consisting of pions and nucleons, one observes that there is a connection between isospin and ordinary spin. Namely, both of them must be integers or half-integers. What is new here is that the isospin assignment of V particles does not respect this rule. The assignments $I = 0$ to Λ and $I = 1$ to Σ are readily acceptable, but the assignment $I = 1/2$ to K^+, K° impies a new feature in that their antiparticles K^-, $\bar{K}^\circ$ form a separate isospin doublet. Two points should be emphasized here. First K^+ and K^- do not belong to the same isospin multiplet. This gives a clue to the understanding of the positive excess mentioned already. Second, we encountered for the first time a neutral boson which is different from its antiparticle. So far we had known only γ and π°, which are identical with their antiparticles. In the beginning I doubted whether such an assignment was right, but after discussing this subject with Nakano 10) I was convinced that this should be the only possibility. The K mesons kept playing the most important rôle in particle physics for many years to come, providing such subjects as $\theta - \tau$ puzzle, CP violation and so on. They entered the history of particle physics as the most important object next only to the hydrogen atom.

Now we come back to the question of the cascade particle. One assigns $I = 1/2$ to Ξ°, Ξ^- although Ξ° had not been observed at that time. The multiplicative selection rule based on V parity failed to account for the metastability of Ξ^-, because of the presence of two decay channels of opposite V parities. Now V parity can be identified with $(-1)^S$. The additive quantum number S can be utilized to formulate a more detailed selection rule than the multiplicative one. Since $S = -2$ for the cascade particles, their instability can be explained by postulating a selection rule

$$\Delta S = 0, \pm 1 \qquad (3)$$

for weak interactions. The V parity selection rule cannot forbid processes obeying $\Delta S \equiv 0 \pmod{2}$.

After completing these isospin assignments to V particles we have learned from Professor Nambu that Professor Gell-Mann was also developing a similar theory 11) . These recollections exhaust what I wanted to say in addition to what Professor Gell-Mann has told us about strangeness.

REFERENCES K. NISHIJIMA

(1) See reports of the session on cosmic ray physics.

(2) Y. Nambu, K. Nishijima and Y. Yamaguchi, Prog. Theor. Phys. 6 (1951) 615, 619.

(3) H. Miyazawa, Prog. Theor. Phys. 6 (1951) 631.

(4) S. Oneda, Prog. Theor. Phys. 6 (1951) 633.

(5) W. B. Fowler, R. P. Schutt, A. M. Thorndike, and W. L. Whittemore, Phys. Rev. 91 (1953) 1287.

(6) A. Pais, Phys. Rev. 86 (1952) 663.

(7) H. Fukuda and Y. Miyamoto, Prog. Theor. Phys. 4 (1949) 392.

(8) P. H. Fowler, M. G. K. Menon, C. F. Powell, and O. Rochat, Phil. Mag. 42 (1951) 1040.

(9) N. Kemmer, report of the session on isospin.

(10) T. Nakano and K. Nishijima, Prog. Theor. Phys. 10(1953) 581.

K. Nishijima, Prog. Theor. Phys. 13(1955) 285.

(11) M. Gell-Mann, Phys. Rev. 92(1953) 833 ; Nuovo Cim. Suppl. 4(1956) 2848.

C. PEYROU.- One has often accused the cosmic rays physicists of making difficulties to the Gell-Mann scheme in not finding associated production. In fact before September 1953 the cosmic rays physicists were asked to verify the Pais theory which predicted the production of Λ_o Λ_o pairs. They said they had no evidence for it and they were right. True associated production was almost impossible to prove in the complicated situation of cosmic rays events. Emulsions could not see $\Lambda_o^{\prime}$s;$K_o^{\prime}$s have only $\frac{1}{3}$ probability to decay in π^+, π^-; K^+ were very difficult to detect in a systematic way. There was in MIT chamber the beginning of an indication that when you see a K° you had good chance of seeing a Λ_o but on a very poor statistics.

O. PICCIONI.- It is interesting to note that the suggestion that a large angular momentum could explain the long life of strange particles, rejected by Fermi as mentioned by Gell-Mann, was the same as the suggestion of Niels Bohr to explain the non capture of muon in carbon (what Bohr called "Pinocchio effect"). There also, the large ℓ should have explained a discrepancy of $\sim 10^{10}$. In the case of the muons Fermi with Teller and Weisskopf showed that the hypothesis of a large angular momentum was untenable.

R. DALITZ.- Yes, as Gell-Mann said, pion physics was indeed the central topic for theoretical physics in the mid 1950s, and that was what the young theoretician was expected to work on. The strange particles were considered generally to be an obscure and uncertain area of phenomena, as some kind of dirt effect which could not have much role to play in the nuclear forces, whose comprehension was considered to be the purpose of our research. Gell-Mann remarked that he spent the major part of his effort on pion physics in that period, and I did the same, although with much less success, of course.

Fashions have always been strong in theoretical physics, and that holds true today as much as ever. The young physicist who is not working on those problems considered central and promising at the time, is at a disadvantage when he seeks a post. This tendency stems from human nature, of course, but it is unfortunate, I think, that the system operates in such a way as to discourage the young physicist from following an independent line of thought.

There is one aspect of Gell-Mann's scheme which I have not heard mentioned here, namely the $\Delta I = 1/2$ rule for strange particle decays, which he proposed at a very early stage [1954, I believe]. This rule gave a simple explanation for one fact which puzzled the early workers, namely that the $\theta^0 \to \pi\pi$ lifetime was about 100 times shorter than the K^+ lifetime, despite the fact that the K^+ meson had so many additional modes of decay. Strictly applied, this rule forbids the θ^+ mode, $K^+ \to \pi^+\pi^0$, since the K^+ meson has $I = 1/2$ and the $\pi^+\pi^0$ system for $J = 0$ has $I = 2$ only, while allowing the θ^0 mode $K^0 \to \pi^+\pi^-$. The observed rate for the θ^+ mode may be due to electromagnetic effects or, more likely, to deviations from a strict $\Delta I = 1/2$ rule in the weak interaction itself. This rule also gave correct predictions for the ratios of the various $K \to 3\pi$ decay modes and for the Λ and Σ hyperon decay amplitudes, as well as for all semi-leptonic decay modes. In fact, the dominance of the $\Delta I = 1/2$ component in the weak interaction has been successful everywhere it has been tested, whereas its theoretical origin has remained quite obscure, as far as the non-leptonic decay modes are concerned. The reason for the validity of this rule is not yet understood.

C8-407

R.A. SALMERON. - The first examples of associated production of (K°, $\bar{K}^\circ$) pairs were obtained in cosmic rays by the Manchester-Jungfraujoch cloud chamber group. In both pictures the V°'s were seen to decay in the cloud chamber and identified as K°'s and the K^+ was identified by its ionization. These were the first examples of production of two neutral kaons with opposite strangeness, as well as that of a positive and a neutral kaon (Il Nuovo Cimento 5 (1957) 1388). Two examples of associated production of ($K^+ K^-$) pairs had been previously reported by two emulsions groups (Il Nuovo Cimento 2 (1955) 666 ; 2 (1955) 828).

Y. YAMAGUCHI.- In 1951, we discussed on these V-particles. It was true that there were few examples of associated production of V's seen in cloud chambers. If you would analyze them statistically as usual, you might find that V-particles production would be dominantly of single production. Nevertheless, since production and decays of V's must be controlled by different interactions —— otherwise we could not understand them at all ——, I firmly insisted upon the idea of pair production of V's. Some cosmic ray theorists (and experimentalists) including S. Hayakawa, however, objected naturally me saying that there were no evidences for pair production from their statistical analyses of cloud chamber photos.

In early 1953, Hayakawa and Nishijima wrote a review article (in Japanese) on strange particles (V-particles) in the monthly journal of the Physical Society of Japan, saying that the pair production of V's has no experimental evidences. Under such a situation, the idea of pair production was hardly acceptable to high energy community.

I may remind you that at that time there was another hot controversy : whether is meson production at high energy mucleon-nucleus collisions multiple production or plural production ? (multiple production : mesons are produced in nucleon-nucleon collisions in the form of multiple production. plural production : meson is singly produced at nucleon-nucleon collision, while cascade processes taking place in nucleon-nucleus collisions will lead "multiple" production of meson for nucleon-nucleus collisions.) At that time it was very difficult to select experimentally these two alternatives for meson-production ! And there were a lot of cosmic-ray experiments and hot discussions on this issue.

At present, it might be very difficult to understand why such a "trivial" issue was so hotly discussed and pursued !

I may conclude that, cosmic-rays brought us a lot of interesting and valuable findings for particle physics, but also sometimes misleading impression because of inherent poor statistics on information obtained by cosmic-rays.

C.N. YANG.- In reference to Murray's interesting account of the history of the concept of strangeness, I remember that in the summer of 1953, I did not like Murray's idea at all. In fact, I convinced everybody at bull sessions at Brookhaven in the early summer of 1953 that Murray's proposal was all wrong. I had two objections. I did not feel that a boson should have half integral isospin, and I had believed that there is only one neutral K. But just to keep the records straight, I was not the referee that Murray mentioned .

J. TIOMNO.- It may be convenient for myself to make at this point an observation related to what Yang has said in that. Althought, the paper on isospin with the classification of the isodoublet for K impressed me very much, I also had this prejudice and then I developped a treatment on a doublet scheme where all baryons (at least those which were known at that time), were isofermion, being fermions. Correspondingly pions and kaons, being bosons, would be isobosons. In 1957, I was really much convinced that this would be usefull when I developped the scheme with O(7) invariance, proposing for the first time (Nuovo Cimento, 6, 69) the unification of the baryon octet and unification of the seven mesons, π and $\bar{K}$. At the Rochester conference in 1957, I submitted this paper, at the same time as Gell-Mann was using the isodoublets for the $\Lambda - \Sigma$ from a different approach (global symmetry in π interactions). Then, as I had mentioned that there was a similarity among the papers by Schwinger, Gell-Mann, and mine, Murray said that someone should point out that they were not quite the same thing - clearly they were not the same thing. Also I like to mention that when Yang was in Rio, he too was thinking on this question and we were studying the possibility of getting a sub-group of O(7) in order to eliminate some unsatisfactory selection rules. We did not work enough to find what Neeman found a few years later, that if you just include the complete set of Γ^A and Γ^{AB} operators you get SU(3) in the octet representation.

M. GELL-MANN.- I think that we can learn from many of these stories a double principle, which is that a good theoretical idea in science often needs to be stripped of unnecessary baggage with which it is accompanied at the beginning, and that then it may need to be taken much more seriously than it was by its original proponent. I said this at the Einstein centenary celebration in Jerusalem and pointed out that in 1905, in the same volume of the Annalen der Physik, Einstein published three articles : one on special relativity, one on the photo-electric effect, and one on Brownian motion. In the Brownian motion article he took seriously the notion of the physical existence of a molecule ; in the article on the photo-electric effect, he took seriously the possibility of the physical existence of a quantum ; and in the article on special relativity he took seriously the physical importance of the symmetry group of the electromagnetic equation.

PHYSICAL REVIEW VOLUME 95, NUMBER 5 SEPTEMBER 1, 1954

Quantum Electrodynamics at Small Distances*

M. GELL-MANN† AND F. E. LOW
Physics Department, University of Illinois, Urbana, Illinois
(Received April 1, 1954)

The renormalized propagation functions D_{FC} and S_{FC} for photons and electrons, respectively, are investigated for momenta much greater than the mass of the electron. It is found that in this region the individual terms of the perturbation series to all orders in the coupling constant take on very simple asymptotic forms. An attempt to sum the entire series is only partially successful. It is found that the series satisfy certain functional equations by virtue of the renormalizability of the theory. If photon self-energy parts are omitted from the series, so that $D_{FC}=D_F$, then S_{FC} has the asymptotic form $A[p^2/m^2]^n[i\gamma\cdot p]^{-1}$, where $A=A(e_1^2)$ and $n=n(e_1^2)$. When all diagrams are included, less specific results are found. One conclusion is that the *shape* of the charge distribution surrounding a test charge in the vacuum does not, at small distances, depend on the coupling constant except through a scale factor. The behavior of the propagation functions for large momenta is related to the magnitude of the renormalization constants in the theory. Thus it is shown that the unrenormalized coupling constant $e_0^2/4\pi\hbar c$, which appears in perturbation theory as a power series in the renormalized coupling constant $e_1^2/4\pi\hbar c$ with divergent coefficients, may behave in either of two ways:

(a) It may really be infinite as perturbation theory indicates;
(b) It may be a finite number independent of $e_1^2/4\pi\hbar c$.

1. INTRODUCTION

IT is a well-known fact that according to quantum electrodynamics the electrostatic potential between two classical test charges in the vacuum is not given exactly by Coulomb's law. The deviations are due to vacuum polarization. They were calculated to first order in the coupling constant α by Serber[1] and Uehling[2] shortly after the first discussion of vacuum polarization by Dirac[3] and Heisenberg.[4] We may express their results by writing a formula for the potential energy be-

* This work was supported by grants from the U. S. Office of Naval Research and the U. S. Atomic Energy Commission.
† Now at Department of Physics and Institute for Nuclear Studies, University of Chicago.

[1] R. Serber, Phys. Rev. 48, 49 (1935).
[2] A. E. Uehling, Phys. Rev. 48, 55 (1935).
[3] P. A. M. Dirac, Proc. Cambridge Phil. Soc. 30, 150(1934).
[4] W. Heisenberg, Z. Physik 90, 209 (1934).

tween two heavy point test bodies, with renormalized charges q and q', separated by a distance r:

$$V(r)=\frac{qq'}{4\pi r}\left\{1+\frac{\alpha}{3\pi}\int_{(2m)^2}^{\infty}\exp\left[-r\left(\frac{M^2c^2}{\hbar^2}\right)^{\frac{1}{2}}\right]\times\left(1+\frac{2m^2}{M^2}\right)\left(1-\frac{4m^2}{M^2}\right)^{\frac{1}{2}}\frac{dM^2}{M^2}+O(\alpha^2)+\cdots\right\}. \quad (1.1)^{5}$$

Here $\alpha=e_1^2/4\pi\hbar c\cong 1/137$ is the renormalized fine structure constant and m is the renormalized (observed) rest mass of the electron.

If $r\ll\hbar/mc$, then (1.1) takes the simple asymptotic form,

$$V(r)=\frac{qq'}{4\pi r}\left\{1+\frac{2\alpha}{3\pi}\left[\ln\left(\frac{\hbar}{mcr}\right)-\tfrac{5}{6}-\ln\gamma\right]+O(\alpha^2)+\cdots\right\}, \quad (1.2)^{5}$$

where $\gamma\cong 1.781$. We shall discuss the behavior of the entire series (1.2), to all orders in the coupling constant, making use of certain simple properties that it possesses in virtue of the approximation $r\ll\hbar/mc$. These properties are intimately connected with the concept of charge renormalization. The relation between (1.2) and charge renormalization can be made clear by the following physical argument:

A test body of "bare charge" q_0 polarizes the vacuum, surrounding itself by a neutral cloud of electrons and positrons; some of these, with a net charge δq, of the same sign as q_0, escape to infinity, leaving a net charge $-\delta q$ in the part of the cloud which is closely bound to the test body (within a distance $\hbar/mc$). If we observe the body from a distance much greater than $\hbar/mc$, we see an effective charge q equal to $(q_0-\delta q)$, the renormalized charge. However, as we inspect more closely and penetrate through the cloud to the core of the test body, the charge that we see inside approaches the bare charge q_0, concentrated in a point at the center. It is clear, then, that the potential $V(r)$, in Eqs. (1.1) and (1.2), must approach $q_0q_0'/4\pi r$ as r approaches zero. Thus, using (1.1), we may write

$$q_0q_0'=qq'\left\{1+\frac{2\alpha}{3\pi}\left[\ln\left(\frac{\hbar/mc}{0}\right)-\tfrac{5}{6}-\ln\gamma\right]+O(\alpha^2)\right\}, \quad (1.3)$$

where the individual terms in the series diverge logarithmically in a familiar way. The occurrence of these logarithmic divergences will play an important role in our work.

Such divergences occur in quantum electrodynamics whenever observable quantities are expressed in terms of the bare charge e_0 and the bare (or mechanical) mass m_0 of the electron. The renormalizability of the theory consists in the fact that, when the observable quantities are re-expressed in terms of the renormalized parameters e_1 and m, no divergences appear, at least when a power series expansion in $e_1^2/4\pi\hbar c$ is used. The proof of renormalizability has been given by Dyson,[6] Salam,[7] and Ward.[8] We shall make particular use in Secs. III and IV of the elegant techniques of Ward.

We shall show that the fact of renormalizability gives considerable information about the behavior of the complete series (1.2). It may be objected to an investigation of this sort that while (1.2) is valid for $r\ll\hbar/mc$, the first few terms should suffice for calculation unless r is as small as $e^{-137}\hbar/mc$, a ridiculously small distance. We have no reason, in fact, to believe that at such distances quantum electrodynamics has any validity whatever, particularly since interactions of the electromagnetic field with particles other than the electron are ignored. However, a study of the mathematical character of the theory at small distances may prove useful in constructing future theories. Moreover, in other field theories now being considered, such as the relativistic pseudoscalar meson theory, conclusions similar to ours may be reached, and the characteristic distance at which they become useful is much greater, on account of the largeness of the coupling constant.

In this paper we shall be mainly concerned with quantum electrodynamics, simply because gauge invariance and charge conservation simplify the calculations to a considerable extent. Actually, our considerations apply to any renormalizable field theory, and we shall from time to time indicate the form they would take in meson theory.

2. REPRESENTATIONS OF THE PROPAGATION FUNCTIONS

The modified Coulomb potential discussed in Sec. I can be expressed in terms of the finite modified photon propagation function $D_{FC}(p^2,e_1^2)$ that includes vacuum polarization effects to all orders in the coupling constant. (Here p^2 is the square of a four-vector momentum p_μ.) The function D_{FC} is calculated by summing all Feynman diagrams that begin and end with a single photon line, renormalizing to all orders. The potential is given by[9]

$$V(r)=\frac{qq'}{(2\pi)^3}\int d^3p\,e^{i\mathbf{p}\cdot\mathbf{r}}D_{FC}(\mathbf{p}^2,e_1^2), \quad (2.1)$$

where $\mathbf{p}$ is a three-dimensional vector.

If we were to sum all the Feynman diagrams that make up D_{FC} without renormalizing the charge we

[5] J. Schwinger, Phys. Rev. 75, 651 (1949).

[6] F. J. Dyson, Phys. Rev. 75, 1756 (1949).
[7] A. Salam, Phys. Rev. 84, 426 (1951).
[8] J. C. Ward, Proc. Phys. Soc. (London) A64, 54 (1951); see also Phys. Rev. 84, 897 (1951).
[9] From this point on, we take $\hbar=c=1$.

would obtain the divergent function $D_F'(p^2,e_0^2)$, which is related to the finite propagation function by Dyson's equations

$$D_F'(p^2,e_0^2)=Z_3D_{FC}(p^2,e_1^2), \tag{2.2}$$

$$e_1^2=Z_3e_0^2, \tag{2.3}$$

where Z_3 is a power series in e_1^2 with divergent coefficients. The bare and renormalized charges of a test body satisfy a relation similar to (2.3):

$$q_1^2=Z_3q_0^2, \tag{2.4}$$

so that Z_3^{-1} is just the bracketed quantity in (1.3).

The function D_{FC} can be represented in the form

$$D_{FC}(p^2,e_1^2)=\frac{1}{p^2-i\epsilon}+\int_0^\infty f\left(\frac{M^2}{m^2},e_1^2\right)\frac{dM^2}{M^2}\frac{1}{p^2+M^2-i\epsilon}, \tag{2.5}$$

where f is real and positive; the quantity Z_3 may be expressed in terms of f through the relation

$$Z_3^{-1}=1+\int_0^\infty f\left(\frac{M^2}{m^2},e_1^2\right)\frac{dM^2}{M^2}. \tag{2.6}$$

These equations have been presented and derived, in a slightly different form by Källén.[10] Their derivation is completely analogous to the derivation given in Appendix A of Eqs. (2.8) and (2.9) for the propagation function of the electron, which is discussed below.

We see from (2.5) and (2.6) that a virtual photon propagates like a particle with a probability distribution of virtual masses. In D_{FC}, the distribution is not normalized, but in the unrenormalized propagation function $D_F'=Z_3D_{FC}$ the probabilities are normalized to 1. The normalization integral is just the formally divergent quantity Z_3^{-1}. In D_{FC}, it is the coefficient of $1/(p^2-i\epsilon)$ that is 1, corresponding to the fact that the potential $V(r)$ in (2.1) at large distances is simply $qq'/4\pi r$.

It has been remarked[10,11] that Z_3^{-1} must be greater than unity, a result that follows immediately from Eq. (2.6). To this property of the renormalization constant there corresponds a simple property of the finite function D_{FC}, to wit, that as $p^2\to\infty$, the quantity p^2D_{FC} approaches Z_3^{-1}. If Z_3^{-1} is in fact infinite, as it appears to be when expanded in a power series, then D_{FC} is more singular than the free photon propagation function $D_F=1/(p^2-i\epsilon)$. In any case, D_{FC} can never be less singular than D_F, nor even smaller asymptotically. This is a general property of existing field theories; it is of particular interest in connection with the hope often expressed that in meson theory the exact modified propagation functions are less singular, or smaller, for large momenta, than the corresponding free-particle propagation functions.

The propagation functions for the electron behave quite similarly to the photon functions we have been discussing. Analogous to D_F' is the divergent electron propagation function $S_F'(p,e_0^2)$. It is obtained by summing all Feynman diagrams beginning and ending in a single electron line, renormalizing the mass of the electron, but not its charge, to all orders. Corresponding to D_{FC} there is the finite function $S_{FC}(p,e_1^2)$, related to S_F' by an equation similar to (2.2):

$$S_F'(p,e_0^2)=Z_2S_{FC}(p,e_1^2). \tag{2.7}$$

The quantity Z_2, like Z_3, appears as a power series in e_1^2 with divergent coefficients. It does not, however, contribute to charge renormalization.

A parametric representation of S_{FC}, resembling Eq. (2.5) for D_{FC}, is derived in Appendix A and reproduced here:

$$S_{FC}(p,e_1^2)=\frac{1}{i\gamma p+m-i\epsilon}+\int_m^\infty\frac{g(M/m,e_1^2)}{i\gamma p+M-i\epsilon}\frac{dM}{M}+\int_m^\infty\frac{h(M/m,e_1^2)}{i\gamma p-M+i\epsilon}\frac{dM}{M}. \tag{2.8}$$

Both g and h are real; in meson theory they are positive, but in quantum electrodynamics they may assume negative values. Z_2 can be expressed in terms of g and h through the relation

$$Z_2^{-1}=1+\int_m^\infty g\left(\frac{M}{m},e_1^2\right)\frac{dM}{M}+\int_m^\infty h\left(\frac{M}{m},e_1^2\right)\frac{dM}{M}. \tag{2.9}$$

Again we have a sort of probability distribution of virtual masses with a formally divergent normalization integral. As before, the modified propagation function is more singular, or at least asymptotically greater, than the free-particle propagation function, since $Z_2^{-1}\geqslant 1$, except possibly in quantum electrodynamics.

Equations (2.8) and (2.9), like (2.5) and (2.6), are similar to ones derived by Källén.[10] However, our notation and approach are perhaps sufficiently different from his to warrant separate treatment.

It should be noted that Källén's paper contains a further equation, (70), which, in our notation expresses the mechanical mass m_0 of the electron in terms of g and h:

$$m_0=\left[m+\int_m^\infty Mg\frac{dM}{M}+\int_m^\infty(-M)h\frac{dM}{M}\right]\times\left[1+\int_m^\infty g\frac{dM}{M}+\int_m^\infty h\frac{dM}{M}\right]^{-1}. \tag{2.10}$$

We see that m_0 is simply the mean virtual mass of the electron.

[10] G. Källén, Helv. Phys. Acta 25, 417 (1952).
[11] J. Schwinger (private communication from R. Glauber).

It may be remarked that a quantity analogous to m_0 can be constructed for the photon field, that is, the mean squared virtual mass of the photon:[12]

$$\mu_0^2=\left[\int_0^\infty fM^2\frac{dM^2}{M^2}\right]\left[1+\int_0^\infty f\frac{dM^2}{M^2}\right]^{-1}. \quad (2.11)$$

While gauge invariance forbids the occurrence of a mechanical mass of the photon in the theory, it is well known that a quadratically divergent quantity that looks like the square of a mechanical mass frequently turns up in calculations and must be discarded. That quantity is just μ_0^2, as given by (2.11). An equation similar to (2.11) holds in pseudoscalar meson theory, where μ_0^2 is really the square of a mechanical mass:

$$\mu_0^2=\left[\mu^2+\int_{9\mu^2}^\infty fM^2\frac{dM^2}{M^2}\right]\left[1+\int_{9\mu^2}^\infty f\frac{dM^2}{M^2}\right]^{-1}. \quad (2.12)$$

Here μ^2 is the observed meson mass. Evidently (2.12) implies that $\mu_0^2>\mu^2$.

3. EXAMPLE: QUANTUM ELECTRODYNAMICS WITHOUT PHOTON SELF-ENERGY PARTS

Before examining the asymptotic forms of the singular functions in the full theory of quantum electrodynamics let us consider a simplified but still renormalizable form of the theory in which all photon self-energy parts are omitted. A photon self-energy part is a portion of a Feynman diagram which is connected to the remainder of the diagram by two and only two photon lines. By omitting such parts, we effectively set

$$D_{FC}(p^2)=D_F(p^2)=1/(p^2-i\epsilon). \quad (3.1)$$

Moreover, there is no charge renormalization left in the theory, so that

$$Z_3=1$$

and

$$e_0^2=e_1^2. \quad (3.2)$$

Although some finite effects of vacuum polarization (such as its contribution to the second-order Lamb shift) have been left out, others (such as the scattering of light by light) are still included.

If the mass of the electron is now renormalized the only divergence remaining in the theory is Z_2. It has been shown by Ward[8] that in the calculation of any observable quantity, such as a cross section, Z_2 cancels out. We shall nevertheless be concerned with Z_2 since it does appear in a calculation of the electron propagation function S_F'.

In order to deal with Z_2, a divergent quantity, we shall make use of the relativistic high-momentum cutoff procedure introduced by Feynman, which consists of replacing the photon function D_F by a modified function $D_{F\lambda}$ defined by

$$D_{F\lambda}(p^2)=D_F(p^2)\lambda^2/(\lambda^2+p^2-i\epsilon).$$

In a given calculation, if λ^2 is large enough, quantities that would be finite in the absence of a cutoff remain unchanged while logarithmically divergent quantities become finite logarithmic functions of λ^2.

Thus, if we calculate $S_F'(p)$ using a Feynman cutoff with $\lambda^2\gg|p^2|$ and $\lambda^2\gg m^2$, and drop terms that approach zero as λ^2 approaches infinity, we must find a relation similar to (2.7):

$$S_{F'\lambda}(p)=z_{2\lambda}S_{FC}(p), \quad (3.3)$$

where the finite function S_{FC} has remained unchanged by the cut-off process, while the infinite constant Z_2 has been converted to the finite quantity $z_{2\lambda}$, which is a function of λ^2/m^2. (The reader who is not impressed with the rigor of these arguments should refer to the next section, where a more satisfactory cut-off procedure is introduced.)

Calculation to the first few orders in the coupling constant indicates that $z_{2\lambda}$ has the form

$$z_{2\lambda}=1+e_1^2\left(a_1+b_1\ln\frac{\lambda^2}{m^2}\right)+e_1^4\left[a_2+b_2\ln\frac{\lambda^2}{m^2}+c_2\left(\ln\frac{\lambda^2}{m^2}\right)^2\right]+\cdots. \quad (3.4)$$

The propagation function S_{FC} may also be calculated to fourth order in e_1; for $|p^2|\gg m^2$ it has the form

$$|p^2|\gg m^2:\ S_{FC}(p)=\frac{1}{i\gamma p}\{1+e_1^2[a_1'+b_1'\ln(p^2/m^2)]+e_1^4[a_2'+b_2'\ln(p^2/m^2)+c_2'(\ln(p^2/m^2))^2]+\cdots\}. \quad (3.5)$$

In order to obtain some understanding of the properties of Eq. (3.3), let us substitute these approximate expressions into that equation. Let us then examine what happens in the limit $m\to0$. We see that in neither of the expressions (3.4) and (3.5) can m be set equal to zero with impunity; that is to say, both factors on the right-hand side of (3.3) contain logarithmic divergences as $m\to0$. Thus we should naively expect that the left-hand side have no limit as $m\to0$. Rather, we should expect to find logarithmic divergences to each order in e_1^2, unless fantastic cancellations, involving the constants a_1, a_1', etc., should happen to occur.

But such cancellations must indeed occur, since a direct calculation of $S_{F\lambda}'$ with m set equal to zero exhibits no divergences whatever. Instead, each Feynman diagram yields a term equal to $1/i\gamma p$ times a finite function of λ^2/p^2. It is clear that this must be so, since λ

[12] We are indebted to Dr. Källén for a discussion of this point.

provides an ultraviolet cutoff for every integral, while p provides an infrared cutoff.

Let us now make use of the remarkable cancellations that we have discussed, but in such a way that we do not rely on the specific forms of (3.4) and (3.5), for which we have so far claimed no validity beyond fourth order in e_1. We shall consider the asymptotic region $\lambda^2 \gg |p^2| \gg m^2$. We may write

$$S_{FC} = (1/i\gamma p) s_C(p^2/m^2) \tag{3.6}$$

and

$$z_{2\lambda} = z_2(\lambda^2/m^2). \tag{3.7}$$

Moreover, in the asymptotic region, we may drop m entirely in $S_{F'\lambda'}$, since a limit exists as m^2/λ^2 and m^2/p^2 approach zero, with $\lambda^2 \gg |p^2|$. Thus we have

$$S_{F'\lambda} \approx (1/i\gamma p) s(\lambda^2/p^2). \tag{3.8}$$

Equation (3.3) then implies the following functional equation:

$$s(\lambda^2/p^2) = z_2(\lambda^2/m^2) \cdot s_C(p^2/m^2). \tag{3.9}$$

The functional equation has the general solution[13]

$$z_2(\lambda^2/m^2) = A(\lambda^2/m^2)^{-n} = A \exp[-n \ln(\lambda^2/m^2)], \tag{3.10}$$

$$s_C(p^2/m^2) = B(p^2/m^2)^n = B \exp[n \ln(p^2/m^2)], \tag{3.11}$$

$$s(\lambda^2/p^2) = AB(\lambda^2/p^2)^{-n} = AB \exp[-n \ln(\lambda^2/p^2)]. \tag{3.12}$$

Here A, B, and n are functions of e_1^2 alone. If all three constants are expanded in power series in e_1^2, then formulas like (3.4) and (3.5) can be seen to be valid to all orders in e_1^2. The constants are given, to second order in e_1, by the equations

$$n = \frac{e_1^2}{16\pi^2} + \cdots, \tag{3.13}$$

$$A = 1 + \frac{e_1^2}{16\pi^2}\left[\frac{3}{2} + \int_{m^2}^{\infty} \frac{dM^2}{M^4(M^2-m^2)}(5m^2M^2 - m^4)\right], \tag{3.14}$$

$$B = 1 - \frac{e_1^2}{16\pi^2}\int_{m^2}^{\infty} \frac{dM^2}{M^4(M^2-m^2)}(5m^2M^2 - m^4) + \cdots. \tag{3.15}$$

It is now apparent that we have glossed over a difficulty, although it turns out to be a minor one. While AB is perfectly finite, A and B separately contain infrared divergences that must be cut off by the introduction of a small fictitious photon mass μ. These divergences are well-known and arise from the requirement that $(i\gamma p + m) S_{FC}(p)$ approach unity as $i\gamma p + m$ tends to zero, while the point $i\gamma p + m = 0$ is in fact a singularity of the function $(i\gamma p + m) S_F'(p)$.

From the asymptotic form for S_{FC},

$$S_{FC}(p) \approx \frac{B}{i\gamma p}\left(\frac{p^2}{m^2}\right)^n, \tag{3.16}$$

we may derive an asymptotic form for the vertex operator[6] $\Gamma_{\mu C}(p,p')$ for equal arguments.

We use Ward's[8] relation

$$\Gamma_{\mu C}(p,p) = -\frac{1}{i}\frac{\partial}{\partial p_\mu}[S_{FC}(p)]^{-1}, \tag{3.17}$$

and obtain

$$\Gamma_{\mu C}(p,p) = B^{-1}(p^2/m^2)^{-n}(\gamma_\mu - 2np_\mu \gamma p/p^2). \tag{3.18}$$

A result similar to this was found by Edwards,[14] who summed a small subset of the diagrams we consider here. We may note that corresponding to the increase in the singularity of S_F there is a decrease in that of Γ_μ. The two are obviously tied together by (3.17). It is therefore highly inadvisable to take seriously any calculation using a modified Γ_μ and unmodified S_F', or vice versa.

It is unfortunate that the inclusion of photon self-energy parts (omitted in this section) invalidates the simple results we have obtained here. In Sec. V we shall derive and solve the functional equations that replace (3.9) in the general case, but the solutions give much less detailed information than (3.16). In order to treat the general case, we must first develop (in Sec. IV) a more powerful cut-off technique than the one we have used so far.

4. WARD'S METHOD[8] USED AS A CUTOFF[15]

The starting point of Ward's method of renormalization is a set of four integral equations derived by summing Feynman diagrams. The equations involve four functions: $S_F'(p)$, $D_F'(k)$, the vertex operator $\Gamma_\mu(p_1,p_2)$, and a function $W_\mu(k)$ defined by

$$W_\mu = (\partial/\partial k_\mu)[D_F'(k)]^{-1}. \tag{4.1}$$

Two of the equations are trivial, following from (4.1) and (3.17), respectively:

$$[S_F'(p)]^{-1} = i\int_0^1 dx(p_\mu - p_\mu') \times \Gamma_\mu(px + p'(1-x),\, px + p'(1-x)) \tag{4.2}$$

and

$$[D_F'(k)]^{-1} = \int_0^1 dy k_\mu W_\mu(ky), \tag{4.3}$$

where p' is a free electron momentum, i.e., after integration p'^2 is to be replaced by $-m^2$ and $i\gamma p'$ by $-m$, where m is the experimental electron mass. No further mass renormalization is necessary.

[13] J. C. Maxwell, Phil. Mag. (Series 4) 19, 19 (1860).

[14] S. F. Edwards, Phys. Rev. 90, 284 (1953).

[15] The purpose of this section is to justify the use of a cutoff when photon self-energy parts are included. The reader who is willing to take this point for granted need devote only the briefest attention to the material between Eqs. (4.3) and (4.6). The remainder of the section contains some simple but important algebraic manipulation of the cut-off propagation functions.

The two remaining equations are nonlinear power-series integral equations for Γ_μ and W_μ in which each term describes an "irreducible" Feynman diagram. An irreducible diagram, as defined by Dyson, is one which contains no vertex or self-energy parts inside itself. When the complete series of irreducible diagrams for Γ_μ or W_μ is written down, and in each one S_F' is substituted for S_F, D_F' for D_F, Γ_μ for γ_μ, and W_μ for $2k_\mu$, then the complete series of *all* diagrams for Γ_μ or W_μ is generated. We give below the first two terms of each of the integral equations. Equation (4.4) corresponds to Fig. 1 and Eq. (4.5) to Fig. 2.

$$\Gamma_\mu(p_1,p_2)\equiv\gamma_\mu+\Lambda_\mu(p_1,p_2)$$
$$=\gamma_\mu+\frac{ie_0^2}{(2\pi)^4}\int\Gamma_\lambda(p_1,p_1-k)$$
$$\times S_F'(p_1-k)\Gamma_\mu(p_1-k,p_2-k)\cdot S_F'(p_2-k)$$
$$\times\Gamma_\lambda(p_2-k,p_2)D_F'(k)d^4k+\cdots,\quad(4.4)$$

$$W_\mu(k)\equiv 2k_\mu+Tk_\mu$$
$$=2k_\mu+\frac{e_0^2}{(2\pi)^4}\cdot\tfrac{1}{3}\,\mathrm{Tr}\int\Gamma_\nu(p,p+k)$$
$$\times S_F'(p+k)\Gamma_\mu(p+k,p+k)S_F'(p+k)$$
$$\times\Gamma_\nu(p+k,p)S_F'(p)d^4p+\cdots.\quad(4.5)$$

The factor of one-third arises in (4.5) because we are interested only in the coefficient of $\delta_{\nu\lambda}$ in the tensor $[\delta_{\nu\lambda}-(k_\nu k_\lambda/k^2)]D_F'(k)$.

The heavy lines and dots have been drawn as a reminder that the complete S_F', D_F', and Γ_μ are to be inserted.

The symbols Λ_μ and T_μ are simply a convenient shorthand for the sum of all the integrals occurring on the right in (4.4) and (4.5), respectively.

The properties of (4.4) and (4.5) that are crucial for the possibility of renormalization are the following:

(i) All divergences that occur in the power series solution of the equations are logarithmic divergences. (W_μ is actually formally linearly divergent but on grounds of covariance the linear divergence will vanish.)

(ii) In (4.4), each terms with coefficient $(e_0^2)^n$ con- exactly nD_F' functions and contains one more Γ_μ function than S_F' functions.

(iii) In (4.5), each term with coefficient $(e_0^2)^n$ contains $(n-1)$ more D_F' functions than W_μ functions, and contains equal numbers of Γ_μ and S_F' functions.

At this point Ward introduces a subtraction procedure which alters Eqs. (4.4) and (4.5) so that the solutions of the new equations are finite functions S_{FC}, D_{FC}, $\Gamma_{\mu C}$, and $W_{\mu C}$. He then shows that the modi-

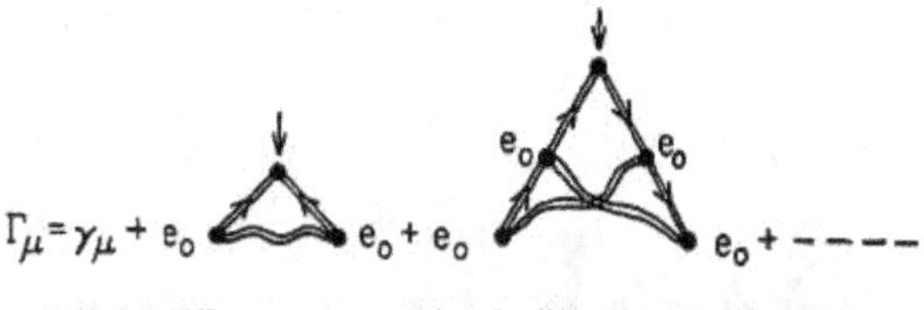

FIG. 1. The sequence of irreducible diagrams for Γ_μ.

fications introduced are equivalent to charge renormalization. The method that we shall present here is a slight generalization of Ward's involving two cut-off parameters λ and λ'. When $\lambda=0$ and $\lambda'=im$, our method reduces to Ward's.

From the right-hand side of (4.4) we subtract $\int dx\Lambda_\mu(l'x+p'(1-x),\,l'x+p'(1-x))$, where p' has the same significance as in (4.2) and l' is a vector parallel to p' but with $(l')^2$ set equal to $(\lambda')^2$ and $\gamma\cdot l'$ set equal to λ' after the integration. This choice of subtraction procedure may appear arbitrarily complicated, so that a remark about our motivation may be in order. Since Λ_μ consists of logarithmically divergent integrals, the quantity

$$\Lambda_\mu{}^x=\Lambda_\mu(p_1,p_2)-\Lambda_\mu(l'x+p'(1-x),\,l'x+p'(1-x))$$

is certainly finite. Therefore so is

$$\Lambda_{\mu l'}=\int_0^1 dx\Lambda_\mu{}^x=\Lambda_\mu(p_1,p_2)$$
$$-\int\Lambda_\mu(l'x+p'(1-x),\,l'x+p'(1-x))dx,$$

which is the quantity of interest. However, referring to (4.2), we see that if we replace Γ_μ by $\gamma_\mu+\Lambda_{\mu l'}$, $S_F'(p)$ will have the value $1/(i\gamma p+m)$ at $p=l'$. This subtraction procedure therefore provides a convenient normalization for the cut-off functions.

Similarly, from the right-hand side of (4.5) we subtract $k_\mu\int_0^1 2ydyT(ly)$, where $l^2=\lambda^2$. (The motivation is the same here as before.) Let us denote the solutions of the modified equations by the symbols $S_{F\lambda'\lambda}$, etc. Like Ward's functions, they are finite to all orders in the coupling constant, the logarithmic divergences having disappeared in the course of the subtraction. In the modified equations, let us everywhere replace the coupling constant e_0^2 by another one, e_2^2. Then we may show that if e_2^2 is a properly chosen function of e_0^2, λ^2, and m^2 the modified functions are multiples of the

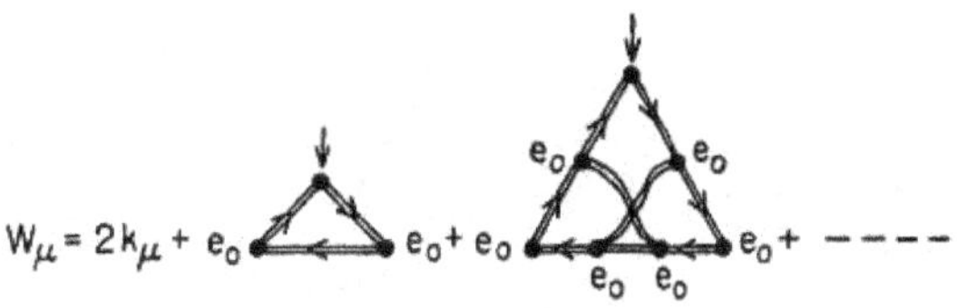

FIG. 2. The sequence of irreducible diagrams for W_μ. Notice that there is no irreducible diagram in fourth order.

original divergent functions. In fact, we have the relations:

$$\Gamma_\mu(p_1,p_2;e_0^2)=Z_2^{-1}(\lambda,\lambda';e_2^2)\Gamma_{\mu\lambda\lambda'}(p_1,p_2;e_2^2), \quad (4.6)$$

$$S_F'(p;e_0^2)=Z_2(\lambda,\lambda';e_2^2)S_{F\lambda\lambda'}(p,e_2^2), \quad (4.7)$$

$$W_\mu(k,e_0^2)=Z_3^{-1}(\lambda;e_2^2)W_{\mu\lambda}(k,e_2^2), \quad (4.8)$$

$$D_F'(k,e_0^2)=Z_3(\lambda;e_2^2)D_{F\lambda}(k,e_2^2), \quad (4.9)$$

$$e_2^2=Z_3(\lambda;e_2^2)e_0^2. \quad (4.10)$$

These relations can be proved by substituting them into the original integral Eqs. (4.2) through (4.5) and making use of the properties (ii) and (iii). Z_2 and Z_3 are then given by

$$\gamma_\mu Z_2(\lambda,\lambda';e_2^2)=\gamma_\mu-\int_0^1 dx\times$$

$$\Lambda_{\mu\lambda\lambda'}(l'x+p'(1-x),\, l'x+p'(1-x);e_2^2) \quad (4.11)$$

and

$$Z_3(\lambda,e_2^2)=1-\int_0^1 ydyT_\lambda(ly,e_2^2). \quad (4.12)$$

Here Λ and T stands for the series of integrals on the right-hand side of (4.4) and (4.5) calculated *with the cut-off functions.*

We note that the photon functions are independent of λ'. This is of course due to the fact that Z_2 cancels out in calculating these functions. Furthermore,

$$[(i\gamma p+m)S_{F\lambda\lambda'}(p)]_{\gamma p=\lambda'}=1, \quad (4.13)$$

and

$$[k^2D_{F\lambda}(k)]_{k^2=\lambda^2}=1. \quad (4.14)$$

It is clear, then, that for $\lambda'=im$ and $\lambda=0$ our modified functions reduce to the usual convergent functions as defined by Ward. Furthermore, when λ and λ' are both infinite our modified functions are (in some sense which we need not worry about) the original divergent functions S_F', D_F', etc. We may now relate our functions to the usual convergent functions very simply. (From now on we shall limit our discussion to S_F' and D_F', since Γ and W are obtained from them by differentiation.)

We rewrite (4.7), (4.9), and (4.10):

$$S_F'(p,e_0^2)=Z_2(\lambda,\lambda';e_2^2)S_{F\lambda\lambda'}(p,e_2^2), \quad (4.7)$$

$$D_F'(k,e_0^2)=Z_3(\lambda,e_2^2)D_{F\lambda}(k,e_2^2), \quad (4.9)$$

$$e_2^2=Z_3(\lambda,e_2^2)e_0^2. \quad (4.10)$$

We obtain the conventional renormalization theory by setting $\lambda=0$, $\lambda'=im$, so that

$$S_F'(p,e_0^2)=Z_2(0,im;e_1^2)S_{FC}(p,e_1^2), \quad (4.7)'$$

$$D_F'(k,e_0^2)=Z_3(0,e_1^2)D_{FC}(k,e_1^2), \quad (4.9)'$$

and

$$e_1^2=Z_3(0,e_1^2)e_0^2, \quad (4.10)'$$

where e_1 is the observed electronic charge.

Dividing the unprimed equations by the primed ones leads to the relations:

$$S_F(\lambda,\lambda',p;e_2^2)=z_2(\lambda,\lambda';e_1^2)S_{FC}(p,e_1^2), \quad (4.7)''$$

$$D_F(\lambda,k;e_2^2)=z_3(\lambda,e_1^2)D_{FC}(k,e_1^2), \quad (4.9)''$$

$$e_1^2=z_3(\lambda,e_1^2)e_2^2. \quad (4.10)''$$

In these last equations all the quantities involved are finite. As λ' and λ approach ∞, however, they approach their original divergent values. We have therefore established a cutoff which is useful in the presence of photon self-energy parts and which has the desired renormalization property for any values of the cut-off parameters. We may call attention to the essential simplicity of the cutoff. For example, if we had used it in Sec. III instead of the Feynman cutoff we would have found $s=(p^2/\lambda^2)^n$ to all orders in Eq. (3.12), i.e., $AB=1$.

For our purposes it is convenient to eliminate the trivial $1/(i\gamma p+m)$ dependence of S_{FC} and the $1/k^2$ dependence of D_{FC}. We therefore set, as in Sec. III:

$$S_{FC}(p)=\frac{1}{i\gamma p+m}s_C(p), \quad (4.15)$$

$$D_{FC}(k)=(1/k^2)d_C(k), \quad (4.16)$$

and

$$S_{F\lambda\lambda'}(p)=\frac{1}{i\gamma p+m}s(\lambda,\lambda',p) \quad (4.15)'$$

$$D_{F\lambda}(k)=(1/k^2)d(\lambda,k). \quad (4.16)'$$

Equations (4.13) and (4.14) are now very useful. Using them together with (4.7)″, (4.9)″, (4.10)″, and the definitions (4.15), (4.16), we find:

$$z_2^{-1}(\lambda,\lambda';e_1^2)=[s_C(p,e_1^2)]_{\gamma p=\lambda'}, \quad (4.17)$$

$$z_3^{-1}(\lambda,e_1^2)=[d_C(k,e_1^2)]_{k^2=\lambda^2}. \quad (4.18)$$

z_2 is thus independent of λ since the right-hand side of (4.17) is independent of λ.

Our final equations are therefore

$$s(\lambda,\lambda',p,e_2^2)=s_C(p,e_1^2)/s_C(\lambda',e_1^2), \quad (4.19)$$

$$d(\lambda,k,e_2^2)=d_C(k,e_1^2)/d_C(\lambda,e_1^2), \quad (4.20)$$

and

$$e_2^2=d_C(\lambda,e_1^2)e_1^2. \quad (4.21)$$

The renormalization constants are seen to be the convergent functions calculated at infinite values of their arguments. This confirms the results of Sec. II and Appendix A.

Before closing this section we might remark that the entire treatment presented here can be very easily transcribed to meson theory. The situation in that case is somewhat more complicated since both Z_2 and Z_5 (the renormalization constant for Γ_5) contribute to charge renormalization, which is present even in the absence of closed loops (as is well known). The renormalization of Γ_5 must be carried through somewhat differently from that of Γ_μ, since we shall want $\Gamma_{5\lambda\lambda'}(p,p;g_2^2)$ to equal γ_5 at $\gamma p=\lambda'$. The equations analogous to (4.19)–(4.21) for pseudoscalar meson theory are:

$$s(\lambda,\lambda',p,g_2^2)=s_C(p,g_1^2)/s_C(\lambda',g_1^2), \qquad (4.19)'$$

$$\delta(\lambda,\lambda',k,g_2^2)=\delta_C(k,g_1^2)/\delta_C(\lambda,g_1^2), \qquad (4.20)'$$

$$\gamma_5(\lambda,\lambda',p,g_2^2)=\gamma_{5C}(p,g_1^2)/\gamma_{5C}(\lambda',g_1^2), \qquad (4.22)$$

$$g_2^2=g_1^2[s_C(\lambda',e_1^2)\gamma_{5C}(\lambda',g_1^2)]^2\delta_C(\lambda,g_1^2), \qquad (4.21)'$$

where

$$\Gamma_{5C}(p,p,g_1^2)=\gamma_5\cdot\gamma_{5C}(p,g_1^2),\ \text{etc.},$$

and where $\Delta=\delta/(k^2+\mu^2)$ is the meson propagation function.

We shall not investigate these equations further; we shall confine our attention to the much simpler case of quantum electrodynamics.

5. ASYMPTOTIC BEHAVIOR OF THE PROPAGATION FUNCTIONS IN QUANTUM ELECTRODYNAMICS

With the aid of the cut-off procedure introduced in the previous section, we may return to the discussion, begun in Sec. III, of the behavior of the propagation functions in the asymptotic region ($|p^2|\gg m^2$). In Sec. III photon self-energy parts were omitted and it was sufficient to use a Feynman cutoff in order to find the functional equation (3.9) satisfied asymptotically by the electron propagation function S_F'. The new cutoff enables us to include all Feynman diagrams. We shall now find a new functional equation for S_F' and one for D_F' as well.

Our starting point is the set of Eqs. (4.19)–(4.21) that express the cut-off propagation functions in terms of renormalized quantities. [In Sec. III, we used Eq. (3.3) instead.] We must first observe, as in Sec. III, that in a power series calculation of the cut-off functions the results remain finite when the electron mass is set equal to zero. The quantities λ and λ' provide, of course, ultraviolet cutoffs, while p provides an infrared cutoff for all Feynman integrals. Thus in the asymptotic region we may drop the electron mass in a calculation of the cut-off functions, which then take the forms

$|p^2|,\ \lambda^2,\ \lambda'^2\gg m^2$:

$$s(p/\lambda,p/\lambda',m^2/p^2,e_2^2)\approx s(p/\lambda,p/\lambda',0,e_2^2), \qquad (5.1)$$

$|k^2|,\ \lambda^2\gg m^2$:

$$d(k^2/\lambda^2,m^2/k^2,e_2^2)\approx d(k^2/\lambda^2,0,e_2^2). \qquad (5.2)$$

It should be noted that the asymptotic form of s depends only on p^2 and not on $i\gamma p$.

Equations (4.19)–(4.21) now give us at once the required functional equations:

$$|p^2|,\ \lambda^2,\ \lambda'^2\gg m^2:\quad s(p^2/\lambda^2,p^2/\lambda'^2,e_2^2)=\frac{s_C(p^2/m^2,e_1^2)}{s_C(\lambda'^2/m^2,e_1^2)} \qquad (5.3)$$

$$|k^2|,\ \lambda^2\gg m^2:\quad d(k^2/\lambda^2,e_2^2)=\frac{d_C(k^2/m^2,e_1^2)}{d_C(\lambda^2/m^2,e_1^2)} \qquad (5.4)$$

$$e_2^2=e_1^2d_C(\lambda^2/m^2,e_1^2). \qquad (5.5)$$

(We have omitted the argument $m^2/p^2=0$ in s and d.)

In Appendix B it is shown that the general nontrivial solution of these equations is given by:

$$e_1^2d_C(k^2/m^2,e_1^2)=F((k^2/m^2)\phi(e_1^2)), \qquad (5.6)$$

$$s_C(p^2/m^2,e_1^2)=A(e_1^2)H((p^2/m^2)\phi(e_1^2)). \qquad (5.7)$$

Here F, H, ϕ, and A are unknown functions of their arguments. (A contains an infrared divergent factor, which is always canceled, in calculations, by a similar factor in the vertex operator $\Gamma_{\mu C}$.) It is evident that we have obtained much less information here than we did in Sec. III. Also, the results of Sec. III are not correct in the general case, since Eq. (3.11) is not a special case of (5.7), but corresponds rather to a trivial solution of the functional equations (5.3)–(5.5), peculiar to the case of no charge renormalization.[16]

At least one striking result has emerged from the work in this section, however. The quantity on the left-hand side of Eq. (5.6) is, as remarked in the introduction, the Fourier transform of the Laplacian of the potential energy of two heavy point charges. It represents, therefore, the Fourier transform of a kind of effective charge density for the cloud of pairs surrounding a test body in the vacuum. Equation (5.6) states, in effect, that the *shape* of the charge distribution, at distances much smaller than $\hbar/mc$, is independent of the coupling constant $e_1^2/4\pi$, which enters only into the *scale factor* $\phi(e_1^2)$.

This result has an important consequence for the magnitude of e_0^2, the square of the bare charge, which is associated with the strength of the singularity at the center of the effective charge distribution. We have learned in Sec. II that e_0^2 is given by

$$e_0^2=e_1^2d_C(\infty,e_1^2) \qquad (5.8)$$

[16] It should be noted that, in any simplified form of the theory in which a restricted class of diagrams is summed, our results are unchanged provided that the conditions for renormalizability, as discussed in Sec. IV, are fulfilled. If, in a renormalizable approximation to the theory, there is charge renormalization, then the results of this section apply; if not, then the results of Sec. III apply.

For example, if the full integral equations (4.4) and (4.5), when renormalized, should turn out not to have solutions, it may be that solutions will exist for the integral equations obtained by cutting off the sequences on the right-hand sides after a finite number of terms and renormalizing. Such a procedure would be renormalizable and would not affect our functional equations.

and that $e_1^2 d_C(k^2/m^2, e_1^2)$ is a positive increasing function of k^2/m^2. Equation (5.6) tells us, then, that $F(\phi(e_1^2)k^2/m^2)$ is an increasing function of k^2/m^2 and that:

(a) If as $k^2/m^2 \to \infty$, $F(\phi k^2/m^2) \to \infty$, then e_0^2 is infinite, and the singularity at the center of the charge distribution is stronger than the δ function that corresponds to a finite point charge. This is the result indicated by perturbation theory.

(b) If, as $k^2/m^2 \to \infty$, $F(\phi k^2/m^2)$ approaches a finite limit, then e_0^2 equals that limit, which is *independent of the value of* e_1^2. The singularity at the center of the effective charge distribution is then a δ function with a strength corresponding to a finite bare charge e_0.

We shall return, at the end of this section, to the discussion of cases (a) and (b). Meanwhile, let us look at the solution of the functional equations from another point of view.

While the functions F and ϕ are unknown, certain of their properties can be deduced from the perturbation expansion of d_C. In the asymptotic region, d_C appears as a double power series in e_1^2 and $\ln(k^2/m^2)$ with finite numerical coefficients. (Of course, the convergence properties of this series are unknown. We have assumed throughout, however, that it defines a function which satisfies the same functional equations that we have derived for the series.) To facilitate comparison of (5.6) with the series, let us make use of the alternative form derived in Appendix B:

$$|k^2| \gg m^2: \quad \ln\frac{k^2}{m^2} = \int_{q(e_1^2)}^{e_1^2 d_C(k^2/m^2, e_1^2)} \frac{dx}{\psi(x)}. \tag{5.9}$$

In Eq. (5.9) both unknown functions q and ψ have power series expansions in their arguments and the first few terms of these series may be determined from the first few orders of perturbation theory. (The functions F and ϕ do not have this property.)

Perturbation theory yields for d_C in the asymptotic region the expansion

$$|k^2| \gg m^2: \quad d_C\left(\frac{k^2}{m^2}, e_1^2\right) = \left[1 - \frac{e_1^2}{12\pi^2}\left(\ln\frac{k^2}{m^2} - \frac{5}{3}\right) - \frac{e_1^4}{64\pi^4}\left(\ln\frac{k^2}{m^2} + c\right) + \cdots\right]^{-1}. \tag{5.10}$$

The fourth-order calculation was performed by Jost and Luttinger,[17] who did not compute c. Comparison of (5.9) and (5.10) yields the expansions of q and ψ:

$$\psi(x) = \frac{1}{12\pi^2}\left[x^2 + \frac{3}{16\pi^2}x^3 + \cdots\right], \tag{5.11}$$

$$q(e_1^2) = e_1^2 - \frac{5}{36\pi^2}e_1^4 + \cdots. \tag{5.12}$$

[17] R. Jost and J. M. Luttinger, Helv. Phys. Acta **23**, 201 (1950).

For the actual value of the coupling constant in quantum electrodynamics, $q(e_1^2)$ is presumably well approximated by e_1^2 and need not concern us very much. The crucial function is $\psi(x)$, which is given by (5.11) for very small x but is needed for large x in order to determine the behavior of the propagation function at very high momenta and to resolve the question of the finiteness of the bare charge.

We can restate, in terms of the properties of ψ, the two possibilities (a) and (b) for the behavior of the theory at high momenta:

(a) The integral $\int dx/\psi(x)$ in (5.9) does not diverge until the upper limit reaches $+\infty$. In that case $\ln(k^2/m^2) = +\infty$ corresponds to $e_1^2 d_C(k^2/m^2, e_1^2) = +\infty$ and the bare coupling constant $e_0^2/4\pi$ is infinite.

(b) For some finite value x_0 of the upper limit, $\int^{x_0} dx/\psi(x)$ diverges; for this to happen, $\psi(x)$ must come down to zero at $x = x_0$. Then $e_1^2 d_C(k^2/m^2, e_1^2) \to x_0$ as $\ln(k^2/m^2) \to \infty$, so that $e_0^2 = x_0$, a finite number independent of e_1^2. Since $e_1^2 < e_0^2$, the theory can exist only for e_1^2 less than some critical value $e_c^2 \leqslant e_0^2$, where $q(e_c^2) = e_0^2$. As $q(e_1^2)$ approaches its maximum value e_0^2, $\psi(q(e_1^2)) \to 0$, so we learn from Eq. (B.26) in Appendix B that the asymptotic form of $e_1^2 d_C(k^2/m^2, e_1^2)$ reduces simply to the constant e_0^2. A constant asymptotic form of $e_1^2 d_C(k^2/m^2, e_1^2)$ means that the weighting function $f(M^2/m^2, e_1^2)$ in Eq. (2.5) must vanish in the asymptotic region to order $1/M^2$. If the bare charge is finite, then the effective coupling at high momenta varies in a strange way with $q(e_1^2)$, increasing at first with increasing $q(e_1^2)$ and then decreasing to zero at $q(e_1^2) = e_0^2$, beyond which point the theory is meaningless.

Since we cannot discriminate between cases (a) and (b), the methods we have developed have not really served to settle fully the question of the asymptotic character of the propagation function D_{FC}. However, it is to be hoped that these methods may be used in the future to obtain more powerful results.

Recently Källén[18] has investigated the question of the finiteness of the bare charge. His result is that of the three renormalization quantities Z_2^{-1}, e_0^2, and m_0, at least one is infinite. Unfortunately, it is not possible to conclude from Källén's work that case (b) must be rejected.

APPENDIX A. CONSTRUCTION OF PARAMETRIC REPRESENTATIONS FOR THE PROPAGATION FUNCTIONS

The function $S_{F'\alpha\gamma}(x-y)$ is given by the matrix element

$$S_{F'\alpha\gamma}(x-y) = \epsilon(x_0 - y_0)(\Psi_0, P[\psi_\alpha(x), \bar{\psi}_\gamma(y)]\Psi_0). \tag{A.1}$$

Here $\psi(x)$ is the electron (or nucleon) field operator at the space-time point x in the Heisenberg representation; $\bar{\psi}(y)$ is the Dirac adjoint $\psi^*(y)\beta$ of $\psi(y)$; P is Dyson's

[18] G. Källén, Kgl. Danske Videnskab. Selskab, Mat.-fys. Medd. **27**, 12 (1953).

time-ordering operator; $\epsilon(t)$ is $t/|t|$; and Ψ_0 is the vacuum state of the coupled electron and photon (or nucleon and meson) fields.

In quantum electrodynamics S_F' is a gauge-variant function and for that reason one may easily be misled in dealing with it. To avoid such difficulties let us first discuss the function in meson theory (say the neutral pseudoscalar theory).

For $x_0 > y_0$ we may write (A.1) in the form

$$S_{F'\alpha\gamma}(x-y) = \sum_{\mathbf{p},M,s,\Pi,n} (\Psi_0,\psi_\alpha(x)\Psi_{\mathbf{p},M,s,\Pi,n}) \times(\Psi_{\mathbf{p},M,s,\Pi,n},\bar{\psi}_\gamma(y)\Psi_0) \quad \text{(A.2)}$$

where the Ψ's are a complete set of eigenstates of the coupled-field Hamiltonian and momentum. The momentum eigenvalue is denoted by $\mathbf{p}$ and the energy eigenvalue is given by $(M^2+p^2)^{\frac{1}{2}}$, so that the system described by the state Ψ has the energy M in its rest frame. For those states which contribute nonvanishing matrix elements to (A.2), the angular momentum of the system in its rest frame is $\frac{1}{2}$ and may have z component $\pm\frac{1}{2}$, denoted by the values ± 1 of the index s. Π is the parity of the system in its rest frame and may be $+1$ or -1. The remaining index n labels all the other (invariant) quantum numbers necessary to specify the state.

We may list some of the simplest types of states that contribute:

(1) one nucleon: $M=m$, the renormalized nucleon mass, and $\Pi=+1$ by convention.

(2) one nucleon, one meson: $M=(m^2+k^2)^{\frac{1}{2}}+(\mu^2+k^2)^{\frac{1}{2}}$, where μ is the renormalized meson mass and k is the relative momentum in the rest frame. The parity Π is $+1$ for a meson in a p state and -1 for a meson in an s state relative to the nucleon. It should be emphasized that in the latter case the matrix element does not vanish. For more complicated systems, more quantum numbers n are needed.

The space-time dependence of the matrix elements in (A.2) is determined by the energy and momentum eigenvalues of the Ψ's and so we have, for $x_0 > y_0$,

$$S_{F'\alpha\gamma}(x-y) = \frac{1}{(2\pi)^3}\int d^3p \sum_M \times\exp\{i[\mathbf{p}\cdot(\mathbf{x}-\mathbf{y})-(M^2+p^2)^{\frac{1}{2}}(x_0-y_0)]\} \times\sum_{n,s,\Pi} u_\alpha(\mathbf{p},M,s,\Pi,n)u_\delta^*(\mathbf{p},M,s,\Pi,n)\beta_{\delta\gamma}. \quad \text{(A.3)}$$

We may consider first only those states with $\mathbf{p}=0$ and later discuss the others by means of Lorentz transformations. For states of zero momentum the sum over spins must give simply

$$\sum_s u_\alpha(0,M,s,\Pi,n)u_\delta^*(0,M,s,\Pi,n) = \tfrac{1}{2}(1+\beta)_{\alpha\delta}U(M,n) \quad \text{if } \Pi=+1$$
$$= \tfrac{1}{2}(1-\beta)_{\alpha\delta}V(M,n) \quad \text{if } \Pi=-1, \quad \text{(A.4)}$$

since for such a state the parity operator is β. That all U's and V's are both real and positive follows from taking the trace of both sides of (A.4).

The generalization of (A.4) to the case of $\mathbf{p}\neq 0$ is easily calculated by transforming both sides with the Lorentz transformation matrix for velocity $-\mathbf{p}/(p^2+M^2)^{\frac{1}{2}}$. We obtain

$$\sum_s u_\alpha(\mathbf{p},M,s,\Pi,n)u_\delta^*(\mathbf{p},M,s,\Pi,n) = \left(\frac{\alpha\cdot\mathbf{p}+\beta M+(p^2+M^2)^{\frac{1}{2}}}{2(p^2+M^2)^{\frac{1}{2}}}\right)_{\alpha\delta} U(M,n) \quad \text{if } \Pi=+1,$$
$$= \left(\frac{\alpha\cdot\mathbf{p}-\beta M+(p^2+M^2)^{\frac{1}{2}}}{2(p^2+M^2)^{\frac{1}{2}}}\right)_{\alpha\delta} V(M,n) \quad \text{if } \Pi=-1. \quad \text{(A.5)}$$

Before substituting these expressions into the formula for S_F', let us make use of the relations

$$\frac{1}{2\pi i}\int_{-\infty}^{\infty}\frac{dp_0\exp(-ip_0t)}{i\gamma_\mu p_\mu+M-i\epsilon} = \frac{\alpha\cdot\mathbf{p}+\beta M+(p^2+M^2)^{\frac{1}{2}}}{2(p^2+M^2)^{\frac{1}{2}}}\beta \times\exp[-i(p^2+M^2)^{\frac{1}{2}}t], \quad (t>0), \quad \text{(A.6)}$$

and

$$\frac{1}{2\pi i}\int_{-\infty}^{\infty}\frac{dp_0\exp(-ip_0t)}{i\gamma_\mu p_\mu-M+i\epsilon} = \frac{\alpha\cdot\mathbf{p}-\beta M+(p^2+M^2)^{\frac{1}{2}}}{2(p^2+M^2)^{\frac{1}{2}}}\beta \times\exp[-i(p^2+M^2)^{\frac{1}{2}}t] \quad (t>0). \quad \text{(A.6')}$$

Here ϵ is a positive infinitesimal quantity.

We may now rewrite (A.3) as follows:

$$S_{F'}(x) = \frac{1}{(2\pi)^4 i}\int d^4p\exp(ip_\mu x_\mu)\sum_M \times\left[\frac{1}{i\gamma_\lambda p_\lambda+M-i\epsilon}\sum_{\substack{n\\ \Pi=1}} U(M,n) + \frac{1}{i\gamma_\lambda p_\lambda-M+i\epsilon}\sum_{\substack{n\\ \Pi=-1}} V(M,n)\right]. \quad \text{(A.7)}$$

Let us separate off the contribution to S_F' of one-nucleon states, which are associated with $M=m$; the coefficient U for that case is the formally divergent constant called Z_2. For all other values of M, let us put

$$\sum_{\substack{n\\ \Pi=+1}} U(M,n) = Z_2 g(M)/M \quad \text{(A.8)}$$

and

$$\sum_{\substack{n\\ \Pi=-1}} V(M,n) = Z_2 h(M)/M. \quad \text{(A.9)}$$

Evidently Z_2, g, and h are all real and positive. We now have

$$S_{F'}(x)=\frac{Z_2}{(2\pi)^4 i}\int d^4p\exp(ip_\mu x_\mu)$$

$$\times\left\{\frac{1}{i\gamma p+m-i\epsilon}+\int_{m+\mu}^{\infty}\frac{g(M)}{i\gamma p+M-i\epsilon}\frac{dM}{M}+\int_{m+\mu}^{\infty}\frac{h(M)}{i\gamma p-M+i\epsilon}\frac{dM}{M}\right\}\quad (x_0>0).\quad (A.10)$$

It follows from the invariance of the theory under charge conjugation that the Eq. (A.10), which we have derived for $x_0>0$, holds also for $x_0<0$.

We must still find an expression for Z_2 in terms of g and h. So far, we have used nothing but relativistic invariance; now we must make use of the anticommutation rules for ψ and ψ^*. Let us calculate the quantity

$$\text{disc}\,S_F'\equiv\lim_{t\to 0^+}[S_F'(\mathbf{x},t)-S_F'(\mathbf{x},-t)].\quad (A.11)$$

We utilize the relation

$$\text{disc}\frac{1}{(2\pi)^4 i}\int\frac{d^4p\exp(ip_\mu x_\mu)}{i\gamma p+M-i\epsilon}=\text{disc}\frac{1}{(2\pi)^4 i}\int\frac{d^4p\exp(ip_\mu x_\mu)}{i\gamma p-M+i\epsilon}=\beta\delta(\mathbf{x}).\quad (A.12)$$

Equation (A.10) then yields

$$\text{disc}\,S_F'=Z_2\beta\delta(\mathbf{x})\times\left(1+\int_{m+\mu}^{\infty}\frac{dM}{M}[g(M)+h(M)]\right).\quad (A.13)$$

Another expression for disc S_F' can be obtained from Eq. (A.1):

$$\text{disc}\,S_F'=(\Psi_0,\{\psi(\mathbf{x},t),\bar\psi(0,t)\}\Psi_0)=\beta\delta(\mathbf{x}).\quad (A.14)$$

Comparison of (A.13) and (A.14) yields

$$Z_2^{-1}=1+\int_{m+\mu}^{\infty}\frac{dM}{M}[g(M)+h(M)].\quad (A.15)$$

With the aid of (A.10) and Dyson's Eq. (2.7) we have for the Fourier transform of the renormalized propagation function the representation

$$S_{FC}(p)=\frac{1}{i\gamma p+m-i\epsilon}+\int_{m+\mu}^{\infty}\frac{g(M)dM/M}{i\gamma p+M-i\epsilon}+\int_{m+\mu}^{\infty}\frac{h(M)dM/M}{i\gamma p-M+i\epsilon}.\quad (A.16)$$

The meson propagation functions Δ_F' and Δ_{FC} can be dealt with by methods entirely analogous to those we have used for the nucleon functions. In place of (A.1) we have

$$\Delta_F'(x-y)-(\Psi_0,P[\phi(x),\phi(y)]\Psi_0).\quad (A.17)$$

Considerations of relativistic invariance lead us to the result

$$\Delta_F'(x)=Z_3\frac{1}{(2\pi)^4 i}\int d^4k\exp(ik_\mu x_\mu)\times\left[\frac{1}{k^2+\mu^2-i\epsilon}+\int_{(3\mu)^2}^{\infty}\frac{dM^2}{M^2}\frac{f(M^2)}{k^2+M^2-i\epsilon}\right]\quad (A.18)$$

corresponding to (A.10). Utilizing the canonical commutation rule for ϕ and $\partial\phi/\partial t$ at equal times, we obtain, in complete analogy with (A.15),

$$Z_3^{-1}=1+\int_{(3\mu)^2}^{\infty}\frac{dM^2 f(M^2)}{M^2}.\quad (A.19)$$

In quantum electrodynamics, S_F' must be calculated, according to Gupta[19] and Bleuler,[20] from the equation

$$S_F'(x-y)=\epsilon(x_0-y_0)(\Psi_0,\eta P[\psi(x),\bar\psi(y)]\Psi_0)\quad (A.20)$$

rather than (A.1). Here $\eta=(-1)^{N_4}$ and N_4 is the operator describing the number of "temporal photons." The only effect of the introduction of η into our previous work is to remove the requirement that g and h be positive.

The photon propagation function D_F' satisfies the relation

$$\langle\Psi_0,\eta P[A_\mu(x),A_\nu(y)]\Psi_0\rangle=\delta_{\mu\nu}D_F'(x-y)+\frac{\partial^2}{\partial x_\mu\partial x_\nu}G(x-y),\quad (A.21)$$

analogous to (A.17). Here G is gauge-variant but D_F' is not. We may determine D_F', moreover, from the transverse part of $A_\mu(x)$ alone, so that the operator η does not disturb us. The results are then identical with (A.18) and (A.19) with $\mu=0$; f is positive as in meson theory.

The functions S_{FC} and D_{FC} in quantum electrodynamics are given, to first order in the coupling constant $\alpha=e_1^2/4\pi$, by

$$S_{FC}(p)=\frac{1}{i\gamma p+m-i\epsilon}+\frac{e_1^2}{16\pi^2}\int_m^{\infty}\frac{dM}{M^3(M^2-m^2)}\times\left[\frac{(M+m)^2(M^2+m^2-4mM)}{i\gamma p+M-i\epsilon}+\frac{(M-m)^2(M^2+m^2+4mM)}{i\gamma p-M+i\epsilon}\right]\quad (A.22)$$

[19] S. W. Gupta, Proc. Phys. Soc. (London) **63**, 681 (1950).
[20] K. Bleuler, Helv. Phys. Acta **23**, 567 (1950).

and

$$D_{FC}(k)=\frac{1}{k^2}+\frac{e_1^2}{12\pi^2}\int_{4m^2}^{\infty}\frac{dM^2}{M^2}\times\frac{(1+2m^2/M^2)(1-4m^2/M^2)^{\frac{1}{2}}}{k^2+M^2-i\epsilon}. \quad \text{(A.23)}$$

It will be seen that the function $g(M)$ in (A.22) is not always positive. This is in contrast with the situation in meson theory. In the scalar symmetric theory with $\mu=0$ we have

$$S_{FC}(p)=\frac{1}{i\gamma p+m-i\epsilon}+\frac{3g_1^2}{16\pi^2}\int_m^{\infty}\times\frac{dM}{M^3(M^2-m^2)}\left[\frac{(M+m)^2(M^2+m^2)}{i\gamma p+M-i\epsilon}+\frac{(M-m)^2(M^2+m^2)}{i\gamma p-M+i\epsilon}\right], \quad \text{(A.24)}$$

while in the pseudoscalar symmetric theory we have

$$S_{FC}(p)=\frac{1}{i\gamma p+m-i\epsilon}+\frac{3g_1^2}{16\pi^2}\int_m^{\infty}\frac{dM}{2M^3}(M^2-m^2)\times\left[\frac{1}{i\gamma p+M-i\epsilon}+\frac{1}{i\gamma p-M+i\epsilon}\right]. \quad \text{(A.25)}$$

APPENDIX B. SOLUTION OF THE FUNCTIONAL EQUATIONS[21]

We may solve Eqs. (5.4) and (5.5) for the photon propagation function without reference to Eq. (5.3). For convenience we reproduce the equations:

$$d(k^2/\lambda^2,e_2^2)=d_C(k^2/m^2,e_1^2)/d_C(\lambda^2/m^2,e_1^2), \quad (5.4)$$

$$e_2^2=e_1^2d_C(\lambda^2/m^2,e_1^2). \quad (5.5)$$

Combining the equations, we obtain

$$e_1^2d_C(k^2/m^2,e_1^2)=e_1^2d_C(\lambda^2/m^2,e_1^2)\times d(k^2/\lambda^2,e_1^2d_C(\lambda^2/m^2,e_1^2)). \quad \text{(B.1)}$$

Giving new names to the left- and right-hand sides of (B.1), we may write

$$g(k^2/m^2,e_1^2)=Q(k^2/\lambda^2,g(\lambda^2/m^2,e_1^2)) \quad \text{(B.2)}$$

or

$$g(x,e_1^2)=Q(x/y,g(y,e_1^2)). \quad \text{(B.3)}$$

Except in trivial cases we may invert the function g and put

$$x=h(g,e_1^2), \quad y=h(g',e_1^2). \quad \text{(B.4)}$$

[21] We would like to thank Dr. T. D. Lee for suggesting the form of the solution to us.

Then (B.3) becomes

$$g=Q(h(g,e_1^2)/h(g',e_1^2),\, g'). \quad \text{(B.5)}$$

If the functional form of Q is nontrivial, then $h(g,e_1^2)/h(g',e_1^2)$ must be independent of e_1^2. Thus

$$h(g,e_1^2)=G(g)/\phi(e_1^2), \quad \text{(B.6)}$$

which implies that

$$g(x,e_1^2)=F(x\phi(e_1^2)), \quad \text{(B.7)}$$

where F is the inverse function to G. In terms of the original labels, we have

$$e_1^2d_C(k^2/m^2,e_1^2)=F((k^2/m^2)\phi(e_1^2)). \quad \text{(B.8)}$$

Substituting (B.8) into (5.5), we obtain

$$e_2^2=F((\lambda^2/m^2)\phi(e_1^2)). \quad \text{(B.9)}$$

Using (B.8) and (B.9) we find

$$F\left(\frac{k^2}{m^2}\phi(e_1^2)\right)=F\left(\frac{k^2}{\lambda^2}\frac{\lambda^2}{m^2}\phi(e_1^2)\right)=F((k^2/\lambda^2)G(e_2^2)). \quad \text{(B.10)}$$

Since F and G are inverse functions, there is only one arbitrary function in (B.10). It is now evident that (B.1) is indeed satisfied, with

$$e_2^2d(k^2/\lambda^2,e_2^2)=F((k^2/\lambda^2)G(e_2^2)). \quad \text{(B.11)}$$

In a power series calculation, e_2^2d appears as a double series in e_2^2 and $\ln(k^2/\lambda^2)$. Let us transform (B.11) so that comparison with the series solution becomes possible:

$$G(e_2^2d)=(k^2/\lambda^2)G(e_2^2), \quad \text{(B.12)}$$

$$\ln k^2/\lambda^2=\ln G(e_2^2d)-\ln G(e_2^2), \quad \text{(B.13)}$$

$$\ln\frac{k^2}{\lambda^2}=\int_{e_2^2}^{e_2^2d}\frac{dx}{\psi(x)}. \quad \text{(B.14)}$$

Here

$$\psi(x)=G(x)(dG/dx)^{-1}. \quad \text{(B.15)}$$

Differentiating both sides of (B.14) with respect to $\ln(k^2/\lambda^2)$, we have

$$1=\frac{1}{\psi(e_2^2d)}\frac{\partial(e_2^2d)}{\partial(\ln(k^2/\lambda^2))}, \quad \text{(B.16)}$$

or

$$\psi(e_2^2d)=\frac{\partial(e_2^2d)}{\partial(\ln(k^2/\lambda^2))}. \quad \text{(B.17)}$$

If in (B.17) we put $\ln(k^2/\lambda^2)=0$, we obtain

$$\psi(e_2^2)=\left.\frac{\partial(e_2^2d)}{\partial(\ln(k^2/\lambda^2))}\right|_{\ln(k^2/\lambda^2)=0}. \quad \text{(B.18)}$$

Evidently the double series expansion of e_2^2d yields a power series expansion of $\psi(e_2^2)$. In fact the entire double series can be rewritten in terms of $\psi(e_2^2)$. If we differentiate both sides of (B.14) with respect to e_2^2, we get

$$0=\frac{1}{\psi(e_2^2d)}\frac{\partial(e_2^2d)}{\partial(e_2^2)}-\frac{1}{\psi(e_2^2)}. \quad \text{(B.19)}$$

Combining this result with (B.17) yields

$$\psi(e_2^2)=\frac{\partial(e_2^2d)}{\partial(\ln(k^2/\lambda^2))}\bigg/\frac{\partial(e_2^2d)}{\partial e_2^2}, \quad \text{(B.20)}$$

whence

$$e_2^2d\left(\frac{k^2}{\lambda^2},e_2^2\right)=\sum_{n=0}^{\infty}\frac{[\ln(k^2/\lambda^2)]^n}{n!}\left[\psi(e_2^2)\frac{d}{de_2^2}\right]^n e_2^2. \quad \text{(B.21)}$$

Representations similar to (B.14) and (B.21) can be found for $e_1^2d_C(k^2/m^2,e_1^2)$. We transform (B.8) as follows:

$$G(e_1^2d_C)=(k^2/m^2)\phi(e_1^2), \quad \text{(B.22)}$$

$$\ln(k^2/m^2)=\ln G(e_1^2d_C)-\ln\phi(e_1^2), \quad \text{(B.23)}$$

$$\ln(k^2/m^2)=\int_{q(e_1^2)}^{e_1^2d_C}dx/\psi(x) \quad \text{(B.24)}$$

where ψ is the same function as before and

$$q(e_1^2)=F(\phi(e_1^2)). \quad \text{(B.25)}$$

A comparison of (B.24) and (B.14) shows that the functional dependence of $e_1^2d_C$ on k^2/m^2 and $q(e_1^2)$ is the same as the dependence of e_2^2d on k^2/λ^2 and e_2^2. Therefore, we have, from Eq. (B.21), the series expansion

$$e_1^2d_C\left(\frac{k^2}{m^2}e_1^2\right)=\sum_{n=0}^{\infty}\frac{[\ln(k^2/m^2)]^n}{n!}\times\left\{\left[\psi(y)\frac{d}{dy}\right]^n y\right\}_{y=q(e_1^2)}. \quad \text{(B.26)}$$

The representation (B.26) may easily be compared with the double series in e_1^2 and $\ln(k^2/m^2)$ obtained from perturbation theory. We see that when $\ln(k^2/m^2)$ is set equal to 0 we obtain for the right-hand side just $q(e_1^2)$, so that the perturbation theory gives a power series expansion for $q(e_1^2)$.

So far our discussion has been confined to the photon propagation function. We must now solve the functional equation (5.3) for the electron propagation function:

$$s(p^2/\lambda^2,p^2/\lambda'^2,e_2^2)=\frac{s_C(p^2/m^2,e_1^2)}{s_C(\lambda'^2/m^2,e_1^2)}. \quad \text{(5.3)}$$

Using (B.9), we can write the left-hand side as

$$s(p^2/\lambda^2,p^2/\lambda'^2,e_2^2)=s(p^2/\lambda^2,p^2/\lambda'^2,F(\phi(e_1^2)\lambda^2/m^2)). \quad \text{(B.27)}$$

By virtue of (5.3) this must be independent of λ^2 and may be written

$$s(p^2/\lambda^2,p^2/\lambda'^2,e_2^2)=R((p^2/m^2)\phi(e_1^2),(\lambda'^2/m^2)\phi(e_1^2)). \quad \text{(B.28)}$$

Since the quotient on the right-hand side of (5.3) depends on its arguments only through $(p^2/m^2)\phi(e_1^2)$ and $(\lambda'^2/m^2)\phi(e_1^2)$, we must have

$$s_C(p^2/m^2,e_1^2)=A(e_1^2)H(p^2/m^2\phi(e_1^2)). \quad \text{(B.29)}$$

PHYSICAL REVIEW VOLUME 97, NUMBER 5 MARCH 1, 1955

Behavior of Neutral Particles under Charge Conjugation

M. GELL-MANN,* *Department of Physics, Columbia University, New York, New York*

AND

A. PAIS, *Institute for Advanced Study, Princeton, New Jersey*

(Received November 1, 1954)

Some properties are discussed of the θ^0, a heavy boson that is known to decay by the process $\theta^0 \rightarrow \pi^+ + \pi^-$. According to certain schemes proposed for the interpretation of hyperons and K particles, the θ^0 possesses an antiparticle $\bar{\theta}^0$ distinct from itself. Some theoretical implications of this situation are discussed with special reference to charge conjugation invariance. The application of such invariance in familiar instances is surveyed in Sec. I. It is then shown in Sec. II that, within the framework of the tentative schemes under consideration, the θ^0 must be considered as a "particle mixture" exhibiting two distinct lifetimes, that each lifetime is associated with a different set of decay modes, and that no more than half of all θ^0's undergo the familiar decay into two pions. Some experimental consequences of this picture are mentioned.

I

IT is generally accepted that the microscopic laws of physics are invariant to the operation of charge conjugation (CC); we shall take the rigorous validity of this postulate for granted. Under CC, every particle is carried into what we shall call its "antiparticle". The principle of invariance under CC implies, among other things, that a particle and its antiparticle must have exactly the same mass and intrinsic spin and must have equal and opposite electric and magnetic moments.

A charged particle is thus carried into one of opposite charge. For example, the electron and positron are each other's antiparticles; the π^+ and π^- and the μ^+ and μ^- mesons are supposed to be pairs of antiparticles; and the proton must possess an antiparticle, the "antiproton".

Neutral particles fall into two classes, according to their behavior under CC:

(a) Particles that transform into themselves, and which are thus their own antiparticles. For instance the photon and the π^0 meson are bosons that behave in this fashion. It is conceivable that fermions, too, may belong to this class. An example is provided by the Majorana theory of the neutrino.

In a field theory, particles of class (a) are represented by "real" fields, i.e., Hermitian field operators. There is an important distinction to be made within this class, according to whether the field takes on a plus or a minus sign under CC. The operation of CC is performed by a unitary operator $\mathcal{C}$. The photon field operator $A_\mu(x)$ satisfies the relation

$$\mathcal{C}A_\mu(x)\mathcal{C}^{-1} = -A_\mu(x), \tag{1}$$

while for the π^0 field operator $\phi(x)$ we have

$$\mathcal{C}\phi(x)\mathcal{C}^{-1} = \phi(x). \tag{2}$$

Equation (1) expresses the obvious fact that the electromagnetic field changes sign when positive and negative charges are interchanged; that the π^0 field must not change sign can be inferred from the observed two-photon decay of the π^0.

We are effectively dealing here with the "charge conjugation quantum number" C, which is the eigenvalue of the operator $\mathcal{C}$, and which is rigorously conserved in the absence of external fields. If only an odd (even) number of photons is present, we have $C = -1(+1)$; if only π^0's are present, $C = +1$; etc. As a trivial example of the conservation of C, we may mention that the decay of the π^0 into an odd number of photons is forbidden.[1]

We may recall that a state of a neutral system composed of charged particles may be one with a definite value of C. For example, the $^1S_0$ state of positronium has $C = +1$; a state of a π^+ and a π^- meson with relative orbital angular momentum l has $C = (-1)^l$; etc.

For fermions, as for bosons, a distinction may be made between "odd" and "even" behavior of neutral fields of class (a) under CC. However, the distinction is then necessarily a relative rather than an absolute one.[2] In other words, it makes no sense to say that a single such fermion field is "odd" or "even", but it does make sense to say that two such fermion fields have the same behavior under CC or that they have opposite behavior.

(b) Neutral particles that behave like charged ones in that: (1) they have antiparticles distinct from themselves; (2) there exists a rigorous conservation law that prohibits virtual transitions between particle and antiparticle states.

A well-known member of this class is the neutron N, which can obviously be distinguished from the antineutron $\bar{N}$ by the sign of its magnetic moment. The law that forbids the virtual processes $N \rightleftarrows \bar{N}$ is the law

* On leave from Department of Physics and Institute for Nuclear Studies, University of Chicago, Chicago, Illinois.

[1] For other consequences of invariance under charge conjugation see A. Pais and R. Jost, Phys. Rev. 87, 871 (1952); L. Wolfenstein and D. G. Ravenhall, Phys. Rev. 88, 279 (1952); L. Michel, Nuovo cimento 10, 319 (1953).

[2] This is due to the fact that fermion fields can interact only bilinearly. For example, one easily sees that the interactions responsible for $P \rightarrow N + e^+ + \nu$ would not lead to physically distinguishable results if ν were either an even or an odd Majorana neutrino.

of conservation of baryons,[3] which is, so far as we know, exact, and which states that n, the number of baryons minus the number of antibaryons, must remain unchanged. Clearly all neutral hyperons likewise belong to this class. Although we know of no "elementary" bosons in the same category, we have no *a priori* reason for excluding their existence. [Note that the H atom is an example of a "non-elementary" boson of class (b).]

Particles in this class are represented by "complex" fields, and the operation of charge conjugation transforms the field operators into their Hermitian conjugates.

It is the purpose of this note to discuss the possible existence of neutral particles that seem, at first sight, to belong neither to class (a) nor to class (b).

II

Recently, attempts have been made to interpret hyperon and K-particle phenomena by distinguishing sharply between strong interactions, to which essentially all production of these particles is attributed, and weak interactions, which are supposed to induce their decay. It is necessary to assume that the strong interactions give rise to "associated production "exclusively.[4]

Certain detailed schemes[5] which meet this requirement lead to further specific properties of particles and interactions. In particular, a suggestion has been made about the θ^0 particle, a heavy boson that is known to decay according to the scheme:

$$\theta^0 \rightarrow \pi^+ + \pi^- + (\sim 215 \text{ Mev}). \tag{3}$$

It has been proposed that the θ^0 possesses an antiparticle $\bar{\theta}^0$ distinct from itself, and that in the absence of the weak decay interactions, there is a conservation law that prohibits the virtual transitions $\theta^0 \rightleftarrows \bar{\theta}^0$. [In our present language, we would say that the θ^0 belongs to class (b) if the weak interactions are turned off.] This conservation law also leads to stability of the θ^0 and $\bar{\theta}^0$; moreover, while it permits the reaction $\pi^- + P \rightarrow \Lambda^0 + \theta^0$ it forbids the analogous process $\pi^- + P \rightarrow \Lambda^0 + \bar{\theta}^0$. In the schemes under consideration this is the same law that forbids the reaction: 2 neutrons$\rightarrow 2\Lambda^0$.

The weak interactions that must be invoked to account for the observed decay (3) evidently cause the conservation law to break down (a fact that is, of course, of little importance for production). This breakdown makes the forbiddenness of the processes $\theta^0 \rightleftarrows \bar{\theta}^0$ no longer absolute, as can be seen from the following argument: In the decay (3) the pions are left in a state with a definite relative angular momentum and therefore with a definite value of the charge-conjugation quantum number C. The charge-conjugate process,

$$\bar{\theta}^0 \rightarrow \pi^+ + \pi^-, \tag{4}$$

must also occur and must leave the pions in the same state; moreover the reverse of (4) must also be possible, at least as a virtual process. Therefore the virtual transition $\theta^0 \rightleftarrows \pi^+ + \pi^- \rightleftarrows \bar{\theta}^0$ is induced by the weak interactions, and we are no longer dealing exactly with case (b).

In order to treat this novel situation, we shall find it convenient to introduce a change of representation. Since the θ^0 and $\bar{\theta}^0$ are distinct, they are associated, in a field theory, with a "complex" field ψ (a non-Hermitian field operator), just as in case (b). Under charge conjugation ψ must transform according to the law:

$$\mathcal{C}\psi\mathcal{C}^{-1} = \psi^+, \quad \mathcal{C}\psi^+\mathcal{C}^{-1} = \psi, \tag{5}$$

where ψ^+ is the Hermitian conjugate of ψ. Let us now define

$$\psi_1 \equiv (\psi + \psi^+)/\sqrt{2} \tag{6}$$

and

$$\psi_2 \equiv (\psi - \psi^+)/\sqrt{2}i, \tag{7}$$

so that ψ_1 and ψ_2 are Hermitian field operators satisfying

$$\mathcal{C}\psi_1\mathcal{C}^{-1} = \psi_1, \tag{8}$$

and

$$\mathcal{C}\psi_2\mathcal{C}^{-1} = -\psi_2. \tag{9}$$

The fields ψ_1 and ψ_2 evidently correspond to class (a); in fact ψ_1 is "even" like the π^0 field and ψ_2 is "odd" like the photon field. Corresponding to these real fields there are quanta, which we shall call θ_1^0 and θ_2^0 quanta. The relationship that these have to the quanta of the complex ψ field, which we have called θ^0 and $\bar{\theta}^0$, may be seen from an example: Let Ψ_1 be the wave-functional representing a single θ_1 quantum in a given state, while Ψ_0 and Ψ_0' describe a θ^0 and a $\bar{\theta}^0$, respectively, in the same state. Then we have

$$\Psi_1 = (\Psi_0 + \Psi_0')/\sqrt{2}.$$

Thus the creation of a θ_1 (or, for that matter, of a θ_2) corresponds physically to the creation, with equal probability and with prescribed relative phase, of either a θ^0 or a $\bar{\theta}^0$. Conversely, the creation of a θ^0 (or of a $\bar{\theta}^0$) corresponds to the creation, with equal probability and prescribed relative phase, of either a θ_1^0 or a θ_2^0.

The transformation (6), (7) to two real fields could equally well have been applied to a complex field of class (b), such as that associated with the neutron. However, this would not be particularly enlightening. It would lead us, for instance, to describe phenomena involving neutrons and antineutrons in terms of "N_1 and N_2 quanta". Now a state with an N_1 (or N_2) quantum is a mixture of states with different values of the quantum number n, the number of baryons minus the number of antibaryons. But the law of conservation of baryons requires this quantity to be a constant of the motion, and so a mixed state can never arise from a pure one. Since in our experience we deal exclusively

[3] Nucleons and hyperons are collectively referred to as baryons.

[4] A. Pais, Phys. Rev. 86, 663 (1952).

[5] M. Gell-Mann, Phys. Rev. 92, 833 (1953); A. Pais, Proc. Nat, Acad. Sci. U. S. 40, 484, 835 (1954); M. Gell-Mann and A. Pais. *Proceedings of the International Conference Glasgow* (Pergamon Press, London, to be published).

with states that are pure with respect to n, the introduction of N_1 and N_2 quanta can only be a mathematical device that distracts our attention from the truly physical particles N and $\bar{N}$.

On the other hand, it can obviously not be argued in a similar way that the θ_1^0 and θ_2^0 quanta are completely unphysical, for the corresponding conservation law in that case is not a rigorous one. Always assuming the correctness of our model of the θ^0, we still have the θ^0 and $\bar{\theta}^0$ as the primary objects in production phenomena. But we shall now show that the decay process is best described in terms of θ_1^0 and θ_2^0.

The weak interactions, in fact, must lead to very different patterns of decay for the θ_1^0 and θ_2^0 into pions and (perhaps) γ rays; any state of pions and/or γ rays that is a possible decay mode for the θ_1^0 is not a possible one for the θ_2^0, and *vice versa*. This is because, according to the postulate of rigorous CC invariance, the quantum number C is conserved in the decay; the θ_1^0 must go into a state that is even under charge conjugation, while the θ_2^0 must go into one that is odd. Since the decay modes are different and even mutually exclusive for the θ_1^0 and θ_2^0, their rates of decay must be quite unrelated. There are thus two independent lifetimes, one for the θ_1^0, and one for the θ_2^0.

An important illustration of the difference in decay modes of the θ_1^0 and θ_2^0 is provided by the two-pion disintegration. We know that reaction (3) occurs; therefore at least one of the two quanta θ_1^0 and θ_2^0, say θ_1^0, must be capable of decay into two charged pions. The final state of the two pions in the θ_1^0 decay is then even under charge conjugation like the θ_1^0 state itself. These two pions are thus in a state of even relative angular momentum and therefore of even parity. So the θ_1^0 must have even spin and even parity. Now we assume that the θ^0 has a definite intrinsic parity, and therefore the parity and spin of the θ_2^0 must be the same as those of the θ_1^0, both even. If the θ_2^0 were to decay into two pions, these would again be in a state of even relative angular momentum and thus even with respect to charge conjugation. However, the θ_2^0 is itself odd under charge conjugation; its decay into two pions is therefore forbidden.

Alternatively, if the θ_2^0 is the one that actually goes into two pions, then the spin and parity of θ_1^0 and the θ_2^0 are both odd, and so the θ_1^0 cannot decay into two pions.

Of the θ_1^0 and the θ_2^0, that one for which the two-pion decay is forbidden may go instead into $\pi^+ + \pi^- + \gamma$ or possibly into three pions (unless the spin and parity of the θ^0 are $0+$), etc.

While we have seen that the θ_1^0 and θ_2^0 may each be assigned a lifetime, this is evidently not true of the θ^0 or $\bar{\theta}^0$. Since we should properly reserve the word "particle" for an object with a unique lifetime, it is the θ_1^0 and θ_2^0 quanta that are the true "particles". The θ^0 and the $\bar{\theta}^0$ must, strictly speaking, be considered as "particle mixtures."

It should be remarked that the θ_1^0 and the θ_2^0 differ not only in lifetime but also in mass, though the mass difference is surely tiny. The weak interactions responsible for decay cause the θ_1^0 and the θ_2^0 to have their respective small level widths and correspondingly must produce small level shifts which are different for the two particles.

To sum up, our picture of the θ^0 implies that it is a particle mixture exhibiting two distinct lifetimes, that each lifetime is associated with a different set of decay modes, and that *not more than half of all θ^0's* can undergo the familiar decay into two pions.[6]

We know experimentally that the lifetime τ for the decay mode (3) (and hence for all decay modes that may compete with this one) is about 1.5×10^{-10} sec. The present qualitative considerations, even if at all correct in their underlying assumptions, do not enable us to predict the value of the "second lifetime" τ' of the θ^0. Nevertheless, the examples given above of decays responsible for the second lifetime lead one to suspect that[7] $\tau' \gg \tau$. As an illustration of the experimental implications of this situation consider the study of the reaction $\pi^- + P \to \Lambda^0 + \theta^0$ in a cloud chamber. If the reaction occurs and subsequently $\Lambda^0 \to P + \pi^-$, $\theta^0 \to \pi^+ + \pi^-$, there should be a reasonable chance to observe this whole course of events in the chamber, as the lifetime for the Λ^0 decay ($\sim 3.5\times10^{-10}$ sec) is comparable to τ. However, if it is true that $\tau' \gg \tau$, it would be very difficult to detect the decay with the second lifetime in the cloud chamber with its characteristic bias for a limited region of lifetime values.[8] Clearly this also means an additional complication in the determination from cloud chamber data as to whether or not production always occurs in an associated fashion. In some such cases the analysis of the reaction $\pi^- + P \to \Lambda^0 + ?$ may still be pushed further, however, if one assumes that besides the Λ^0 only one other neutral object is formed.[9]

At any rate, the point to be emphasized is this: a neutral boson may exist which has the characteristic θ^0 mass but a lifetime $\neq \tau$ and which may find its natural place in the present picture as the second component of the θ^0 mixture.

One of us, (M. G.-M.), wishes to thank Professor E. Fermi for a stimulating discussion.

[6] Note that if the spin and parity of the θ^0 are even, then the θ_1^0 may decay into $2\pi^0$'s as well as into $\pi^+ + \pi^-$.

[7] The process $\theta^0 \to \pi^+ + \pi^- + \gamma$ may occur as a radiative correction to the allowed decay into $\pi^+ + \pi^-$ connected with the lifetime τ; see S. B. Treiman, Phys. Rev. **95**, 1360 (1954). The process may also occur as one of the principal decay modes associated with the second lifetime τ'. The latter case may be distinguished from the former not only by the distinct lifetime but also by a different energy spectrum which probably favors higher γ-ray energies; such a spectrum is to be expected in a case where the emission of the γ ray is not just part of the "infrared catastrophe", but is an integral part of the decay process.

[8] See, e.g., Leighton, Wanlass, and Anderson, Phys. Rev. **89**, 148 (1953), Sec. III.

[9] See Fowler, Shutt, Thorndike, and Whittemore, Phys. Rev. **91**, 1287 (1953).

Field Theory on the Mass Shell*

After the discussion, the chairman called upon *GELL-MANN* to talk on "dispersion relations in pion–nucleon and photon–nucleon scattering." Following is a paraphrase of his talk:

"Goldberger and myself and others started some years ago to look at the dispersion relation for forward scattering of γ-rays from 'things,' in particular, protons. The idea then was to write down a rule, an exact law, which depends on very simple assumptions which might some day be questioned at high energies, but which would be independent of all the details of field theory. We found that just by using microscopic causality and a few other very simple things, we could get dispersion relations expressing the real part of the photon scattering amplitude in the forward direction in terms of total cross sections for photons producing *anything*. This is a relation for the real part at a given frequency in terms of an integral over the cross section at all frequencies, with a dispersion denominator, and is simply the Kramers–Kronig relation. The only thing new was generalizing it to include spin-flip. A great deal of water has gone under the bridge since then. In particular the Chicago group, namely Goldberger and company, have generalized these relations, first to particle with mass, and then, simultaneously with other people, to momentum transfer other than zero (i.e., non-forward scattering). The character of the program has changed. What one had set out to do is no longer the same. As one writes down more and more dispersion relations (more and more exact relations characteristic of amplitudes in field theory), still invoking very simple things such as microscopic causality, Lorentz invariance, reasonable behavior of amplitudes and cross sections at high energies, and the symmetry of field theory under the process of absorbing a particle and creating a particle, and so forth, one begins to come dangerously close to finding all of the results of field theory. At present, at least for simple processes, we seem to have come very close to having prescriptions, which, if taken in expansions in powers of the coupling constant, seem to yield the S-matrix in field theory. It has therefore become rather absurd to say that one is working now with a set of relations based on very simple principles which hardly anyone would challenge, and which do not depend on the details of field theory. So the philosophy must be something quite different. What it is, I don't know. I would like just to talk about the various points of view that one can adopt."

"I would like to write down a few things that one can try to do, depending on who one is. One can try, for example, to carry out Heisenberg's program set forth in 1943. Heisenberg complained that in our usual versions of quantum mechanics we are inclined to calculate in detail the wave functions, or scattering amplitudes on and off the energy shell — in other words, not only the asymtotic forms of wave functions, but wave functions everywhere. In

*Reprinted from High Energy Nuclear Physics, Proceedings of the Sixth Annual Rochester Conference, April 3–7, 1956, pp. III-30–III-36.

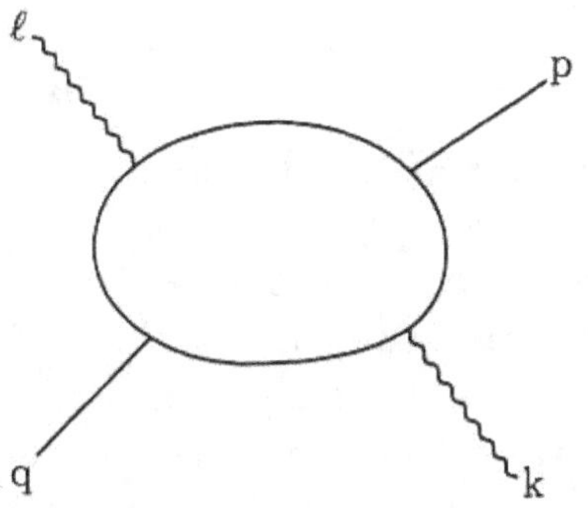

relativistic field theory, in particular in Feynman's version, we calculate matrix elements for processes which we generally indicate by a diagram as shown, where the four-momenta can assume any values, on or off the mass shell. From these we obtain progagators for any values of these four-momenta. Actually, we are only interested in physical processes on the mass shell. The analog of Heisenberg's program here is to ask whether we can write down a set of formulas relating only amplitudes with particles having their physical masses, unlike the field equations, which must first invoke the general amplitudes before going on to the mass shell. Heisenberg had realized that one cannot eliminate all but the physical amplitudes without paying a price, namely, one has then to extend the S-matrix to the region of unphysical energies, energies below the rest masses of particles, for example. Exactly the same sort of thing has to be done in present-day work. One cannot find simple relations among the physical amplitudes without extending them, by certain prescriptions, to the region of the unphysical energies, imaginary momenta, etc. although one is always on the mass shell. In this way, one tries to carry the program through."

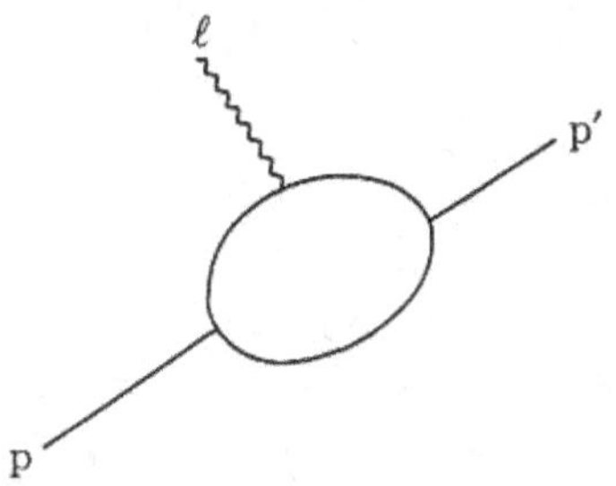

"To sketch the program very briefly, let us talk about the case of a meson absorbed by a nucleon. The diagram for this process is shown, where all momenta are on the mass shell. This process can happen only if the total energy is unphysical, namely, minus a few Mev. The coupling constant is then the probability for this process. Now we go to meson–nucleon scattering, with the diagram shown previously. Here we have only two independent relativistic scalars:

$$(k - \ell)^2 = \Delta^2, \quad \text{the invariant momentum transfer}\,,$$

and

$$p \cdot k = -M\nu - \frac{1}{4}\Delta^2\,,$$

where the frequency ν is closely related to the energy of the meson in a convenient coordinate system as described by Goldberger yesterday. Now we try to describe this process — namely, the four scattering amplitudes corresponding to spin-flip, no spin-flip, charge-flip, no flip — in terms of the two parameters ν and Δ. We find that we must consider, for fixed Δ, *all* value of ν, and not only the physical range of

$$\frac{\sqrt{\mu^2 + \frac{\Delta^2}{4}}\sqrt{M^2 + \frac{\Delta^2}{4}}}{M} \quad \text{to } \infty\,.$$

When ν goes below the physical lower limit, we find ourselves in what Goldberger called the Never-Never Land, where momenta becomes imaginary, cosines of angles become less than -1 and so on. The dispersion relations provide integral relations between the real and imaginary parts of the scattering amplitude, for each value of Δ, provided we consider the amplitude over the whole range of ν, both physical and unphysical. Where the imaginary part of the scattering amplitude appears in the equation, we have to substitute experimental values. In the unphysical region, we may either find an extension procedure by analytic continuation or by extrapolating with a French curve."

"We now probably have almost enough equations to carry through the Heisenberg program. What one would do then is to calculate the scattering in second order. One would calculate the imaginary part of the amplitude in fourth order by squaring that second order amplitude. One would then use the dispersion relation to calculate the real part of the amplitude in fourth order, and so on. In this way, by using dispersion relations and conservation of probability, suitably extended into the Never-Never Land, one may hope to generate the entire S-matrix from more or less fundamental principles. And when this program is completed, it is presumably the same as constructing field theory, with the following exceptions:

(1) All quantities that enter are renormalized. This is then the relativistic analog of the Chew–Low theory.

(2) One has a number of exact relations involving only experimental quantities, for example, the forward direction dispersion relations. If one can suitably extrapolate the non-forward amplitude into the Never-Never Land and substitute them into the dispersion relations, one will then have further relations among physical quantities."

"What to do with these is another question. They may, in a certain sense, be used to check the fundamental assumptions underlying the whole program, which are then the usual assumptions underlying the local relativistic field theory. However, in order for that to be true we have to be sure that the amplitudes and cross sections we feed in are reliable, up to the highest energy important in the dispersion relations. This is, of course, a very serious matter. There may be some hope that some fo these equations are positive-definite. If the experimentalists can manage to dredge together enough terms to feed into the right-hand side of the equations until it equals the left-hand side, we would then have to stop. Either one passes a law not to build higher energy machines, or experimentalists are forced not to find any more cross sections above that point. If we have such a positive-definite relation, where the right-hand side just keeps accumulating as you feed in the cross sections at higher and higher energies — and this comes close to being challenged by experiments — this will be a very exciting situation."

"However, nature may conspire to satisfy all these positive-definite equations exactly, while only the others run into trouble. It seems that it is only those equations that have the firmest footing (those most closely related to macroscopic causality) which have this postive-definite character. This is very unfortunate. So that, as a means of testing our fundamental assumptions, these things seem good, but may prove very, very hard to apply in fact. As a means of generating the S-matrix in place of the usual field equations, they seem very hopeful. They have the advantage that one deals always with quantities that one understands — renormalized quantities, etc. — and the calculations one can perform, as in the Chew–Low work, are presumably better than perturbation calculations."

"Finally, one may have some insight into the question of how it is that these principles of causality, relativistic invariance, reasonable behavior at high energy, etc., keep forcing us back to the same old theories — same old electrodynamics, same old meson theories — that no matter how hard we try to make new theories, we can never find anything that makes sense that does not reduce to these old theories. Maybe this approach will throw some light on how to construct a new theory that is *different*, and which may have some chance of explaining the high energy phenomena which field theory has no chance whatsoever to explain."

The Nature of the Weak Interaction*

Turning to the weak interactions, Gell-Mann discussed attempts to include them in a single scheme devised by himself and by Dallaporta and collaborators. One starts with the familiar triangle proposed to account for processes like β decay, μ decay, and μ absorption in nuclei:

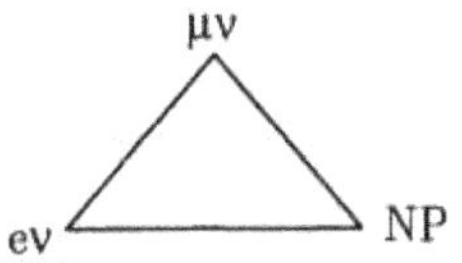

We add to these the strange particles by additional linkages:

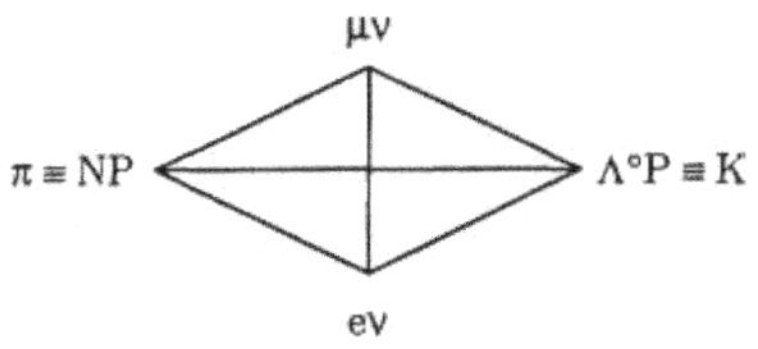

The triple linkages represent strong interactions. The single linkages represent 4-fermion interactions. We say nothing about the character of the weak interactions, but *do* assume that they have roughly the same strength. The other hyperons can be thought to belong to the Λ°P corner. (Of course, this picture may make no sense if one thinks of the Λ° and $\Lambda^{\circ\prime}$, related by C_p. The two parts of Gell-Mann's remarks are disjoint.) Processes such as $K \to \mu + \nu$, $\pi \to \mu + \nu$, etc. can now take place. An attempt, made last year to find a universal form for such weak interactions failed, and did so even without the consideration of strange particles, because of the small value of the ratio $\frac{\pi \to e+\nu}{\pi \to \mu+\nu}$. There are still general consequences of this scheme, however, if one only maintains the assumption of a universal strength for the interaction, but allows its form to vary from linkage to linkage. Two of these consequences can be summarized by the following selection rules for decay:

(1) stronglys → stronglys $\quad |\Delta S| = 1 \quad |\Delta I| = \frac{1}{2}$ or $\frac{3}{2}$

(2) stronglys → stronglys + leptons $\quad |\Delta S| = 1 \quad |\Delta I| = \frac{1}{2}$

The stronglys refer to strongly interacting particles, such as Λ, N, π, and the leptons to μ, ν and e. The second rule refers only to the relation between the strongly interacting particles. The leptons are assumed to carry away no isotopic spin. If one uses rule 1), one may obtain the same result for the τ/τ' ratio as that obtained on the basis of Wentzel's selection

*Reprinted from High Energy Nuclear Physics, Proceedings of the Sixth Annual Rochester Conference, April 3–7, 1956, pp. II-25–II-29.

rule (see above). On the other hand Wentzel's rule would predict an infinite lifetime for the θ^+. The finiteness of the lifetime arises only from electromagnetic corrections of order e^4. Use of the above Gell-Mann selection rules would lead, perhaps more naturally, to a finite lifetime. Further, for processes like $\Lambda^\circ \to P + \pi^-$, $\Lambda^\circ \to n + \pi^\circ$, etc., the above weak rule has no consequences, while Wentzel's rule has very strong consequences. As far as rule 2) above is concerned, it would give essentially a one-to-one partial lifetime for the processes $\theta^+ \to \pi + e + \nu$ and $\theta^\circ \to \pi + e + \nu$. (An exact statement is hard to make in view of the complicated situation of the neutral θ.) This assertion is subject to experimental test. We know the lifetime and branching ratios of the charged K particle and thus can compute the partial lifetime of the process $\theta^+ \to \pi + e + \nu$. We know the lifetime of the θ°. If we can find the branching ratio of the process $\theta^\circ \to \pi + e + \nu$ (events such as the one mentioned by Peyrou above) we can obtain the partial lifetime for this process as well.

Gell-Mann doesn't necessarily advocate this approach to the weak interactions. He merely wished to point out its existence. More detailed rules, involving specific interactions for the various linkages, are hard to state at present, since, except for nuclear β decay, the interactions are not known.

An extensive discussion followed. In connection with Gell-Mann's idea of assuming that C_p commutes with the e.m. field, *Yang* felt that so long as we understand as little as we do about the $\theta - \tau$ degeneracy, it may perhaps be best to keep an open mind on the subject. Pursuing the open mind approach, *Feynman* brought up a question of Block's: Could it be that the θ and τ are different parity states of the same particle which has no definite parity, i.e., that parity is not conserved. That is, does nature have a way of defining right of left-handedness uniquely? *Yang* stated that he and Lee looked into this matter without arriving at any definite conclusions. Wigner (as discussed in some notes prepared by Michel and Wightman) has been aware of the possible existence of two states of opposite parity, degenerate with respect to each other because of space-time transformation properties. So perhaps a particle having both parities could exist. But how could it decay, if one continues to believe that there is absolute invariance with respect to space-time transformations? Perhaps one could say that parity conservation, or else time inversion invariance, could be violated. Perhaps the weak interactions could all come from this same source, a violation of space-time symmetries. The most attractive way out is the nonsensical idea that perhaps a particle is emitted which has no mass, charge, and energy-momentum but only carries away some strange space-time transformation properties. *Gell-Mann* felt that one should also keep an open mind about possibilities like the suggestion by Marshak (see above) that the θ and τ may, without requiring radical assumptions, turn out to be the same particle.

Michel suggested another way out of the difficulty. What is seemingly well known from experiment is the parity of the τ^+ and θ°. If one assumes that the π° emitted in the process $\theta^+ \to \pi^+ + \pi^\circ$ is a "nearly real π°" in the same sense that the pair emitted in the process $\pi^\circ \to \pi + e + e$ is a "nearly real γ," there is nothing to prevent θ^+ from having spin-parity 0^-. This "nearly real π°" would appear only virtually, perhaps due to some selection rules.

The chairman felt that the moment had come to close our minds...

Primakoff returned to the subject of the e.m. interactions with matter — the law of "minimal electromagnetic couplings." One can set limits from experiment to the e.m. couplings associated with other than charge and current interactions. (For example, there is no evidence of the radiative decay of the μ mesons without the emission of neutrinos, or for the (radiative) annihilation of stopped positive μ mesons against electrons in matter.) These couplings, on the basis of present evidence, are very much smaller than the fine structure constant, in fact considerably smaller than the β decay coupling constant, and the limits on them could be pushed even lower with presently available experimental techniques.

Oppenheimer professed himself a believer in the principle of minimal e.m. couplings without understanding it. He suggested: "Perhaps some oscillation between learning from the past and being surprised by the future of this [$\theta - \tau$ dilemma] is the only way to mediate the battle."

Turning to the weak interaction selection rule, Oppenheimer commented that the mass differences of the various pions are large compared to fine structure constant orders of magnitude; and so Wentzel's suggestion for explaining the ratio of θ° and θ^{+} lifetimes is in fact more acceptable than Gell-Mann's argument, based on the magnitude of e.m. interactions, would tend to indicate. *Marshak* stated that he did not believe in the $|\Delta I| = \frac{1}{2}$ selection rule. As a possible explanation of the difference between the θ° and θ^{+} lifetimes, he referred to the theory he worked out with Okuba (see abstract, last Berkeley meeting). They suggest that this difference could be due to an interaction between π^{+} and π^{-} in the $T = 0$ state (the interaction suggested by Dyson and Takeda to account for the second resonance in $\pi = P$ scattering), provided the range of interaction is of order $\frac{\hbar}{\mu c}$, and not $\frac{\hbar}{Mc}$.

Lindenbaum asked if there were any reason for the asymmetry of the strangeness assignments between the heavy ($|S| \leq 2$) and light ($|S| \leq 1$) particles. *Weisskopf* replied that this was a reflection of the non-existence of doubly charged particles — a fact that he hoped reflected another law of nature. *Oppenheimer* expressed the wish, and Weisskopf concurred, that by next year we will also add the law that the maximum spin of fundamental particles in no greater than $\frac{1}{2}$.

PHYSICAL REVIEW VOLUME 109, NUMBER 1 JANUARY 1, 1958

Theory of the Fermi Interaction

R. P. FEYNMAN AND M. GELL-MANN
California Institute of Technology, Pasadena, California
(Received September 16, 1957)

The representation of Fermi particles by two-component Pauli spinors satisfying a second order differential equation and the suggestion that in β decay these spinors act without gradient couplings leads to an essentially unique weak four-fermion coupling. It is equivalent to equal amounts of vector and axial vector coupling with two-component neutrinos and conservation of leptons. (The relative sign is not determined theoretically.) It is taken to be "universal"; the lifetime of the μ agrees to within the experimental errors of 2%. The vector part of the coupling is, by analogy with electric charge, assumed to be not renormalized by virtual mesons. This requires, for example, that pions are also "charged" in the sense that there is a direct interaction in which, say, a π^0 goes to π^- and an electron goes to a neutrino. The weak decays of strange particles will result qualitatively if the universality is extended to include a coupling involving a Λ or Σ fermion. Parity is then not conserved even for those decays like $K\to 2\pi$ or 3π which involve no neutrinos. The theory is at variance with the measured angular correlation of electron and neutrino in He^6, and with the fact that fewer than 10^{-4} pion decay into electron and neutrino.

THE failure of the law of reflection symmetry for weak decays has prompted Salam, Landau, and Lee and Yang[1] to propose that the neutrino be described by a two-component wave function. As a consequence neutrinos emitted in β decay are fully polarized along their direction of motion. The simplicity of this idea makes it very appealing, and considerable experimental evidence is in its favor. There still remains the question of the determination of the coefficients of the scalar, vector, etc., couplings.

There is another way to introduce a violation of parity into weak decays which also has a certain amount of theoretical *raison d'être*. It has to do with the number of components used to describe the electron in the Dirac equation,

$$(i\nabla - \mathbf{A})\psi = m\psi. \qquad (1)$$

Why must the wave function have four components? It is usually explained by pointing out that to describe the electron spin we must have two, and we must also represent the negative-energy states or positrons, requiring two more. Yet this argument is unsatisfactory. For a particle of spin zero we use a wave function of only one component. The sign of the energy is determined by how the wave function varies in space and time. The Klein-Gordon equation is second order and we need both the function and its time derivative to predict the future. So instead of two components for spin zero we use one, but it satisfies a second order equation. Initial states require specification of that one and its time derivative. Thus for the case of spin $\frac{1}{2}$ we would expect to be able to use a simple two-component spinor for the wave function, but have it satisfy a second order differential equation. For example, the wave function for a free particle would look like $U\exp[-i(Et-\mathbf{P}\cdot\mathbf{x})]$, where U has just the two components of a Pauli spinor and whether the particle refers to electron or positron depends on the sign of E in the four-vector $p_\mu=(E,\mathbf{P})$.

In fact it is easy to do this. If we substitute

$$\psi = \frac{1}{m}(i\nabla - \mathbf{A} + m)\chi \qquad (2)$$

in the Dirac equation, we find that χ satisfies

$$(i\nabla - \mathbf{A})^2\chi = [(i\nabla_\mu - A_\mu)\cdot(i\nabla_\mu - A_\mu) - \tfrac{1}{2}\sigma_{\mu\nu}F_{\mu\nu}]\chi = m^2\chi, \qquad (3)$$

where $F_{\mu\nu}=\partial A_\nu/\partial x_\mu - \partial A_\mu/\partial x_\nu$ and $\sigma_{\mu\nu}=\frac{1}{2}i(\gamma_\mu\gamma_\nu-\gamma_\nu\gamma_\mu)$. Now we have a second order equation, but χ still has four components and we have twice as many solutions as we want. But the operator $\gamma_5=\gamma_x\gamma_y\gamma_z\gamma_t$ commutes with $\sigma_{\mu\nu}$; therefore there are solutions of (3) for which $i\gamma_5\chi=\chi$ and solutions for $i\gamma_5\chi=-\chi$. We may select, say, the first set. We always take

$$i\gamma_5\chi = \chi. \qquad (4)$$

Then we can put the solutions of (3) into one-to-one correspondence with the Dirac equation (1). For each ψ there is a unique χ; in fact we find

$$\chi = \tfrac{1}{2}(1+i\gamma_5)\psi \qquad (5)$$

by multiplying (2) by $1+i\gamma_5$ and using (4). The function χ has really only two independent components. The conventional ψ requires knowledge of both χ and its time derivative [see Eq. (2)]. Further, the six $\sigma_{\mu\nu}$ in (3) can be reduced to just the three $\sigma_{xy}, \sigma_{yz}, \sigma_{zx}$. Since $\sigma_{zt}=i\gamma_z\gamma_t=i\sigma_{xy}\cdot i\gamma_5$, Eq. (4) shows that σ_{zt} may be replaced by $i\sigma_{xy}$ when operating on χ as it does in (3)

Let us use the representation

$$\gamma_t=\begin{pmatrix}1 & 0\\ 0 & -1\end{pmatrix},\quad \gamma=\begin{pmatrix}0 & \sigma\\ -\sigma & 0\end{pmatrix},\quad i\gamma_5=-\begin{pmatrix}0 & 1\\ 1 & 0\end{pmatrix},$$

where $\sigma_{x,y,z}$ are the Pauli matrices. If

$$\psi=\begin{pmatrix}a\\ b\end{pmatrix},$$

[1] A. Salam, Nuovo cimento **5**, 299 (1957); L. Landau, Nuclear Phys. 3, 127 (1957); T. D. Lee and C. N. Yang, Phys. Rev. **105**, 1671 (1957).

where a, b are two-component spinors, we find from (5) that

$$\chi=\begin{pmatrix}\varphi\\ -\varphi\end{pmatrix},$$

where $\varphi=\frac{1}{2}(a-b)$. Our Eq. (3) for the two-component spinor φ is

$$[(i\nabla_\mu-A_\mu)^2+\boldsymbol{\sigma}\cdot(\mathbf{B}+i\mathbf{E})]\varphi=m^2\varphi, \qquad (6)$$

where $B_x=F_{yz}$, $E_x=F_{tx}$, etc., which is the equation we are looking for.

Rules of calculation for electrodynamics which involve only the algebra of the Pauli matrices can be worked out on the basis of (6). They, of course, give results exactly the same as those calculated with Dirac matrices. The details will perhaps be published later.

One of the authors has always had a predilection for this equation.[2] If one tries to represent relativistic quantum mechanics by the method of path integrals, the Klein-Gordon equation is easily handled, but the Dirac equation is very hard to represent directly. Instead, one is led first to (3), or (6), and from there one must work back to (1).

For this reason let us imagine that (6) had been discovered first, and (1) only deduced from it later. It would make no difference for any problem in electrodynamics, where electrons are neither created nor destroyed (except with positrons). But what would we do if we were trying to describe β decay, in which an electron is created? Would we use a field operator ψ directly in the Hamiltonian to represent the annihilation of an electron, or would we use φ? Now everything we can do one way, we can represent the other way. Thus if ψ were used it could be replaced by

$$\frac{1}{m}(p\!\!\!/-A\!\!\!/+m)\begin{pmatrix}\varphi\\ -\varphi\end{pmatrix}, \qquad \text{(a)}$$

while an expression in which φ was used could be rewritten by substituting

$$\tfrac{1}{2}(1+i\gamma_5)\psi. \qquad \text{(b)}$$

If φ were really fundamental, however, we might be prejudiced against (a) on the grounds that gradients are involved. That is, an expression for β coupling which does not involve gradients from the point of view of ψ, does from the point of view of φ. So we are led to suggest *φ as the field annihilation operator to be used in β decay without gradients.* If φ is written as in (b), we see this does not conserve parity, but now we know that that is consistent with experiment.

For this reason one of us suggested the rule[3] that the electron in β decay is coupled directly through φ, or, what amounts to the same thing, in the usual four-particle coupling

$$\textstyle\sum_i C_i(\bar{\psi}_n O_i\psi_p)(\bar{\psi}_\nu O_i\psi_e), \qquad (7)$$

we always replace ψ_e by $\frac{1}{2}(1+i\gamma_5)\psi_e$.

One direct consequence is that the electron emitted in β decay will always be left-hand polarized (and the positron right) with polarization approaching 100% as $v\to c$, irrespective of the kind of coupling. That is a direct consequence of the projection operator

$$a=\tfrac{1}{2}(1+i\gamma_5).$$

A priori we could equally well have made the other choice and used

$$\bar{a}=\tfrac{1}{2}(1-i\gamma_5);$$

electrons emitted would then be polarized to the right. We appeal to experiment[4] to determine the sign. Notice that $a^2=a$, $\bar{a}a=0$.

But now we go further, and suppose that the same rule applies to the wave functions of all the particles entering the interaction. We take for the β-decay interaction the form

$$\textstyle\sum C_i(\overline{a\psi_n}O_ia\psi_p)(\overline{a\psi_\nu}O_ia\psi_e),$$

and we should like to discuss the consequences of this hypothesis.

The coupling is now essentially completely determined. Since $\overline{a\psi}=\bar{\psi}\bar{a}$, we have in each term expressions like $\bar{a}O_ia$. Now for S, T, and P we have O_i commuting with γ_5 so that $\bar{a}O_ia=O_i\bar{a}a=0$. For A and V we have $aO_ia=O_ia^2=O_ia$ and the coupling survives. Furthermore, for axial vector $O_i=i\gamma_\mu\gamma_5$, and since $i\gamma_5a=a$, we find $O_ia=\gamma_\mu a$; thus A leads to the same coupling as V:

$$(8)^{\frac{1}{2}}G(\bar{\psi}_n\gamma_\mu a\psi_p)(\bar{\psi}_\nu\gamma_\mu a\psi_e), \qquad (8)$$

the most general β-decay interaction possible with our hypothesis.[5]

This coupling is not yet completely unique, because our hypothesis could be varied in one respect. Instead of dealing with the neutron and proton, we could have made use of the antineutron and antiproton, considering them as the "true particles." Then it would be the wave function $\psi_{\bar{n}}$ of the antineutron that enters with the factor a. We would be led to

$$(8)^{\frac{1}{2}}G(\bar{\psi}_{\bar{p}}\gamma_\mu a\psi_{\bar{n}})(\bar{\psi}_\nu\gamma_\mu a\psi_e). \qquad (9)$$

This amounts to the same thing as

$$(8)^{\frac{1}{2}}G(\bar{\psi}_n\gamma_\mu\bar{a}\psi_p)(\bar{\psi}_\nu\gamma_\mu a\psi_e), \qquad (9')$$

and from the *a priori* theoretical standpoint is just as good a choice as (8).

We have assumed that the neutron and proton are

[2] R. P. Feynman, Revs. Modern Phys. **20**, 367 (1948); Phys. Rev. **84**, 108 (1951).

[3] R. P. Feynman, *Proceedings of the Seventh Annual Rochester Conference on High-Energy Nuclear Physics, 1957* (Interscience Publishers, Inc., New York, 1957).

[4] See, for example, Boehm, Novey, Barnes, and Stech, Phys. Rev. **108**, 1497 (1957).

[5] A universal V, A interaction has also been proposed by E. C. G. Sudarshan and R. E. Marshak (to be published).

either both "particles" or both "antiparticles." We have defined the electron to be a "particle" and the neutrino must then be a particle too.

We shall further assume the interaction "universal," so for example it is

$$(8)^{\frac{1}{2}}G(\bar{\psi}_\mu\gamma_\mu a\psi_\nu)(\bar{\psi}_\nu\gamma_\mu a\psi_e) \quad (10)$$

for μ decay, as currently supposed; the μ^- is then a particle. Here the other choice, that the μ^- is an antiparticle, leads to $(8)^{\frac{1}{2}}G(\bar{\psi}_\nu\gamma_\mu a\psi_{\bar{\mu}})(\bar{\psi}_\nu\gamma_\mu a\psi_e)$, which is excluded by experiment since it leads to a spectrum falling off at high energy (Michel's $\rho=0$).

Since the neutrino function always appears in the form $a\psi_\nu$ only neutrinos with left-hand spin can exist. That is, the two-component neutrino theory with conservation of leptons is valid. Our neutrinos spin oppositely to those of Lee and Yang.[6] For example, a β particle is a lepton and spins to the left; emitted with it is an antineutrino which is an antilepton and spins to the right. In a transition with $\Delta J=0$ they tend to go parallel to cancel angular momentum. This is the angular correlation typical of vector coupling.

We have conservation of leptons and double β decay is excluded.

There is a symmetry in that the incoming particles can be exchanged without affecting the coupling. Thus if we define the symbol

$$(\bar{A}B)(\bar{C}D)=(\bar{\psi}_A\gamma_\mu a\psi_B)(\bar{\psi}_C\gamma_\mu a\psi_D),$$

we have $(\bar{A}B)(\bar{C}D)=(\bar{C}B)(\bar{A}D)$. (We have used anticommuting ψ's; for C-number ψ's the interchange gives a minus sign.[7])

The capture of muons by nucleons results from a coupling $(\bar{n}p)(\bar{\nu}\mu)$. It is already known that this capture is fitted very well if the coupling constant and coupling are the same as in β decay.[8]

If we postulate that the universality extends also to the strange particles, we may have couplings such as $(\bar{\Lambda}^0p)(\bar{\nu}\mu)$, $(\bar{\Lambda}^0p)(\bar{\nu}e)$, and $(\bar{\Lambda}^0p)(\bar{p}n)$. The $(\bar{\Lambda}^0p)$ might be replaced by $(\bar{\Sigma}^-n)$, etc. At any rate the existence of such couplings would account qualitatively for the existence of all the weak decays. Consider, for example, the decay of the K^+. It can go virtually into an anti-Λ^0 and a proton by the fairly strong coupling of strange particle production. This by the weak decay $(\bar{\Lambda}^0p)(\bar{p}n)$ becomes a virtual antineutron and proton. These become, on annihilating, two or three pions. The parity is not conserved because of the a in front of the nucleons in the virtual transition. The theory in which only the neutrino carries the a cannot explain the parity failure for decays not involving neutrinos (the τ-θ puzzle). Here we turn the argument around; both the lack of parity conservation for the K and the fact that neutrinos are always fully polarized are consequences of the same universal weak coupling.

For β decay the expression (8) will be recognized as that for the two-component neutrino theory with couplings V and A with equal coefficients and opposite signs [expression (9) or (9′) makes the coupling $V+A$]. The coupling constant of the Fermi (V) part is equal to G. This constant has been determined[9] from the decay of O^{14} to be $(1.41\pm0.01)\times10^{-49}$ erg/cm^3. In units where $\hbar=c=1$, and M is the mass of the proton, this is

$$G=(1.01\pm0.01)\times10^{-5}/M^2. \quad (11)$$

At the present time several β-decay experiments seem to be in disagreement with one another. Limiting ourselves to those that are well established, we find that the most serious disagreement with our theory is the recoil experiment in He^6 of Rustad and Ruby[10] indicating that the T interaction is more likely than the A. Further check on this is obviously very desirable. Any experiment indicating that the electron is not 100% left polarized as $v\rightarrow c$ for any transition allowed or forbidden would mean that (8) and (9) are incorrect. An interesting experiment is the angular distribution of electrons from polarized neutrons for here there is an interference between the V and A contributions such that if the coupling is $V-A$ there is no asymmetry, while if it is $V+A$ there is a maximal asymmetry. This would permit us to choose between the alternatives (8) and (9). The present experimental results[11] agree with neither alternative.

We now look at the muon decay. The fact that the two neutrinos spin oppositely and the ρ parameter is $\frac{3}{4}$ permitted us to decide that the μ^- is a lepton if the electron is, and determines the order of $(\bar{\mu},\nu)$ which we write in (10). But now we can predict the direction of the electron in the $\pi^-\rightarrow\mu^-+\bar{\nu}\rightarrow e^-+\nu+\bar{\nu}$ sequence. Since the muon comes out with an antineutrino which spins to the right, the muon must also be spinning to the right (all senses of spin are taken looking down the direction of motion of the particle in question). When the muon disintegrates with a high-energy electron the two neutrinos are emitted in the opposite direction. They have spins opposed. The electron emitted must spin to the left, but must carry off the angular momentum of the muon, so it must proceed in the direction opposite to that of the muon. This direction agrees with experiment. The proposal of Lee and Yang predicted

---

[6] This is only because they used S and T couplings in β decay; had they used V and A, their theory would be similar to ours, with left-handed neutrinos.

[7] We can express $(\bar{A}B)(\bar{C}D)$ directly in terms of the two-component spinors φ: $(\bar{A}B)(\bar{C}D)=4(\varphi_A{}^*\varphi_B)(\varphi_C{}^*\varphi_D)-4(\varphi_A{}^*\boldsymbol{\sigma}\varphi_B)\cdot(\varphi_C{}^*\boldsymbol{\sigma}\varphi_D)$. If we put $\varphi_A=\binom{A_1}{A_2}$, etc., where A_1 and A_2 are complex numbers, we obtain $8(A_1{}^*C_2{}^*-A_2{}^*C_1{}^*)(B_1D_2-B_2D_1)$ and the symmetry is evident.

[8] See, for example, J. L. Lopes, Phys. Rev. (to be published); L. Michel, *Progress in Cosmic-Ray Physics*, edited by J. G. Wilson (Interscience Publishers, Inc., New York, 1952), Vol. 1, p. 125.

[9] Bromley, Almquist, Gove, Litherland, Paul, and Ferguson, Phys. Rev. **105**, 957 (1957).

[10] B. M. Rustad and S. L. Ruby, Phys. Rev. **97**, 991 (1955).

[11] Burgy, Epstein, Krohn, Novey, Raboy, Ringo, and Telegdi, Phys. Rev. **107**, 1731 (1957).

the electron spin here to be opposite to that in the case of β decay. Our β-decay coupling is V, A instead of S, T and this reverses the sign. That the electron have the same spin polarization in all decays (β, muon, or strange particles) is a consequence of putting $a\psi_e$ in the coupling for this particle. It would be interesting to test this for the muon decay.

Finally we can calculate the lifetime of the muon, which comes out

$$\tau = 192\pi^3/G^2\mu^5 = (2.26\pm0.04)\times10^{-6}\text{ sec}$$

using the value (11) of G. This agrees with the experimental lifetime[12] $(2.22\pm0.02)\times10^{-6}$ sec.

It might be asked why this agreement should be so good. Because nucleons can emit virtual pions there might be expected to be a renormalization of the effective coupling constant. On the other hand, if there is some truth in the idea of an interaction with a universal constant strength it may be that *the other interactions are so arranged so as not to destroy this constant.* We have an example in electrodynamics. Here the coupling constant e to the electromagnetic field is the same for all particles coupled. Yet the virtual mesons do not disturb the value of this coupling constant. Of course the distribution of charge is altered, so the coupling for high-energy fields is apparently reduced (as evidenced by the scattering of fast electrons by protons), but the coupling in the low-energy limit, which we call the total charge, is not changed.

Using this analogy to electrodynamics, we can see immediately how the Fermi part, at least, can be made to have no renormalization. For the sake of this discussion imagine that the interaction is due to some intermediate (electrically charged) vector meson of very high mass M_0. If this meson is coupled to the "current" $(\bar{\psi}_p\gamma_\mu a\psi_n)$ and $(\bar{\psi}_\mu\gamma_\mu a\psi_\nu)$ by a coupling $(4\pi f^2)^{\frac{1}{2}}$, then the interaction of the two "currents" would result from the exchange of this "meson" if $4\pi f^2 M_0^{-2} = (8)^{\frac{1}{2}}G$. Now we must arrange that the total current

$$J_\mu = (\bar{\psi}_p\gamma_\mu a\psi_n) + (\bar{\psi}_\nu\gamma_\mu a\psi_e) + (\bar{\psi}_\nu\gamma_\mu a\psi_\mu) + \cdots \quad (12)$$

be not renormalized. There are no known large interaction terms to renormalize the $(\bar{\nu}e)$ or $(\bar{\nu}\mu)$, so let us concentrate on the nucleon term. This current can be split into two: $J_\mu = \frac{1}{2}(J_\mu{}^V + J_\mu{}^A)$, where $J_\mu{}^V = \bar{\psi}_p\gamma_\mu\psi_n$ and $J_\mu{}^A = \bar{\psi}_p i\gamma_\mu\gamma_5\psi_n$. The term $J_\mu{}^V = \bar{\psi}\gamma_\mu\tau_+\psi$, in isotopic spin notation, is just like the electric current. The electric current is

$$J_\mu{}^{el} = \bar{\psi}\gamma_\mu(\tfrac{1}{2}+\tau_z)\psi.$$

The term $\frac{1}{2}\bar{\psi}\gamma_\mu\psi$ is conserved, but the term $\bar{\psi}\gamma_\mu\tau_z\psi$ is not, unless we add the current of pions, $i[\varphi^* T_z\nabla_\mu\varphi - (\nabla_\mu\varphi^*)T_z\varphi]$, because the pions are charged. Likewise $\bar{\psi}\gamma_\mu\tau_+\psi$ is not conserved but the sum

$$J_\mu{}^V = \bar{\psi}\gamma_\mu\tau_+\psi + i[\varphi^* T_+\nabla_\mu\varphi - (\nabla_\mu\varphi)^* T_+\varphi] \quad (13)$$

is conserved, and, like electricity, leads to a quantity whose value (for low-energy transitions) is unchanged by the interaction of pions and nucleons. If we include interactions with hyperons and K particles, further terms must be added to obtain the conserved quantity.

We therefore suppose that this conserved quantity be substituted for the vector part of the first term in (12). Then the Fermi coupling constant will be strictly universal, except for small electromagnetic corrections. That is, the constant G from the μ decay, which is accurately $V-A$, should be also the exact coupling constant for at least the vector part of the β decay. (Since the energies involved are so low, the spread in space of $J_\mu{}^V$ due to the meson couplings is not important, only the total "charge.") It is just this part which is determined by the experiment with O^{14}, and that is why the agreement should be so close.

The existence of the extra term in (13) means that other weak processes must be predicted. In this case there is, for example, a coupling

$$(8)^{\frac{1}{2}}Gi(\varphi^*\nabla_\mu T_+\varphi - (\nabla_\mu\varphi)^* T_+\varphi)(\bar{\psi}_\nu\gamma_\mu a\psi_e),$$

by which a π^- can go to a π^0 with emission of $\bar{\nu}$ and e. The amplitude is

$$4G(p_\mu{}^- + p_\mu{}^0)(\bar{\psi}_\nu\gamma_\mu a\psi_e),$$

where p^-, p^0 are the four-momenta of π^- and π^0. Because of the low energies involved, the probability of the disintegration is too low to be observable. To be sure, the process $\pi^-\to\pi^0+e+\bar{\nu}$ could be understood to be qualitatively necessary just from the existence of β decay. For the π^- may become virtually an antiproton and neutron, the neutron decay virtually to a proton, e, and $\bar{\nu}$ by β decay and the protons annihilate forming the π^0. But the point is that by our principle of a universal coupling whose vector part requires no renormalization we can calculate the rate directly without being involved in closed loops, strong couplings, and divergent intervals.

For any transition in which strangeness doesn't change, the current $J_\mu{}^V$ is the total current density of isotopic spin T_+. Thus the vector part gives transitions $\Delta T=0$ with square matrix element $T(T+1)-T_zT_z'$ if we can neglect the energy release relative to the rest mass of the particle decaying. For the nucleon and $K^-\to K^0+e+\bar{\nu}$ the square of the matrix element is 1, for the pion and $\Sigma^-\to\Sigma^0+e+\bar{\nu}$ it is 2. The axial coupling in the low-energy limit is zero between states of zero angular momentum like the π meson or O^{14}, so for both of these we can compute the lifetime knowing only the vector part. Thus the $\pi^-\to\pi^0+e+\bar{\nu}$ decay should have the same ft value as O^{14}. Unfortunately because of the very small energies involved (because isotopic spin is such a good quantum number) none of these decays of mesons or hyperons are fast enough to observe in competition to other decay processes in which T or strangeness changes.

[12] W. E. Bell and E. P. Hincks, Phys. Rev. 84, 1243 (1951).

This principle, that the vector part is not renormalized, may be useful in deducing some relations among the decays of the strange particles.

Now with present knowledge it is not so easy to say whether or not a pseudovector current like $\bar{\psi}i\gamma_5\gamma_\mu\tau_+\psi$ can be arranged to be not renormalized. The present experiments[13] in β decay indicate that the ratio of the coupling constant squared for Gamow-Teller and Fermi is about 1.3 ± 0.1. This departure from 1 might be a renormalization effect.[14] On the other hand, an interesting theoretical possibility is that it is exactly unity and that the various interactions in nature are so arranged that it need not be renormalized (just as for V). It might be profitable to try to work out a way of doing this. Experimentally it is not excluded. One would have to say that the $ft_{\frac{1}{2}}$ value of 1220 ± 150 measured[15] for the neutron was really 1520, and that some uncertain matrix elements in the β decay of the mirror nuclei were incorrectly estimated.

The decay of the π^- into a μ^- and $\bar{\nu}$ might be understood as a result of a virtual process in which the π becomes a nucleon loop which decays into the $\mu+\bar{\nu}$. In any event one would expect a decay into $e+\bar{\nu}$ also. The ratio of the rates of the two processes can be calculated without knowledge of the character of the closed loops. It is $(m_e/m_\mu)^2(1-m_\mu^2/m_\pi^2)^{-2}=13.6\times10^{-5}$. Experimentally[16] no $\pi\rightarrow e+\nu$ have been found, indicating that the ratio is less than 10^{-5}. This is a very serious discrepancy. The authors have no idea on how it can be resolved.

We have adopted the point of view that the weak interactions all arise from the interaction of a current J_μ with itself, possibly via an intermediate charged vector meson of high mass. This has the consequence that any term in the current must interact with all the rest of the terms and with itself. To account for β decay and μ decay we have to introduce the terms in (12) into the current; the phenomenon of μ capture must then also occur. In addition, however, the pairs $e\nu$, $\mu\nu$, and pn must interact with themselves. In the case of the $(\bar{e}\nu)(\bar{\nu}e)$ coupling, experimental detection of electron-neutrino scattering might some day be possible if electron recoils are looked for in materials exposed to pile neutrinos; the cross section[17] with our universal coupling is of the order of 10^{-45} cm^2.

To account for all observed strange particle decays it is sufficient to add to the current a term like $(\bar{p}\Lambda^0)$, $(\bar{p}\Sigma^0)$, or $(\overline{\Sigma^-}n)$, in which strangeness is increased by one as charge is increased by one. For instance, $(\bar{p}\Lambda^0)$ gives us the couplings $(\bar{p}\Lambda^0)(\bar{e}\nu)$, $(\bar{p}\Lambda^0)(\bar{\mu}\nu)$, and $(\bar{p}\Lambda^0)(\bar{n}p)$. A direct consequence of the coupling $(\bar{p}\Lambda^0)(\bar{e}\nu)$ would be the reaction

$$\Lambda^0\rightarrow p+e+\bar{\nu} \tag{14}$$

at a rate 5.3×10^7 sec^{-1}, assuming no renormalization of the constants.[18] Since the observed lifetime of the Λ^0 (for disintegration into other products, like $p+\pi^-$, $n+\pi^0$) is about 3×10^{-10} sec, we should observe process (14) in about 1.6% of the disintegrations. This is not excluded by experiments. If a term like $(\overline{\Sigma^-}n)$ appears, the decay $\Sigma^-\rightarrow n+e^-+\nu$ is possible at a predicted rate 3.5×10^8 sec^{-1} and should occur (for $\tau_{\Sigma^-}=1.6\times10^{-10}$ sec) in about 5.6% of the disintegrations of the Σ^-. Decays with μ replacing the electron are still less frequent. That such disintegrations actually occur at the above rates is not excluded by present experiments. It would be very interesting to look for them and to measure their rates.

These rates were calculated from the formula Rate$=(2G^2W^5c/30\pi^3)$ derived with neglect of the electron mass. Here $W=(M_\Lambda^2-M_p^2)/2M_\Lambda$ is the maximum electron energy possible and c is a correction factor for recoil. If $x=W/M_\Lambda$ it is

$$c=-\tfrac{15}{16}x^{-5}(1-2x)^2\ln(1-2x)$$
$$-\tfrac{5}{8}x^{-4}(1-x)(3-6x-2x^2),$$

and equals 1 for small x, about 1.25 for the Σ decay, and 2.5 for $M_p=0$.

It should be noted that decays like $\Sigma^+\rightarrow n+e^++\nu$ are forbidden if we add to the current only terms for which $\Delta S=+1$ when $\Delta Q=+1$. In order to cause such a decay, the current would have to contain a term with $\Delta S=-1$ when $\Delta Q=+1$, for example $(\overline{\Sigma^+}n)$. Such a term would then be coupled not only to $(\bar{\nu}e)$, but also to all the others, including one like $(\bar{p}\Lambda^0)$. But a coupling of the form $(\overline{\Sigma^+}n)(\bar{\Lambda}^0p)$ leads to strange particle decays with $\Delta S=\pm2$, violating the proposed rule $\Delta S=\pm1$. It is important to know whether this rule really holds; there is evidence for it in the apparent absence of the decay $\Xi^-\rightarrow\pi^-+n$, but so few Ξ particles have been seen that this is not really conclusive. We are not sure, therefore, whether terms like $(\overline{\Sigma^+}n)$ are excluded from the current.

We deliberately ignore the possibility of a neutral current, containing terms like $(\bar{e}e)$, $(\bar{\mu}e)$, $(\bar{n}n)$, etc., and possibly coupled to a neutral intermediate field. No weak coupling is known that requires the existence of such an interaction. Moreover, some of these couplings, like $(\bar{e}e)(\bar{\mu}e)$, leading to the decay of a muon into three electrons, are excluded by experiment.

It is amusing that this interaction satisfies simultaneously almost all the principles that have been

[13] A. Winther and O. Kofoed-Hansen, Kgl. Danske Videnskab. Selskab, Mat.-fys. Medd. (to be published).

[14] This slight inequality of Fermi and Gamow-Teller coupling constants is not enough to account for the experimental results of reference 11 on the electron asymmetry in polarized neutron decay.

[15] Spivac, Sosnovsky, Prokofiev, and Sokolov, *Proceedings of the International Conference on the Peaceful Uses of Atomic Energy, Geneva, 1955* (United Nations, New York, 1956), A/Conf. 8/p/650.

[16] C. Lattes and H. L. Anderson, Nuovo cimento (to be published).

[17] For neutrinos of energy ω (in units of the electron mass m) the total cross section is $\sigma_0\omega^2/(1+2\omega)$, and the spectrum of recoil energies ϵ of the electron is uniform $d\epsilon$. For antineutrinos it is $\sigma_0(\omega/6)[1-(1+2\omega)^{-3}]$ with a recoil spectrum varying as $(1+\omega-\epsilon)^2$. Here $\sigma_0=2G^2m^2/\pi=8.3\times10^{-45}$ cm^2.

[18] R. E. Behrends and C. Fronsdal, Phys. Rev. **106**, 345 (1957).

proposed on simple theoretical grounds to limit the possible β couplings. It is universal, it is symmetric, it produces two-component neutrinos, it conserves leptons, it preserves invariance under CP and T, and it is the simplest possibility from a certain point of view (that of two-component wave functions emphasized in this paper).

These theoretical arguments seem to the authors to be strong enough to suggest that the disagreement with the He^6 recoil experiment and with some other less accurate experiments indicates that these experiments are wrong. The $\pi \to e + \bar{\nu}$ problem may have a more subtle solution.

After all, the theory also has a number of successes. It yields the rate of μ decay to 2% and the asymmetry in direction in the $\pi \to \mu \to e$ chain. For β decay, it agrees with the recoil experiments[19] in A^{35} indicating a vector coupling, the absence of Fierz terms distorting the allowed spectra, and the more recent electron spin polarization[4] measurements in β decay.

[19] Herrmansfeldt, Maxson, Stähelin, and Allen, Phys. Rev. **107**, 641 (1957).

Besides the various experiments which this theory suggests be done or rechecked, there are a number of directions indicated for theoretical study. First it is suggested that all the various theories, such as meson theory, be recast in the form with the two-component wave functions to see if new possibilities of coupling, etc., are suggested. Second, it may be fruitful to analyze further the idea that the vector part of the weak coupling is not renormalized; to see if a set of couplings could be arranged so that the axial part is also not renormalized; and to study the meaning of the transformation groups which are involved. Finally, attempts to understand the strange particle decays should be made assuming that they are related to this universal interaction of definite form.

ACKNOWLEDGMENTS

The authors have profited by conversations with F. Boehm, A. H. Wapstra, and B. Stech. One of us (M. G. M.) would like to thank R. E. Marshak and E. C. G. Sudarshan for valuable discussions.

From: California Institute of Technology Laboratory Report CTSL-20 (1961), 11

THE EIGHTFOLD WAY:

A THEORY OF STRONG INTERACTION SYMMETRY*

Murray Gell-Mann

March 15, 1961

(Second printing: April, 1962)
(Third printing: October, 1963)

(Preliminary version circulated Jan. 20, 1961)

*Research supported in part by the U. S. Atomic Energy Commission Contract No. AT(11-1)-68, and the Alfred P. Sloan Foundation.

Reprint from *The Eightfold Way*, eds. M. Gell-Mann and Y. Ne'eman (W. A. Benjamin, 1964).

We attempt once more, as in the global symmetry scheme, to treat the eight known baryons as a supermultiplet, degenerate in the limit of a certain symmetry but split into isotopic spin multiplets by a symmetry-breaking term. Here we do not try to describe the symmetry violation in detail, but we ascribe it phenomenologically to the mass differences themselves, supposing that there is some analogy to the μ-e mass difference.

The symmetry is called unitary symmetry and corresponds to the "unitary group" in three dimensions in the same way that charge independence corresponds to the "unitary group" in two dimensions. The eight infinitesimal generators of the group form a simple Lie algebra, just like the three components of isotopic spin. In this important sense, unitary symmetry is the simplest generalization of charge independence.

The baryons then correspond naturally to an eight-dimensional irreducible representation of the group; when the mass differences are turned on, the familiar multiplets appear. The pion and K meson fit into a similar set of eight particles, along with a predicted pseudoscalar meson χ^o having $I = 0$. The pattern of Yukawa couplings of π, K, and χ is then nearly determined, in the limit of unitary symmetry.

The most attractive feature of the scheme is that it permits the description of eight vector mesons by a unified theory of the Yang-Mills type (with a mass term). Like Sakurai, we have a triplet ρ of vector mesons coupled to the isotopic spin current and a singlet vector meson ω^o coupled to the hypercharge current. We also have a pair of doublets M and $\overline{M}$, strange vector mesons coupled to strangeness-changing currents that are conserved when the mass differences are turned off. There is only one coupling constant, in the symmetric limit, for the system of eight vector mesons. There is some experimental evidence for the existence of ω^{0} and M, while ρ is presumably the famous $I = 1$, $J = 1$, π-π resonance.

A ninth vector meson coupled to the baryon current can be accommodated naturally in the scheme.

The most important prediction is the qualitative one that the eight baryons should all have the same spin and parity and that the pseudoscalar and vector mesons should form "octets", with possible additional "singlets".

If the symmetry is not too badly broken in the case of the renormalized coupling constants of the eight vector mesons, then numerous detailed predictions can be made of experimental results.

The mathematics of the unitary group is described by considering three fictitious "leptons", ν, e^{-}, and μ^{-}, which may or may not have something to do with real leptons. If there is a connection, then it may throw light on the structure of the weak interactions.

It has seemed likely for many years that the strongly interacting particles, grouped as they are into isotopic multiplets, would show traces of a higher symmetry that is somehow broken. Under the higher symmetry, the eight familiar baryons would be degenerate and form a supermultiplet. As the higher symmetry is broken, the Ξ, Λ, Σ, and N would split apart, leaving inviolate only the conservation of isotopic spin, of strangeness, and of baryons. Of these three, the first is partially broken by electromagnetism and the second is broken by the weak interactions. Only the conservation of baryons and of electric charge are absolute.

An attempt[1,2] to incorporate these ideas in a concrete model was the scheme of "global symmetry", in which the higher symmetry was valid for the interactions of the π meson, but broken by those of the K. The mass differences of the baryons were thus attributed to the K couplings, the symmetry of which was unspecified, and the strength of which was supposed to be significantly less than that of the π couplings.

The theory of global symmetry has not had great success in predicting experimental results. Also, it has a number of defects. The peculiar distribution of isotopic multiplets among the observed mesons and baryons is left unexplained. The arbitrary K couplings (which are not really particularly weak) bring in several adjustable constants. Furthermore, as admitted in Reference 1 and reemphasized recently by Sakurai[3,4] in his remarkable articles predicting vector mesons, the global model makes no direct connection between physical couplings and the currents of the conserved symmetry operators.

In place of global symmetry, we introduce here a new model of the higher symmetry of elementary particles which has none of these faults and a number of virtues.

We note that the isotopic spin group is the same as the group of all unitary 2x2 matrices with unit determinant. Each of these matrices can be written as exp(iA), where A is a hermitian 2x2 matrix. Since there are three independent hermitian 2x2 matrices (say, those of Pauli), there are three components of the isotopic spin.

Our higher symmetry group is the simplest generalization of isotopic spin, namely the group of all unitary 3x3 matrices with unit determinant. There are eight independent traceless 3x3 matrices and consequently the new "unitary spin" has eight components. The first three are just the components of the isotopic spin, the eighth is proportional to the hypercharge Y (which is +1 for N and K, -1 for Ξ and $\bar{K}$, 0 for Λ, Σ, π, etc.), and the remaining four are strangeness-changing operators.

Just as isotopic spin possesses a three-dimensional representation (spin 1), so the "unitary spin" group has an eight-dimensional irreducible representation, which we shall call simply $\underset{\sim}{8}$. In our theory, the baryon supermultiplet corresponds to this representation. When the symmetry is reduced, then $\underset{\sim}{I}$ and Y are still conserved but the four other components of unitary spin are not; the supermultiplet then breaks up into Ξ, Σ, Λ, and N. Thus the distribution of multiplets and the nature of strangeness or hypercharge are to some extent explained.

The pseudoscalar mesons are also assigned to the representation $\underset{\sim}{8}$. When the symmetry is reduced, they become the multiplets K, $\bar{K}$, π, and χ, where χ is a neutral isotopic singlet meson the existence of which we predict. Whether the PS mesons are regarded as fundamental or as bound states, their Yukawa couplings in the limit of "unitary" symmetry are describable in terms of only two coupling parameters.

The vector mesons are introduced in a very natural way, by an extension of the gauge principle of Yang and Mills.[5)] Here too we have a supermultiplet of eight mesons, corresponding to the representation $\underset{\sim}{8}$. In the limit of unitary symmetry and with the mass of these vector mesons "turned off", we have a completely gauge-invariant and minimal theory, just like electromagnetism. When the mass is turned on, the gauge invariance is reduced (the gauge function may no longer be space-time dependent) but the conservation of unitary spin remains exact. The sources of the vector mesons are the conserved currents of the eight components of the unitary spin.[6)]

When the symmetry is reduced, the eight vector mesons break up into a triplet ρ (coupled to the still-conserved isotopic spin current), a singlet ω (coupled to the still-conserved hypercharge current), and a pair of doublets M and $\bar{M}$ (coupled to a strangeness-changing current that is no longer conserved). The particles ρ and ω were both discussed by Sakurai. The ρ meson is presumably identical to the $I = 1$, $J = 1$, π-π resonance postulated by Frazer and Fulco[7)] in order to explain the isovector electromagnetic form factors of the nucleon. The ω meson is no doubt the same as the $I = 1$, $J = 0$ particle or 3π resonance predicted by Nambu[8)] and later by Chew[9)] and others in order to explain the isoscalar form factors of the nucleon. The strange meson M may be the same as the K^* particle observed by Alston *et al.*[10)]

Thus we predict that the eight baryons have the same spin and parity, that K is pseudoscalar and that χ exists, that ρ and ω exist with the properties assigned to them by Sakurai, and that M exists. But besides these qualitative predictions, there are also the many symmetry rules associated with the unitary spin. All of these are broken, though, by whatever destroys the unitary symmetry, and it is a delicate matter to find ways in which these effects of a broken symmetry can be explored experimentally.

Besides the eight vector mesons coupled to the unitary spin, there can be a ninth, which is invariant under unitary spin and is thus not degenerate with the other eight, even in the limit of unitary symmetry. We call this meson B^o. Presumably it exists too and is coupled to the baryon current. It is the meson predicted by Teller[11)] and later by Sakurai[3)] and explains most of the hard-core repulsion between nucleons and the attraction between nucleons and anti-nucleons at short distances.

We begin our exposition of the "eightfold way" in the next Section by discussing unitary symmetry using fictitious "leptons" which may have nothing to do with real leptons but help to fix the physical ideas in a rather graphic way. If there is a parallel between these "leptons" and the real ones, that would throw some light on the weak interactions, as discussed briefly in Section VI.

Section III is devoted to the $\underset{\sim}{8}$ representation and the baryons and Section IV to the pseudoscalar mesons. In Section V we present the theory of the vector mesons.

The physical properties to be expected of the predicted mesons are discussed in Section VII, along with a number of experiments that bear on those properties.

In Section VIII we take up the vexed question of the broken symmetry, how badly it is broken, and how we might succeed in testing it.

II The "Leptons" as a Model for Unitary Symmetry

For the sake of a simple exposition, we begin our discussion of unitary symmetry with "leptons", although our theory really concerns the baryons and mesons and the strong interactions. The particles we consider here for mathematical purposes do not necessarily have anything to do with real leptons, but there are some suggestive parallels. We consider three leptons, ν, e^-, and μ^-, and their antiparticles. The neutrino is treated on the same footing as the other two, although experience suggests that if it is treated as a four-component Dirac field, only two of the components have physical interaction. (Furthermore, there may exist two neutrinos, one coupled to the electron and the other to the muon.)

As far as we know, the electrical and weak interactions are absolutely symmetrical between e^- and μ^-, which are distinguished, however, from ν. The charged particles e^- and μ^- are separated by the mysterious difference in their masses. We shall not necessarily attribute this difference to any interaction, nor shall we explain it in any way. (If one insists on connecting it to an interaction, one might have to consider a coupling that becomes important only at exceedingly high energies and is, for the time being, only of academic interest.) We do, however, guess that the μ-e mass splitting is related to the equally mysterious mechanism that breaks the unitary symmetry of the baryons and mesons and splits the super-multiplets into isotopic multiplets. For practical purposes, we shall put all of these splittings into the mechanical masses of the particles involved.

It is well known that in present quantum electrodynamics, no one has succeeded in explaining the e-ν mass difference as an electromagnetic effect. Without prejudice to the question of its physical origin, we shall proceed with our discussion as if that mass difference were "turned on" along with the charge of the electron.

If we now "turn off" the μ-e mass difference, electromagnetism, and the weak interactions we are left with a physically vacuous theory of three exactly similar Dirac particles with no rest mass and no known couplings. This empty model is ideal for our mathematical purposes, however, and is physically motivated by the analogy with the strongly interacting particles, because it is at the corresponding stage of total unitary symmetry that we shall introduce the basic baryon mass and the strong interactions of baryons and mesons.

The symmetric model is, of course, invariant under all unitary transformations on the three states, ν, e^-, and μ^-.

Let us first suppose for simplicity that we had only two particles ν and e^-. We can factor each unitary transformation uniquely into one which multiplies both particles by the same phase factor and one (with determinant unity) which leaves invariant the product of the phase factors of ν and e^-. Invariance under the first kind of transformation corresponds to conservation of leptons ν and e^-. It may be considered separately from invariance under the class of transformations of the second kind (called by mathematicians the unitary unimodular group in two dimensions).

Each transformation of the first kind can be written as a matrix $e^{i\phi}1$, where 1 is the unit 2x2 matrix. The infinitesimal transformation is $1 + i(\delta\phi)1$ and so the unit matrix is the infinitesimal generator of these transformations. The transformations of the second kind are generated in the same way by the three independent traceless 2x2 matrices, which may be taken to be the three Pauli isotopic spin matrices τ_1, τ_2, τ_3. We thus have

$$1 + i \sum_{k=1}^{3} \delta\theta_k \frac{\tau_k}{2} \tag{2.1}$$

as the general infinitesimal transformation of the second kind. Symmetry under all the transformations of the second kind is the same as symmetry under τ_1, τ_2, τ_3, in other words charge independence or isotopic spin symmetry. The whole formalism of isotopic spin theory can then be constructed by considering the transformation properties of this doublet or spinor (ν, e^-) and of more complicated objects that transform like combinations of two or more such leptons.

The Pauli matrices τ_k are hermitian and obey the rules

$$\begin{aligned} &\mathrm{Tr}\ \tau_i \tau_j = 2\delta_{ij} \\ &\left[\tau_i, \tau_j\right] = 2ie_{ijk}\,\tau_k \\ &\left\{\tau_i, \tau_j\right\} = 2\delta_{ij}\, 1 \end{aligned} \tag{2.2}$$

We now generalize the idea of isotopic spin by including the third object μ^-. Again we factor the unitary transformations on the leptons into those which are generated by the 3x3 unit matrix 1 (and which correspond to lepton conservation) and those that are generated by the eight independent traceless 3x3 matrices (and which form the "unitary unimodular group" in three dimensions). We may construct a typical set of eight such matrices by analogy with the 2x2 matrices of Pauli. We call them $\lambda_1 \ldots\ldots \lambda_8$ and list them in Table I. They are hermitian and have the properties

$$\mathrm{Tr}\,\lambda_i\,\lambda_j = 2\delta_{ij}$$
$$\left[\lambda_i, \lambda_j\right] = 2if_{ijk}\,\lambda_k$$
$$\left\{\lambda_i, \lambda_j\right\} = \frac{4}{3}\,\delta_{ij}\,1 + 2d_{ijk}\,\lambda_k \qquad , \qquad (2.3)$$

where the f_{ijk} are real and totally antisymmetric like the Kronecker symbols e_{ijk} of Eq. (2.2), while the d_{ijk} are real and totally symmetric. These properties follow from the equations

$$\mathrm{Tr}\,\lambda_k \left[\lambda_i, \lambda_j\right] = 4if_{ijk}$$
$$\mathrm{Tr}\,\lambda_k \left\{\lambda_i, \lambda_j\right\} = 4d_{ijk} \qquad (2.4)$$

derived from (2.3).

The non-zero elements of f_{ijk} and d_{ijk} are given in Table II for our choice of λ_i. Even and odd permutations of the listed indices correspond to multiplication of f_{ijk} by ± 1 respectively and of d_{ijk} by +1.

The general infinitesimal transformation of the second kind is, of course,

$$1 + i \sum_i \delta\theta_i \frac{\lambda_i}{2} \qquad (2.5)$$

by analogy with (2.1). Together with conservation of leptons, invariance under the eight λ_i corresponds to complete "unitary symmetry" of the three leptons.

It will be noticed that λ_1, λ_2, and λ_3 correspond to τ_1, τ_2, and τ_3 for ν and e^- and nothing for the muon. Thus, if we ignore symmetry between (ν, e^-) and the muon, we still have conservation of isotopic spin. We also have conservation of λ_8, which commutes with λ_1, λ_2, and λ_3 and is diagonal in our representation. We can diagonalize at most two λ's at the same time and we have chosen them to be λ_3 (the third component of the ordinary isotopic spin) and λ_8, which is like strangeness or hypercharge, since it distinguishes the isotopic singlet μ^- from the isotopic doublet (ν, e^-) and commutes with the isotopic spin.

Now the turning-on of the muon mass destroys the symmetry under λ_4, λ_5, λ_6, and λ_7 (i.e., under the "strangeness-changing" components of the "unitary spin") and leaves lepton number, "isotopic spin", and "strangeness" conserved. The electromagnetic interactions (along with the electron mass) then break the conservation of λ_1 and λ_2, leaving lepton number λ_3, and strangeness conserved. Finally, the weak interactions allow the strangeness to be changed (in muon decay) but continue to conserve the lepton number n_ℓ and the electric charge

$$Q = \frac{e}{2}\left(\lambda_3 + \frac{\lambda_8}{\sqrt{3}} - \frac{4}{3}\, n_\ell\right) \tag{2.6}$$

where n_ℓ is the number of leptons minus the number of antileptons and equals 1 for ν, e^-, and μ^- (i.e., the matrix 1).

We see that the situation is just what is needed for the baryons and mesons. We transfer the symmetry under unitary spin to them and assign them strong couplings and basic symmetrical masses. Then we turn on the mass splittings, and the symmetry under the 4th, 5th, 6th, and 7th components of the unitary spin is lifted, leaving baryon number, strangeness, and isotopic spin conserved. Electromagnetism destroys the symmetry under the 1st and 2nd components of the spin, and the weak interactions destroy strangeness conservation. Finally, only charge and baryon number are conserved.

III Mathematical Description of the Baryons

In the case of isotopic spin $\underset{\sim}{I}$, we know that the various possible charge multiplets correspond to "irreducible representations" of the simple 2x2 matrix algebra described above for (ν, e^-). Each multiplet has $2I + 1$ components, where the quantum number I distinguishes one representation from another and tells us the eigenvalue $I(I + 1)$ of the operator $\sum_{i=1}^{3} I_i^{\,2}$, which commutes with all the elements of the isotopic spin group and in particular with all the infinitesimal group elements $1 + i \sum_{i=1}^{3} \delta\theta_i\, I_i$. The operators I_i are represented, within the multiplet, by hermitian $(2I + 1) \times (2I + 1)$ matrices having the same commutation rules

$$\left[I_i, I_j\right] = ie_{ijk} I_k \tag{3.1}$$

as the 2x2 matrices $\tau_i/2$. For the case of $I = 1/2$, we have just $I_i = \tau_i/2$ within the doublet.

If we start with the doublet representation, we can build up all the others by considering superpositions of particles that transform like the original doublet. Thus, the antiparticles e^+, $-\bar{\nu}$ also form a doublet. (Notice the minus sign on the anti-neutrino state or field.) Taking $\frac{e^+e^- + \bar{\nu}\nu}{\sqrt{2}}$, we obtain a singlet, that is, a one-dimensional representation for which all the I_i are zero. Calling the neutrino and electron e_α with $\alpha = 1, 2$, we can describe the singlet by $\frac{1}{\sqrt{2}}\,\bar{e}_\alpha e_\alpha$ or, more concisely, $\frac{1}{\sqrt{2}}\,\bar{e}e$. The three components of a triplet can be formed by taking $e^+\nu = \frac{1}{2}\,\bar{e}(\tau_1 - i\tau_2)e$, $\frac{e^+e^- - \bar{\nu}\nu}{\sqrt{2}} = \frac{1}{\sqrt{2}}\,\bar{e}\,\tau_3\,e$, and $\nu e^- = \frac{1}{2}\,\bar{e}(\tau_1 + i\tau_2)e$. Rearranging these, we have just $\frac{1}{\sqrt{2}}\,\bar{e}\,\tau_j\,e$ with $j = 1, 2, 3$. Among these three states, the 3x3 matrices I_i^{jk} of the three components of I are given by

$$I_i^{jk} = -ie_{ijk} \quad . \tag{3.2}$$

Now let us generalize these familiar results to the set of three states ν, e^-, and μ^-. Call them ℓ_α with $\alpha = 1, 2, 3$ and use $\bar{\ell}\ell$ to mean $\bar{\ell}_\alpha \ell_\alpha$, etc. For this system we define $F_i = \lambda i/2$ with $i = 1, 2, \ldots, 8$, just as $I_i = \tau i/2$ for isotopic spin. The F_i are the 8 components of the unitary spin operator $\underset{\sim}{F}$ in this case and we shall use the same notation in all representations. The first three components of $\underset{\sim}{F}$ are identical with the three components of the isotopic spin $\underset{\sim}{I}$ in all cases, while F_8 will always be $\frac{\sqrt{3}}{2}$ times the hypercharge Y (linearly related to the strangeness). In all representations, then, the components of $\underset{\sim}{F}$ will have the same commutation rules

$$\left[F_i, F_j\right] = if_{ijk} F_k \qquad (3.3)$$

that they do in the simple lepton representation for which $F_i = \lambda_i/2$. (Compare the commutation rules in Eq. (2.3).) The trace properties and anticommutation properties will not be the same in all representations any more than they are for $\underset{\sim}{I}$. We see that the rules (3.1) are just a special case of (3.3) with indices 1, 2, 3, since the f's equal the e's for these values of the indices.

We must call attention at this point to an important difference between unitary or $\underset{\sim}{F}$ spin and isotopic or $\underset{\sim}{I}$ spin. Whereas, with a simple change of sign on $\bar{\nu}$, we were able to construct from $\bar{e}_\alpha$ a doublet transforming under $\underset{\sim}{I}$ just like e_α, we are not able to do the same thing for the $\underset{\sim}{F}$ spin when we consider the three anti-leptons $\bar{\ell}_\alpha$ compared to the three leptons ℓ_α. True, the anti-leptons do give a representation for $\underset{\sim}{F}$, but it is, in mathematical language, inequivalent to the lepton representation, even though it also has three dimensions. The reason is easy to see: when we go from leptons to anti-leptons the eigenvalues of the electric charge, the third component of $\underset{\sim}{I}$, and the lepton number all change sign, and thus the eigenvalues of F_8 change sign. But they were $\frac{1}{2\sqrt{3}}$, $\frac{1}{2\sqrt{3}}$, and $\frac{-1}{\sqrt{3}}$ for leptons and so they are a different set for anti-leptons and no similarity transformation can change one representation into the other. We shall refer to the lepton representation as 3 and the anti-lepton representation as $\underset{\sim}{\bar{3}}$.

Now let us consider another set of "particles" L_α transforming exactly like the leptons ℓ_α under unitary spin and take their anti-particles $\bar{L}_\alpha$. We follow the same procedure used above for the isotopic spin and the doublet e. We first construct the state $\frac{1}{\sqrt{3}} \bar{L}_\alpha \ell_\alpha$ or $\frac{1}{\sqrt{3}} \bar{L} \ell$. Just as $\frac{\bar{e}e}{\sqrt{2}}$ gave a one-dimensional representation of $\underset{\sim}{I}$ for which all the I_i were zero, so $\frac{\bar{L}\ell}{\sqrt{3}}$ gives a one-dimensional representation of $\underset{\sim}{F}$ for which all the F_i are zero. Call this one-dimensional representation $\underset{\sim}{1}$.

Now, by analogy with $\frac{\bar{e}\ \tau_i\ e}{\sqrt{2}}$ with $i = 1, 2, 3$, we form $\frac{\bar{L}\ \lambda_i\ \ell}{\sqrt{2}}$ with $i = 1, 2, \ldots, 8$. These states transform under unitary spin $\underset{\sim}{F}$ like an irreducible representation of dimension 8, which we shall call $\underset{\sim}{8}$. In this representation, the 8x8 matrices F_i^{jk} of the eight components F_i of the unitary spin are given by the relation

$$F_i^{jk} = -\, i f_{ijk} \qquad , \tag{3.4}$$

analogous to Eq. (3.2).

When we formed an isotopic triplet from two isotopic doublets, in the discussion preceding Eq. (3.2), we had to consider linear combinations of the $\frac{\bar{e}\ \tau_i\ e}{\sqrt{2}}$ in order to get simple states with definite electric charges, etc. We must do the same here. Using the symbol ~ for "transforms like", we define

$$
\begin{aligned}
\Sigma^+ &\sim \tfrac{1}{2}\,\bar{L}(\lambda_1 - i\lambda_2)\ell && \sim D^+\nu \\
\Sigma^- &\sim \tfrac{1}{2}\,\bar{L}(\lambda_1 + i\lambda_2)\ell && \sim D^o e^- \\
\Sigma^o &\sim \tfrac{1}{\sqrt{2}}\,\bar{L}\,\lambda_3\,\ell && \sim \frac{D^o\nu - D^+e^-}{\sqrt{2}} \\
p &\sim \tfrac{1}{2}\,\bar{L}(\lambda_4 - i\lambda_5)\ell && \sim S^+\nu \\
n &\sim \tfrac{1}{2}\,\bar{L}(\lambda_6 - i\lambda_7)\ell && \sim S^+e^- \\
\Xi^o &\sim \tfrac{1}{2}\,\bar{L}(\lambda_6 + i\lambda_7)\ell && \sim D^+\mu^- \\
\Xi^- &\sim \tfrac{1}{2}\,\bar{L}(\lambda_4 + i\lambda_5)\ell && \sim D^o\mu^- \\
\Lambda &\sim \tfrac{1}{\sqrt{2}}\,\bar{L}\,\lambda_8\,\ell && \sim (D^o\nu + D^+e^- - 2S^+\mu^-)/\sqrt{6}\quad . \qquad (3.5)
\end{aligned}
$$

The most graphic description of what we are doing is given in the last column, where we have introduced the notation D^o, D^+, and S^+ for the $\bar{L}$ particles analogous to the $\bar{\ell}$ particles $\bar{\nu}$, e^+, and μ^+ respectively. D stands for doublet and S for singlet with respect to isotopic spin. Using the last column, it is easy to see that the isotopic spins, electric charges, and hypercharges of the multiplets are exactly as we are accustomed to think of them for the baryons listed.

We say, therefore, that the eight known baryons form one degenerate supermultiplet with respect to unitary spin. When we introduce a perturbation that transforms like the μ-e mass difference, the supermultiplet will break up into exactly the known multiplets. (Of course, D will split from S at the same time as e^-, ν from μ^-.)

Of course, another type of baryon is possible, namely a singlet neutral one that transforms like $\frac{1}{\sqrt{3}}\ \bar{L}\ \ell$. If such a particle exists, it may be very heavy and highly unstable. At the moment, there is no evidence for it.

We shall attach no physical significance to the ℓ and $\bar{L}$ "particles" out of which we have constructed the baryons. The discussion up to this point is really just a mathematical introduction to the properties of unitary spin.

IV Pseudoscalar Mesons

We have supposed that the baryon fields N_j transform like an octet $\underset{\sim}{8}$ under $\underset{\sim}{F}$, so that the matrices of $\underset{\sim}{F}$ for the baryon fields are given by Eq. (3.4). We now demand that all mesons transform under $\underset{\sim}{F}$ in such a way as to have $\underset{\sim}{F}$-invariant strong couplings. If the 8 mesons π_i are to have Yukawa couplings, they must be coupled to $\bar{N}\ \theta_i\ N$ for some matrices θ_i, and we must investigate how such bilinear forms transform under $\underset{\sim}{F}$.

In mathematical language, what we have done in Section III is to look at the direct product $\bar{\underset{\sim}{3}} \times \underset{\sim}{3}$ of the representations $\bar{\underset{\sim}{3}}$ and $\underset{\sim}{3}$ and to find that it reduces to the direct sum of $\underset{\sim}{8}$ and $\underset{\sim}{1}$. We identified $\underset{\sim}{8}$ with the baryons and, for the time being, dismissed $\underset{\sim}{1}$. What we must do now is to look at $\bar{\underset{\sim}{8}} \times \underset{\sim}{8}$. Now it is easy to show that actually $\bar{\underset{\sim}{8}}$ is equivalent to $\underset{\sim}{8}$; this is unlike the situation for $\bar{\underset{\sim}{3}}$ and $\underset{\sim}{3}$. (We note that the values of Y, I_3, Q, etc., are symmetrically disposed about zero in the $\underset{\sim}{8}$ representation.) So the anti-baryons transform essentially like the baryons and we must reduce out the direct product $\underset{\sim}{8} \times \underset{\sim}{8}$. Standard group theory gives the result

$$\underset{\sim}{8} \times \underset{\sim}{8} = \underset{\sim}{1} + \underset{\sim}{8} + \underset{\sim}{8} + \underset{\sim}{10} + \underset{\sim}{\overline{10}} + \underset{\sim}{27} \qquad , \qquad (4.1)$$

where $\underset{\sim}{\overline{27}} = \underset{\sim}{27}$ (this can happen only when the dimension is the cube of an integer). The representation $\underset{\sim}{27}$ breaks up, when mass differences are turned on, into an isotopic singlet, triplet, and quintet with $Y = 0$, a doublet and a quartet with $Y = 1$, a doublet and a quartet with $Y = -1$, a triplet with $Y = 2$, and a triplet with $Y = -2$. The representation $\underset{\sim}{10}$ breaks up, under the same conditions, into a triplet with $Y = 0$, a doublet with $Y = -1$, a quartet with $Y = +1$, and a singlet with $Y = +2$. The conjugate representation $\underset{\sim}{\overline{10}}$ looks the same, of course, but with equal and opposite values of Y. None of these much resembles the pattern of the known mesons.

The $\underset{\sim}{8}$ representation, occurring twice, looks just the same for mesons as for baryons and is very suggestive of the known π, K, and $\bar{K}$ mesons plus one more neutral pseudoscalar meson with $I = 0$, $Y = 0$, which corresponds to Λ in the baryon case. Let us call this meson χ^0 and suppose it exists, with a fairly low mass. Then we have identified the known pseudoscalar mesons with an octet under unitary symmetry, just like the baryons. The representations $\underset{\sim}{1}$, $\underset{\sim}{10}$, $\underset{\sim}{\overline{10}}$, and $\underset{\sim}{27}$ may also correspond to mesons, even pseudoscalar ones, but presumably they lie higher in mass, some or all of them perhaps so high as to be physically meaningless.

To describe the eight pseudoscalar mesons as belonging to $\underset{\sim}{8}$, we put (very much as in (3.5))

$$\chi^{o} = \pi_8$$

$$\pi^{+} = (\pi_1 - i\pi_2)/\sqrt{2}$$

$$\pi^{-} = (\pi_1 + i\pi_2)/\sqrt{2}$$

$$\pi^{o} = \pi_3$$

$$K^{+} = (\pi_4 - i\pi_5)/\sqrt{2}$$

$$K^{o} = (\pi_6 - i\pi_7)/\sqrt{2}$$

$$\overline{K^{o}} = (\pi_6 + i\pi_7)/\sqrt{2}$$

$$K^{-} = (\pi_4 + i\pi_5)/\sqrt{2} \qquad (4.2)$$

and we know then that the matrices of $\underset{\sim}{F}$ connecting the π_j are just the same as those connecting the N_j, namely $F_i^{jk} = -if_{ijk}$.

To couple the 8 mesons invariantly to 8 baryons (say by γ_5), we must have a coupling

$$2i\, g_o\, \overline{N}\, \gamma_5\, \theta_i\, N\, \pi_i \qquad (4.3)$$

for which the relation

$$\left[F_i, \theta_j\right] = if_{ijk}\, \theta_k \qquad (4.4)$$

holds. Now the double occurrence of $\underset{\sim}{8}$ in Eq. (4.1) assures us that there are two independent sets of eight 8x8 matrices θ_i obeying (4.4). One of these sets evidently consists of the F_i themselves. It is not hard to find the other set if we go back to the commutators and anti-commutators of the λ matrices in the $\underset{\sim}{3}$ representation (Eq. (2.3)). Just as we formed

$F_i^{jk} = - if_{ijk}$, we define

$$D_i^{jk} = d_{ijk} \tag{4.5}$$

and it is easy to show that the D's also satisfy Eq. (4.4). We recall that where the F matrices are imaginary and antisymmetric with respect to the basis we have chosen, the D's are real and symmetric.

Now what is the physical difference between coupling the pseudoscalar mesons π_i by means of D_i and by means of F_i? It lies in the symmetry under the operation

$$R:\ p \leftrightarrow \Xi^-,\quad n \leftrightarrow \Xi^0,\quad \Sigma^+ \leftrightarrow \Sigma^-,\quad \Sigma^0 \leftrightarrow \Sigma^0,\quad \Lambda \leftrightarrow \Lambda$$

$$K^+ \leftrightarrow \pm K^-,\quad K^0 \leftrightarrow \pm \overline{K^0},\quad \pi^+ \leftrightarrow \pm\pi^-,\quad \pi^0 \leftrightarrow \pm\pi^0,\quad \chi^0 \leftrightarrow \pm\chi^0 \quad , \tag{4.6}$$

which is not a member of the unitary group, but a kind of reflection. In the language of N_i, we may say that R changes the sign of the second, fifth, and seventh particles; we note that λ_2, λ_5, and λ_7 are imaginary while the others are real. From Table II we can see that under these sign changes f_{ijk} is odd and d_{ijk} even.

It may be that in the limit of unitary symmetry the coupling of the pseudoscalar mesons is invariant under R as well as the unitary group. In that case, we choose either the plus sign in (4.6) and the D coupling or else the minus sign and the F coupling. The two possible coupling patterns are listed in Table III.

If only one of the patterns is picked out (case of R-invariance), it is presumably the D coupling, since that gives a large $\Lambda\pi\Sigma$ interaction (while the F coupling gives none) and the $\Lambda\pi\Sigma$ interaction is the best way of explaining the binding of Λ particles in hypernuclei.

In general, we may write the Yukawa coupling (whether fundamental or phenomenological, depending on whether the π_i are elementary or not) in the form

$$L_{int} = 2i\, g_o\, \overline{N}\, \gamma_5 \left[\alpha\, D_i + (1 - \alpha)\, F_i\right] N\, \pi_i \quad . \qquad (4.7)$$

We note that in no case is it possible to make the couplings ΛKN and ΣKN both much smaller than the $N\pi N$ coupling. Since the evidence from photo-K production seems to indicate smaller effective coupling constants for ΛKN and ΣKN than for $N\pi N$ (indeed, that was the basis of the global symmetry scheme), we must conclude that our symmetry is fairly badly broken. We shall return to that question in Section VII.

A simple way to read off the numerical factors in Table III, as well as those in Table IV for the vector mesons, is to refer to the chart in Table V, which gives the transformation properties of mesons and baryons in terms of the conceptual "leptons" and "L particles" of Section III.

An interesting remark about the baryon mass differences may be added at this point. If we assume that they transform like the μ-e mass difference, that is, like the 8th component of the unitary spin, then there are only two possible mass-difference matrices, F_8 and D_8. That gives rise to a sum rule for baryon masses:

$$1/2\ (m_N + m_\Xi) = 3/4\ m_\Lambda + 1/4\ m_\Sigma \quad , \qquad (4.8)$$

which is very well satisfied by the observed masses, much better than the corresponding sum rule for global symmetry.

There is no particular reason to believe, however, that the analogous sum rules for mesons are obeyed.

V Vector Mesons

The possible transformation properties of the vector mesons under $\underset{\sim}{F}$ are the same as those we have already examined in the pseudoscalar case. Again it seems that for low mass states we can safely ignore the representations $\underset{\sim}{27}$, $\underset{\sim}{10}$, and $\underset{\sim}{\overline{10}}$. We are left with 1 and the two cases of $\underset{\sim}{8}$.

A vector meson transforming according to $\underset{\sim}{1}$ would have $Q = 0$, $I = 0$, $Y = 0$ and would be coupled to the total baryon current $i \bar{N} \gamma_\mu N$, which is exactly conserved. Such a meson may well exist and be of great importance. The possibility of its existence has been envisaged for a long time.

We recall that the conservation of baryons is associated with the invariance of the theory under infinitesimal transformations

$$N \to (1 + i\epsilon)N \quad , \tag{5.1}$$

where ϵ is a constant. This is gauge-invariance of the first kind. We may, however, consider the possibility that there is also guage invariance of the second kind, as discussed by Yang and Lee.[12] Then we could make ϵ a function of space-time. In the free baryon Lagrangian

$$L_N = - \bar{N}(\gamma_\alpha \partial_\alpha + m_o)N \tag{5.2}$$

this would produce a new term

$$L_N \to L_N - i \bar{N} \gamma_\alpha N \partial_\alpha \epsilon \tag{5.3}$$

which can be cancelled only if there exists a neutral vector meson field B_α coupled to the current $\bar{N} \gamma_\alpha N$:

$$L_B = - 1/4 \, (\partial_\alpha B_\beta - \partial_\beta B_\alpha)^2$$

$$L_{int} = if_o \bar{N} \gamma_\alpha N B_\alpha \qquad (5.4)$$

and which undergoes the gauge transformation

$$B_\alpha \to B_\alpha + 1/f_o \, \partial_\alpha \epsilon \quad . \qquad (5.5)$$

As Yang and Lee pointed out, such a vector meson is massless and if it existed with any appreciable coupling constant, it would simulate a kind of anti-gravity, for baryons but not leptons, that is contradicted by experiment.

We may, however, take the point of view that there are vector mesons associated with a gauge-invariant Lagrangian plus a mass term, which breaks the gauge invariance of the second kind while leaving inviolate the gauge invariance of the first kind and the conservation law. Such situations have been treated by Glashow,[13)] Salam and Ward,[14)] and others, but particularly in this connection by Sakurai.[3)]

The vector meson transforming according to $\underline{1}$ would then be of such a kind. Teller,[11)] Sakurai,[3)] and others have discussed the notion that such a meson may be quite heavy and very strongly coupled, binding baryons and anti-baryons together to make the pseudoscalar mesons according to the compound model of Fermi and Yang.[15)] We shall leave this possibility open, but not consider it further here. If it is right, then the Yukawa couplings (4.7) must be treated as phenomenological rather than fundamental; from an immediate practical point of view, it may not make much difference.

We go on to consider the $\underset{\sim}{8}$ representation. An octet of vector mesons would break up into an isotopic doublet with $Y = 1$, which we shall call M (by analogy with K -- the symbol L is already used to mean π or μ); the corresponding doublet $\bar{M}$ analogous to $\bar{K}$; a triplet ρ with $Y = 0$ analogous to π; and a singlet ω^o with $Y = 0$ analogous to χ^o.

We may tentatively identify M with the K^* reported by Alston et al.[10)] at 884 MeV with a width $\Gamma \approx 15$ MeV for break-up into $\pi + K$. Such a narrow width certainly points to a vector rather than a scalar state. The vector meson ρ may be identified, as Sakurai has proposed, with the $I = 1$, $J = 1$, π-π resonance discussed by Frazer and Fulco[7)] in connection with the electromagnetic structure of the nucleon. The existence of ω^o has been postulated for similar reasons by Nambu,[8)] Chew,[9)] and others.

In principle, we have a choice again between couplings of the $\underset{\sim}{D}$ and the $\underset{\sim}{F}$ type for the vector meson octet. But there is no question which is the more reasonable theory. The current $i\,\bar{N}\,F_j\,\gamma_\alpha\,N$ is the current of the F-spin for baryons and in the limit of unitary symmetry the total F-spin current is exactly conserved. (The conservation of the strangeness-changing currents, those of F_4, F_5, F_6, and F_7, is broken by the mass differences, the conservation of F_2 and F_3 by electromagnetism, and that of F_3 and F_8 separately by the weak interactions. Of course, the current of the electric charge

$$Q = e\left(F_3 + \frac{F_8}{\sqrt{3}}\right) \tag{5.6}$$

is exactly conserved.)

Sakurai has already suggested that ρ is coupled to the isotopic spin current and ω to the hypercharge current. We propose in addition that the strange vector mesons M are coupled to the strangeness-changing components of the F-spin current and that the whole system is completely invariant under $\underset{\sim}{F}$ before the mass-differences have been turned on, so that the three coupling constants (suitably defined) are approximately equal even in the presence of the mass differences.

Now the vector mesons themselves carry F spin and therefore contribute to the current which is their source. The problem of constructing a nonlinear theory of this kind has been completely solved in the case of isotopic spin by Yang and Mills[5)] and by Shaw.[5)] We have only to generalize their result (for three vector mesons) to the case of F spin and eight vector mesons.

We may remark parenthetically that the Yang-Mills theory is irreducible, in the sense that all the 3 vector mesons are coupled to one another inextricably. We may always make a "reducible" theory by adjoining other, independent vector mesons like the field B_α discussed earlier in connection with the baryon current. It is an interesting mathematical problem to find the set of all irreducible Yang-Mills tricks. Glashow and the author[16)] have shown that the problem is the same as that of finding all the simple Lie algebras, one that was solved long ago by the mathematicians. The possible dimensions are 3, 8, 10, 14, 15, 21, and so forth. Our generalization of the Yang-Mills trick is the simplest one possible.

But let us "return to our sheep", in this case the 8 vector mesons. We first construct a completely gauge-invariant theory and then add a mass term for the mesons. Let us call the eight fields $\rho_{i\alpha}$, just as we denoted the eight pseudoscalar fields by π_i. We may think of the N_i, the π_i, and the $\rho_{i\alpha}$ as vectors in an 8-dimensional space. (The index α here refers to the four space-time components of a vector field.) We use our totally antisymmetric tensor f_{ijk} to define a cross product

$$(\underset{\sim}{A} \times \underset{\sim}{B})_i = f_{ijk} A_j B_k \tag{5.7}$$

The gauge transformation of the second kind analogous to Eqs. (5.1) and (5.5) is performed with an eight-component gauge function $\underset{\sim}{\phi}$:

$$\begin{aligned} \underset{\sim}{N} &\to \underset{\sim}{N} + \underset{\sim}{\phi} \times \underset{\sim}{N} \\ \underset{\sim}{\rho}_\alpha &\to \underset{\sim}{\rho}_\alpha + \underset{\sim}{\phi} \times \underset{\sim}{\rho}_\alpha - (2\gamma_o)^{-1} \partial_\alpha \underset{\sim}{\phi} \\ \underset{\sim}{\pi} &\to \underset{\sim}{\pi} + \underset{\sim}{\phi} \times \underset{\sim}{\pi} \end{aligned} \tag{5.8}$$

We have included the pseudoscalar meson field for completeness, treating it as elementary. We shall not write the π-N and possible π-π couplings in what follows, since they are not relevant and may simply be added in at the end. The bare coupling parameter is γ_o.

We define gauge-covariant field strengths by the relation

$$\underset{\sim}{G}_{\alpha\beta} = \partial_\alpha \underset{\sim}{\rho}_\beta - \partial_\beta \underset{\sim}{\rho}_\alpha + 2\gamma_o \underset{\sim}{\rho}_\alpha \times \underset{\sim}{\rho}_\beta \tag{5.9}$$

and the gauge-invariant Lagrangian (to which a common vector meson mass term is presumably added) is simply

$$\begin{aligned} L = &- \tfrac{1}{4} \underset{\sim}{G}_{\alpha\beta} \cdot \underset{\sim}{G}_{\alpha\beta} - m_o \overline{\underset{\sim}{N}} \cdot \underset{\sim}{N} - \overline{\underset{\sim}{N}} \gamma_\alpha \cdot (\partial_\alpha \underset{\sim}{N} + 2\gamma_o \underset{\sim}{\rho}_\alpha \times \underset{\sim}{N}) \\ &- \tfrac{1}{2} \mu_o^2 \underset{\sim}{\pi} \cdot \underset{\sim}{\pi} - \tfrac{1}{2} (\partial_\alpha \underset{\sim}{\pi} + 2\gamma_o \underset{\sim}{\rho}_\alpha \times \underset{\sim}{\pi}) \cdot (\partial_\alpha \underset{\sim}{\pi} + 2\gamma_o \underset{\sim}{\rho}_\alpha \times \underset{\sim}{\pi}) \ . \end{aligned} \tag{5.10}$$

There are trilinear and quadrilinear interactions amongst the vector mesons, as usual, and also trilinear and quadrilinear couplings with the pseudoscalar mesons. All these, along with the basic coupling of vector mesons to the baryons, are characterized in the limit of no mass differences by the single coupling parameter γ_0. The symmetrical couplings of $\underset{\sim}{\rho}_\alpha$ to the bilinear currents of baryons and pseudoscalar mesons are listed in Table IV. In Section VII, we shall use them to predict a number of approximate relations among experimental quantities relevant to the vector mesons.

As in the case of the pseudoscalar couplings, the various vector couplings will have somewhat different strengths when the mass differences are included, and some couplings which vanish in (5.10) will appear with small coefficients. Thus, in referring to experimental renormalized coupling constants (evaluated at the physical masses of the vector mesons) we shall use the notation $\gamma_{N\Lambda M}$, $\gamma_{NN\rho}$, etc. In the limit of unitary symmetry, all of these that do not vanish are equal.

VI Weak Interactions

So far, the role of the leptons in unitary symmetry has been purely symbolic. Although we introduced a mathematical F spin for ν, e^-, and μ^-, that spin is not coupled to the eight vector mesons that take up the F spin gauge for baryons and mesons. If we take it seriously at all, we should probably regard it as a different spin, but one with the same mathematical properties.

Let us make another point, which may seem irrelevant but possibly is not. The photon and the charge operator to which it is coupled have not so far been explicitly included in our scheme. They must be put in as an afterthought, along with the corresponding gauge transformation, which was the model for the more peculiar gauge transformations we have treated. If the weak interactions are carried[17)] by vector bosons X_α and generated by a gauge transformation[18,19)] of their own, then these bosons and gauges have been ignored as well. Such considerations might cause us, if we are in a highly speculative frame of mind, to wonder about the possibility that each kind of interaction has its own type of gauge and its own set of vector particles and that the algebraic properties of these gauge transformations conflict with one another.

When we draw a parallel between the "F spin" of leptons and the F spin of baryons and mesons, and when we discuss the weak interactions at all, we are exploring phenomena that transcend the scheme we are using. Everything we say in this Section must be regarded as highly tentative and useful only in laying the groundwork for a possible future theory. The same is true of any physical interpretation of the mathematics in Sections II and III.

We shall restrict our discussion to charge - exchange weak currents and then only to the vector part. A complete discussion of the axial vector weak currents may involve more complicated concepts and even new mesons[20)] (scalar and/or axial vector) lying very high in energy.

The vector weak current of the leptons is just $\bar{\nu}\gamma_\alpha e + \bar{\nu}\gamma_\alpha \mu$. If we look at the abstract scheme for the baryons in Eq. (3.5), we see that a baryon current with the same transformation properties under F would consist of two parts: one, analogous to $\bar{\nu}\gamma_\alpha e$, would have $|\Delta \underline{I}| = 1$ and $\Delta S = 0$, while the other, analogous to $\bar{\nu}\gamma_\alpha \mu$, would have $|\Delta \underline{I}| = 1/2$ and

$\Delta S/\Delta Q = +1$. These properties are exactly the ones we are accustomed to associate with the weak interactions of baryons and mesons.

Now the same kind of current we have taken for the leptons can be assigned to the conceptual bosons L of Section III. Suppose it to be of the same strength. Then, depending on the relative sign of the lepton and L weak currents, the matrices in the baryon system may be F's or D's.

Suppose, in the $\Delta S = 0$ case, the relative sign is such as to give F. Then the resulting current is just one component of the isotopic spin current; and the same result will hold for mesons. Thus we will have the conserved vector current that has been proposed[17)] to explain the lack of renormalization of the Fermi constant.

In the $\Delta S = 1$ case, by taking the same sign, we could get the almost-conserved strangeness-changing vector current, the current of $F_4 + iF_5$.

Further speculations along these lines might lead to a theory of the weak interactions.[21)]

VII Properties of the New Mesons

The theory we have sketched is fairly solid only in the realm of the strong interactions, and we shall restrict our discussion of predictions to the interactions among baryons and mesons.

We predict the existence of 8 baryons with equal spin and parity following the pattern of N, Λ, Σ, and Ξ. Likewise, given the π and its coupling constant, we predict a pseudoscalar K and a new particle, the χ^o, both coupled (in the absence of mass differences) as in Eq. (4.7), and we predict pion couplings to hyperons as in the same equation.

Now in the limit of unitary symmetry an enormous number of selection and intensity rules apply. For example, for the reactions PS meson + baryon → PS meson + baryon, there are only 7 independent amplitudes. Likewise, baryon-baryon forces are highly symmetric. However, the apparent smallness of $g_1^2/4\pi$ for $NK\Lambda$ and $NK\Sigma$ compared to $N\pi N$ indicates that unitary symmetry is badly broken, assuming that it is valid at all. We must thus rely principally on qualitative predictions for tests of the theory; in Section VIII we take up the question of how quantitative testing may be possible.

The most clear-cut new prediction for the pseudoscalar mesons is the existence of χ^o, which should decay into 2γ like the π^o, unless it is heavy enough to yield $\pi^+ + \pi^- + \gamma$ with appreciable probability. (In the latter case, we must have $(\pi^+\pi^-)$ in an odd state.) $\chi^o \rightarrow 3\pi$ is forbidden by conservation of I and C. For a sufficiently heavy χ^o, the decay $\chi^o \rightarrow 4\pi$ is possible, but hampered by centrifugal barriers.

Now we turn to the vector mesons, with coupling pattern as given in Table IV. We predict, like Sakurai, the ρ meson, presumably identical with the resonance of Frazer and Fulco, and the ω meson, coupled to the hypercharge. In addition, we predict the strange vector meson M, which may be the same as the K^* of Alston et al.

Some of these are unstable with respect to the strong interactions and their physical coupling constants to the decay products are given by the decay widths. Thus, for $M \rightarrow K + \pi$, we have

$$\Gamma_M = 2 \frac{\gamma^2_{MK\pi}}{4\pi} \frac{k^3}{m_M^2} \quad , \tag{7.1}$$

where k is the momentum of one of the decay mesons. We expect, of course, a $\cos^2\theta$ angular distribution relative to the polarization of M and a charge ratio of 2:1 in favor of $K^o + \pi^+$ or $K^+ + \pi^-$.

For the $I = 1$, $J = 1$, π-π resonance we have the decay $\rho \to 2\pi$ with width

$$\Gamma_\rho = \frac{8}{3} \frac{\gamma^2_{\rho\pi\pi}}{4\pi} \frac{k^3}{m_\rho^2} \tag{7.2}$$

Using a value $m_\rho = 4.5\, m_\pi$, we would have $\Gamma \approx m_\pi \frac{\gamma^2}{4\pi}$ and agreement with the theory of Bowcock <u>et al</u>.[7] would require a value of $\frac{\gamma^2}{4\pi}$ of the order of 2/3. If, now, we assume that the mass of M is really around 880 MeV, then Eq. (7.1) yields $\Gamma_M \approx \frac{\gamma^2}{4\pi} \cdot 50$ MeV. If the width is around 15 MeV, then the two values of $\gamma^2/4\pi$ are certainly of the same order.

We can obtain information about vector coupling constants in several other ways. If we assume, with Sakurai and Dalitz, that the Y^* of Alston <u>et al</u>.[22] (at 1380 MeV with decay $Y^* \to \pi + \Lambda$) is a bound state of $\bar{K}$ and N in a potential associated with the exchange of ω and ρ, then with simple Schrödinger theory we can roughly estimate the relevant coupling strengths. In the Schrödinger approximation (which is fairly bad, of course) we have the potential

$$V(\text{triplet}) \approx -3 \frac{\gamma_{NN\omega}\,\gamma_{KK\omega}}{4\pi} \frac{e^{-m_\omega r}}{r} + \frac{\gamma_{NN\rho}\,\gamma_{KK\rho}}{4\pi} \frac{e^{-m_\rho r}}{r} \quad . \tag{7.3}$$

If ω has a mass of around 400 MeV (as suggested by the isoscalar form factor of the nucleon), then the right binding results with both $\gamma^2/4\pi$ of the order of 2/3.

A most important result follows if this analysis has any element of truth, since the singlet potential is

$$V(\text{singlet}) \approx -3\ \frac{\gamma_{NN\omega}\,\gamma_{KK\omega}}{4\pi}\ \frac{e^{-m_\omega r}}{r} - 3\ \frac{\gamma_{NN\rho}\,\gamma_{KK\rho}}{4\pi}\ \frac{e^{-m_\rho r}}{r} \quad . \qquad (7.4)$$

A singlet version of Y^* should exist considerably below the energy of Y^* itself. Call it Y_s^*. If it is bound by more than 100 MeV or so, it is metastable and decays primarily into $\Lambda + \gamma$, since $\Lambda + \pi$ is forbidden by charge independence. Thus, Y_s^* is a fake Σ^0, with $I = 0$ and different mass, and may have caused some difficulty in experiments involving the production of Σ^0 at high energy. If, because of level shifts due to absorption, Y_s^* is not very far below Y^*, then it should be detectable in the same way as Y^*; one should observe its decay into $\pi + \Sigma$.

Bound systems like Y^* and Y_s^* should occur not only for $\bar{K}N$ but also for $K\Xi$. (In the limit of unitary symmetry, these come to the same thing.)

The vector coupling constants occur also in several important poles. (For the unstable mesons, these are of course not true poles, unless we perform an analytic continuation of the scattering amplitude onto a second sheet, in which case they become poles at complex energies; they behave almost like true poles, however, when the widths of the vector meson states are small.) There is the pole at $q^2 = -m_M^2$ in the reactions $\pi^- + p \to \Lambda + K^0$ and $\pi^- + p \to \Sigma + K$; a peaking of K in the forward direction has already been observed in some of these reactions and should show up at high energies in all of them. Likewise, the pole at $q^2 = -m_\pi^2$ in the reaction $K + N \to M + N$ should be observable at high energies and its strength can be predicted directly from the width of M. In the reactions $\pi + N \to \Lambda + M$ and $\pi + N \to \Sigma + M$, there is a pole at $q^2 = -m_K^2$ and measurement of its strength can determine the coupling constants $g_{NK\Lambda}^2/4\pi$ and $g_{NK\Sigma}^2/4\pi$ for the K meson.

In πN scattering, we can measure the pole due to exchange of the ρ meson. In KN and $\bar{K}N$ scattering, there are poles from the exchange of ρ and of ω; these can be separated since only the former occurs in the charge-exchange reaction. In NN scattering with charge-exchange, there is a ρ meson pole in addition to the familiar pion pole. Without charge exchange, the situation is terribly complicated, since there are poles from π, ρ, ω, χ, and B.

When the pole term includes a baryon vertex for the emission or absorption of a vector meson, we must remember that there is a "strong magnetic" term analogous to a Pauli moment as well as the renormalized vector meson coupling constant.

In a relatively short time, we should have a considerable body of information about the vector mesons.

VIII Violations of Unitary Symmetry

We have mentioned that within the unitary scheme there is no way that the coupling constants of K to both NΛ and NΣ can both be much smaller than 15, except through large violations of the symmetry. Yet experiments on photoproduction of K particles seem to point to such a situation. Even if unitary symmetry exists as an underlying pattern, whatever mechanism is responsible for the mass differences apparently produces a wide spread among the renormalized coupling constants as well. It is true that the binding of Λ particles in hypernuclei indicates a $\pi\Lambda\Sigma$ coupling of the same order of magnitude as the πNN coupling, but the anomalously small renormalized constants of the K meson indicate that a quantitative check of unitary symmetry will be very difficult.

What about the vector mesons? Let us discuss first the ρ and ω fields, which are coupled to conserved currents. For typical couplings of these fields, we have the relations

$$\gamma^2_{\rho\pi\pi} = \gamma_o^2\, Z_3(\rho) \left[V_\pi{}^\rho(0)\right]^{-2} , \qquad (8.1)$$

$$\gamma^2_{\rho NN} = \gamma_o^2\, Z_3(\rho) \left[V_1{}^\rho(0)\right]^{-2} , \qquad (8.2)$$

$$\gamma^2_{\omega NN} = \gamma_o^2\, Z_3(\omega) \left[V_1{}^\omega(0)\right]^{-2} , \qquad (8.3)$$

etc. Here, each renormalized coupling constant is written as a product of the bare constant, a vacuum polarization renormalization factor, and a squared form factor evaluated at zero momentum transfer. The point is that at zero momentum transfer there is no vertex renormalization because the source currents are conserved. To check, for example, the hypothesis that ρ is really coupled to the isotopic spin current, we must check that γ_o^2 in (8.1) is the same as γ_o^2 in (8.2). We can measure (say, by "pole experiments" and by the width of the π-π resonance) the renormalized constants on the left. The quantities V^2 are of the order unity in any case, and their ratios can be measured by studying electromagnetic form factors.[23)]

The experimental check of "universality" between (8.1) and (8.2) is thus possible, but that tests only the part of the theory already proposed by Sakurai, the coupling of ρ to the isotopic spin current. To test unitary symmetry, we must compare (8.2) and (8.3); but then the ratio $Z_3(\rho)/Z_3(\omega)$ comes in to plague us. We may hope, of course, that this ratio is sufficiently close to unity to make the agreement striking, but we would like a better way of testing unitary symmetry quantitatively.

When we consider the M meson, the situation is worse, since the source current of M is not conserved in the presence of the mass differences. For each coupling of M, there is a vertex renormalization factor that complicates the comparison of coupling strengths.

An interesting possibility arises if the vector charge-exchange weak current is really given in the $|\Delta S| = 1$ case by the current of $F_4 \pm iF_5$ just as it is thought to be given in the $\Delta S = 0$ case by that of $F_1 \pm iF_2$ (the conserved current) and if the $\Delta S = 0$ and $|\Delta S| = 1$ currents are of equal strength, like the $e\nu$ and $\mu\nu$ currents. Then the leptonic $|\Delta S| = 1$ decays show renormalization factors that must be related to the vertex renormalization factors for the M meson, since the source currents are assumed to be the same. The experimental evidence on the decay $K \to \pi$ + leptons then indicates a renormalization factor, in the square of the amplitude, of the order of 1/20. In the decays $\Lambda \to p$ + leptons and $\Sigma^- \to n$ + leptons, both vector and axial vector currents appear to be renormalized by comparable factors.

The width for decay of M into $K + \pi$, if it is really about 15 MeV, indicates that the renormalized coupling constant $\gamma^2_{K\pi M}/4\pi$ is <u>not</u> much smaller than $\gamma^2_{\rho\pi\pi}/4\pi \approx 2/3$ and so there is at present no sign of these small factors in the coupling constants of M. It will be interesting, however, to see what the coupling constant $\gamma^2_{N\Lambda M}/4\pi$ comes out, as determined from the pole in $\pi^- + p \to \Lambda + K^0$.

We have seen that the prospect is rather gloomy for a quantitative test of unitary symmetry, or indeed of any proposed higher symmetry that is broken by mass differences or strong interactions. The best hope seems to lie in the possibility of direct study of the ratios of bare constants in experiments involving very high energies and momentum transfers, much larger than all masses.[24)] However, the theoretical work on this subject is restricted to renormalizable theories. At present, theories of the Yang-Mills type with a mass do not seem to be renormalizable,[25)] and no one knows how to improve the situation.

It is in any case an important challenge to theoreticians to construct a satisfactory theory of vector mesons. It may be useful to remark that the difficulty in Yang-Mills theories is caused by the mass. It is also the mass which spoils the gauge invariance of the first kind. Likewise, as in the μ-e case, it may be the mass that produces the violation of symmetry. Similarly, the nucleon and pion masses break the conservation of any axial vector current in the theory of weak interactions. It may be that a new approach to the rest masses of elementary particles can solve many of our present theoretical problems.

IX Acknowledgments

The author takes great pleasure in thanking Dr. S. L. Glashow and Professor R. P. Feynman for their enthusiastic help and encouragement and for numerous ideas, although they bear none of the blame for any errors or defects in the theory. Conversations with Professor R. Block about Lie algebras have been very enlightening.

TABLE I.

A Set of Matrices λ_i.

$$\lambda_1 = \begin{pmatrix} 0 & 1 & 0 \\ 1 & 0 & 0 \\ 0 & 0 & 0 \end{pmatrix} \qquad \lambda_2 = \begin{pmatrix} 0 & -i & 0 \\ i & 0 & 0 \\ 0 & 0 & 0 \end{pmatrix} \qquad \lambda_3 = \begin{pmatrix} 1 & 0 & 0 \\ 0 & -1 & 0 \\ 0 & 0 & 0 \end{pmatrix}$$

$$\lambda_4 = \begin{pmatrix} 0 & 0 & 1 \\ 0 & 0 & 0 \\ 1 & 0 & 0 \end{pmatrix} \qquad \lambda_5 = \begin{pmatrix} 0 & 0 & -i \\ 0 & 0 & 0 \\ i & 0 & 0 \end{pmatrix} \qquad \lambda_6 = \begin{pmatrix} 0 & 0 & 0 \\ 0 & 0 & 1 \\ 0 & 1 & 0 \end{pmatrix}$$

$$\lambda_7 = \begin{pmatrix} 0 & 0 & 0 \\ 0 & 0 & -i \\ 0 & i & 0 \end{pmatrix} \qquad \lambda_8 = \begin{pmatrix} \frac{1}{\sqrt{3}} & 0 & 0 \\ 0 & \frac{1}{\sqrt{3}} & 0 \\ 0 & 0 & \frac{-2}{\sqrt{3}} \end{pmatrix}$$

TABLE II.

Non-zero elements of f_{ijk} and d_{ijk}. The f_{ijk} are odd under permutations of any two indices while the d_{ijk} are even.

ijk	f_{ijk}	ijk	d_{ijk}
123	1	118	$1/\sqrt{3}$
147	1/2	146	1/2
156	-1/2	157	1/2
246	1/2	228	$1/\sqrt{3}$
257	1/2	247	-1/2
345	1/2	256	1/2
367	-1/2	338	$1/\sqrt{3}$
458	$\sqrt{3}/2$	344	1/2
678	$\sqrt{3}/2$	355	1/2
		366	-1/2
		377	-1/2
		448	$-1/(2\sqrt{3})$
		558	$-1/(2\sqrt{3})$
		668	$-1/(2\sqrt{3})$
		778	$-1/(2\sqrt{3})$
		888	$-1/\sqrt{3}$

TABLE III.

Yukawa interactions of pseudoscalar mesons with baryons, assuming pure coupling through D.

$$L_{int}/ig_o = \pi^o \left\{ \bar{p}\gamma_5 p - \bar{n}\gamma_5 n + \frac{2}{\sqrt{3}}\ \overline{\Sigma^o}\gamma_5\Lambda + \frac{2}{\sqrt{3}}\ \bar{\Lambda}\gamma_5\Sigma^o - \overline{\Xi^o}\gamma_5\Xi^o + \overline{\Xi^-}\gamma_5\Xi^- \right\}$$

$$+ \pi^+ \left\{ \sqrt{2}\ \bar{p}\gamma_5 n + \frac{2}{\sqrt{3}}\ \overline{\Sigma^+}\gamma_5\Lambda + \frac{2}{\sqrt{3}}\ \bar{\Lambda}\gamma_5\Sigma^- - \sqrt{2}\ \overline{\Xi^o}\gamma_5\Xi^- \right\}$$

$$+ \text{h.c.}$$

$$+ K^+ \left\{ -\frac{1}{\sqrt{3}}\ \bar{p}\gamma_5\Lambda + \bar{p}\gamma_5\Sigma^o + \sqrt{2}\ \bar{n}\gamma_5\Sigma^- - \frac{1}{\sqrt{3}}\ \bar{\Lambda}\gamma_5\Xi^- + \overline{\Sigma^o}\gamma_5\Xi^- + \sqrt{2}\ \overline{\Sigma^+}\gamma_5\Xi^o \right\}$$

$$+ \text{h.c.}$$

$$+ K^o \left\{ -\frac{1}{\sqrt{3}}\ \bar{n}\gamma_5\Lambda - \bar{n}\gamma_5\Sigma^o + \sqrt{2}\ \bar{p}\gamma_5\Sigma^+ - \frac{1}{\sqrt{3}}\ \bar{\Lambda}\gamma_5\Xi^o - \overline{\Sigma^o}\gamma_5\Xi^o + \sqrt{2}\ \overline{\Sigma^-}\gamma_5\Xi^- \right\}$$

$$+ \text{h.c.}$$

$$+ \chi^o \left\{ -\frac{1}{\sqrt{3}}\ \bar{p}\gamma_5 p - \frac{1}{\sqrt{3}}\ \bar{n}\gamma_5 n - \frac{2}{\sqrt{3}}\ \bar{\Lambda}\gamma_5\Lambda + \frac{2}{\sqrt{3}}\ \overline{\Sigma^+}\gamma_5\Sigma^+ + \frac{2}{\sqrt{3}}\ \overline{\Sigma^o}\gamma_5\Sigma^o + \frac{2}{\sqrt{3}}\ \overline{\Sigma^-}\gamma_5\Sigma^- - \frac{1}{\sqrt{3}}\ \overline{\Xi^o}\gamma_5\Xi^o - \frac{1}{\sqrt{3}}\ \overline{\Xi^-}\gamma_5\Xi^- \right\}$$

TABLE III (cont.)

Yukawa interactions of pseudoscalar mesons with baryons,

assuming pure coupling through F.

$$L_{int}/ig_o = \pi^o (\bar{p}\gamma_5 p - \bar{n}\gamma_5 n + 2\, \overline{\Sigma^+}\gamma_5\Sigma^+ - 2\, \overline{\Sigma^-}\gamma_5\Sigma^- + \overline{\Xi^o}\gamma_5\Xi^o - \overline{\Xi^-}\gamma_5\Xi^-)$$

$$+ \pi^+ (\sqrt{2}\; \bar{p}\gamma_5 n - \sqrt{2}\; \overline{\Xi^o}\gamma_5\Xi^- - 2\, \overline{\Sigma^+}\gamma_5\Sigma^o + 2\, \overline{\Sigma^o}\gamma_5\Sigma^-)$$

$$+ \text{h.c.}$$

$$+ K^+ (- \sqrt{3}\; \bar{p}\gamma_5\Lambda + \sqrt{3}\; \bar{\Lambda}\gamma_5\Xi^- - \bar{p}\gamma_5\Sigma^o - \sqrt{2}\; \bar{n}\gamma_5\Sigma^- + \overline{\Sigma^o}\gamma_5\Xi^- + \sqrt{2}\; \overline{\Sigma^+}\gamma_5\Xi^o)$$

$$+ \text{h.c.}$$

$$+ K^o (- \sqrt{3}\; \bar{n}\gamma_5\Lambda + \sqrt{3}\; \bar{\Lambda}\gamma_5\Xi^o + \bar{n}\gamma_5\Sigma^o - \sqrt{2}\; \bar{p}\gamma_5\Sigma^+ - \overline{\Sigma^o}\gamma_5\Xi^o + \sqrt{2}\; \overline{\Sigma^-}\gamma_5\Xi^-)$$

$$+ \text{h.c.}$$

$$+ \chi^o (\sqrt{3}\; \bar{p}\gamma_5 p + \sqrt{3}\; \bar{n}\gamma_5 n - \sqrt{3}\; \overline{\Xi^o}\gamma_5\Xi^o - \sqrt{3}\; \overline{\Xi^-}\gamma_5\Xi^-)$$

TABLE IV.

Trilinear couplings of ρ's to π's and N's.

$$
\begin{aligned}
L_{int}/i\gamma_o = M_\alpha^+ \Big\{ & -\sqrt{3}\ \bar{p}\gamma_\alpha\Lambda + \sqrt{3}\ \bar{\Lambda}\gamma_\alpha\Xi^- - \bar{p}\gamma_\alpha\Sigma^o - \sqrt{2}\ \bar{n}\gamma_\alpha\Sigma^- + \overline{\Sigma^o}\gamma_\alpha\Xi^- \\
& + \sqrt{2}\ \overline{\Sigma^+}\gamma_\alpha\Xi^o - \sqrt{3}\ K^-\partial_\alpha\chi^o + \sqrt{3}\ \chi^o\partial_\alpha K^- - K^-\partial_\alpha\pi^o \\
& + \pi^o\partial_\alpha K^- - \sqrt{2}\ \overline{K^o}\partial_\alpha\pi^- + \sqrt{2}\ \pi^-\partial_\alpha\overline{K^o}\Big\}
\end{aligned}
$$

+ h.c.

$$
\begin{aligned}
+ M_\alpha^o \Big\{ & -\sqrt{3}\ \bar{n}\gamma_\alpha\Lambda + \sqrt{3}\ \bar{\Lambda}\gamma_\alpha\Xi^o + \bar{n}\gamma_\alpha\Sigma^o - \sqrt{2}\ \bar{p}\gamma_\alpha\Sigma^+ - \overline{\Sigma^o}\gamma_\alpha\Xi^o \\
& + \sqrt{2}\ \overline{\Sigma^-}\gamma_\alpha\Xi^- - \sqrt{3}\ \overline{K^o}\partial_\alpha\chi^o + \sqrt{3}\ \chi^o\partial_\alpha K^o + \overline{K^o}\partial_\alpha\pi^o \\
& - \pi^o\partial_\alpha\overline{K^o} - \sqrt{2}\ K^-\partial_\alpha\pi^+ + \sqrt{2}\ \pi^+\partial_\alpha K^-\Big\}
\end{aligned}
$$

+ h.c.

$$
\begin{aligned}
+ \rho_\alpha^+ \Big\{ & \sqrt{2}\ \bar{p}\gamma_\alpha n - \sqrt{2}\ \overline{\Xi^o}\gamma_\alpha\Xi^- - 2\ \overline{\Sigma^+}\gamma_\alpha\Sigma^o + 2\ \overline{\Sigma^o}\gamma_\alpha\Sigma^- + \sqrt{2}\ K^-\partial_\alpha K^o \\
& - \sqrt{2}\ K^o\partial_\alpha K^- - 2\ \pi^-\partial_\alpha\pi^o + 2\ \pi^o\partial_\alpha\pi^-\Big\}
\end{aligned}
$$

+ h.c.

$$
\begin{aligned}
+ \rho_\alpha^o \Big\{ & \bar{p}\gamma_\alpha p - \bar{n}\gamma_\alpha n + 2\ \overline{\Sigma^+}\gamma_\alpha\Sigma^+ - 2\ \overline{\Sigma^-}\gamma_\alpha\Sigma^- + \overline{\Xi^o}\gamma_\alpha\Xi^o - \overline{\Xi^-}\gamma_\alpha\Xi^- \\
& + K^-\partial_\alpha K^+ - K^+\partial_\alpha K^- - \overline{K^o}\partial_\alpha K^o + K^o\partial_\alpha\overline{K^o} + 2\ \pi^-\partial_\alpha\pi^+ - 2\ \pi^+\partial_\alpha\pi^-\Big\}
\end{aligned}
$$

$$
\begin{aligned}
+ \omega_\alpha^o \Big\{ & \sqrt{3}\ \bar{p}\gamma_\alpha p + \sqrt{3}\ \bar{n}\gamma_\alpha n - \sqrt{3}\ \Xi^o\gamma_\alpha\Xi^o - \sqrt{3}\ \overline{\Xi^-}\gamma_\alpha\Xi^- + \sqrt{3}\ K^-\partial_\alpha K^+ \\
& - \sqrt{3}\ K^+\partial_\alpha K^- + \sqrt{3}\ \overline{K^o}\partial_\alpha K^o - \sqrt{3}\ K^o\partial_\alpha\overline{K^o}\Big\}
\end{aligned}
$$

TABLE V.

Transformation properties of baryons and mesons, assuming pseudoscalar mesons coupled through D.

$$K^{+} \sim \frac{\mu^{+}\nu + S^{+}\overline{D^{0}}}{\sqrt{2}}$$

$$K^{0} \sim \frac{\mu^{+}e^{-} + S^{+}D^{-}}{\sqrt{2}}$$

$$\pi^{+} \sim \frac{e^{+}\nu + D^{+}\overline{D^{0}}}{\sqrt{2}}$$

$$\pi^{0} \sim \frac{\bar{\nu}\nu - e^{+}e^{-} + D^{0}\overline{D^{0}} - D^{+}D^{-}}{2}$$

$$\pi^{-} \sim \frac{\bar{\nu}e^{-} + D^{0}D^{-}}{\sqrt{2}}$$

$$\chi^{0} \sim \frac{\bar{\nu}\nu + e^{+}e^{-} - 2\mu^{+}\mu^{-} + D^{0}\overline{D^{0}} + D^{+}D^{-} - 2S^{+}S^{-}}{\sqrt{12}}$$

$$\overline{K^{0}} \sim \frac{e^{+}\mu^{-} + D^{+}S^{-}}{\sqrt{2}}$$

$$K^{-} \sim \frac{\bar{\nu}\mu^{-} + D^{0}S^{-}}{\sqrt{2}}$$

$p \sim S^{+}\nu$	$n \sim S^{+}e^{-}$
$\Sigma^{+} \sim D^{+}\nu$	$\Sigma^{0} \sim \frac{D^{0}\nu - D^{+}e^{-}}{\sqrt{2}}$
$\Sigma^{-} \sim D^{0}e^{-}$	$\Lambda \sim \frac{D^{0}\nu + D^{+}e^{-} - 2S^{+}\mu^{-}}{\sqrt{6}}$
$\Xi^{0} \sim D^{+}\mu^{-}$	$\Xi^{-} \sim D^{0}\mu^{-}$

TABLE V (cont.)

$$M^+ \sim \frac{\mu^+\nu - S^+\overline{D^0}}{\sqrt{2}}$$

$$M^0 \sim \frac{\mu^+e^- - S^+D^-}{\sqrt{2}}$$

$$\rho^+ \sim \frac{e^+\nu - D^+\overline{D^0}}{\sqrt{2}}$$

$$\rho^0 \sim \frac{\bar{\nu}\nu - e^+e^- - D^0\overline{D^0} + D^+D^-}{2}$$

$$\rho^- \sim \frac{\bar{\nu}e^- - D^0D^-}{\sqrt{2}}$$

$$\omega^0 \sim \frac{\bar{\nu}\nu + e^+e^- - 2\mu^+\mu^- - D^0\overline{D^0} - D^+D^- + 2S^+S^-}{\sqrt{12}}$$

$$\overline{M^0} \sim \frac{e^+\mu^- - D^+S^-}{\sqrt{2}}$$

$$M^- \sim \frac{\bar{\nu}\mu^- - D^0S^-}{\sqrt{2}}$$

REFERENCES

1. M. Gell-Mann, Phys. Rev. 106, 1296 (1957).

2. J. Schwinger, Ann. Phys. 2, 407 (1957).

3. J. J. Sakurai, Ann. Phys. 11, 1 (1960).

4. J. J. Sakurai, "Vector Theory of Strong Interactions", unpublished.

5. C. N. Yang and R. Mills, Phys. Rev. 96, 191 (1954). Also, R. Shaw, unpublished.

6. After the circulation of the preliminary version of this work (January 1961) the author has learned of a similar theory put forward independently and simultaneously by Y. Ne'eman (Nuclear Phys., to be published). Earlier uses of the 3-dimensional unitary group in connection with the Sakata model are reported by Y. Ohnuki at the 1960 Rochester Conference on High Energy Physics. A. Salam and J. Ward (Nuovo Cimento, to be published) have considered related questions. The author would like to thank Dr. Ne'eman and Professor Salam for communicating their results to him.

7. W. R. Frazer and J. R. Fulco, Phys. Rev. 117, 1609 (1960). See also J. Bowcock, W. N. Cottingham, and D. Lurie, Phys. Rev. Letters 5, 386 (1960).

8. Y. Nambu, Phys. Rev. 106, 1366 (1957).

9. G. F. Chew, Phys. Rev. Letters 4, 142 (1960).

10. M. Alston *et al.*, to be published.

11. E. Teller, Proceedings of the Rochester Conference, 1956.

12. C. N. Yang and T. D. Lee, Phys. Rev. 98, 1501 (1955).

13. S. L. Glashow, Nuclear Phys. 10, 107 (1959).

REFERENCES (cont.)

14. A. Salam and J. C. Ward, Nuovo Cimento 11, 568 (1959).

15. E. Fermi and C. N. Yang, Phys. Rev. 76, 1739 (1949).

16. S. L. Glashow and M. Gell-Mann, to be published.

17. R. P. Feynman and M. Gell-Mann, Phys. Rev. 109, 193 (1958).

18. S. Bludman, Nuovo Cimento 9, 433 (1958).

19. M. Gell-Mann and M. Levy, Nuovo Cimento 16, 705 (1960).

20. M. Gell-Mann, talk at Rochester Conference on High Energy Physics, 1960.

21. Earlier attempts to draw a parallel between leptons and baryons in the weak interactions have been made by A. Gamba, R. E. Marshak, and S. Okubo, Proc. Nat. Acad. Sci. 45, 881 (1959), and Y. Yamaguchi, unpublished. Dr. S. L. Glashow reports that Yamaguchi's scheme has much in common with the one discussed in this paper.

22. M. Alston, L. W. Alvarez, P. Eberhard, M. L. Good, W. Graziano, H. K. Ticho, and S. G. Wojcicki, Phys. Rev. Letters 5, 518 (1960).

23. M. Gell-Mann and F. Zachariasen, "Form Factors and Vector Mesons", to be published.

24. M. Gell-Mann and F. Zachariasen, "Broken Symmetries and Bare Coupling Constants", to be published.

25. Kamefuchi and Umezawa, to be published. Salam and Kumar, to be published.

PHYSICAL REVIEW VOLUME 125, NUMBER 3 FEBRUARY 1, 1962

Symmetries of Baryons and Mesons*

MURRAY GELL-MANN
California Institute of Technology, Pasadena, California

(Received March 27, 1961; revised manuscript received September 20, 1961)

The system of strongly interacting particles is discussed, with electromagnetism, weak interactions, and gravitation considered as perturbations. The electric current j_α, the weak current J_α, and the gravitational tensor $\theta_{\alpha\beta}$ are all well-defined operators, with finite matrix elements obeying dispersion relations. To the extent that the dispersion relations for matrix elements of these operators between the vacuum and other states are highly convergent and dominated by contributions from intermediate one-meson states, we have relations like the Goldberger-Treiman formula and universality principles like that of Sakurai according to which the ρ meson is coupled approximately to the isotopic spin. Homogeneous linear dispersion relations, even without subtractions, do not suffice to fix the scale of these matrix elements; in particular, for the nonconserved currents, the renormalization factors cannot be calculated, and the universality of strength of the weak interactions is undefined. More information than just the dispersion relations must be supplied, for example, by field-theoretic models; we consider, in fact, the equal-time commutation relations of the various parts of j_4 and J_4. These nonlinear relations define an algebraic system (or a group) that underlies the structure of baryons and mesons. It is suggested that the group is in fact $U(3)\times U(3)$, exemplified by the symmetrical Sakata model. The Hamiltonian density θ_{44} is not completely invariant under the group; the noninvariant part transforms according to a particular representation of the group; it is possible that this information also is given correctly by the symmetrical Sakata model. Various exact relations among form factors follow from the algebraic structure. In addition, it may be worthwhile to consider the approximate situation in which the strangeness-changing vector currents are conserved and the Hamiltonian is invariant under $U(3)$; we refer to this limiting case as "unitary symmetry." In the limit, the baryons and mesons form degenerate supermultiplets, which break up into isotopic multiplets when the symmetry-breaking term in the Hamiltonian is "turned on." The mesons are expected to form unitary singlets and octets; each octet breaks up into a triplet, a singlet, and a pair of strange doublets. The known pseudoscalar and vector mesons fit this pattern if there exists also an isotopic singlet pseudoscalar meson χ^0. If we consider unitary symmetry in the abstract rather than in connection with a field theory, then we find, as an attractive alternative to the Sakata model, the scheme of Ne'eman and Gell-Mann, which we call the "eightfold way"; the baryons N, Λ, Σ, and Ξ form an octet, like the vector and pseudoscalar meson octets, in the limit of unitary symmetry. Although the violations of unitary symmetry must be quite large, there is some hope of relating certain violations to others. As an example of the methods advocated, we present a rough calculation of the rate of $K^+\to\mu^++\nu$ in terms of that of $\pi^+\to\mu^++\nu$.

I. INTRODUCTION

In connection with the system of strongly interacting particles, there has been a great deal of discussion of possible approximate symmetries,[1] which would be violated by large effects but still have some physical consequences, such as approximate universality of meson couplings, approximate degeneracy of baryon or meson supermultiplets, and "partial conservation" of currents for the weak interactions.

In this article we shall try to clarify the meaning of such possible symmetries, for both strong and weak interactions. We shall show that a broken symmetry, even though it is badly violated, may give rise to certain exact relations among measurable quantities. Furthermore, we shall suggest a particular symmetry group as the one most likely to underlie the structure of the system of baryons and mesons.

We shall treat the strong interactions without approximation, but consider the electromagnetic, weak, and gravitational interactions only in first order.

The electromagnetic coupling is described by the matrix elements of the electromagnetic current operator $ej_\alpha(x)$. Likewise, the gravitational coupling is specified by the matrix elements of the stress-energy-momentum tensor $\theta_{\alpha\beta}(x)$, particularly the component $\theta_{44}=H$, the Hamiltonian density.

The weak interactions of baryons and mesons with leptons are assumed to be given (ignoring possible nonlocality) by the interaction term[2]

$$GJ_\alpha^\dagger J_\alpha^{(l)}/\sqrt{2}+\text{H.c.}, \tag{1.1}$$

where the leptonic weak current $J_\alpha^{(l)}$ has the form

$$J_\alpha^{(l)}=i\bar{\nu}\gamma_\alpha(1+\gamma_5)e+i\bar{\nu}\gamma_\alpha(1+\gamma_5)\mu. \tag{1.2}$$

We shall refer to $J_\alpha(x)$ as the weak current of baryons and mesons. Its matrix elements specify completely the weak interactions with leptons.

It is possible that the full weak interaction may be given simply by the term

$$G(J_\alpha+J_\alpha^{(l)})^\dagger(J_\alpha+J_\alpha^{(l)})/\sqrt{2}, \tag{1.3}$$

although this form provides no explanation of the approximate rule $|\Delta\mathbf{I}|=\frac{1}{2}$ in the nonleptonic decays of strange particles. If we can find no *dynamical* explanation of the predominance of the $|\Delta\mathbf{I}|=\frac{1}{2}$ amplitude in these decays, we may be forced to assume that in addition to (1.3) there is a weak interaction involving the product

$$GL_\alpha^\dagger L_\alpha/\sqrt{2}, \tag{1.4}$$

of charge-retention currents (presumably not involving leptons); or else we may be compelled to abandon (1.3)

* Research supported in part by U. S. Atomic Energy Commission and Alfred P. Sloan Foundation. A report of this work was presented at the La Jolla Conference on Strong and Weak Interactions, June, 1961.

[1] For example, see the "global symmetry" scheme of M. Gell-Mann, Phys. Rev. **106**, 1296 (1957) and J. Schwinger, Ann. Phys. **2**, 407 (1957).

[2] We use $\hbar=c=1$. The Lorentz index α takes on the values 1, 2, 3, 4. For each value of α, the Dirac matrix γ_α is Hermitian; so is the matrix γ_5.

altogether. In any case, we shall define the weak current J_α by the coupling to leptons.

We shall assume microcausality and hence the validity of dispersion relations for the matrix elements of the various currents and densities. In addition, we shall sometimes require the special assumption of highly convergent dispersion relations.

Our description of the symmetry group for baryons and mesons is most conveniently given in the framework of standard field theory, where the Lagrangian density L of the strong interactions is expressed as a simple function of a certain number of local fields $\psi(x)$, which are supposed to correspond to the "elementary" baryons and mesons. Recently this type of formalism has come under criticism[3]; it is argued that perhaps none of the strongly interacting particles is specially distinguished as "elementary," that the strong interactions can be adequately described by the analyticity properties of the S matrix, and that the apparatus of field theory may be a misleading encumbrance.

Even if the criticism is justified, the field operators $j_\alpha(x)$, $\theta_{\alpha\beta}(x)$, and $J_\alpha(x)$ may still be well defined (by all their matrix elements, including analytic continuations thereof) and measurable in principle by interactions with external electromagnetic or gravitational fields or with lepton pairs. Since the Hamiltonian density H is a component of $\theta_{\alpha\beta}$, it can be a physically sensible quantity.

In order to make our description of the symmetry group independent of the possibly doubtful details of field theory, we shall phrase it ultimately in terms of the properties of the operators H, j_α, and J_α. In introducing the description, however, we shall make use of field-theoretic models. Moreover, in describing the behavior of a particular group, we shall refer extensively to a special example, the symmetrical Sakata model of Ohnuki *et al.*,[4] Yamaguchi,[5] and Wess.[6]

The order of presentation is as follows: We treat first the hypothesis of highly convergent dispersion relations for the matrix elements of currents; and we show that the notion of a meson being coupled "universally" or coupled to a particular current or density means simply that the meson state dominates the dispersion relations for that current or density at low momenta. Next we discuss the universality of strength of the currents themselves; evidently it cannot be derived from homogeneous linear dispersion relations for the matrix elements of the currents. We show that equal-time commutation relations for the currents fulfill this need (or most of it), and that, in a wide class of model field theories, these commutation rules are simple and reflect the existence of a symmetry group, which underlies the structure of the baryon-meson system even though some of the symmetries are badly violated. We present the group properties in an abstract way that does not involve the details of field theory.

Next, it is asked what group is actually involved. The simplest one consistent with known phenomena is the one suggested. It is introduced, for clarity, in connection with a particular field theory, the symmetrical Sakata model, in which baryons and mesons are built up of fundamental objects with the properties of n, p, and Λ. For still greater simplicity, we discuss first the case in which Λ is absent.

We then return to the question of broken symmetry in the strong interactions and show how some of the symmetries in the group, if they are not too badly violated, would reveal themselves in approximately degenerate supermultiplets. In particular, there should be "octets" of mesons, each consisting of an isotopic triplet with $S=0$, a pair of doublets with $S=\pm1$, and a singlet with $S=0$. In the case of pseudoscalar mesons, we know of π, K, and $\bar{K}$; these should be accompanied by a singlet pseudoscalar meson χ^0, which would decay into 2γ, $\pi^++\pi^-+\gamma$, or 4π, depending on its mass.

In Sec. VIII, we propose, as an alternative to the symmetrical Sakata model, another scheme with the same group, which we call the "eightfold way." Here the baryons, as well as mesons, can form octets and singlets, and the baryons N, Λ, Σ, and Ξ are supposed to constitute an approximately degenerate octet.

In Sec. IX, some topics are suggested for further investigation, including the possibility of high energy limits in which non-conserved quantities become conserved, and we give, as an example of methods suggested here, an approximate calculation of the rate of $K^+\rightarrow\mu^++\nu$ decay from that of $\pi^+\rightarrow\mu^++\nu$ decay.

II. MESONS AND CURRENTS

To introduce the connection between meson states and currents or densities, let us review the derivation[7] of the Goldberger-Treiman relation[8] among the charged pion decay amplitude, the strength of the axial vector weak interaction in the β decay of the nucleon, and the pion-nucleon coupling constant.

The axial vector term in J_α with $\Delta S=0$, $|\Delta\mathbf{I}|=1$, $GP=-1$, can be written as $P_{1\alpha}+iP_{2\alpha}$, where $\mathbf{P}_\alpha$ is an axial vector current that transforms like an isotopic vector. We have, for nucleon β decay,

$$\langle N|\mathbf{P}_\alpha|N\rangle=\bar{u}_f[i\gamma_\alpha F_{\mathrm{ax}}(s)+k_\alpha\beta(s)]\gamma_5(\boldsymbol{\tau}/2)u_i, \quad (2.1)$$

where u_i and u_f are the initial and final spinors, k_α is the four-momentum transfer, and $s=-k^2=-k_\alpha k_\alpha$. At

[3] G. F. Chew, Talk at La Jolla Conference on Strong and Weak Interactions, June, 1961 (unpublished).

[4] M. Ikeda, S. Ogawa, and Y. Ohnuki, Progr. Theoret. Phys. (Kyoto) **22**, 715 (1959); Y. Ohnuki, *Proceedings of the 1960 Annual International Conference on High-Energy Physics at Rochester* (Interscience Publishers, Inc., New York, 1960).

[5] Y. Yamaguchi, Progr. Theoret. Phys. (Kyoto) Suppl. No. 11, 1 (1959).

[6] J. Wess, Nuovo cimento **10**, 15 (1960).

[7] J. Bernstein, S. Fubini, M. Gell-Mann, and W. Thirring, Nuovo cimento **17**, 757 (1960). See also Y. Nambu, Phys. Rev. Letters **4**, 380 (1960); and Chou Kuang-Chao, Soviet Phys.—JETP **12**, 492 (1961).

[8] M. Goldberger and S. Treiman, Phys. Rev. **110**, 1478 (1958).

$s=0$ we have just

$$F_{\alpha x}(0)=-G_A/G, \quad (2.2)$$

the axial vector renormalization constant.

The axial vector current is not conserved; its divergence $\partial_\alpha \mathbf{P}_\alpha$ has the same quantum numbers as the pion ($J=0^-$, $I=1$). Between nucleon states we have

$$\langle N|\partial_\alpha \mathbf{P}_\alpha|N\rangle=\bar{u}_f i\gamma_5(\boldsymbol{\tau}/2)u_i[2m_N F_{\alpha x}(s)+s\beta(s)]. \quad (2.3)$$

We may compare this matrix element with that between the vacuum and a one-pion state

$$\langle 0|\partial_\alpha \mathbf{P}_\alpha|\pi\rangle=m_\pi^2(2f_\pi)^{-1}\boldsymbol{\phi}, \quad (2.4)$$

where $\boldsymbol{\phi}$ is the pion wave function and the constant f_π (or at least its square) may be measured by the rate of $\pi^+\rightarrow\mu^++\nu$:

$$\Gamma_\pi=G^2 m_\pi m_\mu^2(1-m_\mu^2/m_\pi^2)^2(f_\pi^2/4\pi)^{-1}(64\pi^2)^{-1}. \quad (2.5)$$

It is known that the matrix element (2.3) has a pole at $s=m_\pi^2$ corresponding to the virtual emission of a pion that undergoes leptonic decay. The strength of the pole is given by the product of m_π^2/f_π and the pion-nucleon coupling constant $g_{NN\pi}$. If we assume that the expression in brackets vanishes at large s, we have an unsubtracted dispersion relation for it consisting of the pole term and a branch line beginning at $(3m_\pi)^2$, the next lowest mass that can be virtually emitted:

$$2m_N F_{\alpha x}(s)+s\beta(s)=(g_{NN\pi}/f_\pi)m_\pi^2(m_\pi^2-s)^{-1}$$
$$+\int\sigma_{\alpha x}(M^2)M^2dM^2\,(M^2-s-i\epsilon)^{-1}. \quad (2.6)$$

At $s=0$, we have, using (2.2), the sum rule

$$2m_N(-G_A/G)=g_{NN\pi}/f_\pi+\int\sigma_{\alpha x}(M^2)dM^2. \quad (2.7)$$

Now if the dispersion relation (2.6) is not only convergent but dominated at low s by the term with the lowest mass, then we have the approximate Goldberger-Treiman relation

$$2m_N(-G_A/G)\approx g_{NN\pi}/f_\pi, \quad (2.8)$$

which agrees with experiment to within a few percent.

The success of the relation suggests that other matrix elements of $\partial_\alpha \mathbf{P}_\alpha$ may also obey unsubtracted dispersion relations dominated at low s by the one-pion term. For example, if we consider the matrix element between Λ and Σ, we should arrive at the relation

$$(m_\Lambda+m_\Sigma)(-G_A^{\Lambda\Sigma}/G)\approx g_{\Lambda\Sigma\pi}/f_\pi, \quad (2.9)$$

if Λ and Σ have the same parity, or an analogous relation if they have opposite parity.

If such a situation actually obtains, then it may be said that the pion is, to a good approximation, coupled "universally" to the divergence of the axial vector current. To calculate any g approximately, we multiply the universal constant f_π, the sum of the initial and final masses, and the renormalization factor for the axial vector current.

Now let us turn to the case of a current that is conserved, say the isotopic spin current $\mathfrak{J}_\alpha$ with quantum numbers $J=1^-$, $I=1$. Acting on the vacuum, the operator $\mathfrak{J}_\alpha$ does not lead to any stable one-meson state, but it does lead to the unstable vector meson state ρ at around 750 Mev, which decays into 2π or 4π. For simplicity, let us ignore the rather large width ($\Gamma_\rho\sim 100$ Mev) of the ρ state and treat it as stable. The mathematical complications resulting from the instability are not severe and have been discussed elsewhere.[9,10]

In place of (2.4), then, we have the definition

$$\langle 0|\mathfrak{J}_\alpha|\rho\rangle=m_\rho^2(2\gamma_\rho)^{-1}\boldsymbol{\phi}_\alpha, \quad (2.10)$$

of the constant γ_ρ, where $\boldsymbol{\phi}_\alpha$ is the wave function of the ρ meson. In place of (2.1) or (2.3), we consider the matrix element between nucleon states of the isotopic spin current:

$$\langle N|\mathfrak{J}_\alpha|N\rangle=\bar{u}_f i\gamma_\alpha(\boldsymbol{\tau}/2)u_i F_1^V(s)+\text{magnetic term}, \quad (2.11)$$

where $F_1^V(s)$ is the familiar isovector form factor of the electric charge of the nucleon, since the electromagnetic current has the form

$$j_\alpha=\mathfrak{J}_{3\alpha}+\text{isoscalar term}. \quad (2.12)$$

If we continue to ignore the width of ρ, we get a dispersion relation like (2.6) with a pole term at m_ρ^2:

$$F_1^V(s)=(\gamma_{NN\rho}/\gamma_\rho)m_\rho^2(m_\rho^2-s)^{-1}$$
$$+\int\sigma_1^V(M^2)dM^2\,M^2(M^2-s-i\epsilon)^{-1}. \quad (2.13)$$

Here $\gamma_{NN\rho}$ is the coupling constant of ρ to $\bar{u}_f i\boldsymbol{\tau}\gamma_\alpha u_i$, just as $g_{NN\pi}$ is the coupling constant of π to $\bar{u}_f i\boldsymbol{\tau}\gamma_5 u_i$. In this case, we have used an unsubtracted dispersion relation just for convenience.

Since the current is conserved, there is no renormalization and we have

$$F_1^V(0)=1, \quad (2.14)$$

giving, in place of (2.7), the sum rule

$$1=\gamma_{\rho NN}/\gamma_\rho+\int\sigma_1^V(M^2)dM^2. \quad (2.15)$$

If the dispersion relation is dominated at low s by the ρ term, then we obtain the analog of the Goldberger-Treiman formula:

$$1\approx\gamma_{\rho NN}/\gamma_\rho. \quad (2.16)$$

[9] G. F. Chew, University of California Radiation Laboratory Report No. UCRL-9289, 1960 (unpublished).
[10] M. Gell-Mann and F. Zachariasen, Phys. Rev. 124, 953 (1961).

Now the same reasoning may be applied to the isovector electric form factor of another particle, for example the pion:

$$\langle\pi|\mathfrak{J}_\alpha|\pi\rangle=[i\phi_f{}^*\times\partial_\alpha\phi_i-i\partial_\alpha\phi_f{}^*\times\phi_i]F_\pi(s), \quad (2.17)$$

$$F_\pi(s)=(\gamma_{\rho\pi\pi}/\gamma_\rho)m_\rho{}^2(m_\rho{}^2-s)^{-1}$$
$$+\int\sigma_\pi(M^2)dM^2\,M^2(M^2-s-i\epsilon)^{-1}, \quad (2.18)$$

and

$$1=\gamma_{\rho\pi\pi}/\gamma_\rho+\int\sigma_\pi(M^2)dM^2. \quad (2.19)$$

If this dispersion relation, too, is dominated by the ρ pole at low s, then we find

$$1\approx\gamma_{\rho\pi\pi}/\gamma_\rho. \quad (2.20)$$

To the extent that the ρ pole gives most of the sum rule in each case, we have ρ coupled *universally* to the isotopic spins of nucleon, pion, etc., with coupling parameter $2\gamma_\rho$. Such universality was postulated by Sakurai,[11] within the framework of a special theory, in which ρ is treated as an elementary vector meson described by a Yang-Mills field. It can be seen that whether or not such a field description is correct, the *effective* universality ($\gamma_{\rho\pi\pi}\approx\gamma_{\rho NN}\approx\gamma_{\rho KK}$, etc.) is an approximate rule the validity of which depends on the domination of (2.15), (2.19), etc., by the ρ term.

The various coupling parameters $\gamma_{\rho\pi\pi}$, $\gamma_{\rho NN}$, etc., can be determined from the contribution of the ρ "pole" to various scattering processes, for example $\pi+N\rightarrow\pi+N$. But the factors $\gamma_{\rho\pi\pi}/\gamma_\rho$, $\gamma_{\rho NN}/\gamma_\rho$, etc., can also be measured, using electromagnetic interactions.[10]

An approximate determination of $\gamma_{\rho NN}/\gamma_\rho$ was made by Hofstadter and Herman[12] as follows The masses M^2 in the integral in Eq. (2.13) are taken to be effectively vary large, so that (2.13) becomes approximately

$$F_1{}^V(s)\approx(\gamma_{NN\rho}/\gamma_\rho)m_\rho{}^2(m_\rho{}^2-s)^{-1}$$
$$+1-(\gamma_{\rho NN}/\gamma_\rho). \quad (2.21)$$

Fitting the experimental data on $F_1{}^V(s)$ with such a formula and using $m_\rho\approx750$ Mev, we obtain $\gamma_{\rho NN}/\gamma_\rho\approx1.4$. (Hofstadter and Herman, with a smaller value of m_ρ, found 1.2.)

III. EQUAL-TIME COMMUTATION RELATIONS

The dispersion relations for the matrix elements of weak or electromagnetic currents are linear and homogeneous. For example, Eq. (2.6) may be thought of as an expression for the matrix element of $\mathbf{P}_\alpha$ between the vacuum and a nucleon-antinucleon pair state. On the right-hand side, the pole term contains the product of the matrix element of $\mathbf{P}_\alpha$ between the vacuum and a one-pion state multiplied by the transition amplitude for the transition from π to $N\bar{N}$ by means of the strong interactions. The weight function $\sigma_{ax}(M^2)$ is just the sum of such products over many intermediate states (such as 3π, 5π, etc.) with total mass M.

Now such linear, homogeneous equations may determine the dependence of the current matrix elements on variables such as s, but they cannot fix the scale of these matrix elements; constants like $-G_A/G$ cannot be calculated without further information. A field theory of the strong interactions, with explicit expressions for the currents, somehow contains more than these dispersion relations. In what follows, we shall extract some of this additional information in the form of equal-time commutation relations between components of the currents. Since these are nonlinear relations, they can help to fix the scale of each matrix element. Moreover, these relations may be the same for the lepton system and for the baryon-meson system, so that universality of strength of the weak interactions, for example, becomes meaningful.[13]

Let us begin our discussion of equal-time commutation relations with a familiar case—that of the isotopic spin $\mathbf{I}$. Its components I_i obey the well-known commutation relations

$$[I_i,I_j]=ie_{ijk}I_k. \quad (3.1)$$

In terms of the components $\mathfrak{J}_{i\alpha}$ of the isotopic spin current, we have

$$I_i=-i\int\mathfrak{J}_{i4}d^3x, \quad (3.2)$$

and the conservation law

$$\partial_\alpha\mathfrak{J}_{i\alpha}=0 \quad (3.3)$$

tells us that

$$\dot{I}_i=\int\partial_\alpha\mathfrak{J}_{i\alpha}d^3x=0, \quad (3.4)$$

at all times.

Now the commutator of $\mathfrak{J}_{i4}(\mathbf{x},t)$ and $\mathfrak{J}_{j4}(\mathbf{x}',t)$ must vanish for $\mathbf{x}\neq\mathbf{x}'$, in accorance with microcausality. (Note we have taken the times equal.) If the commutator is not more singular than a delta function, then (3.1) and (3.2) give us the relation

$$[\mathfrak{J}_{i4}(\mathbf{x},t),\mathfrak{J}_{j4}(\mathbf{x}',t)]=-ie_{ijk}\mathfrak{J}_{k4}(\mathbf{x},t)\delta(\mathbf{x}-\mathbf{x}'), \quad (3.5)$$

which can also be obtained in any simple field theory by explicit commutation.[14]

In discussing the various parts of the weak current J_α, we shall have to deal with currents like $\mathbf{P}_\alpha$ that are not

---

[11] J. J. Sakurai, Ann. Phys. **11**, 1 (1960).

[12] R. Hofstadter and R. Herman, Phys. Rev. Letters **6**, 293 (1961). See also S. Bergia, A. Stanghellini, S. Fubini, and C. Villi, Phys. Rev. Letters **6**, 367 (1961).

[13] M. Gell-Mann, *Proceedings of the 1960 Annual International Conference on High-Energy Physics at Rochester* (Interscience Publishers, Inc., New York, 1960).

[14] In some cases explicit commutation may be ambiguous and misleading. For example, a superficial consideration of $[j_i(\mathbf{x},t), j_4(\mathbf{x}',t)]$ for $i=1, 2, 3$ may lead to the conclusion that the expression vanishes. Yet the vacuum expectation value of the commutator can be shown to be a nonzero quantity times $\partial_i\delta(\mathbf{x}-\mathbf{x}')$, and that result is confirmed by more careful calculation. See J. Schwinger, Phys. Rev. Letters **3**, 296 (1959).

conserved.[15] Here, too, we may define a quantity analogous to **I**:

$$D_i = -i\int P_{i4}d^3x, \tag{3.6}$$

but D_i is *not* independent of time:

$$\dot{D}_i = \int \partial_\alpha P_{i\alpha}d^3x \neq 0. \tag{3.7}$$

For the moment, let us restrict our attention to the currents $\mathfrak{J}_\alpha$ and $\mathbf{P}_\alpha$ and the operators **I** and $\mathbf{D}(t)$. Since **D** is an isovector, we have the relations

$$[I_i,D_j]=[D_i,I_j]=ie_{ijk}D_k, \tag{3.8}$$

but what is the commutator of two components of **D**? Since $\mathbf{P}_\alpha$ is a physical quantity, so is **D** and the question is one with direct physical meaning. We shall give both a general and a specific answer.

In general, we may take the commutators of D's (divided by i), the components of **I** and **D**, the commutators of all of these with one another (divided by i), etc., until we obtain a system of Hermitian operators that is closed under commutation. Any of these operators can be written as a linear combination of N linearly independent Hermitian operators $R_i(t)$, where N might be infinite, and where the commutator of any two R_i is a linear combination of the R_i:

$$[R_i(t),R_j(t)]=ic_{ijk}R_k(t), \tag{3.9}$$

with c_{ijk} real. Such a system is called an algebra by the mathematicians. If we consider the set of infinitesimal unitary operators $1+i\epsilon R_i(t)$ and all possible products of these, we obtain an N-parameter continuous group of unitary transformations. We can refer to (3.9) as the algebra of the group. It is a physically meaningful statement to specify what group or what algebra is generated in this way by the currents $\mathfrak{J}_\alpha$ and $\mathbf{P}_\alpha$. Since a commutation relation like (3.9) is left invariant by a unitary transformation such as $\exp(-it\int Hd^3x)$, the numbers c_{ijk} are independent of time.

A second mathematical statement is also in order, i.e., the specification of the transformation properties of the Hamiltonian density $H(\mathbf{x},t)$ under the group or the algebra. Those R_i for which $[R_i(t),H(\mathbf{x},t)]=0$ are independent of time, but some of them, like D_i, do not commute with H. If all of the R_i commuted with H, then H would belong to the trivial one-dimensional representation of the group. In fact, H behaves in a more complicated way. By commuting all of the $R_i(t)$ with $H(\mathbf{x},t)$, we obtain a linear set of operators, containing H, that form a representation of the group; it may be broken up into the direct sum of irreducible representations. We want to know, then, what group is generated by **I** and **D** and to what irreducible representations of this group H belongs. Suggested are specific answers to both questions.

Let us look at the vector and axial vector weak currents for the leptons. For the time being, we shall consider only ν and e, ignoring the muon. (In the same way, we shall, in this section, ignore strange particles, and consider only baryons and mesons with $S=0$.) The vector weak current $i\bar{\nu}\gamma_\alpha e$ and the axial current $i\bar{\nu}\gamma_\alpha\gamma_5 e$ can be regarded formally as components of two "isotopic vector" currents for the leptons:

$$\mathfrak{J}_\alpha^{(l)}=i\bar{\xi}\boldsymbol{\tau}\gamma_\alpha\xi/2, \quad \mathbf{P}_\alpha^{(l)}=i\bar{\xi}\boldsymbol{\tau}\gamma_\alpha\gamma_5\xi/2, \tag{3.10}$$

where ξ stands for (ν,e). We can also form the mathematical analogs of **I** and **D**:

$$\mathbf{I}^{(l)}=-i\int\mathfrak{J}_4^{(l)}d^3x, \quad \mathbf{D}^{(l)}=-i\int\mathbf{P}_4^{(l)}d^3x. \tag{3.11}$$

Now in this leptonic case we can easily compute the commutation rules of $\mathbf{I}^{(l)}$ and $\mathbf{D}^{(l)}$:

$$\begin{aligned}
&[I_i^{(l)},I_j^{(l)}]=ie_{ijk}I_k^{(l)},\\
&[I_i^{(l)},D_j^{(l)}]=[D_i^{(l)},I_j^{(l)}]=ie_{ijk}D_k^{(l)},\\
&[D_i^{(l)},D_j^{(l)}]=ie_{ijk}I_k^{(l)}.
\end{aligned} \tag{3.12}$$

Another way to phrase these commutation rules is to put

$$\begin{aligned}
\mathbf{I}^{(l)}&=\mathbf{L}_+^{(l)}+\mathbf{L}_-^{(l)},\\
\mathbf{D}^{(l)}&=\mathbf{L}_+^{(l)}-\mathbf{L}_-^{(l)},
\end{aligned} \tag{3.13}$$

and to notice that $\mathbf{L}_+^{(l)}$ and $\mathbf{L}_-^{(l)}$ are two commuting angular momenta [essentially $\boldsymbol{\tau}(1+\gamma_5)/4$ and $\boldsymbol{\tau}(1-\gamma_5)/4$]. The weak current $i\bar{\nu}\gamma_\alpha(1+\gamma_5)e$ is just a component of the current of $\mathbf{L}_+^{(l)}$.

We now suggest that the algebraic structure of **I** and **D** is exactly the same in the case of baryons and mesons. To (3.1) and (3.8), we add the rule[16,17]

$$[D_i,D_j]=ie_{ijk}I_k, \tag{3.14}$$

which closes the system and makes $\mathbf{I}^+\equiv(\mathbf{I}+\mathbf{D})/2$ and $\mathbf{I}^-\equiv(\mathbf{I}-\mathbf{D})/2$ two commuting angular momenta. Again, we make the weak current a component of the current of $\mathbf{I}^+$. Evidently the statement that $(\mathbf{I}+\mathbf{D})/2$ is an angular momentum and not some factor times an angular momentum, fixes the scale of the weak current. It makes universality of strength between baryons and leptons meaningful, and it specifies, together with the dispersion relations, the value of such constants as $-G_A/G$.

The simplest way to realize the algebraic structure under discussion in a field-theory model of baryons and mesons is to construct the currents $\mathfrak{J}_\alpha$ and $\mathbf{P}_\alpha$ out of p and n fields just as $\mathfrak{J}_\alpha^{(l)}$ and $\mathbf{P}_\alpha^{(l)}$ are made out of ν and e fields:

$$\mathfrak{J}_\alpha=i\bar{N}\boldsymbol{\tau}\gamma_\alpha N/2, \quad \mathbf{P}_\alpha=i\bar{N}\boldsymbol{\tau}\gamma_\alpha\gamma_5 N/2, \tag{3.15}$$

[15] We assume that the vector weak current with $\Delta S=0$ is just a component of the isotopic spin current $\mathfrak{J}_\alpha$ and thus conserved.

[16] F. Gursey, Nuovo cimento **16**, 230 (1960).

[17] M. Gell-Mann and M. Lévy, Nuovo cimento **16**, 705 (1960).

where N means (p,n). We then obtain not only the commutation rules (3.1), (3.8), and (3.14), but the stronger rule (3.5) and its analogs:

$$[\mathfrak{J}_{i4}(\mathbf{x},t),P_{j4}(\mathbf{x}',t)]=-ie_{ijk}P_{k4}(\mathbf{x},t)\delta(\mathbf{x}-\mathbf{x}'),$$
$$[P_{i4}(\mathbf{x},t),P_{j4}(\mathbf{x}',t)]=-ie_{ijk}\mathfrak{J}_{k4}(\mathbf{x},t)\delta(\mathbf{x}-\mathbf{x}'). \quad (3.16)$$

Next we want to use a field-theory model to suggest an answer to the second question—how H behaves under the group or, what is the same thing, under the algebra consisting of $\mathbf{I}$ and $\mathbf{D}$ or of $\mathbf{I}^+$ and $\mathbf{I}^-$. Since $\mathbf{I}^+$ and $\mathbf{I}^-$ are two commuting angular momenta, any irreducible representation of the algebra is specified by a pair of total angular momentum quantum numbers: i_+ for $\mathbf{I}^+$ and i_- for $\mathbf{I}^-$. The total isotopic spin quantum number I is associated with $\mathbf{I}^++\mathbf{I}^-=\mathbf{I}$.

Now we want the vector weak current $\mathfrak{J}_\alpha$ to be the isotopic spin current and to be conserved. Thus H must commute with $\mathbf{I}$; it transforms as an isoscalar, with $I=0$. In order to couple to zero, i_+ and i_- must be equal. So H can consist of terms with $(i_+,i_-)=(0,0)$, $(\frac{1}{2},\frac{1}{2})$, $(1,1)$, $(\frac{3}{2},\frac{3}{2})$, etc. Which of these are in fact present?

The simplest model in which the total isotopic current is given by just (3.15) is the Fermi-Yang[18] model, in which the pion is a composite of nucleon and antinucleon. To write an explicit Lagrangian, it must be decided what form the binding interaction takes. Since a direct four-fermion coupling leads to unpleasant singularities, we whall use a massive neutral vector meson field B^0 coupled to the nucleon current, as proposed by Teller[18] and Sakurai[11]; the exchange of a B^0 gives attraction between nucleon and antinucleon, permitting binding, and it also gives repulsion between nucleons, contributing to the "hard core." The model Lagrangian is then[19]

$$L=-\bar{N}\gamma_\alpha\partial_\alpha N-(\partial_\alpha B_\beta-\partial_\beta B_\alpha)^2/4$$
$$-\mu_0^2B_\alpha B_\alpha/2-ih_0B_\alpha\bar{N}\gamma_\alpha N-m_0\bar{N}N. \quad (3.17)$$

If the mass term for the nucleon were absent, then both $\mathfrak{J}_\alpha$ and $\mathbf{P}_\alpha$ would be conserved; $\mathbf{I}$ and $\mathbf{D}$ would both commute with L and with H. Thus,

$$H=H(0,0)-u_0, \quad (3.18)$$

where $H(0,0)$ transforms according to $(i_+,i_-)=(0,0)$ and the noninvariant term u_0 is just $-m_0\bar{N}N$. To what representation does it belong?

It is easy to see that the field B^0 belongs to $(0,0)$, while $N_L\equiv(1+\gamma_5)N/2$ belongs to $(\frac{1}{2},0)$ and $N_R\equiv(1-\gamma_5)N/2$ belongs to $(0,\frac{1}{2})$. One can thus verify that all terms of (3.17) except the last belong to $(0,0)$, since $\bar{N}\gamma_\alpha N$ or $\bar{N}\gamma_\alpha\partial_\alpha N$ couples $\bar{N}_L$ to N_L and $\bar{N}_R$ to N_R. But the Dirac matrix β, unlike $\beta\gamma_\alpha$, anticommutes with γ_5, so that the last term $-m_0\bar{N}N$ couples $\bar{N}_L$ to N_R and $\bar{N}_R$ to N_L. Thus u_0 belongs to $(\frac{1}{2},\frac{1}{2})$. We have $H=H(0,0)+H(\frac{1}{2},\frac{1}{2})$.

There are four components to the representation $(\frac{1}{2},\frac{1}{2})$ to which $u_0=-H(\frac{1}{2},\frac{1}{2})$ belongs. By commuting $\mathbf{D}$ with u_0, we generate the other three easily and see that they are proportional to $-i\bar{N}\boldsymbol{\tau}\gamma_5N$. In fact $\mathbf{D}$ acts like $\boldsymbol{\tau}\gamma_5/2$, $\mathbf{I}$ like $\boldsymbol{\tau}/2$, u_0 like β, and the other three components like $-i\beta\gamma_5\boldsymbol{\tau}$. Denoting the three new components by v_i, we have

$$[I_i,u_0]=0, \quad [D_i,u_0]=-iv_i,$$
$$[I_i,v_j]=ie_{ijk}v_k, \quad [D_i,v_j]=i\delta_{ij}u_0. \quad (3.19)$$

In the model, there are the even stronger relations for the densities

$$[\mathfrak{J}_{i4}(\mathbf{x},t),u_0(\mathbf{x}',t)]=0, \quad [P_{i4}(\mathbf{x},t),u_0(\mathbf{x}',t)]$$
$$=-iv_i(\mathbf{x},t)\delta(\mathbf{x}-\mathbf{x}'), \text{ etc.} \quad (3.20)$$

The noninvariant term u_0 is what prevents the axial vector current from being conserved. Thus one can express the divergence $\partial_\alpha\mathbf{P}_\alpha$ of the current in terms of the commutator of $\mathbf{D}$ with u_0. The conditions for this relation to hold are treated in the appendix and are applicable to all models we discuss. We find simply

$$\partial_\alpha\mathbf{P}_\alpha=-i[\mathbf{D},H]=i[\mathbf{D},u_0]=\mathbf{v}, \quad (3.21)$$

and, of course,

$$\partial_\alpha\mathfrak{J}_\alpha=-i[\mathbf{I},H]=0. \quad (3.22)$$

It is precisely the operator $\mathbf{v}$, then, that we used in a dispersion relation in order to obtain the Goldberger-Treiman relation in Sec. II. Acting on the vacuum, it leads mostly to the one-pion state, so that the pion is effectively coupled universally to the divergence of the axial vector current. Thus $\mathbf{v}$ is a sort of effective pion field operator for the Fermi-Yang theory, which has no explicit pion field.

If we insist on a model in which there is a field variable $\boldsymbol{\pi}(x,t)$ then we must complicate the discussion. The total isotopic spin current is no longer given by just (3.15); there is a pion isotopic current term as well. In order to preserve the same algebraic structure of $\mathbf{I}$ and $\mathbf{D}$, one must then modify $\mathbf{P}_\alpha$ as well. Such a theory was described by Gell-Mann and Lévy,[17] who called it the "σ-model".[20] Along with the field $\boldsymbol{\pi}$, we must introduce a scalar, isoscalar field σ' in such a way that $\boldsymbol{\pi}$, σ' transform under the group like $\mathbf{v}$, u_0. Then, just as $\mathfrak{J}_\alpha$ has an additional term quadratic in $\boldsymbol{\pi}$, $\mathbf{P}_\alpha$ requires an additional term bilinear in $\boldsymbol{\pi}$ and σ'.

As we shall see in the next section, the introduction of

[18] E. Fermi and C. N. Yang, Phys. Rev. 76, 1739 (1949); E. Teller, *Proceedings of the Sixth Annual Rochester Conference on High-Energy Nuclear Physics, 1956* (Interscience Publishers, Inc., New York, 1956).

[19] Conceivably a massive B^0 meson can be described by (3.17) even with $\mu_0=0$. [J. Schwinger, lectures at Stanford University, summer, 1961 (unpublished)]. In that case the noninvariant term in (3.17) is just equal to $\theta_{\alpha\alpha}$ and the traceless part of $\theta_{\alpha\beta}$ commutes with the group elements at equal times. In any case, whether μ_0 is zero or not, the off-diagonal terms in $\theta_{\alpha\beta}$ commute with the group.

[20] In the σ model, explicit commutation of u_0 and $\mathbf{v}$ at equal times gives zero, while in the Fermi-Yang model this is not so; if we take these results seriously, they give us definite physical distinctions among models.

strange particles makes the group much larger. The term u_0 is then a member of a much larger representation, with eighteen components. Thus if a pion field is introduced, fifteen more components are needed as well. Such a theory is too complicated to be attractive; we shall therefore ignore it and concentrate on the simplest generalization of the Fermi-Yang model to strange particles, namely the symmetrical Sakata model.

IV. SYMMETRICAL SAKATA MODEL AND UNITARY SYMMETRY

In the previous section, we proceeded inductively. We showed that starting from physical currents like $\mathfrak{J}_\alpha$ and $\mathbf{P}_\alpha$ we may construct a group and its algebra and that it is physically meaningful to specify the group and also the transformation properties of H under the group. We chose the algebraic structure by analogy with the case of leptons and we saw that the simplest field theory model embodying the structure is just the Fermi-Yang model, in which p and n fields are treated just like the ν and e fields for the leptons, except that they are given a mass and a strong "gluon" coupling. The transformation properties of H were taken from the model; H consists, then, of an invariant part $H_{0,0}$ plus a term $(-u_0)$, where u_0 and a pseudoscalar isovector quantity $\mathbf{v}$ belong to the representation $(\frac{1}{2},\frac{1}{2})$ of the group. We then have the commutation rules (3.1), (3.8), (3.14), and (3.19). Microcausality with the assumption of commutators that are not too singular, or else direct inspection of the model, gives the stronger commutation rules (3.5), (3.16), and (3.20) for the densities. The model also gives specific equal-time commutation rules for u_0 and $\mathbf{v}$, which we did not list. All of these properties can be abstracted from the model and considered on their own merits as proposed relations among the currents and the Hamiltonian density.

Now, to argue deductively, we want to include the strange particles and all parts of the weak current J_α and the electromagnetic current j_α. We generalize the Fermi-Yang description to obtain the symmetrical Sakata model and abstract from it as many physically meaningful relations as possible.

It has long been recognized that the qualitative properties of baryons and mesons could be understood in terms of the Sakata model,[21] in which all strongly interacting particles are made out of N, Λ, $\bar{N}$, and $\bar{\Lambda}$ (or at least out of basic fields with the same quantum numbers as these particles).

We write the Lagrangian density for the Sakata model as a generalization of (3.17):

$$\begin{aligned}L=&-\bar{p}\gamma_\alpha p-\bar{n}\gamma_\alpha\partial_\alpha\bar{n}-\bar{\Lambda}\gamma_\alpha\partial_\alpha\Lambda-\tfrac{1}{4}(\partial_\alpha B_\beta-\partial_\beta B_\alpha)^2\\&-\tfrac{1}{2}\mu_0^2B_\alpha B_\alpha-ih_0(\bar{p}\gamma_\alpha p+\bar{n}\gamma_\alpha n+\bar{\Lambda}\gamma_\alpha\Lambda)B_\alpha\\&-m_{0N}(\bar{n}n+\bar{p}p)-m_{0\Lambda}\bar{\Lambda}\Lambda.\end{aligned}\quad(4.1)$$

According to this picture, the baryons present a striking parallel with the leptons,[22] for which we write the Lagrangian density

$$L_l=-\bar{\nu}\gamma_\alpha\nu-\bar{e}\gamma_\alpha\partial_\alpha e-\bar{\mu}\gamma_\alpha\partial_\alpha\mu-0\cdot(\bar{\nu}\nu+\bar{e}e)-m_\mu\bar{\mu}\mu,\quad(4.2)$$

if we turn off the electromagnetic and weak couplings, along with the ν-e mass difference. Here it is assumed there is only one kind of neutrino.

The only real difference between baryons and leptons in (4.1) and (4.2), respectively, is that the baryons are coupled, through the baryon current, to the field B. It is tempting to suppose that the weak current of the strongly interacting particles is just the expression.

$$i\bar{p}\gamma_\alpha(1+\gamma_5)n+i\bar{p}\gamma_\alpha(1+\gamma_5)\Lambda,\quad(4.3)$$

analogous to Eq. (1.2) for the leptonic weak current $J_\alpha^{(l)}$. Now (4.3) is certainly a reasonable expression, qualitatively, for weak currents of baryons and mesons. As Okun has emphasized,[23] the following properties of the weak interactions, often introduced as postulates, are derivable from (1.1), (1.2), (4.1), and (4.3):

(a) The conserved vector current.[24] In the model under discussion, as in that of Fermi and Yang, $i\bar{p}\gamma_\alpha n$ is a component of the total isotopic spin current.

(b) The rules $|\Delta S|=1$, $\Delta S/\Delta Q=+1$, and $|\Delta\mathbf{I}|=\frac{1}{2}$ for the leptonic decays of strange particles.[25]

(c) The invariance under GP of the $\Delta S=0$ weak current.[26]

(d) The rules $|\Delta S|=1$, $|\Delta\mathbf{I}|=\frac{1}{2}$ or $\frac{3}{2}$ in the nonleptonic decays of strange particles; along with $|\Delta S|=1$, we have the absence of a large K_1^0-K_2^0 mass difference.

The quantitative facts that the effective coupling constants for $|\Delta S|=1$ leptonic decays are smaller than those for $|\Delta S|=0$ leptonic decays and that in nonleptonic decays of strange particles the $|\Delta\mathbf{I}|=\frac{1}{2}$ amplitude greatly predominates over the $|\Delta\mathbf{I}|=\frac{3}{2}$ amplitude are not explained in any fundamental way.[27]

[21] S. Sakata, Progr. Theoret. Phys. (Kyoto) **16**, 686 (1956).

[22] A. Gamba, R. E. Marshak, and S. Okubo, Proc. Natl. Acad. Sci. U. S. **45**, 881 (1959).

[23] L. Okun, Ann. Rev. Nuclear Sci. **9**, 61 (1959).

[24] R. P. Feynman and M. Gell-Mann, Phys. Rev. **109**, 193 (1958). See also S. S. Gershtein and J. B. Zeldovich, Soviet Phys.—JETP **2**, 576 (1957).

[25] M. Gell-Mann, *Proceedings of the Sixth Annual Rochester Conference on High-Energy Nuclear Physics, 1956* (Interscience Publishers, Inc., New York, 1956). These rules were in fact suggested on the basis of the idea that N and Λ are fundamental. Should the rules prove too restrictive (for example should $\Delta S/\Delta Q=+1$ be violated), then we would try a larger group; in the language of the field-theoretic model, we would assume more fundamental fields. For a discussion of possible larger groups, see M. Gell-Mann and S. Glashow, Ann. Phys. **15**, 437 (1961) and S. Coleman and S. Glashow (to be published).

[26] S. Weinberg, Phys. Rev. **112**, 1375 (1958).

[27] A possible dynamical explanation of the predominance of $|\Delta\mathbf{I}|=\frac{1}{2}$ is being investigated by Nishijima (private communication). For example, consider the decay $\Lambda\rightarrow N+\pi$. A dispersion relation without subtractions is written for the matrix element of $J_\alpha^\dagger J_\alpha$ between the vacuum and a state containing $N+\bar{\Lambda}+\pi$. The parity-violating part leads to intermediate pseudoscalar states with $S=+1$ and with $|\Delta\mathbf{I}|=\frac{1}{2}$ or $\frac{3}{2}$. In the case of $|\Delta\mathbf{I}|=\frac{1}{2}$, there is an intermediate K particle, which may give a large contribution, swamping the term with $|\Delta\mathbf{I}|=\frac{3}{2}$, which has no one-meson state. For the same argument to apply to the parity-conserving part, we need the K' meson of Table III.

TABLE I. A set of matrices λ_i.

$$\lambda_1=\begin{pmatrix}0&1&0\\1&0&0\\0&0&0\end{pmatrix}\quad \lambda_2=\begin{pmatrix}0&-i&0\\i&0&0\\0&0&0\end{pmatrix}\quad \lambda_3=\begin{pmatrix}1&0&0\\0&-1&0\\0&0&0\end{pmatrix}$$

$$\lambda_4=\begin{pmatrix}0&0&1\\0&0&0\\1&0&0\end{pmatrix}\quad \lambda_5=\begin{pmatrix}0&0&-i\\0&0&0\\i&0&0\end{pmatrix}\quad \lambda_6=\begin{pmatrix}0&0&0\\0&0&1\\0&1&0\end{pmatrix}$$

$$\lambda_7=\begin{pmatrix}0&0&0\\0&0&-i\\0&i&0\end{pmatrix}\quad \lambda_8=\begin{pmatrix}1/\sqrt{3}&0&0\\0&1/\sqrt{3}&0\\0&0&-2/\sqrt{3}\end{pmatrix}$$

The electromagnetic properties of baryons and leptons are not exactly parallel in the Sakata model. The electric current (divided by e), which are denoted by j_α, is given by

$$i\bar{p}\gamma_\alpha p \tag{4.4}$$

for the baryons and mesons and by

$$-i(\bar{e}\gamma_\alpha e+\bar{\mu}\gamma_\alpha\mu) \tag{4.5}$$

for the leptons.

Now, we return to the Lagrangian (4.1) and separate it into three parts:

$$L=\bar{L}+L'+L'', \tag{4.6}$$

where $\bar{L}$ stands for everything except the baryon mass terms, while L' and L'' are given by the expressions

$$\begin{aligned}L'&=(2m_{0N}+m_{0\Lambda})(\bar{N}N+\bar{\Lambda}\Lambda)/3,\\ L''&=(m_{0N}-m_{0\Lambda})(\bar{N}N-2\bar{\Lambda}\Lambda)/3.\end{aligned} \tag{4.7}$$

If we now consider the Lagrangian with the mass-splitting term L'' omitted, we have a theory that is completely symmetrical in p, n, and Λ. We may perform any unitary linear transformation (with constant coefficients) on these three fields and leave $\bar{L}+L'$ invariant. Thus in the absence of the mass-splitting term L'' the theory is invariant under the three-dimensional unitary group $U(3)$; we shall refer to this situation as "unitary symmetry."

If we now turn on the mass-splitting, the symmetry is reduced. The only allowed unitary transformations are those involving n and p alone or Λ alone. The group becomes $U(2)\times U(1)$, which corresponds, as we shall see, to the conservation if isotopic spin, strangeness, and baryon number.

For simplicity, let us return briefly to the simpler case in which there is no Λ. The symmetry group is then just $U(2)$, the set of unitary transformations on n and p. We can factor each unitary transformation uniquely into one which multiplies both fields by the same phase factor and one (with determinant unity) which leaves invariant the product of the phase factors of p and n. Invariance under the first kind of transformation corresponds to conservation of nucleons n and p; it may be considered separately from invariance under the class of transformations of the second kind [called by mathematicians the unitary unimodular group $SU(2)$ in two dimensions]. In mathematical language, we can factor $U(2)$ into $U(1)\times SU(2)$.

Each transformation of the first kind can be written as a matrix $1\exp i\phi$, where 1 is the unit 2×2 matrix. The infinitesimal transformation is $1+i1\delta\phi$, and so the unit matrix is the infinitesimal generator of these transformations. Those of the second kind are generated in the same way by the three independent traceless 2×2 matrices, which may be taken to be the Pauli isotopic spin matrices τ_1, τ_2, and τ_3. We thus have

$$N\to(1+i\sum_{k=1}^{3}\delta\theta_k\tau_k/2)N, \tag{4.8}$$

as the general infinitesimal transformation of the second kind. Symmetry under all the transformations of the second kind is the same as symmetry under isotopic spin rotations. The whole formalism of isotopic spin theory can then be constructed by considering the transformation properties of the doublet or spinor (p,n) and of more complicated objects that transform like combinations of two or more such nucleons (or antinucleons).

The Pauli matrices τ_k are Hermitian and obey the rules

$$\begin{aligned}&\mathrm{Tr}\,\tau_i\tau_j=2\delta_{ij},\\ &[\tau_i,\tau_j]=2ie_{ijk}\tau_k,\\ &\{\tau_i,\tau_j\}=2\delta_{ij}1.\end{aligned} \tag{4.9}$$

The invariance under the group $SU(2)$ of isotopic spin rotations corresponds to conservation of the isotopic spin current

$$\mathfrak{J}_\alpha=i\bar{N}\boldsymbol{\tau}\gamma_\alpha N/2,$$

while the invariance under transformations of the first kind corresponds to conservation of the nucleon current $i\bar{N}\gamma_\alpha N/2=n_\alpha$.

Defining the total isotopic spin $\mathbf{I}$ as in (3.2), we obtain for I_i the commutation rules (3.1), which are the same as those for $\tau_i/2$. Likewise the nucleon number is defined as $-i\int n_4 d^3x$ and commutes with $\mathbf{I}$.

We now generalize the idea of isotopic spin by including the third field Λ. Again we factor the unitary transformations on baryons into those which are generated by the 3×3 unit matrix 1 (and which correspond to baryon conservation) and those which are generated by the eight independent traceless 3×3 matrices [and which form the unitary unimodular group $SU(3)$ in three dimensions]. We may construct a typical set of eight such matrices by analogy with the 2×2 matrices of Pauli. We call then $\lambda_1\cdots\lambda_8$ and list them in Table I. They are Hermitian and have the properties

$$\begin{aligned}&\mathrm{Tr}\,\lambda_i\lambda_j=2\delta_{ij},\\ &[\lambda_i,\lambda_j]=2if_{ijk}\lambda_k,\\ &\{\lambda_i,\lambda_j\}=2d_{ijk}\lambda_k+\tfrac{4}{3}\delta_{ij}1,\end{aligned} \tag{4.10}$$

where f_{ijk} is real and totally antisymmetric like the

TABLE II. Nonzero elements of f_{ijk} and d_{ijk}. The f_{ijk} are odd under permutations of any two indices while the d_{ijk} are even.

ijk	f_{ijk}	ijk	d_{ijk}
123	1	118	$1/\sqrt{3}$
147	1/2	146	1/2
156	−1/2	157	1/2
246	1/2	228	$1/\sqrt{3}$
257	1/2	247	−1/2
345	1/2	256	1/2
367	−1/2	338	$1/\sqrt{3}$
458	$\sqrt{3}/2$	344	1/2
678	$\sqrt{3}/2$	355	1/2
...	...	366	−1/2
...	...	377	−1/2
...	...	448	$-1/(2\sqrt{3})$
...	...	558	$-1/(2\sqrt{3})$
...	...	668	$-1/(2\sqrt{3})$
...	...	778	$-1/(2\sqrt{3})$
...	...	888	$-1/\sqrt{3}$

Kronecker symbol e_{ijk} of Eq. (4.9), while d_{ijk} is real and totally symmetric. These properties follow from the equations

$$\mathrm{Tr}\lambda_k[\lambda_i,\lambda_j]=4if_{ijk},\qquad \mathrm{Tr}\lambda_k\{\lambda_i,\lambda_j\}=4d_{ijk}, \tag{4.11}$$

derived from (4.10).

The nonzero elements of f_{ijk} and d_{ijk} are given in Table II for our choice of λ_i. Even and odd permutations of the listed indices correspond to multiplication of f_{ijk} by ± 1, respectively, and of d_{ijk} by $+1$.

The general infinitesimal transformation of the second kind on the three basic baryons b is, of course,

$$b \to (1+i\sum_{i=1}^{8}\delta\theta_i\lambda_i/2)b, \tag{4.12}$$

by analogy with (4.8). Together with conservation of baryons, invariance under these transformations corresponds to complete "unitary symmetry" of the three baryons. We have factored $U(3)$ into $U(1)\times SU(3)$.

The invariance under transformations of the first kind gives us conservation of the baryon current

$$i\bar{b}\gamma_\alpha b = i\bar{n}\gamma_\alpha n + i\bar{p}\gamma_\alpha p + i\bar{\Lambda}\gamma_\alpha\Lambda, \tag{4.13}$$

while invariance under the second class of transformations would give us conservation of the eight-component "unitary spin" current

$$\mathfrak{F}_{i\alpha}=i\bar{b}\lambda_i\gamma_\alpha b/2 \quad (i=1,\cdots,8). \tag{4.14}$$

Now in fact L'' is not zero and so not all the components of $\mathfrak{F}_{i\alpha}$ are actually conserved. This does not prevent us from defining $\mathfrak{F}_{i\alpha}$ as in (4.14), nor does it affect the commutation rules of the unitary spin density. The total unitary spin F_i is defined by the relation

$$F_i=-i\int \mathfrak{F}_{i4}d^3x, \tag{4.15}$$

at any time and at equal times the commutation rules for F_i follow those for $\lambda_i/2$

$$[F_i,F_j]=if_{ijk}F_k. \tag{4.16}$$

The baryon number, of course, commutes with all components F_i.

It will be noticed that λ_1, λ_2, and λ_3 agree with τ_1, τ_2, and τ_3 for p and n and have no matrix elements for Λ. Thus the first three components of the unitary spin are just the components of the isotopic spin. The matrix λ_8 is diagonal in our representation and has one eigenvalue for the nucleon and another for the Λ. Thus F_8 is just a linear combination of strangeness and baryon number. It commutes with the isotopic spin.

The matrices λ_4, λ_5, λ_6, and λ_7 connect the nucleon and Λ. We see that the components F_4, F_5, F_6, and F_7 of the unitary spin change strangeness by one unit and isotopic spin by a half unit. When the mass-splitting term L'' is "turned on," it is these components that are no longer conserved, while the conservation of F_1, F_2, F_3, F_8, and baryon number remains valid.

V. VECTOR AND AXIAL VECTOR CURRENTS

We may unify the mathematical treatment of the baryon current and the unitary spin current if we define a ninth 3×3 matrix

$$\lambda_0=(\tfrac{2}{3})^{\frac{1}{2}}1, \tag{5.1}$$

so that the *nine* matrices λ_i obey the rules

$$\begin{aligned}[\lambda_i,\lambda_j]&=2if_{ijk}\lambda_k \quad (i=0,\cdots,8),\\ \{\lambda_i,\lambda_j\}&=2d_{ijk}\lambda_k \quad (i=0,\cdots,8),\\ \mathrm{Tr}\lambda_i\lambda_j&=2\delta_{ij} \quad (i=0,\cdots,8).\end{aligned} \tag{5.2}$$

Here, f_{ijk} is defined as before, except that it vanishes when any index is zero; d_{ijk} is also defined as before, except that it has additional nonzero matrix elements equal to $(\frac{2}{3})^{\frac{1}{2}}$ whenever any index is zero and the other two indices are equal. The baryon current is now $(\frac{3}{2})^{\frac{1}{2}}\mathfrak{F}_{0\alpha}$.

The definitions (4.15) and the equal-time commutation relations (4.16) now hold for $i=0,\cdots,8$. Moreover, there are the equal-time commutation relations

$$[\mathfrak{F}_{i4}(\mathbf{x},t),\mathfrak{F}_{j4}(\mathbf{x}',t)]=-if_{ijk}\mathfrak{F}_{k4}(\mathbf{x},t)\delta(\mathbf{x}-\mathbf{x}') \tag{5.3}$$

for the densities.

The electric current j_α is then

$$j_\alpha=(\sqrt{2}\mathfrak{F}_{0\alpha}+\mathfrak{F}_{8\alpha}+\sqrt{3}\mathfrak{F}_{3\alpha})/2\sqrt{3}, \tag{5.4}$$

while the vector weak current is

$$\mathfrak{F}_{1\alpha}+i\mathfrak{F}_{2\alpha}+\mathfrak{F}_{4\alpha}+i\mathfrak{F}_{5\alpha}. \tag{5.5}$$

We now wish to set up the same formalism for the axial vector currents. We recall that the presence of the symmetry-breaking term L'' did not prevent us from defining the $\mathfrak{F}_{i\alpha}$ and obtaining the commutation rules (5.3) characteristic of the unitary symmetry group $U(3)$.

In the same way, we now remark that if both L''

and L' are "turned off," we have invariance under the infinitesimal unitary transformations

$$b\rightarrow(1+i\sum_{i=0}^{8}\delta\psi_i\gamma_5\lambda_i/2)b, \tag{5.6}$$

as well as the infinitesimal transformations

$$b\rightarrow(1+i\sum_{i=0}^{8}\delta\theta_i\lambda_i/2)b \tag{5.7}$$

we have used before.[28] Thus the axial vector currents

$$\mathfrak{F}_{i\alpha}{}^5=i\bar{b}\lambda_i\gamma_\alpha\gamma_5 b/2 \tag{5.8}$$

would be conserved if both L' and L'' were absent. Even in the presence of these terms, we have the commutation rules

$$[\mathfrak{F}_{i4}{}^5(\mathbf{x},t),\mathfrak{F}_{j4}(\mathbf{x}',t)]=-if_{ijk}\mathfrak{F}_{k4}{}^5(\mathbf{x},t)\delta(\mathbf{x}-\mathbf{x}') \tag{5.9}$$

and

$$[\mathfrak{F}_{i4}{}^5(\mathbf{x},t),\mathfrak{F}_{j4}{}^5(\mathbf{x}',t)]=-if_{ijk}\mathfrak{F}_{k4}(\mathbf{x},t)\delta(\mathbf{x}-\mathbf{x}') \tag{5.10}$$

at equal times, We may use the definition

$$F_i{}^5(t)\equiv -i\int \mathfrak{F}_{i4}{}^5 d^3x, \tag{5.11}$$

along with (4.15).

Just as we put $\mathbf{I}=\mathbf{I}_+ + \mathbf{I}_-$ and $\mathbf{D}=\mathbf{I}_+ - \mathbf{I}_-$ in the discussion following Eq. (3.16), so we now write

$$\begin{aligned}F_i(t)&=F_i{}^+(t)+F_i{}^-(t),\\ F_i{}^5(t)&=F_i{}^+(t)-F_i{}^-(t),\end{aligned} \tag{5.12}$$

and it is seen that $F_i{}^+$ and $F_i{}^-$ separately obey the commutation rules

$$[F_i{}^\pm,F_j{}^\pm]=if_{ijk}F_k{}^\pm, \tag{5.13}$$

while they commute with each other:

$$[F_i{}^\pm,F_j{}^\mp]=0. \tag{5.14}$$

Thus we are now dealing with the group $U(3)$ taken twice: $U(3)\times U(3)$. Factoring each $U(3)$ into $U(1)\times SU(3)$, we have[29] $U(1)\times U(1)\times SU(3)\times SU(3)$. Thus we have defined a left- and a right-handed baryon number and a left- and right-handed unitary spin.

The situation is just as in Sec. III, where we defined a left- and a right-handed isotopic spin and we could have defined a left- and a right-handed nucleon number.

The left- and right-handed quantities are connected to each other by the parity operation P:

$$PF_i{}^\pm P^{-1}=F_i{}^\mp. \tag{5.15}$$

Now that we have constructed the mathematical apparatus of the group $U(3)\times U(3)$ and its algebra, we may inquire how the Hamiltonian density H behaves under the group, i.e., under commutation with the algebra.

In the model, there is, corresponding to (4.6), the formula

$$H=\bar{H}-L'-L'', \tag{5.16}$$

where $\bar{H}$ is the Hamiltonian density derived from the Lagrangian density $\bar{L}$ and is completely invariant under the group. Instead of defining u_0 as in Sec. III, let us put

$$u_0=L'\propto\bar{b}\lambda_0 b. \tag{5.17}$$

We can easily see that by commutation of u_0 with F_i and $F_i{}^5$ ($i=0,\cdots,8$) at equal times we obtain a set of eighteen quantities:

$$\begin{aligned}u_i&\propto\bar{b}\lambda_i b,\\ v_i&\propto -i\bar{b}\lambda_i\gamma_5 b.\end{aligned} \tag{5.18}$$

In fact F_i acts like $\lambda_i/2$, $F_i{}^5$ like $\lambda_i\gamma_5/2$, u_i like $\beta\lambda_i$, and v_i like $-i\beta\gamma_5\lambda_i$. Thus we have at equal times[30]

$$\begin{aligned}[F_i,u_j]&=if_{ijk}u_k,\\ [F_i,v_j]&=if_{ijk}v_k,\\ [F_i{}^5,u_j]&=-id_{ijk}v_k,\\ [F_i{}^5,v_j]&=id_{ijk}u_k,\end{aligned} \tag{5.19}$$

and the stronger relations

$$[\mathfrak{F}_{i4}(\mathbf{x},t),u_j(\mathbf{x}',t)]=-f_{ijk}u_k(\mathbf{x},t)\delta(\mathbf{x}-\mathbf{x}'),\ \text{etc.} \tag{5.20}$$

for the densities. All indices run from 0 to 8.

Note that we can now express not only L' (which is defined to be u_0) but L'' as well, since by (4.7) it is proportional to u_8. We have, then,

$$H-\bar{H}=u_0-cu_8, \tag{5.21}$$

where c is of the order $(m_{0N}-m_{0\Lambda})/m_{0N}$ in the model.

We may now make a series of abstractions from the model. First, we suppose that currents $\mathfrak{F}_{i\alpha}$ and $\mathfrak{F}_{i\alpha}{}^5$ are defined, with commutation rules (5.3), (5.9), and (5.10), and with the weak current given by the analog of (5.5)[31]:

$$J_\alpha=\mathfrak{F}_{1\alpha}+\mathfrak{F}_{1\alpha}{}^5+i\mathfrak{F}_{2\alpha}+i\mathfrak{F}_{2\alpha}{}^5 +\mathfrak{F}_{4\alpha}+\mathfrak{F}_{4\alpha}{}^5+i\mathfrak{F}_{5\alpha}+i\mathfrak{F}_{5\alpha}{}^5, \tag{5.22}$$

---

[28] Actually the Lagrangian (4.1) without the nucleon mass terms is invariant under a larger continuous group of transformations than the one $[U(3)\times U(3)]$ that we treat here. For example, there are infinitesimal transformations in which the baryon fields b acquire small terms in $\bar{b}$. Invariance under these is associated with the conservation of currents carrying baryon number 2. The author wishes to thank Professor W. Thirring for a discussion of these additional symmetries and of conformal transformations, which give still more symmetry.

[29] The groups $U(1)$, $SU(3)$, and $SU(2)$ cannot be further factored in this fashion. They are called *simple*.

[30] Note that even if we use just F_i and $F_i{}^5$ for $i=1,\cdots,8$, or $SU(3)\times SU(3)$ only, we still generate all eighteen u's and v's. [In the two-dimensional case described in Sec. III the situation is different. Using $SU(2)\times SU(2)$, we generate from u_0 only itself and v_1, v_2, v_3; if we then bring in $F_0{}^5$ as well, we obtain three more u's and one more v.] This remark is interesting because the group that gives currents known to be physically interesting is just $U(1)\times SU(3)\times SU(3)$; there is no known physical coupling to $\mathfrak{F}_{0\alpha}{}^5$, the axial vector baryon current.

[31] Note that the *total* weak current, whether for baryons and mesons or for leptons, is just a component of the current of an angular momentum. See reference 13.

while the electric current is given by (5.4). Next, we may take the Hamiltonian density to be of the form (5.21), with $\bar{H}$ invariant and u_i and v_i transforming as in (5.20). Then, if the theory is of the type described in Appendix A, we can calculate the divergences of the currents in terms of the equal-time commutators

$$\begin{aligned}\partial_\alpha \mathfrak{F}_{i\alpha} &= i[F_i, u_0] + ic[F_i, u_8],\\ \partial_\alpha \mathfrak{F}_{i\alpha}{}^5 &= i[F_i{}^5, u_0] + ic[F_i{}^5, u_8],\end{aligned} \tag{5.23}$$

or, explicitly,

$$\begin{aligned}\partial_\alpha \mathfrak{F}_{i\alpha} &= 0, \quad (i=0,1,2,3,8)\\ \partial_\alpha \mathfrak{F}_{4\alpha} &= (\tfrac{3}{2})^{\frac{1}{2}} u_5, \text{ etc.},\\ \partial_\alpha \mathfrak{F}_{0\alpha}{}^5 &= (\tfrac{2}{3})^{\frac{1}{2}} v_0 + (\tfrac{2}{3})^{\frac{1}{2}} c v_8,\\ \partial_\alpha \mathfrak{F}_{1\alpha}{}^5 &= [(\tfrac{2}{3})^{\frac{1}{2}} + (\tfrac{1}{3})^{\frac{1}{2}} c] v_1, \text{ etc.},\\ \partial_\alpha \mathfrak{F}_{4\alpha}{}^5 &= [(\tfrac{2}{3})^{\frac{1}{2}} - (\tfrac{1}{12})^{\frac{1}{2}} c] v_4, \text{ etc.},\\ \partial_\alpha \mathfrak{F}_{8\alpha}{}^5 &= [(\tfrac{2}{3})^{\frac{1}{2}} - (\tfrac{1}{3})^{\frac{1}{2}} c] v_8 + (\tfrac{2}{3})^{\frac{1}{2}} c v_0.\end{aligned} \tag{5.24}$$

Finally, if we taken the model really seriously, we may abstract the equal-time commutation relations of the u_i and v_i as obtained by explicit commutation in the model.

The relations of Sec. III are all included in those of this section, except that what was called u_0 there is now called $u_0 + cu_8$ and what was called v_i is now called $[(\frac{2}{3})^{\frac{1}{2}} + (\frac{1}{3})^{\frac{1}{2}} c] v_i$ for $i=1, 2, 3$.

All of the relations used here are supposed to be exact and are not affected by the symmetry-breaking character of the non-invariant term in the Hamiltonian. In the next section, we discuss what happens if c can be regarded as small in any sense. We may then expect to see some trace of the symmetry under $U(3)$ that would obtain if c were 0 and L'' disappeared. In this limit, N and Λ are degenerate, and all the components F_i of the unitary spin are conserved. The higher symmetry would show up particularly through the existence of degenerate baryon and meson supermultiplets, which break up into ordinary isotopic multiplets when L'' is turned on. These supermultiplets have been discussed previously for baryons and pseudoscalar mesons[4,6] and then for vector mesons.[32–34]

We shall not discuss the case in which both L' and L'' are turned off; that is the situation, still more remote from reality, in which all the axial vector currents are conserved as well as the vector ones.

VI. BROKEN SYMMETRY—MESON SUPERMULTIPLETS

We know that because of isotopic spin conservation the baryons and mesons form degenerate isotopic multiplets, each corresponding to an irreducible representation of the isotopic spin algebra (3.1). Each multiplet has $2I+1$ components, where the quantum number I distinguishes one representation from another and gives us the eigenvalue $I(I+1)$ of the operator $\sum_{i=1}^{3} I_i^2$, which commutes with all the elements of the isotopic spin group. The operators I_i are represented, within the multiplet, by Hermitian $(2I+1)\times(2I+1)$ matrices having the commutation rules (3.1) of the algebra.

If we start from the doublet representation, we can build up all the others by considering combinations of particles that transform like the original doublet. Just as (p,n) form a doublet representation for which the I_i are represented by $\tau_i/2$, the antiparticles $(\bar{n}, -\bar{p})$ also form a doublet representation that is equivalent. (Notice the minus sign on the antiproton state or field.) Now, if we put together a nucleon and an antinucleon, we can form the combination

$$\bar{N}N = \bar{p}p + \bar{n}n,$$

which transforms like an isotopic singlet, or the combinations

$$\bar{N}\tau_i N, \quad (i=1,2,3)$$

which form an isotopic triplet. The direct product of nucleon and antinucleon doublets gives us a singlet and a triplet. Any meson that can dissociate virtually into nucleon and antinucleon must be either a singlet or a triplet. For the singlet state, the components I_i are all zero, while for the three triplet states the three 3×3 matrices, $I_i{}^{jk}$ of the components I_i, are given by

$$I_i{}^{jk} = -ie_{ijk}. \tag{6.1}$$

Now let us generalize these familiar results to the unitary spin and the three basic baryons b (comprising n, p, and Λ). These three fields or particles form a three-dimensional irreducible representation of the unitary spin algebra (4.16) from which all the other representations may be constructed.

For example, consider a meson that can dissociate into b and $\bar{b}$. It must transform either like

$$\bar{b}b = \bar{p}p + \bar{n}n + \bar{\Lambda}\Lambda,$$

a unitary singlet, or else like

$$\bar{b}\lambda_i b, \quad (i=1, \cdots, 8)$$

a unitary octet.

The unitary singlet is evidently neutral, with strangeness $S=0$, and forms an isotopic singlet. But how does the unitary octet behave with respect to isotopic spin? We form the combinations

$$\left.\begin{aligned}\bar{b}(\lambda_1 - i\lambda_2)b/2 &= \bar{n}p,\\ \bar{b}\lambda_3 b/\sqrt{2} &= (\bar{p}p - \bar{n}n)/\sqrt{2},\\ \bar{b}(\lambda_1 + i\lambda_2)b/2 &= \bar{p}n,\end{aligned}\right\} \quad I=1,\ S=0$$

$$\left.\begin{aligned}\bar{b}(\lambda_4 - i\lambda_5)b/2 &= \bar{\Lambda}p,\\ \bar{b}(\lambda_6 - i\lambda_7)b/2 &= \bar{\Lambda}n,\end{aligned}\right\} \quad I=\tfrac{1}{2},\ S=+1 \tag{6.2}$$

$$\left.\begin{aligned}\bar{b}(\lambda_4 + i\lambda_5)b/2 &= \bar{p}\Lambda,\\ \bar{b}(\lambda_6 + i\lambda_7)b/2 &= \bar{n}\Lambda,\end{aligned}\right\} \quad I=\tfrac{1}{2},\ S=-1$$

$$\bar{b}\lambda_8 b/\sqrt{2} = (\bar{p}p + \bar{n}n - 2\bar{\Lambda}\Lambda)/\sqrt{6}, \quad I=0,\ S=0,$$

---

[32] M. Gell-Mann, California Institute of Technology Synchrotron Laboratory Report No. CTSL-20, 1961 (unpublished).

[33] Y. Ne'eman, Nuclear Phys. **26**, 222 (1961).

[34] A. Salam and J. C. Ward, Nuovo cimento **20**, 419 (1961).

and we see immediately that the unitary octet comprises an isotopic triplet with $S=0$, a pair of isotopic doublets with $S=\pm 1$, and an isotopic singlet with $S=0$. All these are degenerate only in the limit of unitary symmetry ($L''=0$); when the mass-splitting term is turned on, the singlet, the triplet, and the pair of doublets should have three somewhat different masses.

The known pseudoscalar mesons (π, K, and $\bar{K}$) fit very well into this picture, provided there is an eighth pseudoscalar meson to fill out the octet. Let us call the hypothetical isotopic singlet pseudoscalar meson χ^0. Since it is pseudoscalar, it cannot dissociate (virtually or really) into 2π. It has the value $+1$ for the quantum number G, so that it cannot dissociate into an odd number of pions either. Thus in order to decay by means of the strong interactions, it must have enough energy to yield 4π. It would then appear as a 4π resonance. The decay into 4π is, however, severely hampered by centrifugal barriers.

If the mass of χ^0 is too low to permit it to decay readily into 4π, then it will decay electromagnetically. If there is sufficient energy, the decay mode $\chi^0\to\pi^++\pi^-+\gamma$ is most favorable; otherwise[34a] it will decay into 2γ like π^0.

Let us now turn to the vector mesons. The best known vector meson is the $I=1$, $J=1^-$ resonance of 2π, which we shall call ρ. It has a mass of about 750 Mev. According to our scheme, it should belong, like the pion, to a unitary octet. Since it occupies the same position as the π ($I=1$, $S=0$), we denote it by the succeeding letter of the Greek alphabet.

The vector analog of χ^0 we shall call ω^0 (skipping the Greek letter ψ). It must have $I=0$, $J=1^-$, and $G=-1$ and so it is capable of dissociation into $\pi^++\pi^-+\pi^0$. Presumably it is the 3π resonance found experimentally[35] at about 790 Mev.

In order to complete the octet, we need a pair of strange doublets analogous to K and $\bar{K}$. In the vector case, we shall call them M and $\bar{M}$ (skipping the letter L). Now there is a known $K\pi$ resonance with $I=\frac{1}{2}$ at about 884 Mev. If it is a p-wave resonance, then it fits the description of M perfectly.

In the limit of unitary symmetry, we can have, besides the unitary octet of vector mesons, a unitary singlet. The hypothetical B^0 that we discussed in Sec. III would have such a character. If B^0 exists, then the turning-on of the mass-splitting term L'' mixes the states B^0 and ω^0, which are both *isotopic* singlets.

Other mesons may exist besides those discussed, for example, scalar and axial vector mesons. All those that can associate into $b+\bar{b}$ should form unitary octets or singlets or both, with each octet splitting into isotopic multiplets because of the symmetry-breaking term L''.

TABLE III. Possible meson octets and singlets.

Unitary spin	Isotopic spin	Strangeness	Pseudoscalar	Vector	Scalar	Axial vector
Octet	1	0	π	ρ	π'	ρ'
	1/2	+1	K	M	K'	M'
	1/2	−1	$\bar{K}$	$\bar{M}$	$\bar{K}'$	$\bar{M}'$
	0	0	χ	ω	χ'	ω'
Singlet	0	0	A	B	A'	B'

A list of some possible meson states is given in Table III, along with suggested names for the mesons.

It is interesting that we can predict not only the degeneracy of an octet in the limit $L''\to 0$ but also a sum rule[32] that holds in first order in L'':

$$(m_K+m_{\bar{K}})/2=(3m_\chi+m_\pi)/4,$$
$$(m_M+m_{\bar{M}})/2=(3m_\omega+m_\rho)/4. \tag{6.3}$$

If M is at about 884 Mev and ρ at about 750 Mev, then ω should lie at about 930 Mev according to the sum rule; since it is actually at 790 Mev, the sum rule does not seem to give a good description of the splitting. Perhaps an important effect is the repulsion between the ω^0 and B^0 levels, pushing ω^0 down and B^0 up. For what it is worth, (6.3) gives a χ^0 mass of around 610 Mev.

In the limit of unitary symmetry, not only are the supermultiplets degenerate but their effective couplings are symmetrical. For example, the effective coupling of the unitary pseudoscalar octet to N and Λ takes the form

$$ig_1\{\bar{N}\boldsymbol{\tau}\gamma_5N\cdot\boldsymbol{\pi}+\bar{N}\gamma_5\Lambda K+\bar{\Lambda}\gamma_5N\bar{K}$$
$$+3^{-\frac{1}{2}}\bar{N}\gamma_5N\chi-2\times 3^{-\frac{1}{2}}\bar{\Lambda}\gamma_5\Lambda\chi\}, \tag{6.4}$$

in terms of renormalized "fields." Now, as the term L'' is turned on, the various coupling constants become unequal; instead of calling them all g_1, we refer to them as $g_{NN\pi}$, $g_{N\Lambda K}$, $g_{NN\chi}$, and $g_{\Lambda\Lambda\chi}$, respectively, each of these constants being the measurable renormalized coupling parameter at the relevant pole.

We have written the effective coupling (6.4) as if there were renormalized fields for all the particles involved, but that is only a matter of notation; the mesons can perfectly well be composite. We may simplify the notation still further by constructing a traceless 3×3 matrix Π containing the pseudoscalar "fields" in such a way that (6.4) becomes

$$ig_1\bar{b}\Pi\gamma_5 b. \tag{6.5}$$

We may now write, in a trivial way, other effective couplings in the limit of unitary symmetry. We define a traceless 3×3 matrix W_α containing the "fields" for the vector meson octet just as Π^* contains those for the pseudoscalar octet. We then have the invariant

[34a] *Note added in proof.* H. P. Duerr and W. Heisenberg (preprint) have pointed out the importance of the decay mode $\chi^0\to 3\pi$ induced by electromagnetism. For certain χ masses, it may be a prominent mode.

[35] B. C. Maglić, L. W. Alvarez, A. H. Rosenfeld, and M. L. Stevenson, Phys. Rev. Letters 7, 178 (1961).

effective coupling

$$i\gamma_1 \operatorname{Tr} W_\alpha(\Pi\partial_\alpha\Pi - \partial_\alpha\Pi\Pi)/2 \quad (6.6)$$

in the symmetric limit. When the asymmetry is turned on, the single coupling parameter γ_1 is replaced by the set of different parameters $\gamma_{\rho\pi\pi}$, $\gamma_{\rho KK}$, $\gamma_{\omega KK}$, $\gamma_{MK\pi}$, and $\gamma_{MK\chi}$.

In the same way, we have another effective coupling

$$ih_1 \operatorname{Tr}\Pi(\partial_\alpha W_\beta - \partial_\beta W_\alpha)(\partial_\gamma W_\delta - \partial_\delta W_\gamma)e_{\alpha\beta\gamma\delta} \quad (6.7)$$

in the symmetric limit; in the actual asymmetric case, we define the distinct constants $h_{\pi\omega\rho}$, $h_{\pi MM}$, $h_{\chi\omega\omega}$, $h_{\chi\rho\rho}$, $h_{\chi MM}$, $h_{KM\rho}$, and $h_{KM\omega}$. All of these constants can be measured, in principle, in "pole" experiments, except that for the broad resonances like ρ the poles are well off the physical sheet.

We have generalized the definitions of constants like $g_{NN\pi}$ and $\gamma_{\rho\pi\pi}$, as used in Sec. II, to other particles. The constants γ_ρ and f_π of Sec. II also have analogs, of course, and we define γ_ω, f_K, etc., in the obvious way. In the limit of unitary symmetry, of course, we would have $f_\pi = f_K = f_\chi$ and $\gamma_\rho = \gamma_\omega = \gamma_M$. Likewise, the constant $-G_A/G$ for nucleon β decay would equal the corresponding quantity $-G_A{}^{\Lambda N}/G$ for the β decay of Λ.

VII. BROKEN SYMMETRY—BARYON SUPERMULTIPLETS

What has been done in the previous section may be described mathematically as follows. We considered a three-dimensional representation of the unitary spin algebra (4.16) or of the group $SU(3)$ that is generated by the algebra. It is the representation to which b belongs (that is, n, p, and Λ) and we may denote it by the symbol **3**.

The antiparticles $\bar{b}$ belong to the conjugate representation **3***, which is inequivalent[36] to **3**. We have then taken the direct product **3×3*** and found it to be given by the rule

$$\mathbf{3}\times\mathbf{3}^* = \mathbf{8}+\mathbf{1}, \quad (7.1)$$

where **8** is the octet representation and **1** the singlet representation of unitary spin. Each of these is its own conjugate; that is a situation that occurs only when the dimension is the cube of an integer.

There are, of course, more complicated representations to which mesons might belong that are incapable (in the limit of unitary symmetry) of dissociation into $b+\bar{b}$ but capable of dissociation into $2b+2\bar{b}$ or higher configurations. But we might guess that at least the mesons of lowest mass would correspond to the lowest configurations.

Now we want to examine the simplest configurations for baryons, apart from just b. Evidently the next simplest is $2b+\bar{b}$, which poses the problem of reducing the direct product **3×3×3***; the result is the following:

$$\mathbf{3}\times\mathbf{3}\times\mathbf{3}^* = \mathbf{3}\times\mathbf{1}+\mathbf{3}\times\mathbf{8} = \mathbf{3}+\mathbf{3}+\mathbf{6}+\mathbf{15}. \quad (7.2)$$

The six-dimensional representation **6** is composed of an isotopic triplet with $S=-1$, a doublet with $S=0$, and a singlet with $S=+1$; the fifteen-dimensional representation **15** is composed of a doublet with $S=-2$, a singlet and a triplet with $S=-1$, a doublet and a quartet with $S=0$, and a triplet with $S=+1$.

According to the scheme, then, Ξ should belong to **15**. Where are the other members of the supermultiplet? For $S=-1$ and $S=0$, there are many known resonances, some of which might easily have the same spin and parity as Ξ. For $S=+1$, $I=1$, however, no resonance has been found so far (in K^+-p scattering, for example).

The hyperon Σ should also be placed in a supermultiplet, which may or may not be the same one to which Ξ belongs; we do not know if the spin and parity of Σ and Ξ are the same, with K taken to be pseudoscalar. If Σ belongs to **6** in the limit of unitary symmetry, then there should be a KN resonance in the $I=0$ state.

It is difficult to say at the present time if the baryon states can be reconciled with the model. Further knowledge of the baryon resonances is required.

One curious possibility is that the fundamental objects b are hidden and that the physical N and Λ, instead of belonging to **3**, belong, along with Σ and Ξ, to the representation **15** in the limit of unitary symmetry. That would require the spins and parities of N, Λ, Σ, and Ξ to be equal, and it would require a πN resonance in the $p_{\frac{1}{2}}$, $I=\frac{3}{2}$ state as well as a KN resonance in the $p_{\frac{1}{2}}$, $I=1$ state to fill out the supermultiplet.

VIII. THE "EIGHTFOLD WAY"

Unitary symmetry may be applied to the baryons in a more appealing way if we abandon the connection with the symmetrical Sakata model and treat unitary symmetry in the abstract. (An abstract approach is, of course, required if there are no "elementary" baryons and mesons.) Of all the groups that could be generated by the vector weak currents, $SU(3)$ is still the smallest and the one that most naturally gives rise to the rules $|\Delta\mathbf{I}| = \frac{1}{2}$ and $\Delta S/\Delta Q = 0, +1$.

There is no longer any reason for the baryons to belong to the **3** representation or the other spinor representations of the group $SU(3)$; the various irreducible spinor representations are those obtained by reducing direct products like **3×3×3***, **3×3×3×3*×3***, etc.

Instead, the baryons may belong, like the mesons, to representations such as **8** or **1** obtained by reducing the direct products of equal numbers of **3**'s and **3***'s. It is then natural to assign the stable and metastable baryons N, Λ, Σ, and Ξ to an octet, degenerate in the limit of unitary symmetry. We thus obtain the scheme

[36] In other words, no unitary transform can convert the representations 3 and 3* into each other. That is easy to see, since the eigenvalues of λ_8 are opposite in sign for the two representations, and changing the signs changes the set of eigenvalues. In the case of the group $SU(2)$ of isotopic spin transformations, the basic spinor representation $I=\frac{1}{2}$ is equivalent to the corresponding antiparticle representation.

of Gell-Mann[32] and Ne'eman[33] that we call the "eightfold way." The component F_8 of the unitary spin is now $(\sqrt{3}/2)Y$, where Y is the hypercharge (equal to strangeness plus baryon number).

The baryons of the octet must have the same spin and parity (treating K as pseudoscalar). To first order in the violation of unitary symmetry, the masses should obey the sum rule analogous to (6.3):

$$(m_N+m_\Xi)/2=(3m_\Lambda+m_\Sigma)/4, \qquad (8.1)$$

which agrees surprisingly well with observations, the two sides differing by less than 20 Mev.

To form mesons that transform like combinations of these baryons and their antiparticles, we reduce the direct product $\mathbf{8}\times\mathbf{8}$ (remembering that $\mathbf{8}=\mathbf{8}^*$) and obtain

$$\mathbf{8}\times\mathbf{8}=\mathbf{1}+\mathbf{8}+\mathbf{8}+\mathbf{10}+\mathbf{10}^*+\mathbf{27}, \qquad (8.2)$$

where $\mathbf{1}$ and $\mathbf{8}$ are the singlet and octet representations already discussed; $\mathbf{10}$ consists of an isotopic triplet with $Y=0$, a doublet with $Y=-1$, a quartet with $Y=+1$, and a singlet with $Y=-2$; $\mathbf{10}^*$ has the opposite behavior with respect to Y; and $\mathbf{27}$ consists of an isotopic singlet, triplet, and quintet with $Y=0$, a pair of doublets with $Y=\pm1$, a pair of quartets with $Y=\pm1$, and a pair of triplets with $Y=\pm2$. Evidently the known mesons are to be assigned to octets and perhaps singlets, as in Sec. VI. The meson-nucleon scattering resonances must then also be assigned representations among those in (8.2); the absence so far of any observed structure in K-N scattering makes it difficult to place the $I=3/2$, $J=3/2$, π-N resonance in a supermultiplet.

The fact that $\mathbf{8}$ occurs twice in Eq. (8.2) means that there are two possible forms of symmetrical Yukawa coupling of a meson octet to the baryon octet in the limit of unitary symmetry. As in Sec. VI for the mesons, we form a 3×3 traceless matrix out of the formal "fields" of the baryon octet; call it $\mathcal{B}$. The effective symmetrical coupling of pseudoscalar mesons may then be written as

$$ig_1\alpha \operatorname{Tr}(\bar{\mathcal{B}}\pi\gamma_5\mathcal{B}+\pi\bar{\mathcal{B}}\gamma_5\mathcal{B})/2 \\ +ig_1(1-\alpha)\operatorname{Tr}(\bar{\mathcal{B}}\pi\gamma_5\mathcal{B}-\pi\bar{\mathcal{B}}\gamma_5\mathcal{B})/2. \qquad (8.3)$$

The two types of coupling differ in their behavior under the operation R that exchanges N and Ξ, K and $\bar{K}$, M and $\bar{M}$, etc.; the first term is symmetric while the second is antisymmetric under R. The parameter α just specifies how much of each effective coupling is presented in the limit of unitary symmetry. When we take into account violations of the symmetry, we must define separate coupling constants $g_{NN\pi}$, $g_{NK\Lambda}$, etc., in a suitable way.

Likewise the vector mesons have the general symmetrical coupling

$$i\gamma_1\beta \operatorname{Tr}(\bar{\mathcal{B}}W_\alpha\gamma_\alpha\mathcal{B}+W_\alpha\bar{\mathcal{B}}\gamma_\alpha\mathcal{B}) \\ +i\gamma_1(1-\beta)\operatorname{Tr}(\bar{\mathcal{B}}W_\alpha\gamma_\alpha\mathcal{B}-W_\alpha\bar{\mathcal{B}}\gamma_\alpha\mathcal{B}), \qquad (8.4)$$

where we ignore Pauli moment terms for simplicity. To the extent that the vector meson octet W_α dominates the dispersion relations for the unitary spin current $\mathfrak{F}_{i\alpha}$, then the mesons of W_α are coupled effectively to the components of $\mathfrak{F}_{i\alpha}$, and we have $\beta=0$ in (8.5). Then ρ is effectively coupled to the isotopic spin current, ω to the hypercharge current, and M to the strangeness-changing vector current. The first two of these currents are conserved, and so we have the approximate universality of ρ and ω couplings proposed by Sakurai[11] and discussed in Sec. II. In the limit of unitary symmetry, under the assumptions just mentioned, ρ is effectively coupled to the current of $2\gamma_1\mathbf{I}$ and ω to the current of $2\gamma_1F_8=\sqrt{3}\gamma_1Y$.

The electromagnetic current is now given by the formula

$$j_\alpha=\mathfrak{F}_{3\alpha}+3^{-\frac{1}{2}}\mathfrak{F}_{8\alpha} \qquad (8.5)$$

instead of (5.4), while the weak vector current is still described by Eq. (5.5). If we are to treat the vector and axial vector currents by means of $SU(3)\times SU(3)$, as we did earlier, then the entire weak current is given by (5.22) and we have the commutation rules (5.3), (5.9), and (5.10) for the various components of the currents. The question of the behavior of H under the group $SU(3)\times SU(3)$ should, however, be re-examined for the eightfold way; we shall not go into that question here. But let us consider how the baryon octet transforms in the limit of conserved vector *and* axial vector currents [invariance under $SU(3)\times SU(3)$]. In the Sakata model, the left-handed baryons transformed under (F_i^+,F_j^-) like $(\mathbf{3},\mathbf{1})$, while the right-handed baryons transformed according to $(\mathbf{1},\mathbf{3})$. For the eightfold way, there are two simple possibilities for these transformation properties. Either we have $(\mathbf{8},\mathbf{1})$ and $(\mathbf{1},\mathbf{8})$ or else we adjoin a ninth neutral baryon (which need not be degenerate with the other eight in the limit of conserved *vector* currents and which need not have the same parity) and use the transformation properties $(\mathbf{3},\mathbf{3}^*)$ and $(\mathbf{3}^*,\mathbf{3})$. In the first case, the baryons transform like the quantities $\mathfrak{F}_{i\alpha}$ and $\mathfrak{F}_{i\alpha}{}^5$ $(i=1,\cdots,8)$ and in the second case they transform like u_i and v_i $(i=0,\cdots,8)$ of Sec. V.

IX. REMARKS AND SUGGESTIONS

Our approach to the problem of baryon and meson couplings leads to a number of suggestions for new investigations, both theoretical and experimental.

First, the equal-time commutation relations for currents and densities lead to exact sum rules for the weak and electromagnetic matrix elements. As an example, take the commutation rules (3.5) for the isotopic spin current. These do not, of course, depend on any higher symmetry, but they can be used to illustrate the results that can be obtained from the more general relations like (5.3).

Consider the electromagnetic form factor $F_\pi(s)$ of the charged pion, which is just the form factor of the isotopic spin current between one-pion states. Let p and p' be the initial and final pion four-momenta, with $s=-(p-p')^2$. Let K be any four-momentum with $K^2=-m_\pi^2$. Then, taking the matrix element of (3.5) between one-pion states, we obtain the result

$$\begin{aligned}2(p_0+p_0')K_0F_\pi(-(p-p')^2)\\ =(p_0+K_0)(p_0'+K_0)F_\pi(-(p-K)^2)F_\pi(-(p'-K)^2)\\ -(p_0-K_0)(p_0'-K_0)F_\pi(-(p+K)^2)F_\pi(-(p'+K)^2)\\ +\text{inelastic terms},\quad (9.1)\end{aligned}$$

where the inelastic terms come from summing over bilinear forms in the inelastic matrix elements of the current. We see that if there is no inelasticity the form factor is unity. Thus the departure from unity of $F_\pi(s)$ is related to the amount of inelasticity.

A similar relation is familiar in nonrelativistic quantum mechanics:

$$\langle e^{i(\mathbf{p}-\mathbf{p}')\cdot\mathbf{x}}\rangle_{00}=\sum_n\langle e^{i(\mathbf{p}-\mathbf{k})\cdot\mathbf{x}}\rangle_{0n}\langle e^{i(\mathbf{k}-\mathbf{p}')\cdot\mathbf{x}}\rangle_{n0}. \quad (9.2)$$

If we apply relations like (9.1) to the matrix elements of non-conserved currents like $\mathbf{P}_\alpha$, along with the linear homogeneous dispersion relations for these matrix elements, we can in principle determine constants like $-G_A/G$.

A second line of theoretical investigation is suggested by the vanishing at high momentum transfer of matrix elements of divergences of non-conserved currents, like $\partial_\alpha\mathbf{P}_\alpha$. We should try to find limits involving high energies and high momentum transfers in which we can show that the conservation of helicity, unitary spin, etc., becomes valid. A preliminary effort in this direction has been made by Gell-Mann and Zachariasen.[37]

A third topic of study is the testing of broken symmetry at low energy. Do the mesons fall into unitary octets and singlets? An experimental search for χ^0 is required and also a determination of the spin of K^* at 884 Mev to see if it really is our M meson.

Let us discuss briefly the properties of χ^0. An $I=0$ state of 4π can have two types of symmetry: either totally symmetric (partition [4]) in both space and isotopic spin or else the symmetry of the partition [2+2] in space and in isotopic spin. For a pseudoscalar state, the first type of wave function in momentum space is very complicated. If $\mathbf{p}$, $\mathbf{q}$, and $\mathbf{r}$ are the three momentum differences, it must look like

$$\mathbf{p}\cdot\mathbf{q}\times\mathbf{r}(E_1-E_2)(E_2-E_3)(E_3-E_4)\times(E_1-E_3)(E_1-E_4)(E_2-E_4),$$

times a symmetric function of the energies E_1, E_2, E_3, E_4 of the four pions. On the basis of any reasonable dynamical picture of χ^0, such a wave function should have a very small amplitude. In particular, dispersion theory suggests that the wave function of χ^0 should have large contributions from virtual dissociation into 2ρ, which gives a wave function with [2+2] symmetry.

[37] M. Gell-Mann and F. Zachariasen, Phys. Rev. 123, 1065 (1961).

If [2+2] predominates, then the charge ratio in decay is 2:1 in favor of $2\pi^0+\pi^++\pi^-$ over $2\pi^++2\pi^-$, with $4\pi^0$ absent. If virtual dissociation into 2ρ actually predominates, then the matrix element of the 4π configuration is easily written down and the spectrum of the decay $\chi^0\to 4\pi$ can be calculated.

If χ^0 is lighter than 4π, it will, of course, decay electromagnetically. Even if it is above threshold for 4π, however, the matrix element for decay contains so many powers of pion momenta that electromagnetic decay should be appreciable over a large range of masses. The branching ratio $(\pi^++\pi^-+\gamma)/(4\pi)$ is approximately calculable by dispersion methods. In both cases χ^0 first dissociates into 2ρ. Then either both virtual ρ mesons decay into 2π, or else (in the case where both are neutral) one may decay into $\pi^++\pi^-$, while the other turns directly into γ. If we draw a diagram for such a process, then the constant $\gamma_{\rho\pi\pi}$ is inserted whenever we have a $\rho\pi\pi$ vertex and the constant $em_\rho^2/2\gamma_\rho$ at a ρ-γ junction.[10]

If the meson spectrum is consistent with broken unitary symmetry, we should examine the baryons, and see whether the various baryon states fit into the representations **3**, **6**, and **15** (or the representations **1**, **8**, **10**, **10***, and **27** that arise in the alternative form of unitary symmetry).

If some states are lacking in a given supermultiplet, it does not necessarily prove that the broken symmetry is wrong, but only that it is badly violated. We assume that baryon isobars like the πN $\frac{3}{2}$, $\frac{3}{2}$ resonance are dynamical in nature; there may be some attractive and some repulsive forces in this channel, and the attractive ones have won out, producing the resonance. In the KN channel with $I=1$, for example, it is conceivable that the repulsive ones are stronger (because of symmetry violation), and the analogous $p_{\frac{3}{2}}$ resonance disappears. In such a situation, the concept of broken symmetry at low energies is evidently of little value.

Suppose, however, that the idea of broken unitary symmetry is confirmed for both mesons and baryons, say according to the Sakata picture, in which N and Λ belong to the representation **3** in the limit of unitary symmetry. There are, nevertheless, gross violations of unitary symmetry, and the elucidation of these, both theoretical and experimental, is a fourth interesting subject.

If unitary symmetry were exact, then not only would m_K/m_π equal unity, instead of about 3.5, but f_K^2/f_π^2 would be 1 instead of about 6, and $3G_A^2+G_V^2$ for the β decay of Λ would be equal to $3G_A^2+G_V^2$ for the nucleon instead of being 1/15 as large. All these huge departures from unity represent very serious violations of unitary symmetry.

Yet the relatively small mass difference of N and Λ compared to their masses would seem to indicate, if our model is right, that the constant c in Eq. (5.21) is considerably smaller than unity. It is conceivable that the large mass ratio of K to π comes about because the total mass of the system is so small. It is possible that even with a fairly small c (say $\sim -\frac{1}{10}$) we might explain the gross violations of unitary symmetry. We might try to interpret the large values of $g_{NN\pi}^2/g_{NK\Lambda}^2$, f_K^2/f_π^2, etc., in terms of the large value of m_K^2/m_π^2.

An example of such a calculation, and one that illustrates the various methods suggested in this article, is the following. We try to calculate f_K^2/f_π^2 in terms of m_K^2/m_π^2.

Consider the following vacuum expectation value, written in parametric representation:

$$\langle[\mathfrak{F}_{1\alpha}{}^5(x),\partial_\beta\mathfrak{F}_{1\beta}{}^5(x')]\rangle_0=i/(2\pi)^3\int d^4K\, e^{iK\cdot(x-x')}$$
$$\times K_\alpha\epsilon(K)\int dM^2/M^2\delta(K^2+M^2)\rho(M^2). \quad (9.3)$$

Here x and x' are arbitrary space-time points. In terms of (9.3), we have

$$\langle[\partial_\alpha\mathfrak{F}_{1\alpha}{}^5(x),\partial_\beta\mathfrak{F}_{1\beta}{}^5(x')]\rangle=1/(2\pi)^3\int d^4K\, e^{iK\cdot(x-x')}$$
$$\times\epsilon(K)\int dM^2\,\delta(K^2+M^2)\rho(M^2). \quad (9.4)$$

Now the contribution of the one-pion intermediate state is easily obtained in terms of the constant f_π^2:

$$\rho(M^2)=\delta(M^2-m_\pi^2)m_\pi^4/4f_\pi^2+\text{higher terms.} \quad (9.5)$$

If $\int\rho(M^2)dM^2/M^2$ converges and if the one-pion term dominates, we have

$$\int\rho(M^2)dM^2/M^2\approx m_\pi^2/4f_\pi^2. \quad (9.6)$$

But from (9.3) we can extract the expectation value of the equal-time commutator of the fourth component of $\mathfrak{F}_{1\alpha}{}^5$ with $\partial_\beta\mathfrak{F}_{1\beta}{}^5$; making use of (5.20) and (5.24), we can express the result in terms of $\langle u_0\rangle$ and $\langle u_8\rangle$. Thus we find

$$\int\rho(M^2)dM^2/M^2=[(2/3)^{\frac{1}{2}}+(1/3)^{\frac{1}{2}}c]$$
$$\times[(2/3)^{\frac{1}{2}}\langle u_0\rangle_0+(1/3)^{\frac{1}{2}}\langle u_8\rangle_0], \quad (9.7)$$

assuming convergence.

Now we can do exactly the same thing for $\mathfrak{F}_{4\alpha}{}^5$ and the K meson, obtaining, in place of the formula

$$m_\pi^2/4f_\pi^2\approx[(\tfrac{2}{3})^{\frac{1}{2}}+(\tfrac{1}{3})^{\frac{1}{2}}c][(\tfrac{2}{3})^{\frac{1}{2}}\langle u_0\rangle_0+(\tfrac{1}{3})^{\frac{1}{2}}\langle u_8\rangle_0], \quad (9.8)$$

the analogous result

$$m_K^2/4f_K^2\approx[(\tfrac{2}{3})^{\frac{1}{2}}-(\tfrac{1}{12})^{\frac{1}{2}}c][(\tfrac{2}{3})^{\frac{1}{2}}\langle u_0\rangle_0-(\tfrac{1}{12})^{\frac{1}{2}}\langle u_8\rangle_0]. \quad (9.9)$$

If c is really small, presumably $\langle u_8\rangle_0$ is also small compared to $\langle u_0\rangle_0$. Then we can, roughly, set (9.8) equal to (9.9), obtaining

$$f_K^2/f_\pi^2\approx m_K^2/m_\pi^2. \quad (9.10)$$

The left-hand side is about 6 and the right-hand side about 10. Thus we can, in a crude approximation, calculate the rate of $K^+\to\mu^++\nu$ in terms of that for $\pi^+\to\mu^++\nu$ and explain one large violation of symmetry in terms of another.

The Goldberger-Treiman formula relating f_π, $g_{NN\pi}$, and $(-G_A/G)$ can also be used for the K particle to give a relation among f_K, $g_{N\Lambda K}$, and $(-G_A/G)$ for the β decay of Λ. Of course, the K-particle pole is much closer to the branch line beginning at $(m_K+2m_\pi)^2$ than the pion pole is to the branch line beginning at $9m_\pi^2$; thus the Goldberger-Treiman formula may be quite bad for the K meson. Still, we may try to use it to discuss the coupling of N and Λ to K and to leptons. We have

$$(m_N+m_\Lambda)(-G_A{}^{N\Lambda}/G)\approx g_{N\Lambda K}/f_K, \quad (9.11)$$

by analogy with (2.8). Comparing the two formulas, we have

$$(-G_A{}^{N\Lambda}/G)^2(-G_A/G)^{-2}$$
$$\approx g_{N\Lambda K}{}^2g_{NN\pi}{}^{-2}(2m_Nf_\pi)^2[(m_\Lambda+m_N)f_K]^{-2}. \quad (9.12)$$

The ratio of g^2 factors is thought to be $\sim$0.1 from photoproduction of K, while the remaining factor on the right is also $\sim$0.1, so that the Goldberger-Treiman relation leads us to expect a very small axial vector β-decay rate for the Λ, much smaller than the observed one. The observed β decay would be nearly all vector; this prediction of the Goldberger-Treiman formula can easily be checked by observing the electron-neutrino angular correlation in the β decay of Λ, using bubble chambers.

We should, of course, try to predict the value of $g_{N\Lambda K}{}^2g_{NN\pi}{}^{-2}$ in terms of m_K^2/m_π^2 just as we did above for f_K^2/f_π^2; however, it is a much harder problem.

When we know more about the coupling constants of the vector mesons (strong coupling constants such as $\gamma_{\omega NN}$, $h_{\omega\pi\rho}$, etc., and coupling strengths of currents such as γ_ω, γ_M, etc.) we will be able to make a survey of the pattern of coupling constants as well as the pattern of masses and see whether the higher symmetry has any relevance. Also it should become clear how well the approximation of dominant low-mass states works, in terms of universality of meson couplings and Goldberger-Treiman relations.[38]

In summary, then, we suggest the use of the equal-time commutators to predict sum rules, attempts to derive high-energy conservation laws and to check them

[38] An interesting relation of the Goldberger-Treiman type is one that holds if the trace $\theta_{\alpha\alpha}$ of the stress-energy-momentum tensor has matrix elements obeying highly convergent dispersion relations. Because of the vanishing of the self-stress, the expectation value of $\theta_{\alpha\alpha}$ in the state of a particle at rest gives the mass of the particle. Rewriting the matrix element as one between the vacuum and a one-pair state, we see that the dispersion relation involves intermediate states with $I=0$, $J=0^+$, $G=+1$. If there is a resonance or quasi-resonance in this channel (like the χ' meson of Table III) and if that resonance dominates the dispersion relation at low momentum transfers, then the coupling of the resonant state to different particles is roughly proportional to their masses. That is just the situation discussed by Schwinger in reference 1 and by Gell-Mann and Lévy in reference 17 for the "σ meson."

experimentally, the search for broken symmetry at low energies, attempts to calculate some violations in terms of others, and efforts to check the highly convergent dispersion relations dominated by low-mass states.

Nowhere does our work conflict with the program of Chew *et al.* of dynamical calculation of the S matrix for strong interactions, using dispersion relations. If something like the Sakata model is correct, then most of the mesons are dynamical bound states or resonances, and their properties are calculable according to the program. Those particles for which there are fundamental fields (like n, p, Λ, and B^0 in the specific field-theoretic model) would presumably occur as CDD poles or resonances in the dispersion relations.[39]

If there are no fundamental fields and no CDD poles, all baryons and mesons being bound or resonant states of one another, models like that of Sakata will fail; the symmetry properties that we have abstracted can still be correct, however. This situation would presumably differ in two ways[10] from the one mentioned above. First, all the masses and coupling constants could be calculated from coupled dispersion relations. Second, certain scattering amplitudes at high energies would show different behavior, corresponding to different kinds of subtractions in the dispersion relations. The second point should be investigated further, as it could lead to experimental tests of the "fundamental" character of various particles.[10,40]

ACKNOWLEDGMENTS

It is a pleasure to thank R. P. Feynman, S. L. Glashow, and R. Block for many stimulating discussions of symmetry, and to acknowledge the great value of conversations with G. F. Chew, S. Frautschi, R. Haag, R. Schroer, and F. Zachariasen about the explanation of approximate universality in terms of highly convergent dispersion relations.

APPENDIX

The field theories of the Fermi-Yang and Sakata models, given by Eqs. (3.17) and (4.1), respectively, belong to a general class of theories, which we now describe.

The Lagrangian density L is given as a function of a number of fields ψ_A and their gradients. The kinetic part of the Lagrangian (consisting of those terms containing gradients) is invariant under a set of infinitesimal unitary transformations generated by N independent Hermitian operators R_i, which may depend on the time. Under the transformations, the various fields ψ_A undergo linear recombinations:

$$\begin{aligned}\psi_A(\mathbf{x},t) \to \psi_A(\mathbf{x},t) &- i\Lambda_i[R_i(t),\psi_A(\mathbf{x},t)] \\ &= \psi_A(\mathbf{x},t)+i\Lambda_i \textstyle\sum_B M_i^{AB}\psi_B(\mathbf{x},t), \qquad \text{(A1)}\end{aligned}$$

where Λ_i is the infinitesimal gauge constant associated with the ith transformation. The equal-time commutation rules of the R_i are the same as those of the matrices M_i. Moreover, the set of R_i and linear combinations of R_i is algebraically complete under commutation; in other words, we have an algebra. The matrices M_i are the basis of a representation of the algebra (in general, a reducible representation). It is convenient to take the matrices of the basis to be orthonormal,

$$\mathrm{Tr} M_i M_j = (\mathrm{const})\delta_{ij}, \qquad \text{(A2)}$$

redefining the R_i accordingly. The structure constants c_{ijk} in the commutation rules

$$\begin{aligned}[M_i,M_j]&=ic_{ijk}M_k,\\ [R_i(t),R_j(t)]&=ic_{ijk}R_k(t),\end{aligned} \qquad \text{(A3)}$$

are now real and totally antisymmetric in i, j, and k. We may still perform real rotations in the N-dimensional space of the R_i or the M_i. Suppose, after performing such a rotation, that we can split the R_i into two sets that commute with each other. Then our algebra is the direct sum of two commuting algebras. We continue this process until no further splitting is possible, even after performing rotations. The algebra has then been expressed as the direct sum of *simple* algebras. All the simple algebras have been listed by Cartan.[40] Besides the trivial one-dimensional algebra of $U(1)$ (which is not included by the mathematicians), there are the three-dimensional algebra of $SU(2)$, the eight-dimensional algebra of $SU(3)$, and so forth.

Now let us construct the currents of the operators R_i. We consider the gauge transformation of the second kind

$$\psi_A \to \psi_A(\mathbf{x},t) - i\Lambda_i(\mathbf{x},t)[R_i(t),\psi_A(\mathbf{x},t)], \qquad \text{(A4)}$$

and ask what change it induces in the Lagrange density L. There will be a term in Λ_i and a term in $\partial_\alpha\Lambda_i$, so adjusted[17] that the total change is just the divergence of a four-vector:

$$L \to L(\mathbf{x},t) - \partial_\alpha\Lambda_i(\mathbf{x},t)R_{i\alpha}(\mathbf{x},t) - \Lambda_i(\mathbf{x},t)\partial_\alpha R_{i\alpha}(\mathbf{x},t). \qquad \text{(A5)}$$

We define $R_{i\alpha}$ to be the current of R_i. It can be shown that R_i is in fact given by the relation

$$R_i = -i\int R_{i4}d^3x. \qquad \text{(A6)}$$

Now if, for constant Λ_i, the whole Lagrangian is invariant under R_i, then the term in Λ_i in (A4) must vanish; we have $\partial_\alpha R_{i\alpha}=0$. In other words, if there is exact symmetry under R_i, the current $R_{i\alpha}$ is conserved.

If there is a noninvariant part of L with respect to the symmetry operation R_i, then the current will not be conserved. By hypothesis, the noninvariant term (call it u) contains no gradients. Therefore, the effect

[39] L. Castillejo, R. H. Dalitz, and F. J. Dyson, Phys. Rev. 101, 453 (1956).

[40] S. C. Frantschi, M. Gell-Mann, and F. Zachariasen (to be published).

[40] E. Cartan, *Sur la Structure des groupes de transformations finis et continus*, thèse (Paris, 1894; 2nd ed., 1933).

of the transformation (A3) for *constant* Λ_i will be simply to add a term $-i\Lambda_i[R_i,u]$ to the Lagrangian density. We have, then, using (A4), the result

$$\partial_\alpha R_{i\alpha}(\mathbf{x},t)=i[R_i(t),u(\mathbf{x},t)]. \qquad \text{(A7)}$$

Since u contains no gradients, it is not only the noninvariant term in the Lagrangian density, but also the negative of the noninvariant term in the Hamiltonian density. The invariant part of H evidently commutes with R_i. Thus we have

$$\partial_\alpha R_{i\alpha}(\mathbf{x},t)=-i[R_i(t),H(\mathbf{x},t)]. \qquad \text{(A8)}$$

By considering the transformation properties of H under commutation with the algebra, we generate the divergences of all the currents. The formula obtained by integrating (A6) over space is, of course, very familiar:

$$\dot{R}_i=\int \partial_\alpha R_{i\alpha}d^3x=-i\left[R_i,\int Hd^3x\right]. \qquad \text{(A9)}$$

Prediction of the Ω^- Particle

Gell-Mann: If we take the unitary symmetry model with baryon and meson octets, with first order violation giving rise to mass differences, we obtain some rules for supermultiplets. The broken symmetry picture is hard to interpret on any fundamental theoretical basis, but I hope that such a justification may be forthcoming on the basis of analytic continuation of resonant states in isotopic spin and strangeness. Instead of constructing just the inverse Regge function $E\ (J)$, we can consider surfaces $E\ (J,\ I,\ Y,$ etc.). Certainly the dynamical equations are as smooth in I and Y as they are in J.

Anyway, we may look at the success of the mass rules:

$$\frac{m_N+m_\Xi}{2}=\frac{3m_\Lambda}{4}+\frac{m_\Sigma}{4}$$

and

$$m_K^2=\frac{3m_\chi^2}{4}+\frac{m_\pi^2}{4}$$

work fine, while

$$m_M^2=\frac{3m_\omega^2}{4}+\frac{m_\rho^2}{4}$$

does not work quite so well if $M=K^*$.

Suppose, now we try to incorporate the 3/2-3/2 nucleon resonance into the scheme. The only supermultiplet that does not lead to non-existent resonances in the $K-N$ channels is the *10* representation, which gives 4 states:

$$I=3/2,\quad S=0$$
$$I=1\ ,\quad S=-1$$
$$I=1/2,\quad S=-2$$
$$I=0\ ,\quad S=-3$$

The mass rule is stronger here and yields *equal spacing* of these states. Starting with the resonance at 1238 MeV, we may conjecture that the Y_1^*, at 1385 MeV and the Ξ^* at 1535 MeV might belong to this supermultiplet. Certainly they fulfil the requirement of equal spacing. If $J=3/2^+$ is really right for these two cases, then our speculation might have some value and we should look for the last particle, called, say, Ω^- with $S=-3$, $I=0$. At 1685 MeV, it would be metastable and should decay by the weak interactions into $K^-+\Lambda$, $\pi^-+\Xi^0$, or $\pi^0+\Xi^-$. Perhaps it would explain the old Eisenberg event. A beam of K^- with momentum $\gtrsim$3,5 GeV/c could yield Ω^- by means of $K^-+p\rightarrow K^++K^0+\Omega^-$.

Reprinted from 1962 International Conference on High Energy Physics 1962, p. 805.

VOLUME 9, NUMBER 6 PHYSICAL REVIEW LETTERS SEPTEMBER 15, 1962

ELEMENTARY PARTICLES OF CONVENTIONAL FIELD THEORY AS REGGE POLES

M. Gell-Mann
California Institute of Technology, Pasadena, California

and

M. L. Goldberger
Palmer Physical Laboratory, Princeton University, Princeton, New Jersey

(Received August 29, 1962)

Composite states in nonrelativistic scattering theory lie on Regge trajectories[1] corresponding to poles in the angular momentum plane that move with varying energy. Simple approximations[2-5] indicate that composite particles in relativistic field theory have the same behavior. According to the Regge pole hypothesis,[6-9] particles like the nucleon, that have customarily been treated as elementary in field theory, also lie on Regge trajectories. Is that in accord with describing such particles by ordinary perturbation theory?

It has been thought that elementary particles behave in perturbation theory as objects of fixed angular momentum.[7-9] In reference 9 the *PS-PS* theory of pions and nucleons is taken as an example. (For simplicity, let us ignore here the isotopic spin of the pion.) Writing the Feynman scattering amplitude as usual in the form $A - i\gamma \cdot qB$, we have in second order

$$A = 0,$$
$$B = -g^2/(m^2-u) + g^2/(m^2-s), \qquad (1)$$

where s and u are the Mandelstam variables. The first term in the expression for B corresponds to the nucleon, with $\alpha = J - \frac{1}{2} = 0$. (It is proportional to s^0.) The second term, although it represents the nucleon singularity in the s channel, corresponds in the u channel to an infinite sequence of angular momentum poles with $\alpha = -1, -2, -3$, etc., just like the Born approximation in nonrelativistic scattering by a Yukawa potential. Now in fourth order B acquires terms that vary as $s^{-1}\ln s$ (for large s and fixed u) and others that vary as s^0, but none that varies as $s^0 \ln s$. Thus the subsidiary angular momentum poles at $\alpha = -1, -2, \cdots$ may be beginning to move, as in potential scattering. For example, if $\alpha = -1$ becomes $\alpha = -1 + g^2F(\sqrt{u}) \cdots$, then $s^{-1+g^2F(\sqrt{u})+\cdots}$ appears in perturbation theory as $s^{-1} + g^2F(\sqrt{u})s^{-1} \times \ln s + \cdots$; but in this order the elementary nucleon pole continues to have $\alpha = 0$.

The situation is different, however, if we replace the virtual pion in the radiative correction by a virtual neutral vector meson with mass λ and coupling parameter γ. The amplitude B then acquires terms that go, for large s and fixed u, like $s^0 \ln s$. We suggest that the nucleon pole now moves too.

The variation of α for the nucleon in perturbation theory can then be studied as follows: The contribution of a Regge pole with the parity of the nucleon is given in Eq. (4.21) of reference 9. However, in order to satisfy the symmetry condition of MacDowell[10] and Frautschi and Walecka,[11] there must be a related Regge pole with the opposite parity. The two α's become coincident at $u = 0$ and complex conjugates of each other for u negative.[12,13]

Using the two equations, we have for the complete contribution[14] at large s

$$A \cong \frac{b(\sqrt{u})(\sqrt{u}-m)}{2\sin\pi\alpha(\sqrt{u})}\left(\frac{s}{s_0}\right)^{\alpha(\sqrt{u})}\{1+\exp[-i\pi\alpha(\sqrt{u})]\} + \frac{b(-\sqrt{u})(-\sqrt{u}-m)}{2\sin\pi\alpha'-\sqrt{u})}\left(\frac{s}{s_0}\right)^{\alpha(-\sqrt{u})}\{1+\exp[-i\pi\alpha(-\sqrt{u})]\};$$

$$B \cong \frac{-b(\sqrt{u})}{2\sin\pi\alpha(\sqrt{u})}\left(\frac{s}{s_0}\right)^{\alpha(\sqrt{u})}\{1+\exp[-i\pi\alpha(\sqrt{u})]\} - \frac{b(-\sqrt{u})}{2\sin\pi\alpha(-\sqrt{u})}\left(\frac{s}{s_0}\right)^{\alpha(-\sqrt{u})}\{1+\exp[-i\pi\alpha(-\sqrt{u})]\}. \quad (2)$$

Now with α equal to zero plus a correction of order γ^2 and b of order $\gamma^2 g^2$, we can write

$$A \cong [b(\sqrt{u})/\pi\alpha(\sqrt{u})](\sqrt{u}-m)[1+\alpha(\sqrt{u})\ln(s/s_0)+\cdots] + [b(-\sqrt{u})/\pi\alpha(-\sqrt{u})](-\sqrt{u}-m) \times [1+\alpha(-\sqrt{u})\ln(s/s_0)+\cdots],$$

$$B \cong [-b(\sqrt{u})/\pi\alpha(\sqrt{u})][1+\alpha(\sqrt{u})\ln(s/s_0)+\cdots] - [b(-\sqrt{u})/\pi\alpha(-\sqrt{u})][1+\alpha(-\sqrt{u})\ln(s/s_0)+\cdots]. \quad (3)$$

We compare these expressions with the results of perturbation theory at large s for fixed u:

$$A \cong 0 + (\gamma^2 g^2/8\pi^2) m I_0(u)\ln(s/s_0),$$

$$B \cong -[g^2/(m^2-u)] + (\gamma^2 g^2/8\pi^2)[I_0(u) - I_1(u)]\ln(s/s_0), \quad (4)$$

where

$$I_n(u) \equiv \int_0^1 \frac{x^n dx}{\lambda^2(1-x)+(m^2-u)x+ux^2-i\epsilon}. \quad (5)$$

We obtain

$$\alpha(\sqrt{u}) = 0 + (\gamma^2/8\pi^2)(\sqrt{u}-m)[\sqrt{u}I_0(u) - (\sqrt{u}+m)I_1(u)] + \cdots,$$

$$b(\sqrt{u}) = -(\gamma^2 g^2/16\pi)\{I_0(u) - [(\sqrt{u}+m)/\sqrt{u}]I_1(u)\} + \cdots. \quad (6)$$

It is easy to verify that these quantities obey a number of rules characteristic of the Regge pole formalism. Both α and b are real between $u = 0$ and the first physical threshold at $u = (m+\lambda)^2$, where they become complex; they both obey dispersion relations in $\sqrt{u}$. The imaginary part of α is positive. The real part of α increases through $\sqrt{u} = m$ up to threshold.

Some properties are unphysical, but these seem to be attributable to perturbation theory. The real part of α increases to $+\infty$ at threshold. Above threshold the imaginary part decreases from $+\infty$ to a constant as $u \to \infty$, while the real part decreases from a finite value and goes logarithmically to $-\infty$ as $u \to \pm\infty$.[15] The infinities at threshold are characteristic of this order of perturbation theory in the case of the Schrödinger equation with a Yukawa potential [where $\alpha = -1 - i\gamma^2/(\text{energy})^{1/2}$, see reference 6]; they should not be taken seriously.

If the nucleon has really turned into a Regge pole as a result of vector meson radiative corrections, then a study of higher order corrections can confirm the fact. The coefficients of powers of $\ln(s/s_0)$ to each order in γ^2 must be such as to agree with the expansion of $\exp[\alpha(\sqrt{u})\ln(s/s_0)]$, where $\alpha(\sqrt{u})$ is a power series in γ^2. We shall assume that the exponential character of the higher corrections will be confirmed, for example by the method of Sudakov.[16]

In ordinary quantum electrodynamics of electrons and photons (but with massive photons), the electron must then be a Regge pole, with corrections to $\alpha = 0$ of order e^2, much as in Eq. (6). The leading terms in the amplitude at large s and fixed u are[17]

$$-e^2\gamma_\nu[(-i\gamma\cdot P+m)/(m^2-u)]\gamma_\mu + (e^4/8\pi^2)\ln(s/s_0)\gamma_\nu\{-i\gamma\cdot P[I_0(u) - I_1(u)] + mI_0(u)\}\gamma_\mu, \quad (7)$$

where P is the initial electron four-momentum minus the final photon four-momentum, while ν and μ are the initial and final photon polarizations respectively.

In the simpler case of the electrodynamics of spin-zero particles (still with massive photons), the lowest order term $4e^2P_\mu P_\nu(m^2-u)^{-1}$ acquires the radiative correction $-(e^4/2\pi^2)P_\mu P_\nu[I_0(u) - 2I_1(u) + I_2(u)]\ln(s/s_0)$, so that we have (with $\alpha = J$) the result

$$\alpha(u) = 0 + (e^2/8\pi^2)(u-m^2)[I_0(u) - 2I_1(u) + I_2(u)] + \cdots,$$

$$b = -(e^4/2\pi)[I_0(u) - 2I_1(u) + I_2(u)] + \cdots. \quad (8)$$

Again these are reasonable forms for α and b except for the wild misbehavior at threshold (and possibly for the logarithmic dependence of $\text{Re}\,\alpha$ at infinity). At threshold $\text{Im}\,\alpha$ should go

VOLUME 9, NUMBER 6 PHYSICAL REVIEW LETTERS SEPTEMBER 15, 1962

like $p^{3+2\,\mathrm{Re}\alpha}$ (where p is the barycentric momentum in the u channel), b should be essentially constant, and the threshold unitarity relation,

$$\mathrm{Im}\alpha \cong \frac{-p^{3+2\alpha}}{8\pi^2} b \frac{(m+\lambda)}{\lambda^2} \frac{\sqrt{\pi}\,\Gamma(\alpha+1)}{(2\alpha+1)\Gamma(\alpha+\frac{1}{2})}, \qquad (9)$$

should hold. Presumably, the higher corrections fix up the situation at threshold, as in potential theory.

As the photon mass, λ, tends to zero all of our expressions become infrared divergent. This comes as no surprise, of course, since the divergence of the Compton amplitude in fourth order must be present in order to cancel the corresponding soft photon emission in the double Compton cross section. We have not fully analyzed the requisite modifications of the Regge formalism for this limiting situation and hope to return to the question elsewhere.

For finite photon mass, the photon is presumably on a Regge trajectory if the electron is. If we consider the correction to the amplitude for the exchange of a photon between two electrons (order e^2) arising from the exchange of three photons (order e^6), the motion of the photon pole should be apparent. Again infrared divergences may arise as $\lambda \to 0$.

The S-matrix or dispersion relation approach to relativistic quantum theory employs dispersion and unitarity formulas abstracted from conventional field theory. If it is really true that the Regge pole boundary conditions are also contained in conventional field theory, as the present work suggests, then there is no evidence for any conflict between the two points of view.

One of us (M.G.-M.) would like to acknowledge the great value of conversations with Professor Gribov, Professor Pomeranchuk, and Professor Feynman.

[1]T. Regge, Nuovo cimento 14, 951 (1959); 18, 947 (1960).

[2]B. Lee and R. Sawyer (to be published).

[3]D. Amati, S. Fubini, and A. Stanghellini (to be published).

[4]S. Mandelstam, Proceedings of the 1962 International Conference on High-Energy Physics at CERN (to be published).

[5]V. N. Gribov and I. Ya. Pomeranchuk, Proceedings of the 1962 International Conference on High-Energy Physics at CERN (to be published).

[6]R. Blankenbecler and M. L. Goldberger, Phys. Rev. 126, 766 (1962).

[7]G. F. Chew and S. C. Frautschi, Phys. Rev. Letters 7, 394 (1961).

[8]G. F. Chew and S. C. Frautschi, Phys. Rev. Letters 8, 41 (1962).

[9]S. C. Frautschi, M. Gell-Mann, and F. Zachariasen, Phys. Rev. 126, 2204 (1962).

[10]S. W. MacDowell, Phys. Rev. 116, 774 (1959).

[11]S. C. Frautschi and J. D. Walecka, Phys. Rev. 120, 1486 (1960).

[12]V. N. Gribov and I. Ya. Pomeranchuk (to be published).

[13]N. Dombey (private communication).

[14]Here s_0 is an arbitrary constant with the dimensions of mass squared; in reference 9 it was set equal to $2m_\pi m_N$.

[15]Thus for large s and then large u, the correction to the lowest order result goes like $(\ln s)(\ln u)$. In the u channel, we have the motion of a Regge pole with $\Delta\alpha(u) \propto \ln u$ for large u. In the s channel, a subsidiary Regge pole is likewise moving with $\Delta\alpha(s) \propto \ln s$ for large s.

[16]V. V. Sudakov, J. Exptl. Theoret. Phys. (U.S.S.R.) 30, 87 (1956) [translation: Soviet Phys. – JETP 3, 65 (1956)].

[17]R. P. Feynman (private communication). In the limit $\lambda \to 0$, this result is implicit in the calculation of radiative corrections to the Compton effect by L. M. Brown and R. P. Feynman, Phys. Rev. 85, 231 (1952).

VOLUME 10, NUMBER 1 PHYSICAL REVIEW LETTERS 1 JANUARY 1963

ERRATUM

ELEMENTARY PARTICLES OF CONVENTIONAL FIELD THEORY AS REGGE POLES. Murray Gell-Mann and Marvin L. Goldberger [Phys. Rev. Letters 9, 275 (1962)].

Our expression for the contribution of a Regge pole in the u channel to the scattering of pseudoscalar mesons by nucleons (with isotopic spin ignored) is not quite correct. In reference 9, the sign of B should be changed in Eqs. (4.17), (4.18), (4.20), and (4.21); the same is true of Eqs. (2) and (3) of our Letter.

Let us denote by P the initial nucleon four-momentum minus the final meson four-momentum and write $\not{P} = -i\gamma \cdot P$, with $\not{P}^2 = u$. The contribution to the scattering amplitude of a Regge pole of positive signature in the u channel can then be written, for large s and fixed u, as

$$\gamma_5 \frac{b(\sqrt{u})(u^{1/2}+\not{P})}{2\sin\pi\alpha(\sqrt{u})}\left(\frac{s}{s_0}\right)^{\alpha(\sqrt{u})}\{1+\exp[-i\pi\alpha(\sqrt{u})]\}\gamma_5$$

$$+\gamma_5 \frac{b(-\sqrt{u})(-u^{1/2}+\not{P})}{2\sin\pi\alpha(-\sqrt{u})}\left(\frac{s}{s_0}\right)^{\alpha(-\sqrt{u})}\{1+\exp[-i\pi\alpha(-\sqrt{u})]\}\gamma_5,$$

in place of Eq. (2). Since $(u^{1/2}+\not{P})(2\sqrt{u})^{-1}$ and $(u^{1/2}-\not{P})(2\sqrt{u})^{-1}$ are projection operators, we may simplify and obtain

$$\gamma_5 \not{P} b(\not{P})\left(\frac{s}{s_0}\right)^{\alpha(\not{P})}\left\{\frac{1+\exp[-i\pi\alpha(\not{P})]}{\sin\pi\alpha(\not{P})}\right\}\gamma_5.$$

We now introduce a neutral vector meson of mass λ coupled to the nucleon current with coupling constant $\gamma^2/4\pi$; perturbation theory to orders g^2 and $\gamma^2 g^2$ then gives, for large s and fixed u, the result

$$\gamma_5\left\{\frac{g^2}{\not{P}-m}+\frac{\gamma^2}{8\pi^2}g^2\ln\left(\frac{s}{s_0}\right)\right.$$

$$\left.\times\int_0^1\frac{dx[\not{P}(1-x)+m]}{\lambda^2(1-x)+(m^2-\not{P}^2)x+\not{P}^2x^2-i\epsilon}\right\}\gamma_5.$$

[Thus, in our Eq. (4), in the formula for A, I_0 should be replaced by I_1.] If we identify the logarithmic term as part of a series transforming the Born approximation term into one representing a Regge pole, then we conjecture that higher orders of perturbation theory in γ^2, with the highest power of $\ln s$ retained in each order, will yield

$$\gamma_5\frac{g^2}{\not{P}-m}\exp\left\{\frac{\gamma^2}{8\pi^2}\ln\left(\frac{s}{s_0}\right)(\not{P}-m)\right.$$

$$\left.\times\int_0^1\frac{dx[\not{P}(1-x)+m]}{\lambda^2(1-x)+(m^2-\not{P}^2)x+\not{P}^2x^2-i\epsilon}\right\}\gamma_5.$$

If the conjecture is right, then we have, correcting Eq. (6), the following lowest-order expressions for α and b:

$$\alpha(\not{P})=\frac{\gamma^2}{8\pi^2}(\not{P}-m)\int_0^1\frac{dx[\not{P}(1-x)+m]}{\lambda^2(1-x)+(m^2-\not{P}^2)x+\not{P}^2x^2-i\epsilon}+\cdots,$$

$$b(\not{P})=\pi g^2\alpha(\not{P})[2\not{P}(\not{P}-m)]^{-1}+\cdots.$$

Volume 8, number 3 PHYSICS LETTERS 1 February 1964

A SCHEMATIC MODEL OF BARYONS AND MESONS *

M. GELL-MANN
California Institute of Technology, Pasadena, California

Received 4 January 1964

If we assume that the strong interactions of baryons and mesons are correctly described in terms of the broken "eightfold way" [1-3], we are tempted to look for some fundamental explanation of the situation. A highly ɔromised approach is the purely dynamical "boo! trap" model for all the strongly interacting particles within which one may try to derive isotopic spin and strangeness conservation and broken eightfold symmetry from self-consistency alone [4]. Of course, with only strong interactions, the orientation of the asymmetry in the unitary space cannot be specified; one hopes that in some way the selection of specific components of the F-spin by electromagnetism and the weak interactions determines the choice of isotopic spin and hypercharge directions.

Even if we consider the scattering amplitudes of strongly interacting particles on the mass shell only and treat the matrix elements of the weak, electromagnetic, and gravitational interactions by means of dispersion theory, there are still meaningful and important questions regarding the algebraic properties of these interactions that have so far been discussed only by abstracting the properties from a formal field theory model based on fundamental entities [3] from which the baryons and mesons are built up.

If these entities were octets, we might expect the underlying symmetry group to be SU(8) instead of SU(3); it is therefore tempting to try to use unitary triplets as fundamental objects. A unitary triplet t consists of an isotopic singlet s of electric charge z (in units of e) and an isotopic doublet (u, d) with charges z+1 and z respectively. The anti-triplet $\bar{t}$ has, of course, the opposite signs of the charges. Complete symmetry among the members of the triplet gives the exact eightfold way, while a mass difference, for example, between the isotopic doublet and singlet gives the first-order violation.

For any value of z and of triplet spin, we can construct baryon octets from a basic neutral baryon singlet b by taking combinations $(b t \bar{t})$, $(b t t \bar{t} \bar{t})$, etc. **. From $(b t \bar{t})$, we get the representations **1** and **8**, while from $(b t t \bar{t} \bar{t})$ we get **1**, **8**, **10**, $\overline{\mathbf{10}}$, and **27**. In a similar way, meson singlets and octets can be made out of $(t \bar{t})$, $(t t \bar{t} \bar{t})$, etc. The quantum number $n_t - n_{\bar{t}}$ would be zero for all known baryons and mesons. The most interesting example of such a model is one in which the triplet has spin $\frac{1}{2}$ and $z = -1$, so that the four particles d^-, s^-, u^0 and b^0 exhibit a parallel with the leptons.

A simpler and more elegant scheme can be constructed if we allow non-integral values for the charges. We can dispense entirely with the basic baryon b if we assign to the triplet t the following properties: spin $\frac{1}{2}$, $z = -\frac{1}{3}$, and baryon number $\frac{1}{3}$. We then refer to the members $u^{\frac{2}{3}}$, $d^{-\frac{1}{3}}$, and $s^{-\frac{1}{3}}$ of the triplet as "quarks" [6] q and the members of the anti-triplet as anti-quarks $\bar{q}$. Baryons can now be constructed from quarks by using the combinations $(q q q)$, $(q q q q \bar{q})$, etc., while mesons are made out of $(q \bar{q})$, $(q q \bar{q} \bar{q})$, etc. It is assuming that the lowest baryon configuration $(q q q)$ gives just the representations **1**, **8**, and **10** that have been observed, while the lowest meson configuration $(q \bar{q})$ similarly gives just **1** and **8**.

A formal mathematical model based on field theory can be built up for the quarks exactly as for p, n, Λ in the old Sakata model, for example [3] with all strong interactions ascribed to a neutral vector meson field interacting symmetrically with the three particles. Within such a framework, the electromagnetic current (in units of e) is just

$$\mathrm{i}\{\tfrac{2}{3}\,\bar{u}\,\gamma_\alpha\,u - \tfrac{1}{3}\,\bar{d}\,\gamma_\alpha\,d - \tfrac{1}{3}\,\bar{s}\,\gamma_\alpha\,s\}$$

or $\mathscr{F}_{3\alpha} + \mathscr{F}_{8\alpha}/\sqrt{3}$ in the notation of ref. [3]. For the weak current, we can take over from the Sakata model the form suggested by Gell-Mann and Lévy [7], namely $\mathrm{i}\,\bar{p}\,\gamma_\alpha(1+\gamma_5)(n\cos\theta + \Lambda\sin\theta)$, which gives in the quark scheme the expression ***

$$\mathrm{i}\,\bar{u}\,\gamma_\alpha(1+\gamma_5)(d\cos\theta + s\sin\theta)$$

* Work supported in part by the U.S. Atomic Energy Commission.

** This is similar to the treatment in ref. [1]. See also ref. [5].

*** The parallel with $\mathrm{i}\,\bar{\nu}_e\,\gamma_\alpha(1+\gamma_5)\,e$ and $\mathrm{i}\,\bar{\nu}_\mu\,\gamma_\alpha(1+\gamma_5)\mu$ is obvious. Likewise, in the model with d^-, s^-, u^0, and b^0 discussed above, we would take the weak current to be $\mathrm{i}(\bar{b}^0\cos\theta + \bar{u}^0\sin\theta)\,\gamma_\alpha(1+\gamma_5)\,s^- + \mathrm{i}(\bar{u}^0\cos\theta - \bar{b}^0\sin\theta)\,\gamma_\alpha(1+\gamma_5)\,d^-$. The part with $\Delta(n_t - n_{\bar{t}}) = 0$ is just $\mathrm{i}\,\bar{u}^0\,\gamma_\alpha(1+\gamma_5)(d^-\cos\theta + s^-\sin\theta)$.

Volume 8, number 3 PHYSICS LETTERS 1 February 1964

or, in the notation of ref. 3),

$$[\mathscr{F}_{1\alpha} + \mathscr{F}^5_{1\alpha} + i(\mathscr{F}_{2\alpha} + \mathscr{F}^5_{2\alpha})] \cos\theta$$
$$+ [\mathscr{F}_{4\alpha} + \mathscr{F}^5_{4\alpha} + i(\mathscr{F}_{5\alpha} + \mathscr{F}^5_{5\alpha})] \sin\theta .$$

We thus obtain all the features of Cabibbo's picture [8] of the weak current, namely the rules $|\Delta I| = 1$, $\Delta Y = 0$ and $|\Delta I| = \frac{1}{2}$, $\Delta Y/\Delta Q = +1$, the conserved $\Delta Y = 0$ current with coefficient $\cos\theta$, the vector current in general as a component of the current of the F-spin, and the axial vector current transforming under SU(3) as the same component of another octet. Furthermore, we have [3] the equal-time commutation rules for the fourth components of the currents:

$$[\mathscr{F}_{j4}(x) \pm \mathscr{F}^5_{j4}(x),\ \mathscr{F}_{k4}(x') \pm \mathscr{F}^5_{k4}(x')] =$$
$$- 2 f_{jkl}\, [\mathscr{F}_{l4}(x) \pm \mathscr{F}^5_{l4}(x)]\, \delta(x-x') ,$$

$$[\mathscr{F}_{j4}(x) \pm \mathscr{F}^5_{j4}(x),\ \mathscr{F}_{k4}(x') \mp \mathscr{F}^5_{k4}(x')] = 0 ,$$

$i = 1, \ldots 8$, yielding the group SU(3) × SU(3). We can also look at the behaviour of the energy density $\theta_{44}(x)$ (in the gravitational interaction) under equal-time commutation with the operators $\mathscr{F}_{j4}(x') \pm \mathscr{F}_{j4}{}^5(x')$. That part which is non-invariant under the group will transform like particular representations of SU(3) × SU(3), for example like $(3, \bar{3})$ and $(\bar{3}, 3)$ if it comes just from the masses of the quarks.

All these relations can now be abstracted from the field theory model and used in a dispersion theory treatment. The scattering amplitudes for strongly interacting particles on the mass shell are assumed known; there is then a system of linear dispersion relations for the matrix elements of the weak currents (and also the electromagnetic and gravitational interactions) to lowest order in these interactions. These dispersion relations, unsubtracted and supplemented by the non-linear commutation rules abstracted from the field theory, may be powerful enough to determine all the matrix elements of the weak currents, including the effective strengths of the axial vector current matrix elements compared with those of the vector current.

It is fun to speculate about the way quarks would behave if they were physical particles of finite mass (instead of purely mathematical entities as they would be in the limit of infinite mass). Since charge and baryon number are exactly conserved, one of the quarks (presumably $u^{\frac{2}{3}}$ or $d^{-\frac{1}{3}}$) would be absolutely stable *, while the other member of the doublet would go into the first member very slowly by β-decay or K-capture. The isotopic singlet quark would presumably decay into the doublet by weak interactions, much as Λ goes into N. Ordinary matter near the earth's surface would be contaminated by stable quarks as a result of high energy cosmic ray events throughout the earth's history, but the contamination is estimated to be so small that it would never have been detected. A search for stable quarks of charge $-\frac{1}{3}$ or $+\frac{2}{3}$ and/or stable di-quarks of charge $-\frac{2}{3}$ or $+\frac{1}{3}$ or $+\frac{4}{3}$ at the highest energy accelerators would help to reassure us of the non-existence of real quarks.

These ideas were developed during a visit to Columbia University in March 1963; the author would like to thank Professor Robert Serber for stimulating them.

References

1) M. Gell-Mann, California Institute of Technology Synchrotron Laboratory Report CTSL-20 (1961).
2) Y. Ne'eman, Nuclear Phys. 26 (1961) 222.
3) M. Gell-Mann, Phys. Rev. 125 (1962) 1067.
4) E.g.: R.H. Capps, Phys. Rev. Letters 10 (1963) 312; R.E. Cutkosky, J. Kalckar and P. Tarjanne, Physics Letters 1 (1962) 93;
E. Abers, F. Zachariasen and A.C. Zemach, Phys. Rev. 132 (1963) 1831;
S. Glashow, Phys. Rev. 130 (1963) 2132;
R.E. Cutkosky and P. Tarjanne, Phys. Rev. 132 (1963) 1354.
5) P. Tarjanne and V.L. Teplitz, Phys. Rev. Letters 11 (1963) 447.
6) James Joyce, Finnegan's Wake (Viking Press, New York, 1939) p. 383.
7) M. Gell-Mann and M. Lévy, Nuovo Cimento 16 (1960) 705.
8) N. Cabibbo, Phys. Rev. Letters 10 (1963) 531.

* There is the alternative possibility that the quarks are unstable under decay into baryon plus anti-di-quark or anti-baryon plus quadri-quark. In any case, some particle of fractional charge would have to be absolutely stable.

* * * * *

Introductory Session

CURRENT TOPICS IN PARTICLE PHYSICS

Murray Gell-Mann

It's a great pleasure and a great honor to be here and to address such a distinguished audience and so many old friends, but I'm not at all clear about what the subject should be. It was implied that I had chosen the task of summarizing the Conference in advance; that is hardly the case. I've had lots of advice from many people about what to do. Some people have said, "I hope that you can tell us everything that's going to be important so we don't have to go to the sessions." Others have insisted that it would be absurd to try to summarize the Conference in advance and what I should do is to give some general philosophical statements about the progress of high energy physics and the meaning of high energy physics. Other people have told me that I am far too young and far too involved in the subject to be able to give any general philosophical pronouncements and that I should concentrate on some discussion of what's going on in the field. I think the last is probably the most reasonable. I'll try to say something about my personal prejudices about the field. The theme, let us say, is what I am eager to hear about at the Conference in the next few days on the basis of all the rumors since the preceding Conference. Now, if I don't mention something that you have done that's not at all because I don't consider it important or because I'm not anxious to hear about it but only because there isn't time to talk about everything here. And if I mention very few names, it will be simply because I don't want to make the mistake of leaving out any. I may mention some for purposes of identification or to quote those people who modestly left themselves off the invitation list.

Now, what sorts of things can we really expect to hear? We've been told by Professor McMillan about some of the profound changes that have taken place in the field since the first Rochester Conference. What has happened to theory in all the time since then? Then we were making unreliable calculations of the deuteron structure based on the exchange of one pion. Now we have gone to the stage where we make unreliable calculations of 50 or 60 bound states on the basis of exchanging 50 or 60 particles, and the progress is amazing. At that time, the experimental people were still debating about whether strange particles existed, and now we know they exist but not why.

In some respects, it's rather humbling to think about how little progress we've made in the last 15 years; but if we actually look at the data accumulated and the theoretical analysis, it's clear that we are much further on our way toward understanding the particles. It's rare that, at a meeting like this, one sees the synthesis actually forming before one's eyes, but what I think happens more often is that going home afterwards one has a lot of ideas that have seeped in during the meeting and that form a coherent picture in the mind.

What can we look forward to hearing? Something which we can certainly look forward to hearing, although not necessarily with pleasure, is a lot of discussion among the different kinds of theorists about whether one should work with "S-matrix theory" or "field theory" or "Lagrangian field theory" or "abstract field theory," and I would like to suggest, as a way of settling this once and for all, that we recognize how remarkable it is that field theory works at all, that so far we have not had to abandon our basic theoretical tool for understanding particles, namely relativistic quantum mechanics. It comes equipped with all its attendant details, which some people want to refer to as microcausality in space-time and other people as analyticity in momentum-space, but as far as anyone knows, these are very similar. It would be better if all the efforts that we expend on the discussions on which form of field theory one should use were devoted to arguing for a higher-energy accelerator so that we can do more experiments over the next generation and really learn more about the basic structure of matter.

The experiments at the highest energies now available at CERN and Brookhaven are certainly exciting. A number of quite refined experiments are now available on high-energy scattering. It's true that you can do experiments with large momentum transfer and experiments with the production of a great many particles, but those have not proved terribly easy to analyze so far; they present a challenge to theory. (If I seem to emphasize theory a great deal in this discussion, it is only because that is the way I make my living.) The experiments at high energies with small momentum transfer seem to be susceptible of rather detailed analysis, and I expect what we hear on the subject to be quite interesting. Some analysis has been carried out using optical model methods and some analysis using Regge poles.

Regge poles, as you recall, were very popular at the last-Conference-but-one for this purpose, and at the most recent Conference they were somewhat muted. To some extent, they seem to work again.

Now, the difficulties that have accumulated with them have been both theoretical and experimental. The experimental one was simply a failure of one or two trajectories to explain the data at high energies, but we now know that there are many mesons and it's not unreasonable that there should be about an equal number of trajectories, and so an analysis with several trajectories is by no means absurd. The theoretical difficulties were more serious; it was discovered that poles apparently implied other kinds of singularity in the complex plane of angular momentum. Some sort of horrible essential singularity seemed to develop which could move up to rather high J and therefore up to great prominence in high energy reactions. Imagine a hole in the wall that we cover with an ugly picture; when we want to take down the ugly picture we remember that we put it up in order to cover the hole in the plaster. So we have cuts in the angular momentum plane which were introduced in order to paper over the essential singularity; their existence is strongly suggested by theory, although not rigorously proved, and they are a real nuisance, as we shall see in a moment. It is still possible that further study may render them more innocuous.

If only poles are used, one finds the famous expression for their contributions to a high-energy scattering amplitude for a particle a going into b and c going into d, as a function of the invariant

Reprinted from *Proceedings of the XIII Int. Conf. on High-Energy Physics* (University of California Press, 1967), pp. 3–9.

energy variable s and momentum transfer variable t:

$$\sum_n \frac{\beta_{abn}(t)\beta_{cdn}(t)}{\sin \pi(\alpha_n(t)-v)} \left\{1 \pm e^{-i\pi[\alpha_n(t)-v]}\right\} s^{\alpha_n(t)-\Delta}, \quad (1)$$

where the index n labels the trajectory, Δ is the number of units of helicity flip, and v is zero for meson trajectories and 1/2 for baryon trajectories. The angular momentum J of the Regge pole is $\alpha_n(t)$. The number ± 1 is called the signature of the trajectory; a meson trajectory of positive signature may have along it, for $t > 0$, meson states of even spin, while a negative-signature trajectory may have states of odd spin. Likewise a baryon trajectory of even signature can have particles of spin 3/2, 7/2, etc., along it, while a negative-signature baryon trajectory can have particles of spin 1/2, 5/2, etc., along it for $t > 0$. These particles correspond to poles in t of Expression 1. In high-energy scattering, however, we are concerned with the region $t < 0$, and we have no poles for particles of negative mass squared!

We can, in various reactions involving the exchange of particular quantum numbers, try to analyze the scattering amplitudes in terms of one or more leading trajectories (with the largest α_n) having those quantum numbers, using Eq. 1. Unfortunately, the most obnoxious cuts that are thought to exist in the angular momentum plane give contributions that, at $t = 0$, are smaller than the leading pole contributions only by a factor of $\ln s$ at large s, and for $t < 0$, are larger than the leading pole contributions. These dominant cuts can be roughly described as giving terms that go as $s^{\alpha(0)}/\ln s$, where α is the leading trajectory. Nevertheless, a description in terms of poles alone usually gives a good description of the data. Are the dominant cuts absent because of some hole in the theoretical arguments for them? Or are they present, but with small numerical coefficients? If the latter is true, then for $t < 0$ the higher we go in energy the more the waters will be muddied by the dominant cuts, a most unpleasant situation. With these remarks, let me forget cuts and go on with an account of the description of high-energy two-body reactions in terms of poles.

In any reaction in which strangeness, isotopic spin, or baryon number is exchanged, the analysis in terms of one or two leading trajectories works beautifully. Thus we find trajectories on which the nucleon and $\Delta(1240)$ could lie, meson trajectories on which the ρ and "A_2" could lie, and so forth.

From forward amplitudes or total cross sections, we find rather precise values for $\alpha(0)$. Also, there are many interesting theoretical properties of trajectories at $t = 0$, some of which will undoubtedly be discussed at this meeting. Trajectories with different signatures and different parities and different charge-conjugation behavior can be connected at $t = 0$ and have α's that are equal or that differ by integers. The reason is that if we solve a bound-state problem at $t = 0$, we can describe the system as having energy and momentum equal to zero, so that there is four-dimensional angular invariance, much higher symmetry than usual.

Another kind of place where interesting information is available, both theoretically and experimentally, is where a trajectory passes through certain half-integral or integral values of α. The contribution of the trajectory to particular scattering amplitudes or to all amplitudes may have to vanish at such a point. The various kinds of zeroes give dips in the angular distributions of scattering processes at characteristic values of t, and a number of such dips seem to have been identified in the data. For the "ρ trajectory," for example, we now have good evidence that $\alpha = 0$ at $t \approx -0.6\ (\text{BeV})^2$, as well as that $\alpha(0) \approx 0.5$ at $t = 0$. For the "N trajectory," we have fair evidence that $\alpha = -1/2$ at $t \approx -0.2\ (\text{BeV})^2$, as well as that $\alpha(0) \approx -0.3$ at $t = 0$.

In this way a set of provisional trajectories has been plotted, using information from scattering at $t \leq 0$ and information from the existence of known particles at certain values of $t > 0$. These trajectories are all nearly linear in t over a considerable range and with roughly the same slope, around 1 $(\text{BeV})^{-2}$. For example, the $t \leq 0$ portion of the nucleon trajectory just mentioned connects smoothly with the points corresponding to $\text{Re}\,\alpha = 1/2$ at $(0.94\ \text{BeV})^2$, $\text{Re}\,\alpha = 5/2$ at $(1.68\ \text{BeV})^2$, and with suspected $9/2^+$ and $13/2^+$ resonances lying higher. The Δ trajectory has $\alpha(0) \approx -0.1$, passes through $\alpha = 3/2$ at $(1.24\ \text{BeV})^2$, $7/2$ at $(1.92\ \text{BeV})^2$, and may pass through higher suspected resonances with $11/2^+$ and $15/2^+$.

Now let us look at meson trajectories. The ρ trajectory is rather straight, and suggests that states with $J = 3^-, 5^-$, etc., should show up at rather well-defined masses; there is even some slight experimental evidence that this is so, as we shall no doubt hear. The "A_2" trajectory with opposite parity and signature appears to lie very close by, passing through the 2^+ meson at about $(1.31\ \text{BeV})^2$ and suggesting the existence of $J = 4^+$, etc., lying higher. The rough coincidence of trajectories and residues of opposite signature and parity has been called "exchange degeneracy" because it is exact in two-body systems with no exchange forces.

A very simple picture of the 1^- and 2^+ nonets of mesons then emerges, in which they lie on roughly straight, parallel, and exchange-degenerate trajectories, but we must then rearrange somewhat our ideas about elastic scattering amplitudes in which no quantum numbers are exchanged, like the sum of pp and $\bar{p}p$ amplitudes, or the sum of π^-p and π^+p amplitudes. Here the main effect experimentally is diffraction scattering, dominated by the exchange of the "Pomeranchuk pole" with $\alpha(0) = 1$ or nearly so, and it was thought that this pole had a trajectory passing through the 2^+ meson f^0 at $(1.25\ \text{BeV})^2$. It is now much more natural to say that the next highest trajectory, with $\alpha(0) \approx 0.4$, passes through this meson, while a still lower trajectory passes through the 2^+ meson $f^{0\prime}$ at $(1.5\ \text{BeV})^2$. The Pomeranchuk pole is left high and dry, with no known mesons on its trajectory, which is also rather flat, since the diffraction scattering peak does not show much shrinking in t as s gets larger. It may be that the Pomeranchuk pole is a fixed one, with $\alpha(t) = 1$. If that is so, then describing the leading term in diffraction scattering by means of the coupling to this pole is the same as describing it by an "optical model," and may be capable of describing not only the diffraction data at small negative values of t but also the data up to very high values. The leading term has, of course, the form $s\, f_{ab}(t)\, f_{cd}(t)$, where, for elastic scattering, $a = b$ and $c = d$.

Let us assume that the high-energy data are correctly described in terms of Regge poles (plus perhaps a fixed pole and whatever associated cuts and other singularities in the complex plane are required). Now consider the bulk of the experimental results at lower energies, described in terms of the formation and production of resonances and, to some extent, in terms of the exchange of bound states and resonances (the peripheral model). An immense amount of information is now being gathered about hundreds of bound and resonant states of the meson and baryon systems. We are learning from the data first of all their masses and quantum numbers. We are learning about the baryon-baryon-meson and meson-meson-meson coupling constants among all the states; these govern the strong decays of resonances and the exchanges of meson and baryon states. We are also learning from electromagnetic production and decays some of the matrix elements of the electromagnetic current between states; in some cases, electron scattering experiments tell us form factors. The much more difficult neutrino experiments are beginning to tell us a little bit about the matrix elements of the weak current and their form factors. All of this information, together with the high-energy work on Regge trajectories, is beginning to fit into a coherent picture of the mesons and baryons.

We discuss first the masses and quantum numbers of the meson and baryon states and the strong coupling constants among them. These can be incorporated, along with our knowledge of trajectories, into a unified systematics of hadron states. We have seen that the trajectories $\alpha(t)$ are studied for $t \leqslant 0$ in high-energy experiments, and that when $t > 0$ the same trajectories give us the hadron states. For example, any meson of mass μ and spin J corresponds to a situation in which a trajectory n with the right quantum numbers and with signature $(-1)^J$ has $\mathrm{Re}\,\alpha = J$ at $t = \mu^2$ and $\mathrm{Im}\,\alpha$ proportional to the width of the meson state, while the coupling parameter $\beta_{abn}(t)$ of the trajectory to two hadron states a and b gives, at $t = \mu^2$, the coupling constant of the meson state to a and b. The same kind of thing is true of baryon states and baryon trajectories. The rough linearity in t of trajectories suggests not only that we will have long series of rotational levels, as in nuclear physics, but also that these will be rather narrow, since the widths are proportional to $\mathrm{Im}\,\alpha$ and a "real analytic function" that is nearly linear does not have much of an imaginary part.

Now it is not only in the relation between mass M and angular momentum J that we have such simplicity in the systematics. We know that if we think of M as a function of I, Y, J, and so forth, we can define families of particles or trajectories for which the variation of M in all these variables is very smooth. That brings us to the subject of approximate symmetries of the hadrons and their strong interaction. The main features of what we now know about the approximate symmetries of the hadron spectrum are most simply described by the "quark model."

We consider three hypothetical and probably fictitious spin 1/2 quarks, falling into an isotopic doublet, u and d (for "up" and "down"), with charges +2/3 and -1/3 respectively, and an isotopic singlet s with charge -1/3. Corresponding to these quarks q we take the three kinds of antiquarks $\bar{q}$. We make the known bound and resonant states of the mesons formally out of $\bar{q}q$ and the known bound and resonant baryon states out of qqq.

For the mesons, we obtain roughly degenerate nonets, including the ^1S nonet that gives pseudoscalar mesons and the ^3S nonet that gives vector mesons. The series $^3S_1$, $^3P_2$, $^3D_3$, $^3F_4$, etc., gives the two trajectories on which the known vector and tensor mesons lie. We expect to find, near the tensor mesons, also the $^3P_1$ nonet (ordinary axial vector mesons) and the $^3P_0$ nonet (scalar mesons). The singlet configuration should give trajectories including not only the pseudoscalar nonet and its expected rotational excitations (^1S, ^1D, ^1G, etc.), but also the ^1P, $^1F, \cdots$ nonets, where ^1P corresponds to axial vector mesons with opposite charge conjugation, that should lie in the region just above 1 BeV as do the other P states. We shall hear, I am sure, how well the theoretical $^1P_1$, $^3P_0$, and $^3P_1$ nonets fit in with the observed bumps in this region of mass. I understand that the evidence is consistent with the theoretical picture, but not conclusive.

For the baryons, we start with the lowest configurations of qqq, corresponding to the $J = 1/2^+$ octet and the $J = 3/2^+$ decimet. These correspond to $^2S_{1/2}$ and $^4S_{3/2}$, where the spin-unitary-spin wave function is symmetrical. The simplest assumption is that the quarks have a totally symmetric wave function altogether (unlike real fermions) and that the ground state is an overall s state, as in a problem with mostly ordinary forces. (It is also possible, of course, that the ground state is a complicated S state made of two internal p waves and that the quarks act like fermions, but I shall not pursue that further.) The next likely configurations include only one with negative parity, namely a P state that transforms, as do the internal coordinates $\underset{\sim}{x}_1 - \underset{\sim}{x}_2$, $(\underset{\sim}{x}_1 + \underset{\sim}{x}_2)/2 - \underset{\sim}{x}_3$, according to the Young diagram

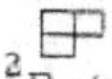

, and gives $^2P_{1/2}(\underset{\sim}{1})$, $^2P_{3/2}(\underset{\sim}{1})$, $^2P_{1/2}(\underset{\sim}{8})$, $^2P_{3/2}(\underset{\sim}{8})$, $^2P_{1/2}(\underset{\sim}{10})$, $^2P_{3/2}(\underset{\sim}{10})$, $^4P_{1/2}(\underset{\sim}{8})$, $^4P_{3/2}(\underset{\sim}{8})$, and $^4P_{5/2}(\underset{\sim}{8})$. In the language of "SU(6)," we have for the ground state a $\underset{\sim}{56}$ with $L = 0^+$, and for the first negative parity family a $\underset{\sim}{70}$ with $L = 1^-$. The observational evidence is, I understand, in reasonable agreement with this picture, but there is also evidence for low-lying excited configurations of positive parity, which may be $\underset{\sim}{56}$, $L = 0^+$, $\underset{\sim}{56}$, $L = 2^+$, and perhaps some others.

In general, the spectrum looks like the solution of a wave equation with a rather simple "potential." For example, for the mesons, the situation for M^2 resembles that of the energy in a three-dimensional harmonic oscillator with perturbations giving spin-orbit splitting, octet-singlet splitting for the spin singlet, and simple SU(3) breaking. The harmonic oscillator would give trajectories with J exactly linear in M^2.

Now what is going on? What are these quarks? It is possible that real quarks exist, but if so they have a high threshold for copious production, many BeV; if this threshold comes from their rest mass, they must be very heavy and it is hard to see how deeply bound states of such heavy real quarks could look like $q\bar{q}$, say, rather than a terrible mixture of $q\bar{q}$, $qq\bar{q}\bar{q}$, and so on. Even if there are light real quarks, and the threshold comes from a very high barrier, the idea that mesons and baryons are made primarily of quarks is difficult to believe, since we know that, in the sense of dispersion theory, they are mostly, if not entirely, made up out of one another. The probability that a meson consists of a real quark pair rather than two mesons or a baryon and antibaryon must be quite small. Thus it seems to me that whether or not real quarks exist, the q and $\bar{q}$ we have been talking about are mathematical; in particular, I would guess that they are mathematical entities that arise when we construct representations of current algebra, which we shall discuss later on. Their effective masses, to the extent that these have meaning, seem to be of the order of one-third the nucleon mass. One may think of mathematical quarks as the limit of real light quarks confined by a barrier, as the barrier goes to an infinitely high one.

If the mesons and baryons are made of mathematical quarks, then the quark model may perfectly well be compatible with the bootstrap hypothesis, that hadrons are made up out of one another.

Experimentally, it is a very interesting question whether all reasonably well-defined excited meson and baryon states fit in with the $\bar{q}q$ and qqq assignments, which allow only nonets for the mesons and only $\underline{1}$, $\underline{8}$, and $\underline{10}$ representations of SU(3) for the baryons, or whether there are resonances that must be assigned to higher configurations, like $\bar{q}q\bar{q}q$ or $qqqq\bar{q}$. If the latter is the case, then we must find resonances with exotic I and Y values. It will be interesting to find out how the evidence for such states stands at present. This matter, although important, is, of course, not absolutely fundamental, since we know that continua with such quantum numbers exist, and it is a dynamical question whether or not well-defined resonances are found.

The quark model, if correctly interpreted, should give information not only about the hadron spectrum, but also about the strong coupling constants or, more generally, the couplings of pairs of particles to trajectories. Some successes have been achieved by the so-called $[U(6)]_W$ symmetry, and it is to be hoped that the relation of this approximate symmetry to the quark model and to current algebra will soon be much better understood. A striking success in interpreting high-energy scattering data comes from a very simple assumption of symmetry and universality of meson and baryon couplings to the highest meson Regge poles, including the "Pomeranchuk pole" that governs diffraction scattering. This assumption amounts merely to "quark counting," and in its most primitive form says that total meson-baryon and baryon-baryon cross sections are in the ratio 2:3, since the meson is made of two quarks and the baryon of three. It will be fascinating to see how well the various relations work that have come out of "quark counting." I should say that one can apply this method to the high energy data even if the Regge pole hypothesis should collapse.

Now we have spoken of the relations of current algebra and I should like to go on now to discuss them, as well as other sets of presumably exact relations (to all orders in the strong interaction) that we use as theoretical tools in describing the hadrons. Let us start with the new superconvergence relations, which I hope will be presented to the Conference, the work of Fubini and collaborators. These pertain to hadron scattering amplitudes without currents.

Consider a hadron scattering amplitude $A(s,t)$, without kinematic singularities and involving Δ units of helicity flip. We have seen in Eq. 1 and the discussion following that the asymptotic behavior of $A(s,t)$ is (at $t = 0$ and, in the worst case of cuts, for $t \leq 0$) $s^{\alpha(0)-\Delta}$, where α is the leading exchanged trajectory. (Even if not derived theoretically, such power laws can be simply taken from experiment.) Take, for convenience, an amplitude for which $\mathrm{Re}\,A(\nu)$ is odd in $\nu \equiv s-u$. If $\alpha(0) - \Delta < 1$ (and it is always ≤ 1), then such an amplitude obeys an unsubtracted dispersion relation in ν for fixed t:

$$A(\nu) = \frac{2\nu}{\pi} \int d\nu' \, \frac{\mathrm{Im}\,A(\nu')}{\nu'^2 - \nu^2 - i\epsilon} \, . \tag{2}$$

Now suppose $\alpha(0) - \Delta < -1$; we get an additional sum rule or "superconvergence relation",

$$\int d\nu' \, \mathrm{Im}\,A(\nu') = 0. \tag{3}$$

To take a concrete example, consider nucleon-antinucleon exchange scattering, so that objects like the deuteron are exchanged. In these two-baryon channels, the trajectories must lie very low; moreover, in the nucleon-antinucleon system we can get Δ as high as 2. Thus there are numerous superconvergence rules like Eq. 2. They express exactly the kind of physics that one attempted to express 25 years ago by saying that there must be several mesons, giving nucleon-nucleon forces with cancelling singularities, so that the deuteron would have a binding energy that is finite and even small. Approximating all the $A = 0$ states by discrete mesons M, Eq. 3 describes such a cancellation,

$$\sum_M g^2_{NNM} f(m_N^2, m_M^2) = 0, \tag{4}$$

where f is a kinematic function.

Now this approximation of integration over complicated continua by summation over resonant states is our most powerful method of approximation in hadron physics. Very difficult problems in analytic functions of several variables are replaced by algebraic problems. It is a challenge at present to algebraize, in a systematic way, all the superconvergence relations for hadrons and then to try to find representations of the resulting algebraic system in terms of discrete meson and baryon states.

Meanwhile, one adopts the temporary expedient of a much more drastic approximation, which involves not only converting the integrals into sums, but also making these finite sums with a small number of terms. In this way one gets interesting but less reliable formulae; for example, in π-ρ scattering there is a superconvergence relation that can be drastically approximated as

$$g^2_{\pi\rho\omega}(m_\omega^2 - m_\rho^2 - m_\pi^2) + g^2_{\pi\rho\phi}(m_\phi^2 - m_\rho^2 - m_\pi^2) + \cdots = 0 . \tag{5}$$

This is amusing, because experimentally $m_\omega^2 - m_\rho^2 - m_\pi^2$ is nearly zero in the first term, while $g^2_{\pi\rho\phi}$ is nearly zero in the second.

Many of the approximate relations among the hadron couplings and masses that have been found in investigations of the bootstrap hypothesis come directly out of this kind of truncation applied to the superconvergence relations.

For the other kinds of relations, we need the weak and electromagnetic currents, local operators that are sandwiched between states of hadrons on the mass shell to give matrix elements for photon emission and absorption and form factors for weak or electromagnetic interaction with lepton pairs.

We can obtain a direct analog of the superconvergence relations by replacing one of the incoming or outgoing particles in a two-body scattering amplitude by a current. The resulting relations can be approximated much as in Eq. 4, giving us, for example,

$$\sum_M g_{\gamma\pi M}\, g_{\pi\pi M}\, \phi(m_\pi^2, m_M^2) = 0 \tag{6}$$

for the photopion effect on pions.

Now let us go further and replace two of the four vertices by currents. Then we no longer have superconvergent dispersion relations, and the right-hand side of our equation is no longer zero. Instead, we pick up, on the right-hand side, a matrix element of the equal-time commutator of the two currents. Now we are talking about the relations of current algebra.

The assumptions of current algebra can be arranged in a hierarchy of credibility. We start with the most believable one and then add further assumptions. The charge operators $F_i = \int \mathcal{F}_{i0}\, d^3x$ of the vector currents $\mathcal{F}_{i\alpha}$ $(i = 0, 1, \cdots 8)$ of hadrons occurring in the electromagnetic and weak interactions are assumed to obey the equal-time commutation rules of the algebra of U(3):

$$[F_i, F_j] = i f_{ijk} F_k. \quad (7)$$

Here, F_0 is proportional to the baryon number; F_1, F_2, and F_3 are the components of the isotopic spin; F_8 is proportional to the hypercharge Y, related to strangeness; and F_4, F_5, F_6, and F_7 have $|\Delta Y| = 1$ and $|\Delta I| = 1/2$ and are not conserved by the strong interaction, hence time-dependent. We may then treat also the axial vector currents $\mathcal{F}_{i\alpha}^5$ ($i = 0, 1, \cdots 8$) of hadrons occurring in the weak interaction and suppose that they form an octet and singlet under the algebra of the F_i, so that their charges $F_i^5 = \int \mathcal{F}_{i0}^5 d^3x$ satisfy

$$[F_i, F_j^5] = i f_{ijk} F_k^5. \quad (8)$$

Then we can assume that the algebra of the F_i and F_i^5 closes in the simplest possible way to form the algebra of U(3)×U(3):

$$[F_i^5, F_j^5] = i f_{ijk} F_k. \quad (9)$$

Finally, we know from microcausality that the charge densities $\mathcal{F}_{i0}(\underline{x})$, $\mathcal{F}_{i0}^5(\underline{x})$ must commute with each other at nonvanishing spatial separations, so that we have such relations as

$$[\mathcal{F}_{i0}(\underline{x}), \mathcal{F}_{j0}(\underline{x}')] = i f_{ijk} \mathcal{F}_{k0}(\underline{x})\, \delta(\underline{x} - \underline{x}')$$
$$+ \text{ gradient terms},$$

where the gradient terms involve a finite number of derivatives of δ functions and integrate to zero. We may assume that these gradient terms vanish and obtain the "local current algebra" at equal times:

$$[\mathcal{F}_{i0}(\underline{x}), \mathcal{F}_{j0}(\underline{x}')] = i f_{ijk} \mathcal{F}_{k0}\, \delta(\underline{x} - \underline{x}'),$$

$$[\mathcal{F}_{i0}(\underline{x}), \mathcal{F}_{j0}^5(\underline{x}')] = i f_{ijk} \mathcal{F}_{k0}^5\, \delta(\underline{x} - \underline{x}'), \quad (10)$$

$$[\mathcal{F}_{i0}^5(\underline{x}), \mathcal{F}_{j0}^5(\underline{x}')] = i f_{ijk} \mathcal{F}_{k0}\, \delta(\underline{x} - \underline{x}').$$

That these equal-time relations do not contradict the principles of relativistic quantum mechanics, no matter how badly approximate symmetries are broken, we can see by constructing a formal mathematical Lagrangian field theory with coupled "quark fields" q (or u, d, and s) and seeing that these relations are true in such a theory; the theory may then be thrown away and the commutation relations retained. Such a formal quark-field approach can be used also to exhibit the universality of strength and form of the weak interaction.

In the limit of zero range for the weak interaction (and we have not yet detected a finite range) it appears to have the form

$$\frac{G}{\sqrt{2}} (J_\alpha^{tot})^+ J_\alpha^{tot},$$

where G is the universal Fermi constant and

$$J_\alpha^{tot} = \bar{\nu}_e \gamma_\alpha (1 + \gamma_5) e + \bar{\nu}_\mu \gamma_\alpha (1 + \gamma_5) \mu^- + J_\alpha. \quad (11)$$

Here, J_α is the hadronic part of the weak current and, in the formal quark field picture, corresponds to a term just like the first two terms above,

$$\bar{u} \gamma_\alpha (1 + \gamma_5)(d \cos\theta + s \sin\theta), \quad (12)$$

where θ is the curious angle between the strangeness-preserving and strangeness-changing weak couplings of hadrons, and has a value around 15°. The currents $\mathcal{F}_{i\alpha}$ and $\mathcal{F}_{i\alpha}^5$ correspond formally to $\bar{q}(\lambda_i/2)\gamma_\alpha q$ and $\bar{q}(\lambda_i/2)\gamma_\alpha\gamma_5 q$ respectively, where the λ_i are 3×3 isotopic matrices such that $\lambda_i/2$ represents the algebra of U(3).

Now let us return to the applications of current algebra, as exhibited in Eq. 10. Serious applications have been delayed from 1961, when the relations were proposed, until 1965, when the trick was suggested of sandwiching the relations between hadron states of infinite momentum (say, P_z equal to a fixed value that goes to infinity, with P_x and P_y finite). The state of motion of the hadron states on both sides of the operator relations is relativistically meaningful, since we have already fixed a Lorentz frame by talking about equal-time commutation relations as a way of presenting relations between operators at points spacelike to each other. Thus, equal-time commutators between states with $P_z = \infty$ are a covariant notion, and can be made "manifestly covariant" if we wish. As a general technique, working at $P_z = \infty$ is very useful. Thus the method of doing relativistic calculations prevalent before the Rochester Conferences were started (taken from Heitler's book, for example) becomes very similar to the method of Feynman diagrams. Running by a system at velocity c, we can no longer tell the difference between Heitler and Feynman. Likewise, we cannot distinguish Low from Goldberger, since the Low equations at $P_z = \infty$ are essentially the dispersion relations.

From local current algebra at $P_z = \infty$, we get, as we said earlier, sum rules that can be written in a manner resembling Eqs. 3, 4, 5, and 6, but with two vertices of the four-legged diagram corresponding to currents and with a right-hand side given by the matrix element of the equal-time commutator. We get equations such as

$$\int d\nu' [\text{Im}\, A_{ij}(\nu', t, k_1^2, k_2^2) - \text{Im}\, A_{ji}(\nu', t, k_1^2, k_2^2)]$$
$$= i f_{ijk} F_k(t), \quad (13)$$

where ν' is, as before, a relativistic energy variable, t is the invariant momentum transfer variable between the initial and final currents, and k_1^2 and k_2^2 are the invariant momentum transfers squared for the initial and final currents respectively. On the right-hand side, we have a form factor for the current that appears on the right-hand side of the equal-time commutator.

Suppose that in Eq. 13 we are dealing with a commutator between vector charge densities. Then the left-hand side will have poles at $k_1^2 = -\mu^2$ and $k_2^2 = -\mu^2$, where μ is the mass of a vector meson. The right-hand side has no such poles. Thus if we look at the coefficient of the compound pole at $k_1^2 = -\mu^2$ and $k_2^2 = -\mu^2$, we recover a superconvergence relation like Eq. 3 for the scattering of a vector meson. If we look at the coefficient of a single pole, say at $k_1^2 = -\mu^2$, we recover a relation like Eq. 6 describing the superconvergence of an amplitude in which an electromagnetic current produces a vector meson.

The new predictions of current algebra are those in which the right-hand side plays a role. For example, we may take the first moment in t of a relation of the type Eq. 13 obtained by sandwiching the commutators between nucleon states and obtain, for $k_1^2 = k_2^2 = 0$, the formula

$$(\mu_A^V)^2 + 1/2\pi^2\alpha \int ds'/(s' - m_N^2)$$
$$\times [2\sigma_{1/2}^V(s') - \sigma_{3/2}^V(s')] = \langle r^2 \rangle_1^V/3, \quad (14)$$

where μ_A^V is the isovector anomalous moment of the nucleon, σ_I^V is the total photonucleon cross section for isovector photons going into hadrons with isotopic spin I, and $\langle r^2\rangle_1^V$ is the mean-square radius corresponding to the isovector Dirac form factor F_1^V of the nucleon. The comparison of this relation with experiment suggests fair agreement, but much more information is needed on the amplitudes for the photoeffect above $s = (1.5 \text{ BeV})^2$.

Using the commutator of two axial-vector charge densities at $k_1^2 = 0$, $k_2^2 = 0$, and $t = 0$, we have the Adler-Weisberger relation

$$\left(\frac{G_A}{G_V}\right)^2 + \int ds' \phi(s')[\sigma(\nu \to e^-, s') - \sigma(\bar{\nu} \to e^+, s')] = 1 , \qquad (15)$$

where G_A is the axial vector coupling constant of the nucleon and G_V the vector coupling constant, $\phi(s')$ is a known kinematical function, and the cross sections are for the sum of all processes in which the lepton comes out with zero momentum transfer when a neutrino or antineutrino strikes a proton.

This relation cannot be tested by observation until we have much better data on neutrino-induced reactions, but a trick can be used to obtain still another kind of sum rule, this time an approximate one. We make the so-called "PCAC approximation," that the matrix elements of the divergence of the isovector axial vector current at $k^2 = 0$ are given roughly by the contribution of the one-pion pole at $k^2 = m_\pi^2$. Combining PCAC with the sum rule Eq. 15 we get a rule that is verifiable by experiment:

$$\left(\frac{G_A}{G_V}\right)^2 + \int ds' \psi(s')[\sigma_{\pi^- p}(s') - \sigma_{\pi^+ p}(s')] = 1, \qquad (16)$$

where ψ is another known kinematic function inversely proportional to the pion decay rate into leptons and the σ's are total π-N cross sections. This rule works well; an optimist would say it verifies the algebra and the PCAC approximation, while a pessimist could say that both principles are in error but the errors cancel.

Current algebra and PCAC have been applied with some success also to leptonic weak decays, for example relating $K \to \pi$ + leptons to $K \to 2\pi$ + leptons. The method is also capable of giving some information about strong amplitudes such as π-π scattering lengths, as we shall probably hear at this meeting.

Still another twist has been given to current algebra sum rules in a contribution to the Conference that discusses matrix elements of commutators between vacuum and two-particle states, rather than between one-particle states. Sum rules emerge that resemble Eq. 13, but with the integration in the left-hand side performed (for example) on $k_1^2 + k_2^2$, with $k_1^2 - k_2^2$, ν, and t fixed. From the assumed convergence of such rules one can, by looking at the coefficients of particle poles, extract superconvergence relations for form factors. Now, high-energy electron scattering experiments have suggested that certain electromagnetic form factors, like $F_2^V(t)$ for the nucleon isovector anomalous magnetic moment, tend to zero faster than $1/t$, so that in the unsubtracted dispersion relation

$$F_2^V(t) = \frac{1}{\pi} \int \frac{dt'}{t' - t - i\epsilon} \operatorname{Im} F_2^V(t') \qquad (17)$$

we have

$$\int dt' \operatorname{Im} F_2^V(t') = 0. \qquad (18)$$

Now we have some theoretical support for such a form factor superconvergence relation. By the way, this Conference should tell us some new experimental results on the nucleon form factors at very large t.

Before leaving the subject of current algebra, let me say that it is tempting to try two theoretical constructions based on such an algebra. One is to write down a set of densities, including the vector and axial vector charge densities, with known equal-time commutation rules and to try to express the total hadronic stress-energy-momentum tensor $\theta_{\mu\nu}$ (the lowest order coupling to gravity) in terms of such currents. Since the integrals $P_\mu = \int \theta_{\mu 0} d^3x$ give the space and time displacement operators (momentum and energy), such a formulation would give "equations of motion" that would completely describe the hadron system. Such a program might possibly be successful, with a relatively simple form for $\theta_{\mu\nu}$, and might even be equivalent to the bootstrap theory.

A less ambitious but more immediate objective is to try to approximate the systems of well-defined baryon and meson levels (idealized to an infinite set of sharp levels going up to infinite energy) as small relativistic representations of the algebra of V and A charge densities at $P_z = \infty$, small in the sense that the huge number of variables of relativistic quantum mechanics would be replaced approximately by a few, for example two spins, a relative coordinate, and two isotopic spins to describe the meson levels. In this way it might be possible to arrive at a relativistic quark model, with a mathematical $\bar{q}q$ description of the meson levels. Now we have lingered long enough on current algebra.

The most exciting topic, of course, is the chapter that was opened two years ago at the Conference in Dubna with the report by Fitch, Cronin, Christensen, and Turlay of CP violation. That chapter is one that is only begun, no doubt. More experiments have confirmed the original result; more investigations have taken place to study the same effect, which is K_2^0 decay into two pions. There are only two more parameters really to be gotten out of that, the phase and the ratio of the $2\pi^0$ to the $\pi^+ + \pi^-$ mode. Some progress has been made with each of those but they're not finished yet. In the meantime, the possible implications for all other kinds of experiments are discussed in some very interesting work; the three hypotheses that are most popular are the following.

One is the "super-weak coupling," that is, a new CP-violating interaction with $|\Delta Y| = 2$, very, very weak (like weak interactions squared), and exploiting the tiny mass difference between K_1^0 and K_2^0 to give the Fitch-Cronin effect. That's my favorite hypothesis because it predicts no new observable effects at all, given present technology. In the Fitch-Cronin effect, it does predict the phase of the K_2^0 into 2π and the charge ratio, but it means that in other processes one is extremely unlikely, for a long time, to see any effects because it's hard to find another process with such a tiny energy denominator of 10^{-5} electron volt, that could show up the super-weak interaction.

Another possibility is that the CP-violating force is 10^{-2} or 10^{-3} of the weak interaction (or comparable to the weak interaction with some inefficiency factor) and gives the effect directly. With this kind of theory, one should certainly expect to see a number of other consequences. In particular, one should find a nonzero value of the neutron electric dipole moment, which has been measured down to the level of weak interactions already and is going to be measured, they say, far more accurately during the next few months or years, using walking neutrons from Oak Ridge or Brookhaven, or perhaps elsewhere.

The third kind of theory is of a totally different nature, saying that the interaction is 10^{-2} or 10^{-3} of the strong interaction, is just C-violating, and combines with the P-violating weak interaction to give the effect. Now, in this case, the experimental implications, of course, are tremendous. Since the strength is 10^{-2} or 10^{-3}, why, you can look for the new force anyplace; you can see it behind every shutter. In particular, when there is a process like η decay, which occurs only through electromagnetism, then competing decay through the new thing will be of a similar order of magnitude and you can look for considerable asymmetry. We can look forward to some enjoyable arguments during this meeting, over whether, in fact, an η-decay asymmetry has been found. I don't know if it will reach the height of comedy achieved in Siena two years ago, but it may be pretty good. Perhaps after a few months we will really know whether there is any asymmetry or not.

If the third explanation should be right, then one can think of the possibility, particularly emphasized by Lee and collaborators, that the new interaction is in fact simply electromagnetic and corresponds to the existence of an electromagnetic current with two terms in it, the ordinary term with $C = -1$ (where C is the strong charge conjugation) and another unusual term with $C = +1$. Now, such a theory poses a very difficult dilemma. If the new current has no charge--in other words, if the charge operator for the $C = +1$ part of the current is zero--then the current doesn't follow the charge in its quantum numbers. But we tend to expect that it should, from Ampere's law that the currents consist completely of the motion of charges. We would at least expect that quantum numbers of the currents should follow the quantum numbers of the charge, what I have sometimes called the idea of minimal electromagnetic interaction. Now, this would be completely wrong, if the new current exists and has a charge operator which is zero. An alternative possibility is that the $C = +1$ charge is not zero. Then there must exist some entirely new charged particle, for example something called $\kappa+$, which is taken into itself by C. Of course CPT still takes $\kappa+$ into $\kappa-$, so you do have a $\kappa-$ particle, but nevertheless C takes $\kappa+$ into itself. Well, that's also extremely interesting, if there exists an entirely new class of strongly interacting particles which are charged but are taken into themselves by strong charge conjugation. That would be as exciting as the violation of minimum electromagnetic interactions.

What I'm really looking forward to hearing at this meeting is the theoretical proposal of a new hypothetical particle that is a triplet with respect to SU(3) like the quark, is an intermediate boson for the weak interactions, is carried into itself by C although charged, obeys parastatistics and has a single magnetic pole. For this particle I suggest the name "chimeron," and I look forward not only to hearing it proposed but also to hearing its existence partially confirmed.

In the meantime the old questions are still with us. Is quantum electrodynamics violated a little bit before we get to the level of the hadron vacuum polarization which we know must be there? Some experimental evidence, I think, will be discussed here and we'll have to see whether there is any good evidence for violation at present or not. What about the weak interaction? The current-current picture for the weak interaction works extremely well. In fact, by combining that picture with current commutation rules and PCAC, theorists have made a lot of progress in relating the different weak decays to each other. This is true of the leptonic weak decays, where decays with no pions, one pion, two pions, and so on, have been related to each other. It is also true, even more strikingly, of the nonleptonic weak decays in which $K_1^0 \to 2\pi$ and $K \to 3\pi$ have been related to each other, and the baryon decays $\Lambda \to N + \pi$, $\Sigma \to N + \pi$, and $\Xi \to \Lambda + \pi$ have been related to each other. The latter is good only for S waves, I might say for the benefit of my theoretical friends; the P wave still resists explanation to some extent. Still, the current-current picture looks good. Another question that has been around for a long time, about 12 years, concerns the $\Delta I = 1/2$ rule or what may now be called octet dominance for the nonleptonic weak interaction. Is it really just a dynamical enhancement or does it come from some basic interaction which has to be added to the charged current times its Hermitian conjugate? In other words, is there an extra current peculiar to hadrons, that makes the $|\Delta I| = 1/2$ rule true by symmetry, or is it simply that the formula $J_\alpha^+ J_\alpha$ is right and that the $\Delta I = 1/2$ rule comes from a dynamical enhancement of the octet part of $J_\alpha^+ J_\alpha$? I would say that a lot of the recent theoretical work indicates that the dynamical enhancement idea may be right, but the issue is by no means settled.

How about the structure of the weak interaction itself, though? Assuming the current-current form is right, what's the range of the interaction? Is there an intermediate boson that's responsible for the exchange? Nobody knows. How do we do calculations in higher order in the weak interaction? Nobody really knows yet. And then there is a still deeper question to which nobody has any answer at all. Why are there hadrons and leptons, strongly interacting particles and particles with no strong interactions? Why are there the gravitational, electromagnetic, and weak interactions? Why is there this funny lepton spectrum with the e and the μ, and the symmetry between the two, and the two neutrinos? As Rabi once asked, "Who ordered that?" My colleague Feynman for many years had on his blackboard in a little box the question "Why does the muon weigh?" It's been erased. Why is there the funny angle that we spoke of before, the Cabibbo angle, between the strangeness-changing and the strangeness-preserving currents? Nobody knows. I think that it's not discouraging, but rather that it's marvelous not to know all of these fundamental things. We still have problems to work on. Thank you.

PHYSICAL REVIEW VOLUME 175, NUMBER 5 25 NOVEMBER 1968

Behavior of Current Divergences under $SU_3\times SU_3$*†

MURRAY GELL-MANN
California Institute of Technology, Pasadena, California 91109

AND

R. J. OAKES
Northwestern University, Evanston, Illinois 60201

AND

B. RENNER‡
California Institute of Technology, Pasadena, California 91109

(Received 22 July 1968)

We investigate the behavior under $SU_3\times SU_3$ of the hadron energy density and the closely related question of how the divergences of the axial-vector currents and the strangeness-changing vector currents transform under $SU_3\times SU_3$. We assume that two terms in the energy density break $SU_3\times SU_3$ symmetry; under SU_3 one transforms as a singlet, the other as the member of an octet. The simplest possible behavior of these terms under chiral transformations is proposed: They are assigned to a single $(3,3^*)+(3^*,3)$ representation of $SU_3\times SU_3$ and parity together with the current divergences. The commutators of charges and current divergences are derived in terms of a single constant c that describes the strength of the SU_3-breaking term relative to the chiral symmetry-breaking term. The constant c is found not to be small, as suggested earlier, but instead close to the value $(-\sqrt{2})$ corresponding to an $SU_2\times SU_2$ symmetry, realized mainly by massless pions rather than parity doubling. Some applications of the proposed commutation relations are given, mainly to the pseudoscalar mesons, and other applications are indicated.

I. INTRODUCTION

We assume here the correctness of the $SU_3\times SU_3$ algebra proposed for equal-time commutators of the vector and axial-vector charge operators F_i and F_i^5 for hadrons.[1] As is well known, there is some experimental evidence in confirmation of it,[2] and especially of the $SU_2\times SU_2$ subalgebra. The corresponding local commutation rules proposed for the charge densities may also be correct. At infinite momentum, all these commutation relations fall into the "good-good" class[3] and yield sum rules that amount to unsubtracted dispersion relations in the variable s.

We investigate in this paper the behavior under $SU_3\times SU_3$ of the hadron energy density θ_{00}, and the closely related question of how the divergences of the axial-vector currents and strangeness-changing vector currents transform under $SU_3\times SU_3$. The commutators involved here fall into the "good-bad" category at infinite momentum, and some of them are tractable in deriving sum rules.

We do not consider here the equal-time commutators of divergences with each other, which give "bad-bad" relations at $P_z=\infty$ and which no one has succeeded in using in a straightforward way at $P_z=\infty$ or in deriving low-energy theorems, although these commutators may be useful when acting on the vacuum state.

Our aim is to explore further the original proposal[1] (as suggested, for example, by a quark model) for commutation relations between charges and current divergences. These relations arise from an energy density θ_{00} in which the $SU_3\times SU_3$-violating part consists of two terms; the first breaks the chiral symmetry $SU_3\times SU_3$ but not SU_3 itself (and corresponds in a quark model to a common quark mass), and the second breaks SU_3 (and corresponds in a quark model to a mass-splitting between isotopic singlet and doublet). The proposed behavior of these terms under $SU_3\times SU_3$ is the simplest possible: They and all the current divergences belong to a single representation of $SU_3\times SU_3$ and parity. This theory, because of its simplicity, contains a single universal parameter c that describes the strength of the SU_3 symmetry-breaking term relative to the chiral symmetry-breaking term and determines the commutators of charges with divergences.

If the whole $SU_3\times SU_3$ violation were abolished, we would have a world in which all sixteen vector and axial-vector currents were conserved. We suppose that the conservation of the vector currents would be achieved through the exact degeneracy of SU_3 multiplets and the conservation of the axial-vector current through the existence and coupling of eight massless pseudoscalar mesons. The chiral symmetry violation raises the masses of the pseudoscalar mesons to finite values, and the SU_3 violation splits the SU_3 multiplets.

---

* Work supported in part by the U. S. Atomic Energy Commission. Prepared under Contract No. AT(11-1)-68 for the San Francisco Operations Office, U. S. Atomic Energy Commission.

† The research described here was performed at the Institute for Advanced Study, Princeton, N. J. The authors would like to thank Professor Carl Kaysen for his hospitality at the Institute.

‡ On leave from Gonville and Caius College, Cambridge, England.

[1] M. Gell-Mann, Phys. Rev. **125**, 1067 (1962).

[2] For a review of the literature, see S. L. Adler and R. F. Dashen, *Current Algebras* (W. A. Benjamin, Inc., New York, 1968) and B. Renner, *Current Algebras and Their Applications* (Pergamon Press, London, 1968).

[3] R. F. Dashen and M. Gell-Mann, in *Proceedings of the Fourth Coral Gables Conference on Symmetry Principles at High Energy, 1966* (W. H. Freeman and Co., San Francisco, 1967); S. Fubini, G. Segrè, and J. D. Walecka, Ann. Phys. (N. Y.), **39**, 381 (1966).

We note that SU_3 mass splittings are always of the same order as the masses of the pseudoscalar octet. This observation suggests that it would not be unreasonable for the strengths of the two symmetry-violating terms to be comparable.

Employing some simple approximate assumptions, we conclude that a consistent picture of several experimental results can be obtained and the crucial constant c estimated. This constant is not small, as originally suggested,[1] but rather is close to the value $-\sqrt{2}$. If it were exactly equal to $-\sqrt{2}$, we would have exact $SU_2 \times SU_2$ invariance and massless pions. (In the quark scheme, which we employ here only as a mnemonic, this value corresponds to a zero mass for the isotopic doublet and a finite mass for the isotopic singlet.)

Many authors have taken steps in the same direction,[2] and in particular we should mention the work of Glashow and Weinberg[4] as being the most closely related to ours. Still, we find that our approach leads to results and to a formulation of the situation that, to the best of our knowledge, have not yet been reported in the literature.

In Sec. II, we discuss the commutators of charges with current divergences; in Sec. III, we estimate the constant c and give some applications.

II. TRANSFORMATION PROPERTIES OF CURRENT DIVERGENCES

The algebraic behavior of current divergences is closely related to the properties of the energy operator H under the $SU_3 \times SU_3$ algebra through the equation[5]

$$\frac{d}{dt}\int j_0(\mathbf{x},t)d^3x = -\int \partial_\mu j_\mu(\mathbf{x},t)d^3x = i\left[H, \int j_0(\mathbf{x},t)\right], \quad (2.1)$$

where

$$H = \int \theta_{00}(\mathbf{x},t)d^3x. \quad (2.2)$$

The local generalization[1] of Eq. (2.1) to

$$\partial_\mu j_\mu(\mathbf{x},t) = -i\left[\mathcal{H}'(\mathbf{x},t), \int j_0(\mathbf{y},t)d^3y\right] \quad (2.3)$$

rests on the assumption that the tensor part in $\theta_{00}(x)$ commutes with the charges, and only a Lorentz scalar part $\delta_{00}\mathcal{H}'(x)$ (such as a mass term in certain Lagrangian models) breaks the symmetry. We make this our first assumption.[6] A decomposition of $\mathcal{H}'(x)$ into terms transforming according to irreducible representations of $SU_3 \times SU_3$ specifies completely the commutators of charges and current divergences.

For simplicity, we make our second assumption: We will not admit operators of isospin or hypercharge 2 into the multiplet of the current divergences. This assumption implies the vanishing of certain commutators, for example,

$$[F_i{}^5 + iF_2{}^5, \partial_\mu \mathfrak{F}_{1\mu}{}^5 + i\partial_\mu \mathfrak{F}_{2\mu}{}^5] = 0, \quad (2.4)$$

$$[F_4{}^5 + iF_5{}^5, \partial_\mu \mathfrak{F}_{4\mu}{}^5 + i\partial_\mu \mathfrak{F}_{5\mu}{}^5] = 0. \quad (2.5)$$

These have been found consistent with experimental information.[7] In fact, this assumption allows components of $\mathcal{H}'(x)$ only in representations $(3,3^*)+(3^*,3)$ and $(1,8)+(8,1)$ of $SU_3 \times SU_3$.

The final simplification is achieved if we make our last assumption, which is that the $(1,8)+(8,1)$ part of $\mathcal{H}'$ vanishes and that the SU_3 singlet and octet parts of $\mathcal{H}'$ belong to the same $(3,3^*)+(3^*,3)$ representation of $SU_3 \times SU_3$. We have arrived at the simplest theory of the behavior of $\mathcal{H}'$, because it introduces the least number of new operators to complete the multiplet of the current divergences. [We shall argue at the end of the article that if there is a $(1,8)+(8,1)$ contribution, its effects on our deductions are unlikely to be large.] We then obtain the formula

$$\mathcal{H}' = -u_0 - cu_8 \quad (2.6)$$

of Ref. 1, where the parameter c expresses the relative scale of the two parts of $\mathcal{H}'$ and is uniquely defined by the transformation properties of the scalar and pseudoscalar nonets, u_i and v_i, in the representation $(3,3^*)+(3^*,3)$ of $SU_3 \times SU_3$:

$$\begin{aligned} [F_i, u_j(y)] &= i f_{ijk} u_k(y), \\ [F_i, v_j(y)] &= i f_{ijk} v_k(y), \\ [F_i{}^5, u_j(y)] &= -i d_{ijk} v_k(y), \\ [F_i{}^5, v_j(y)] &= i d_{ijk} u_k(y), \end{aligned} \quad (2.7)$$

where $i = 1 \cdots 8$ and $j, k = 0 \cdots 8$. We do not consider an operator $F_0{}^5$, because we find that no such operator is "partially conserved" to anywhere near the same degree as the operators we do consider. Furthermore, such an operator is not known to play a role in the physical interactions of hadrons.

The current divergences follow from (2.3):

$$\partial_\mu \mathfrak{F}_{i\mu} = c f_{i8k} u_k, \quad (2.8)$$

$$\partial_\mu \mathfrak{F}_{i\mu}{}^5 = -d_{i0k} v_k - c d_{i8k} v_k = -W_i(c)\delta_{ik} v_k - (\sqrt{\tfrac{2}{3}})c\delta_{i8} v_0, \quad (2.9)$$

[4] S. L. Glashow and S. Weinberg, Phys. Rev. Letters **20**, 224 (1968).

[5] The notation of Ref. 1 will be adopted throughout this paper, in that we use spacelike metric $g_{00} = -1$ and the pion decay constant $(1/\sqrt{2})\langle 0|\partial_\mu \mathfrak{F}_{1\mu}{}^5(0) - i\partial \mathfrak{F}_{2\nu}{}^5(0)|\pi^+\rangle = m_\pi{}^2/2f_\pi$. As in M. Gell-Mann [Physics **1**, 63 (1964)] $f_K{}^{-1}$ and $f_\pi{}^{-1}$ are constants that yield kaon and pion decay amplitudes when multiplied by $\sin\theta$ and $\cos\theta$, respectively.

[6] For a related discussion, see D. J. Gross and R. Jackiw, Phys. Rev. **163**, 1689 (1967).

[7] M. A. Ahmed, Nucl. Phys. **B1**, 679 (1967); D. K. Elias, Oxford University Report, 1967 (unpublished); G. Furlan and C. Rossetti, Phys. Letters **23**, 499 (1966); G. Soliani, University of Torino Report, 1967 (unpublished).

where

$$W_i(c)=(\sqrt{2}+c)/\sqrt{3} \quad \text{for} \quad i=1,2,3,$$
$$(\sqrt{2}-\tfrac{1}{2}c)/\sqrt{3} \quad \text{for} \quad i=4,5,6,7, \tag{2.10}$$
$$(\sqrt{2}-c)/\sqrt{3} \quad \text{for} \quad i=8.$$

Commutation of charges with current divergences can be read off by combining Eqs. (2.7)–(2.10).

All these assumptions and conclusions are as in Ref. 1. However, it was assumed there that c was small compared to unity. If we consider the formal analogy of $-u_0-cu_8$ with the mechanical masses of an isotopic doublet and singlet forming a fundamental triplet, then the old assumption about c amounts to saying that the doublet-singlet splitting is small compared to the average mechanical mass. If we imagine that, in the limit of conservation of all sixteen currents, the masses of the baryons (for example, the $\frac{1}{2}^+$ octet) go to zero, then this might be a reasonable idea, with c corresponding roughly to the ratio of baryon mass splitting to baryon mass. But if it is the masses of the eight pseudoscalar mesons of the 0^- octet that go to zero as we disregard u_0+cu_8, then that is no longer reasonable. In fact, it seems that the real world of hadrons is not too far from a world in which we have eight massless pseudoscalar mesons, SU_3 degeneracy, and conservation of all sixteen currents. We are even closer to a world in which there are massless pions and the algebra of $SU_2\times SU_2$ is conserved. In that limit, we would have $c=-\sqrt{2}$, since the combination $u_0-\sqrt{2}u_8$ commutes with F_i and F_i^5 for $i=1, 2, 3$. (This combination corresponds to zero mass for the conceptual fundamental isospin doublet and a finite mass for the isospin singlet.) Since, in fact, the pion is nearly massless, we should expect that in fact the value of c is close to $(-\sqrt{2})$. We shall see that a number of experimental results can be understood with the aid of the assumed behavior of $\mathcal{H}'$ in Eq. (2.6), a value of c something like (-1.2), and a few approximate assumptions about the matrix elements of currents between SU_3 multiplets and about meson pole dominance.

III. APPLICATIONS

The applications given in this paper will include the estimate of c, a discussion of the approximate equality of the $\pi_{\mu2}$ and $K_{\mu2}$ decay constants, a demonstration of the consistency of the squared-mass formula, and some results for $K_{\mu3}$ decays.

We shall make approximations of two kinds:

(A) Pole dominance for axial current divergences through π, K, and η mesons. We note that although the hypothesis of partially conserved axial-vector current (PCAC) for K mesons appears uncertain, there is no definite evidence known against it, and the recent estimates of K-meson Yukawa coupling constants are compatible with generalized Goldberger-Treiman relations.[8] η-meson PCAC is not testable, but one expects it to be on a similar footing as K-meson PCAC, at least as long as $(\eta\eta')$ mixing may be neglected.

(B) Application of approximate SU_3 symmetry to vertices of certain operator octets involving multiplets with small mixing. We shall apply SU_3 only to form factors and there only at points which are far enough away from important singularities so that differences in their distance due to SU_3 violations are negligible. At small momentum transfers, form factors describing the transverse parts of vector and axial-vector currents (not the induced scalar or pseudoscalar parts) and form factors of scalar densities (u_i) may be estimated by SU_3, while matrix elements of pseudoscalar densities (v_i) are usually too strongly distorted because of the closeness of pseudoscalar-meson poles. No such distortion by the position of singularities affects matrix elements of octet operators between the vacuum and particle octets. We shall find that the vacuum is approximately SU_3 invariant and that these matrix elements are always close to their symmetry values.

Before employing these approximations in our applications of the algebraic properties of current divergences, we give an independent example to show they are reasonable by estimating the observed SU_3 violations in the decay $Y_0^*(1405)\to\Sigma\pi$ as compared with the virtual transition $Y_0^*\to N\bar{K}$, with Y_0^*, a $(\frac{1}{2})^-$ SU_3 singlet. As discussed in (B), we apply SU_3 to the matrix elements of the transverse part of the axial-vector currents

$$\langle Y_0^*|\mathfrak{F}_{\pi\mu}^5|\Sigma\rangle = ig_\pi(t)\bar{u}_Y\gamma_\mu u_\Sigma+\cdots, \tag{3.1a}$$

$$\langle Y_0^*|\mathfrak{F}_{K\mu}^5|N\rangle = ig_K(t)\bar{u}_Y\gamma_\mu u_N+\cdots, \tag{3.1b}$$

by approximating $g_\pi(0)\approx g_K(0)$. Using generalized Goldberger-Treiman relations for the axial divergences, as discussed in (A), we find

$$g_\pi(0)(m_Y-m_\Sigma)\approx(1/2f_\pi)g_{Y_0^*\Sigma\pi} \tag{3.2a}$$

and

$$g_K(0)(m_Y-m_N)\approx(1/2f_K)g_{Y_0^*N\bar{K}}. \tag{3.2b}$$

Therefore,

$$g_{Y_0^*N\bar{K}}/g_{Y_0^*\Sigma\pi}\approx\frac{(m_Y-m_N)}{(m_Y-m_\Sigma)}\frac{f_K}{f_\pi}\approx 2. \tag{3.3}$$

Experimentally, this ratio is estimated[9] to be about 3, while exact SU_3 symmetry predicts 1. Because of the parity change $(\frac{1}{2}^-\to\frac{1}{2}^+)$ the SU_3 corrections enter through the mass differences and are significantly larger than in cases considered previously[10] where there is no parity change $(\frac{1}{2}^+\to\frac{1}{2}^+)$ and the corrections enter

---

[8] J. K. Kim, Phys. Rev. Letters **19**, 1079 (1967); C. H. Chan and F. T. Meiere, *ibid.* **20**, 568 (1968); H. T. Neih, *ibid.* **20**, 1254 (1968).

[9] C. Weil, Phys. Rev. **161**, 1682 (1967).

[10] K. Raman, Phys. Rev. **149**, 1122 (1966); **152**, 1517(E) (1966); R. H. Graham, S. Pakvasa, and K. Raman, *ibid.* **163**, 1774 (1967).

through the sum of the masses. Further corrections would result from singlet-octet mixing.

We now return to our objective and start by considering the vertex of $\mathcal{H}'(0)$ between members of the pseudoscalar-meson octet $|P_i\rangle$ in the low-energy limit

$$\begin{aligned}-\lim_{p\to 0}\langle P_i(p)|u_0+cu_8|P_i(p')\rangle &\approx -2if_i\langle 0|[F_i^5, u_0+cu_8]|P_i(p')\rangle \\ &= 2f_i\langle 0|\partial_\mu \mathfrak{F}_{i\mu}{}^5|P^i(p')\rangle = m_i^2. \quad (3.4)\end{aligned}$$

The application of low-energy limits to matrix elements involving parts of $\mathcal{H}(0)$ may appear questionable, and it certainly is misleading in some theories, but it has been our first assumption that we are not dealing with such cases in that the explicitly momentum-dependent tensor parts in $\theta_{00}(0)$ are $SU_3\times SU_3$ symmetric and do not enter into $u_0(0)+cu_8(0)$. A free-meson theory may serve as an illustration of what we mean; there we have

$$\langle P_i(p)|\theta_{00}(0)|P_i(p')\rangle = p_0p_0'+\mathbf{p}\mathbf{p}'+m_i^2, \quad (3.5)$$

with $(p_0p_0'+\mathbf{p}\mathbf{p}')$ being contributed by the chirally symmetric kinetic tensor term (for which the low-energy limit is a bad approximation) and the symmetry-breaking mass term whose effects are not distorted in the virtual low-energy limit.

Equation (3.4) illustrates the role of the pseudoscalar-meson masses in chiral symmetry breaking and their vanishing in the symmetry limit. It gives us a way of separating the effects of u_0 from the chirally invariant part in $\mathcal{H}$. Only for the pseudoscalar octet, with the help of PCAC, do we know how to isolate u_0; for the other multiplets, this separation is yet to be achieved.

Now we consider a more general scalar vertex with pseudoscalar mesons and apply SU_3 symmetry [see (B)]. Neglecting $(\eta\eta')$ mixing, we have

$$\begin{aligned}-\langle P_i(p)|u_j|P_k(p')\rangle &= \alpha(t)\delta_{j0}\delta_{ik}+\beta(t)d_{ijk}, \\ i, k &= 1\cdots 8;\quad j=0\cdots 8,\end{aligned} \quad (3.6)$$

where $\alpha(0)$ and $\beta(0)$ are related to the pseudoscalar-meson masses through the SU_3 mass formula

$$-\langle P_i(p)|u_0+cu_8|P_i(p)\rangle \approx m_i^2 \approx \langle m^2\rangle_{\rm av}+d^{ii8}\Delta m^2 \quad (3.7a)$$

by

$$\alpha(0)=\langle m^2\rangle_{\rm av}-(\sqrt{\tfrac{2}{3}})\Delta m^2/c \quad \text{and} \quad \beta(0)=\Delta m^2/c. \quad (3.7b)$$

Taking low-energy limits and neglecting the dependence of α and β on t $[\alpha(m_i^2)\approx\alpha(0)\equiv\alpha,\ \beta(m_i^2)\approx\beta(0)\equiv\beta]$, we have

$$\begin{aligned}\alpha\delta_{j0}\delta_{ik}+\beta d_{ijk} &\approx -2f_i d_{ijl}\langle 0|v_l|P_k\rangle \\ &\approx -2f_k d_{ljk}\langle P_i|v_l|0\rangle. \quad (3.8)\end{aligned}$$

Using different values for (i,j,k), we find

$$\begin{aligned}&(1,4,6):\ \beta\approx -2f_\pi\langle 0|v_K|K\rangle \approx -2f_K\langle\pi|v_\pi|0\rangle, \\ &(1,1,8):\ \beta/\sqrt{3}\approx -2f_\pi\langle 0|v_8|\eta\rangle/\sqrt{3}-2f_\pi\langle 0|v_0|\eta\rangle\sqrt{\tfrac{2}{3}} \\ &\qquad\qquad \approx -2f_\eta\langle\pi|v_\pi|0\rangle/\sqrt{3}, \\ &(4,4,8):\ -\beta/2\sqrt{3}\approx 2f_K\langle 0|v_8|\eta\rangle/2\sqrt{3} \\ &\qquad\qquad -2f_K\langle 0|v_0|\eta\rangle\sqrt{\tfrac{2}{3}} \\ &\qquad\qquad \approx 2f_\eta\langle K|v_K|0\rangle/2\sqrt{3}, \\ &(8,8,8):\ -\beta/\sqrt{3}\approx 2f_\eta\langle 0|v_8|\eta\rangle/\sqrt{3}-2f_\eta\langle 0|v_0|\eta\rangle\sqrt{\tfrac{2}{3}}, \\ &(1,0,1):\ \alpha+\beta\sqrt{\tfrac{2}{3}}\approx -2f_\pi\langle 0|v_\pi|\pi\rangle\sqrt{\tfrac{2}{3}}.\end{aligned} \quad (3.9)$$

From these equations, we find

$$f_\pi\approx f_K\approx f_\eta, \quad (3.10)$$

$$\langle 0|v_i|P_i\rangle \approx -\beta/2f, \quad \text{independently of } i, \quad (3.11a)$$

$$\langle 0|v_0|\eta\rangle\approx 0, \quad (3.11b)$$

$$\alpha\approx 0;\quad \langle m^2\rangle_{\rm av}\approx(\sqrt{\tfrac{2}{3}})\Delta m^2/c \Rightarrow c\approx -1.25. \quad (3.12)$$

Equation (3.10) states the equality of the $\pi_{\mu 2}$ and $K_{\mu 2}$ decay constants.[11] Experimentally they differ by less than 25%, and we may take this as an indication of the accuracy of our estimates. Equation (3.11a) reproduces the squared meson mass formula when combined with Eqs. (2.9), (3.10), and (3.11b). This is not particularly astonishing since we began with an equivalent assumption that the pseudoscalar-meson states form an approximately pure octet, yet it appears interesting to us that the simplest ansatz gave a consistent result starting with the squared meson mass formula. The vanishing of $\langle 0|v^0|\eta\rangle$ corresponds to our neglecting of (η,η') mixing; if we had a significant amount of mixing we would not have the octet mass formula for pseudoscalar mesons and η dominance for $\partial_\mu\mathfrak{F}_{\mu 8}{}^5$. In our approach we have to insist on keeping $m_{\eta'}{}^2$ still large in the limit of $SU_3\times SU_3$ symmetry, with no conservation for a hypothetical current of the form $\mathfrak{F}_{\alpha 0}{}^5$, because otherwise we could not explain the large splitting of singlet and octet; the term $\mathcal{H}'$ in the energy density, as given in Eq. (2.6), is clearly not responsible for that large splitting.

The value $c\approx -1.25$ suggests the closeness of our theory to the $SU_2\times SU_2$ symmetric limit with $m_\pi=0$ and $c=-\sqrt{2}$. In fact, we find that in the construction of the commutators (Sec. II), if we take the limit $m_\pi=0$, we could replace our second and third assumptions by the requirements of $SU_2\times SU_2$ symmetry and singlet and octet behavior of $\mathcal{H}(0)$, except that some (1,8) and (8,1) would still be allowed. We see $SU_3\times SU_3$ symmetry (with eight pseudoscalar Nambu-Goldstone bosons)

[11] Compare C. S. Lai, Phys. Rev. Letters 20, 509 (1968) and R. J. Oakes, *ibid.* 20, 513 (1968).

broken in two chains[12]:

$$\begin{array}{ccccc} & & SU_3 & & \\ & \nearrow & & \searrow & \\ SU_3 \times SU_3 & & & & SU_2. \\ & \searrow & & \nearrow & \\ & & SU_2 \times SU_2 & & \end{array} \qquad (3.13)$$

At least for the low-lying multiplets, chiral symmetry appears realized by strong coupling to massless pseudoscalar mesons, rather than by parity doubling. Fubini[13] has recently emphasized that this is a quantitative matter in the real world; if a resonant state of opposite parity and the same spin lies closer in mass than m_π from a given state, then approximate chiral symmetry can manifest itself more in the manner of degenerate multiplets than in the Nambu-Goldstone manner.

The vacuum state should be expected to be approximately invariant under SU_3 but not $SU_3 \times SU_3$. That is consistent with Eq. (3.11), as we shall show when estimating the vacuum expectation values $\langle 0|u_0|0\rangle$ and $\langle 0|u_8 0\rangle$. The limit $p \to 0$ in Eq. (3.11) gives

$$\begin{aligned} \langle 0|v_i|P_i\rangle &= -m_i^2/(2f_i)W_i(c) \approx -\beta/2f \\ &\approx -(2f_i)(d_{ii0}\langle 0|u_0|0\rangle + d_{ii8}\langle 0|u_8|0\rangle). \end{aligned} \qquad (3.14)$$

To the accuracy of Eq. (3.14), we find $\langle 0|u_8|0\rangle \approx 0$, which indeed corresponds to an SU_3-invariant vacuum, and $\langle 0|u_0|0\rangle \approx \frac{3}{2}\langle m^2\rangle_{\rm av}/(2f)^2$.

At this point, we want to argue that the effects of a possible (1,8)+(8,1) admixture in $\mathcal{H}'$ on our deductions are in fact small. We denote the scalar and pseudoscalar members of the multiplet by $g_j(0)$ and $h_j(0)$, respectively. They transform as follows:

$$\begin{aligned} [F_i, g_j] &= i f_{ijk} g_k, \\ [F_i, h_j] &= i f_{ijk} h_k, \\ [F_i^5, g_j] &= i f_{ijk} h_k, \\ [F_i^5, h_j] &= i f_{ijk} g_k. \end{aligned} \qquad (3.15)$$

We note that the h_k behave oppositely under charge conjugation to the v_k. Making an ansatz similar to (3.6) for a possible contribution g_8 in $\mathcal{H}'(0)$, we have

$$\begin{aligned} \langle P_i|g_j|P_k\rangle &= d_{ijk}\gamma(t) \\ &\approx -f_{ijk}\langle 0|h^k|P^k\rangle/2f_i. \end{aligned} \qquad (3.16)$$

We see that, to the accuracy of low-energy limits, $\gamma(0) \approx 0$ and $\langle 0|h^k|P^k\rangle \approx 0$. Roughly speaking, we would find effects of a possible (1,8)+(8,1) admixture only in the order or corrections to PCAC in our results.

This possibility should be kept in mind when trying to use the commutators of currents and divergences in connection with multiplets other than the pseudoscalar octet. Such tests are very desirable for confirming the universality of c. An obvious possibility is the study of the so-called σ terms in the scattering of pseudoscalar mesons at low energies.[14] In πN scattering, the low-energy value of the isospin-symmetric amplitude is given by the formula[15]

$$\begin{aligned} \langle \pi N|T|\pi N\rangle|_{q_\pi=0} &= \tfrac{1}{3}(\sqrt{2}+c)(4f_\pi^2) \\ &\quad \times \langle N|(\sqrt{2}u_0 + u_8)|N\rangle. \end{aligned} \qquad (3.17)$$

A precise estimate is made difficult for three reasons. The first is the occurrence of the factor $(\sqrt{2}+c)$. The second is the lack of knowledge about $\langle N|u_0|N\rangle$; it is not possible to decide *a priori* how much of the nucleon mass should be attributed to chiral symmetry violation. The third reason is that the application of PCAC to a four-point function is a delicate problem. The extreme smallness of the sum of π^+p and π^-p scattering lengths tends to suggest that in fact the quantity $\langle N|u_0|N\rangle$ is not a large fraction of the nucleon mass. In K^+p scattering, we find

$$\begin{aligned} \langle K^+p|T|K^+p\rangle|_{q_K=0} &= \tfrac{1}{3}(\sqrt{2}-\tfrac{1}{2}c)(4f_K^2) \\ &\quad \times \langle p|\sqrt{2}u_0 + \tfrac{1}{2}\sqrt{3}u_3 - \tfrac{1}{2}u_8|p\rangle \end{aligned} \qquad (3.18)$$

and two of these difficulties are still present, with the PCAC trouble aggravated. Attempts at numerical estimates may be considered elsewhere.

Within the meson system, further applications can be made to $K_{\mu 3}$ decays. We found the low-energy theorems emerging from the commutators (2.7) compatible with and, to our accuracy, hardly distinguishable from the Callan-Treiman relation.[16] The details will be given elsewhere, along with applications to hard-meson calculations of the $K_{\mu 3}$ vertex.

In conclusion, let us reemphasize that we have so far investigated the compatibility of the formula $\mathcal{H}' = -u_0 - cu_8$ (with c near $-\sqrt{2}$) only with certain data about hadrons and that further tests are desirable to establish whether a single $(3,3^*)$ and $(3^*,3)$ representation is indeed what is involved and also to decide whether there is an additional term of the form g_8 belonging to (1,8) and (8,1).

ACKNOWLEDGMENT

We wish to thank Professor R. P. Feynman for interesting discussions on the subject of this paper.

[12] Compare G. S. Guralnik, V. S. Mathur, and L. K. Pandit, Phys. Letters **20**, 64 (1966).

[13] S. Fubini and G. Furlan, Massachusetts Institute of Technology Report, 1968 (unpublished).

[14] K. Kawarabayashi and W. W. Wada, Phys. Rev. **146**, 1209 (1966); K. T. Mahanthappa and Riazuddin, Nuovo Cimento **45A**, 252 (1966); R. W. Griffith (to be published), has also emphasized the connections between σ terms and the fraction of the nucleon mass arising from chiral symmetry violation.

[15] We use the notation $S = 1 + i(2\pi)^4\delta(p^i - p^f)T$ and covariant normalization for the states; see Footnote 5.

[16] C. G. Callan and S. B. Treiman, Phys. Rev. Letters **16**, 153 (1966).

9

Light Cone Current Algebra*

HARALD FRITZSCH** and MURRAY GELL-MANN†

PREFACE

THIS TALK follows by a few months a talk by the same authors on nearly the same subject at the Coral Gables Conference. The ideas presented here are basically the same, but with some amplification, some change of viewpoint, and a number of new questions for the future. For our own convenience, we have transcribed the Coral Gables paper, but with an added ninth section, entitled "Problems of light cone current algebra", dealing with our present views and emphasizing research topics that require study.

1. INTRODUCTION

We should like to show that a number of different ideas of the last few years on broken scale invariance, scaling in deep inelastic electron-nucleon scattering, operator product expansions on the light cone, "parton" models, and generalizations of current algebra, as well as some new ideas, form a coherent picture. One can fit together the parts of each approach that make sense and obtain a consistent view of scale invariance, broken by certain terms in the energy density, but restored in operator commutators on the light cone.

We begin in the next section with a review of the properties of the dilation operator D obtained from the stress-energy-momentum tensor $\theta_{\mu\nu}$ and the behavior of operators under equal-time

* Work supported in part by the U.S. Atomic Energy Commission under contract AT(11-1)-68, San Francisco Operations Office.

** Max-Planck-Institut für Physik und Astrophysik, München, Germany. Present address (1971–1972): CERN, Geneva, Switzerland.

† California Institute of Technology, Pasadena, California. Present address. (1971–1972): CERN, Geneva, Switzerland.

commutation with D, which is described in terms of physical dimensions l for the operators. We review the evidence on the relation between the violation of scale invariance and the violation of $SU \times SU_3$ invariance.

Next, in Section 3, we describe something that may seem at first sight quite different, namely the Bjorken scaling of deep inelastic scattering cross sections of electrons on nucleons and the interpretation of this scaling in terms of the light cone commutator of two electromagnetic current operators. We use a generalization of Wilson's work,[1] the light-cone expansion emphasized particularly by Brandt and Preparata[2] and Frishman.[3] A different definition l of physical dimension is thus introduced and the scaling implies a kind of conservation of l on the light cone. On the right-hand side of the expansions, the operators have $l = -J - 2$, where J is the leading angular momentum contained in each operator and l is the leading dimension.

In Section 4, we show that under simple assumptions the dimensions l and l are essentially the same, and that the notions of scaling and conservation of dimension can be widely generalized. The essential assumption of the whole approach is seen to be that the dimension l (or l) of any symmetry-breaking term in the energy (whether violating scale invariance or $SU_3 \times SU_3$) is *higher* than the dimension, -4, of the completely invariant part of the energy density. The conservation of dimension on the light cone then assigns a lower singularity to symmetry-breaking terms than to symmetry-preserving terms, permitting the light-cone relations to be completely symmetrical under scale, $SU_3 \times SU_3$, and perhaps other symmetries.

In Section 5, the power series expansion on the light cone is formally summed to give bilocal operators (as briefly discussed by Frishman) and it is suggested that these bilocal light-cone operators may be very few in number and may form some very simple closed algebraic system. They are then the basic mathematical entities of the scheme.

It is pointed out that several features of the Stanford experiments, as interpreted according to the ideas of scaling, resemble the behavior on the light cone of free field theory or of interacting field theory with naïve manipulation of operators, rather than the behavior of renormalized perturbation expansions of renormalizable field theories. Thus free field theory models may be studied for the purpose of abstracting algebraic relations that might be true on the light cone in the real world of hadrons. (Of course, matrix elements of operators in the real world would not in general resemble matrix elements in free field theory.) Thus in Section 6 we study the light-cone behavior of local and bilocal operators in free quark theory, the simplest interesting case. The relevant bilocal operators turn out to be extremely simple, namely just $i/2(\bar{q}(x)\lambda_i\gamma_\alpha q(y))$ and $i/2(\bar{q}(x)\lambda_i\gamma_\alpha\gamma_5 q(y))$, bilocal generalizations of V and A currents. The algebraic system to which they belong is also very simple.

In Section 7 we explore briefly what it would mean if these algebraic relations of free quark theory were really true on the light cone for hadrons. We see that we obtain, among other things, the sensible features of the so-called "parton" picture of Feynman[4] and of Bjorken and Paschos,[5] especially as formulated more exactly by Landshoff and Polkinghorne,[6] Llewellyn Smith,[7] and others. Many symmetry relations are true in such a theory, and can be checked by deep inelastic experiments with electrons and with neutrinos. Of course, some alleged results of the "parton" model depend not just on light cone commutators but on detailed additional assumptions about matrix elements, and about such results we have nothing to say.

The abstraction of free quark light cone commutation relations becomes more credible if we can show, as was done for equal time charge density commutation relations, that certain kinds of nontrivial interactions of quarks leave the relations undisturbed, according to the method of naïve manipulation of operators, using equations of motion. There is evidence that in fact this is so, in a theory with a neutral scalar or pseudoscalar "gluon" having a Yukawa interaction with the quarks. (If the "gluon" is a vector boson, the commutation relations on the light cone might be disturbed for all we know.)

A special case is one in which we abstract from a model in which there are only quarks, with some unspecified self-interaction, and no "gluons". This corresponds to the pure quark case of the "parton" model. One additional constraint is added, namely the identification of the traceless part of $\theta_{\mu\nu}$ with the analog of the traceless part of the symmetrized $\bar{q}\gamma_\mu\partial_\nu q$. This constraint leads to an additional sum rule for deep inelastic electron and neutrino experiments, a rule that provides a real test of the pure quark case.

We do not, in this paper, study the connection between scaling in electromagnetic and neutrino experiments on hadrons on the one hand and scaling in "inclusive" reactions of hadrons alone on the other hand. Some approaches, such as the intuition of the "parton" theorists, suggest such a connection, but we do not explore that idea here. It is worth reemphasizing, however, that any theory of pure hadron behavior that limits transverse momenta of particles produced at high energies has a chance of giving the Bjorken scaling when electromagnetism and weak interactions are introduced. (This point has been made in the cut-off models of Drell, Levy, and Yan.[8])

2. DILATION OPERATOR AND BROKEN SCALE INVARIANCE[9]

We assume that gravity theory (in first order perturbation approximation) applies to hadrons on a microscopic scale, although no way of checking that assertion is known. There is then a symmetrical, conserved, local stress-energy-momentum tensor $\theta_{\mu\nu}(x)$ and in terms of it the translation operators P_μ, obeying for any operator $\mathcal{O}\cdots(x)$, the relation

$$[\mathcal{O}\cdots(x), P_\mu] = \frac{1}{i}\,\partial_\mu\mathcal{O}\cdots(x), \tag{2.1}$$

are given by

$$P_\mu = \int \theta_{\mu o} d^3x. \tag{2.2}$$

Now we want to define a differential dilation operator $D(t)$ that corresponds to our intuitive notions of such an operator, i.e., one that on equal-time commutation with a local operator $\mathcal{O}\cdots$ of definite physical dimension l_o, gives

$$[\mathcal{O}\cdots(x), D(t)] = ix_\mu\partial_\mu\mathcal{O}\cdots(x) - il_o\mathcal{O}\cdots(x). \tag{2.3}$$

We suppose that gravity selects a $\theta_{\mu\nu}$ such that this dilation operator D is given by the expression

$$D = -\int x_\mu\theta_{\mu o} d^3x. \tag{2.4}$$

It is known that for any renormalizable theory this is possible, and Callan, Coleman, and Jackiw have shown that in such a case the matrix elements of this $\theta_{\mu\nu}$ are finite. From (2.4) we see that the violation of scale invariance is connected with the non-vanishing of $\theta_{\mu\mu}$, since we have

$$\frac{dD}{dt} = -\int \theta_{\mu\mu} d^3x. \tag{2.5}$$

Another version of the same formula says that

$$[D, P_0] = -iP_0 - i\int \theta_{\mu\mu} d^3x \tag{2.6}$$

and we see from this and (2.3) that the energy denisty has a main scale-invariant term $\bar{\bar{\theta}}_{00}$ (under the complete dilation operator D) with $l = -4$ (corresponding to the mathematical dimension of energy density) and other terms w_n with other physical dimensions l_n. The simplest assumption (true of most simple models) is that these terms are world scalars, in which case we obtain

$$-\theta_{\mu\mu} = \sum_n (l_n + 4) w_n \tag{2.7}$$

along with the definition

$$\theta_{00} = \bar{\bar{\theta}}_{00} + \sum_n w_n. \tag{2.8}$$

We note that the breaking of scale invariance prevents D from being a world scalar and that equal-time commutation with D leads to a non-covariant break-up of operators into pieces with different dimensions l.

To investigate the relation between the violations of scale invariance and of chiral invariance, we make a still further simplifying assumption (true of many simple models such as the quark-gluon Langrangian model), namely that there are two q-number w's, the first violating scale invariance but not chiral invariance (like the gluon mass) and the second violating both (like the quark mass):

$$\theta_{00} = \bar{\bar{\theta}}_{00} + \delta + u + \text{const.}, \tag{2.9}$$

with δ transforming like $(\mathbf{1}, \mathbf{1})$ under $SU_3 \times SU_3$. Now how does

u transform? We shall start with the usual theory that it all belongs to a single $(3, \bar{3}) + (\bar{3}, 3)$ representation and that the smallness of m_π^2 is to be attributed, in the spirit of PCAC, to the small violation of $SU_2 \times SU_2$ invariance by u. In that case we have

$$u = -u_0 - cu_8, \tag{2.10}$$

with c not far from $-\sqrt{2}$, the value that gives $SU_2 \times SU_2$ invariance and $m_\pi^2 = 0$ and corresponds in a quark scheme to giving a mass only to the s quark. A small amount of u_3 may be present also, if there is a violation of isotopic spin conservation that is not directly electromagnetic; an expression containing u_0, u_3, and u_8 is the most general canonical form of a CP-conserving term violating $SU_3 \times SU_3$ invariance and transforming like $(3, \bar{3}) + (\bar{3}, 3)$.

According to all these simple assumptions, we have

$$-\theta_{\mu\mu} = (l_\delta + 4)\delta + (l_u + 4)(-u_0 - cu_8) + 4\,(\text{const.}) \tag{2.11}$$

and, since the expected value of $(-\theta_{\mu\mu})$ is $2m^2$, we have

$$O = (l_\delta + 4)\langle \text{vac} | \delta | \text{vac} \rangle + (l_u + 4)\langle \text{vac} | u | \text{vac} \rangle + 4(\text{const.}), \tag{2.12}$$

$$2m_i^2(PS\,8) = (l_\delta + 4)(PS_i | \delta | PS_i) + (l_u + 4)\langle PS_i | u | PS_i \rangle, \tag{2.13}$$

etc.

The question has often been raised whether δ could vanish. Such a theory is very interesting, in that the same term u would break chiral and conformal symmetry. But is it possible?

It was pointed out a year or two ago[10] that for this idea to work, something would have to be wrong with the final result of von Hippel and Kim,[11] who calculated approximately the "σ terms" in meson-baryon scattering and found, using our theory of $SU_3 \times SU_3$ violation, that $\langle N | u | N \rangle$ was very small compared to $2m_N^2$. Given the variation of $\langle B | u_8 | B \rangle$ over the $1/2^+$ baryon octet, the ratio of $\langle \Xi | u | \Xi \rangle$ to $\langle N | u | N \rangle$ would be huge if von Hippel and Kim were right, and this disagrees with the value m_Ξ^2/m_N^2 that obtains if $\delta = 0$.

Now, Ellis[12] has shown that in fact the method of von Hippel and Kim should be modified and will produce different results, provided there is a dilation. A dilation is a neutral scalar meson that dominates the dispersion relations for matrix elements of $\theta_{\mu\mu}$ at low frequency, just as the pseudoscalar octet is supposed to dominate the relations for $\partial_\alpha \mathscr{F}^5_{i\alpha}$. We are dealing in the case of the dilaton, with PCDC (partially conserved dilation current) along with PCAC (partially conserved axial vector current). If we have PCAC, PCDC, *and* $\delta = 0$, we may crudely describe the situation by saying that as $u \to 0$ we have chiral and scale invariance of the energy, the masses of a pseudoscalar octet and a scalar singlet go to zero, and the vacuum is not invariant under either chiral or scale transformations (though it is probably SU_3 invariant). With the dilation, we can have masses of other particles non-vanishing as $u \to 0$, even though that limit is scale invariant.

Dashen and Cheng[13] have just finished a different calculation of the "σ terms" not subject to modification by dilation effects, and they find, using our description of the violation of chiral invariance, that $\langle N|u|N\rangle$ at rest is around $2m_N^2$, a result perfectly compatible with the idea of vanishing δ and yielding in that case a value $l_u \approx -3$ (as in a naive quark picture, where u is a quark mass term!).

An argument was given last year [10] that if $\delta = 0$, the value of l_u would have to be -2 in order to preserve the perturbation theory approach for $m^2(PS\,\mathbf{8})$, $m^2(PS\,\mathbf{8}) \propto u$, which gives the right mass formula for the pseudoscalar octet. Ellis, Weisz, and Zumino[14] have shown that this argument can be evaded if there is a dilation.

Thus at present there is nothing known against the idea that $\delta = 0$, with l_u probably equal to -3. However, there is no strong evidence in favor of the idea either. Theories with non-vanishing δ operators and various values of l_δ and l_u are not excluded at all (although even here a dilaton would be useful to explain why $\langle N|u|N\rangle$ is so large). It is a challenge to theorists to propose experimental means of checking whether the δ operator is there or not.

It is also possible that the simple theory of chiral symmetry violation may be wrong. First of all, the expression $-u_0 + \sqrt{2}u_8$

could be right for the $SU_2 \times SU_2$-conserving but $SU_3 \times SU_3$-violating part of θ_{00}, while the $SU_2 \times SU_2$ violation could be accomplished by something quite different from $(-c-\sqrt{2})u_8$. Secondly, there can easily be an admixture of the eighth component g_8 of an octet belonging to $(\mathbf{1}, \mathbf{8})$ and $(\mathbf{8}, \mathbf{1})$. Thirdly, the whole idea of explaining $m_\pi^2 \approx 0$ by near-conservation of $SU_2 \times SU_2$ might fail, as might the idea of octet violation of SU_3; it is those two hypotheses that give the result that for $m_\pi^2 = 0$ we have only $u_o - \sqrt{2}u_8$ with a possible admixture of g_8. Here again there is a challenge to theoreticians to propose effective experimental tests of the theory of chiral symmetry violation.

3. LIGHT CONE COMMUTATORS AND DEEP INELASTIC ELECTRON SCATTERING

We want ultimately to connect the above discussion of physical dimensions and broken scale invariance with the scaling described in connection with the Stanford experiments on deep inelastic electron scattering.[15] We must begin by presenting the Stanford scaling in suitable form. For the purpose of doing so, we shall assume for convenience that the experiments support certain popular conclusions, even though uncertainties really prevent us from saying more than that the experiments are consistent with such conclusions:

(1) that the scaling formula of Bjorken is really correct, with no logarithmic factors, as the energy and virtual photon mass go to infinity with fixed ratio;

(2) that in this limit the neutron and proton behave differently;

(3) that in the limit the longitudinal cross section for virtual photons goes to zero compared to the transverse cross section.

All these conclusions are easy to accept if we draw our intuition from certain field theories without interactions or from certain field theories with naïve manipulation of operators. However, detailed calculations using the renormalized perturbation expansion in renormalizable field theories do not reveal any of these forms of behavior, unless of course the sum of all orders of perturbation theory somehow restores the simple situation. If we accept the conclusions, therefore, we should probably not think in terms of the renormalized perturbation expansion, but rather conclude, so to speak, that Nature reads books on free field theory, as far as the Bjorken limit is concerned.

To discuss the Stanford results, we employ a more or less conventional notation. The structure functions of the nucleon are defined by matrix elements averaged over nucleon spin,

$$\frac{1}{4\pi}\int d^4x\langle N,p|[j_\mu(x),j_\nu(y)]|N,p\rangle e^{-iq\cdot(x-y)}$$

$$= \left(\delta_{\mu\nu} - \frac{q_\mu q_\nu}{q^2}\right)W_1(q^2,p\cdot q)$$

$$+\left(p_\mu - \frac{p\cdot q}{q^2}q_\mu\right)\left(p_\nu - \frac{p\cdot q}{q^2}q_\nu\right)W_2(q^2,p\cdot q) \qquad (3.1)$$

$$= \left(\delta_{\mu\nu} - \frac{q_\mu q_\nu}{q^2}\right)\left(W_1 - \frac{(p\cdot q)^2}{q^2}W_2\right.$$

$$\left. + \frac{\delta_{\mu\nu}(p\cdot q)^2 + p_\mu p_\nu q^2 - (p_{\mu\nu} + q_\mu p_\nu)p\cdot q}{q^2}W_2\right),$$

where p is the nucleon four-momentum and q the four-momentum of the virtual photon. As q^2 and $q\cdot p$ become infinite with fixed ratio, averaging over the nucleon spin and assuming $\sigma_L/\sigma_T\mathscr{F}\to 0$, we can write the Bjorken scaling in the form

$$\frac{1}{4\pi}\int d^4x\langle N,p|[j_\mu(x),j_\nu(y)]|N,p\rangle e^{-iq\cdot(x-y)}$$

$$\to \frac{(p_\mu q_\nu + p_\nu q_\mu)p\cdot q - \delta_{\mu\nu}(p\cdot q)^2 - p_\mu p_\nu q^2}{q^2(q\cdot p)}F_2(\xi), \qquad (3.2)$$

where $\xi = -q^2/2p\cdot q$ and $F_2(\xi)$ is the scaling function in the deep inelastic region.

In coordinate space, this limit is achieved by approaching the light cone $(x-y)^2 = 0$,[15] and we employ a method, used by Frishman[3] and by Brandt and Preparata,[2] generalizing earlier work of Wilson, that starts with an expansion for commutators or operator products valid near $(x-y)^2 = 0$. (The symbol $\hat{=}$ will be employed for equality in the vicinity of the light cone.) After the expansion is made, then the matrix element is taken between nucleons. To simplify matters, let us introduce the "barred product" of two operators, which means that we average over the mean position $R \equiv (x+y)/2$, leaving a function of $z \equiv x - y$ only (as appropriate for matrix elements with no change of momentum) and that we retain in the expansion only totally symmetric Lorentz tensor operators (as appropriate for matrix elements averaged over spin). Then the assumed light-cone expansion of the barred commutator $[\overline{j_\mu(x), j_\nu(y)}]$ tells us that we have, as $z^2 \to 0$,

$$[\overline{j_\mu(x), j_\nu(y)}] \hat{=} t_{\mu\nu\rho\sigma}\{\varepsilon(z_0)\delta(z^2)(\mathcal{O}_{\rho\sigma} + \frac{1}{2!} z_\alpha z_\beta \mathcal{O}_{\rho\sigma\alpha\beta} + \cdots)\}$$

$$+ (\partial_\mu \partial_\nu - \delta_{\mu\nu}\partial^2)\{\varepsilon(z_0)\delta(z^2)(U + \frac{1}{2!} z_\alpha z_\beta U_{\alpha\beta} + \cdots)\}, \tag{3.3}$$

where

$$t_{\mu\nu\rho\sigma} = \frac{1}{\pi i} \times$$

$$\frac{2\delta_{\mu\nu}\partial_\rho\partial_\sigma - \delta_{\rho\mu}\partial_\nu\partial_\sigma - \delta_{\rho\nu}\partial_\mu\partial_\sigma - \delta_{\sigma\mu}\partial_\nu\partial_\rho - \delta_{\sigma\nu}\partial_\mu\partial_\sigma - \delta_{\mu\sigma}\delta_{\nu\rho}\partial^2 - \partial_{\nu\sigma}\delta_{\mu\rho}\partial^2}{\partial^2}$$

and the second term, the one that gives σ_L, will be ignored for simplicity in our further work.

In order to obtain the Bjorken limit, we have only to examine the matrix elements between $|Np\rangle$ and itself of the operators $\mathcal{O}_{\alpha\beta}$, $\mathcal{O}_{\alpha\beta\gamma\delta}$, $\mathcal{O}_{\alpha\beta\gamma\delta\varepsilon_\rho}$, etc. The leading tensors in the matrix elements have the form $c_2 p_\alpha p_\beta$, $c_4 p_\alpha p_\beta p_\gamma p_\delta$, etc., where the c's are dimensionless constants. The lower tensors, such as $\delta_{\alpha\beta}$, have coefficients that are positive powers of masses, and these tensors give negligible contributions in the Bjorken limit. All we need is the very weak assumption that c_2, c_4, c_8, etc., are not all zero, and we obtain the Bjorken limit.

We define the function

$$\tilde{F}(p \cdot z) = c_2 + \frac{1}{2!} \cdot c_4 (p \cdot z)^2 + \cdots . \tag{3.4}$$

Taking the Fourier transform of the matrix element of (3.3), we get in the Bjorken limit

$$
\begin{aligned}
W_2 &\to \frac{1}{2\pi^2 i}\int d^4z e^{-iq\cdot z}\tilde{F}(p\cdot z)\varepsilon(z_0)\delta(z^2) \\
&= \frac{1}{2\pi^2 i}\int F(\xi)d\xi \int d^4 z e^{-i(q+\xi p)\cdot z}\varepsilon(z_0)\delta(z^2) \\
&= 2\int F(\xi)d\xi\varepsilon(-q\cdot p)\delta(q^2+2q\cdot p\xi) \\
&= \frac{1}{-q\cdot p}F(\xi)
\end{aligned}
\tag{3.5}
$$

where ξ is $-q^2/2q\cdot p$ and $F(\xi)$ is the Fourier transform of $\tilde{F}(p\cdot z)$:

$$F(\xi) = \frac{1}{2\pi}\int e^{i\xi(p\cdot z)}\tilde{F}(p\cdot z)d(p\cdot z). \tag{3.6}$$

The function $F(\xi)$ is therefore the Bjorken scaling function in the deep inelastic limit and is defined only for $-1 < \xi < 1$. We can write (3.6) in the from

$$F(\xi) = c_2\cdot\delta(\xi) - c_4\frac{1}{2!}\delta''(\xi) + c_6\frac{1}{4!}\delta''''(\xi) - \cdots \tag{3.7}$$

The dimensionless numbers c_i defined by the matrix elements of the expansion operators can be written as

$$c_2 = \int_{-1}^{1} F(\xi)d\xi, \quad c_4 = -\int_{-1}^{1} F(\xi)\xi^2 d\xi \cdots \tag{3.8}$$

This shows the connection between the matrix elements of the expansion operators and the moments of the scaling function. The Bjorken limit is seen to be a special case (the matrix element between single nucleon states of fixed momentum) of the light cone expansion.[17]

Now the derivation of the Bjorken limit from the light cone expansion can be described in terms of a kind of physical dimension l for operators. (We shall see in the next section that these dimensions l are essentially the same as the physical dimensions l we described in Section 2.) We define the expansion to conserve dimension on the light cone and assign to each current

$l = -3$ while counting each power of z as having an l-value equal to the power. We see then that on the right-hand side we are assigning to each J-th rank Lorentz tensor (with maximum spin J) the dimension $l = -J - 2$. Furthermore, the physical dimension equals the mathematical dimension in all of these cases.

4. GENERALIZED LIGHT CONE SCALING AND BROKEN SCALE INVARIANCE

We have outlined a situation in which scale invariance is broken by a non-vanishing $\theta_{\mu\mu}$ but restored in the most singular terms of current commutators on the light cone. There is no reason to suppose that such a restoration is restricted to commutators of electromagnetic currents. We may extend the idea to all the vector currents $\mathscr{F}_{i\mu}$ and axial vector currents $\mathscr{F}_{i\mu}{}^5$, to the scalar and pseudoscalar operators u_i and v_i that comprise the $(3, \bar{3})$ and $(\bar{3}, 3)$ representation thought to be involved in chiral symmetry breaking, to the whole stress-energy momentum tensor $\theta_{\mu\nu}$, to any other local operator of physical significance, and finally to all the local operators occurring in the light cone expansions of commutators of all these quantities with one another. Let us suppose that in fact conservation of dimension applies to leading terms in the light cone in the commutators of all these quantities and that finally a closed algebraic system with an infinite number of local operators is attained, such that the light cone commutator of any two of the operators is expressible as a linear combination of operators in the algebra. We devote this section and the next one to discussing such a situation.

If there is to be an analog of Bjorken scaling in all these situations, then on the right-hand side of the light cone commutation relations we want operators with $l = -J - 2$, as above for electromagnetic current commutators, so that we get leading matrix elements between one-particle states going like $cp_\alpha p_\beta \cdots$, where the c's are dimensionless constants.

Of course, there might be cases in which, for some reason, all the c's have to vanish, and the next-to-leading term on the light cone becomes the leading term. Then the coefficients would have the dimensions of positive powers of mass. We want to avoid, however, situations in which coefficients with the dimension of negative powers of mass occur; that means on the right-hand side we want $l \leqq -J-2$ in any case, and $l = -J-2$ when there is nothing to prevent it.

This idea might have to be modified, as in a quark model with a scalar or pseudoscalar "gluon" field, to allow for a single operator ϕ, with $l = -1$ and $J = 0$, that can occur in a barred product, but without a sequence of higher tensors with $l = -J-1$ that could occur in such a product; gradients of ϕ would, of course, average out in a barred product. However, even this modification is probably unnecessary, since preliminary indications are that, in the light cone commutator of any two (physically interesting operators, the operator ϕ with $l = -1$ would not appear on the right-hand side.

Now, on the left-hand side, we want the non-conserved currents among $\mathscr{F}_{i\mu}$ and $\mathscr{F}_{i\mu}^5$ to act as if they have dimension -3 just like the conserved ones, as far as leading singularities on the light cone are concerned, even though the non-conservation implies the admixture of terms that may have other dimensions l, dimensions that become $l-1$ in the divergences, and correspond to dimensions $l-1$ in the $SU_3 \times SU_3$ breaking terms in the energy density. But the idea of conservation of dimension on the light cone tells us that we are dealing with lower singularities when the dimensions of the operators on the left are greater. What is needed, then, is for the dimensions l to be > -3, i.e., for the chiral symmetry breaking terms in $\theta_{\mu\nu}$ to have dimension > -4. Likewise, if we want the stress-energy-momentum tensor itself to obey simple light cone scaling, we need to have the dimension of all scale breaking parts of $\theta_{\mu\nu}$ restricted to values > -4. In general, we can have symmetry on the light cone if the symmetry breaking terms in $\theta_{\mu\nu}$ have dimension greater than -4. (See Appendix I.)

Now we can have $\mathscr{F}_{i\mu}$ and $\mathscr{F}_{i\mu}^5$ behaving, as far as leading singularities on the light cone are concerned, like conserved currents with $l = -3$, $\theta_{\mu\nu}$ behaving like a chiral and scale invariant quantity with $l = -4$, and so forth. To pick out the subsidiary dimensions associated with the non-conservation of $SU_3 \times SU_3$ and dilation, we can study light cone commutators involving,

$\partial_\alpha \mathscr{F}_{i\alpha}$, $\partial_\alpha \mathscr{F}^5_{i\alpha}$, and $\theta_{\mu\mu}$. (If the $(3, \bar{3}) + (\bar{3}, 3)$ hypothesis is correct, that means studying commutators involving u's and v's and also δ, if $\delta \neq 0$.)

In our enormous closed light cone algebra, we have all the operators under consideration occuring on the left-hand side, the ones with $l = -J-2$ on the right-hand side, and coefficients that are functions of z behaving like powers according to the conservation of dimension. But are there restrictions on these powers? And are there restrictions on the dimensions occurring among the operators?

If, for example, the functions of z have to be like powers of z^2 (or $\delta(z^2)$, $\delta'(z^2)$, etc.) multiplied by tensors $z_\alpha z_\beta z_\gamma \cdots$, and if $l+J$ for some operators is allowed to be non-integral or even odd integral, then we cannot always have $l = -J-2$ on the right, i.e., the coefficients of all such operators would vanish in certain commutators, and for those commutators we would have to be content with operators with $l < -J-2$ on the right, and coefficients of leading tensors that act like positive powers of a mass.

Let us consider the example:

$$[\theta_{\mu\nu}(x), u(y)] \mathrel{\hat{=}} E_{\mu\nu}(z) \cdot (\mathcal{O}(y) + z_\rho \mathcal{O}_\rho(y) + \cdots) + \cdots,$$

where $u(y)$ has the dimension -3. In this case we cannot have the Bjorken scaling. Because of the relation

$$[D(O), u(O)] = -3iu(O),$$

the operator $\mathcal{O}(y)$ has to be proportional to $u(y)$. The operator series fulfilling the condition $l = -J-2$ is forbidden in this case on the right-hand side.

We have already emphasized that Nature seems to imitate the algebraic properties of free field theory rather than renormalized perturbation theory. (We could also say that Nature is imitating a super-renormalizable theory, even though no sensible theory of that kind exists, with the usual methods of renormalization, in four dimensions.) This suggests that we should have in our general expansion framework finite equal-time commutators for all possible operators and their time derivatives.

Such a requirement means that all functions of z multiplying operators in a light cone expansion must have the behavior described just above, i.e., the scalar functions involved behave like integral forces of z^2 or like derivatives of delta functions with z^2 as the argument. The formula

$$\frac{1}{(z^2+i\varepsilon)^\alpha} - \frac{1}{(z^2-i\varepsilon)^\alpha} \xrightarrow[z_0 \to 0]{} \text{const. } z_0^{-2\alpha+3}\,\delta(z)$$

shows the sort of thing we mean. It also shows that α must not be too large. That can result in lower limits on the tensorial rank of the first operator in the light cone expansion in higher and higher tensors; to put it differently, the first few operators in a particular light cone expansion may have to be zero in order to give finiteness of equal time commutators with all time derivatives.

Now, on the right-hand side of a light cone commutator of two physically interesting operators, when rules such as we have just discussed do not forbid it, we obtain operators with definite $SU_3 \times SU_3$ and other symmetry properies, of various tensor ranks, and with $l = -J-2$. Now, for a given set of quantum numbers, how many such operators are there? Wilson[1] suggested a long time ago that there may be very few, sometimes only one, and others none. Thus no matter what we have on the left, we always would get the same old operators on the right (when not forbidden and less singular terms with dimensional coefficients occuring instead). This is very important, since the matrix elements of these universal $l = -J-2$ operators are then natural constants occurring in many problems. Wilson presumably went a little too far in guessing that the only Lorentz tensor operator in the light cone expansion of $[j_\mu(x),\ j_\nu(y)]$ would be the stress-energy-momentum tensor $\theta_{\mu\nu}$, with no provision for an accompanying octet of $l = -4$ tensors. That radical suggestion, as shown by Mack,[17] would make $\int F_2^{en}(\xi)\, d\xi$ equal to $\int F_2^{ep}(\xi) d\xi$, which does not appear to be the case. However, it is still possible that one singlet and one octet of tensors may do the job. (See the discussion in Section 7 of the "pure quark" case.)

If we allow z_0 to approach zero in a light cone commutator, we obtain an equal time commutator. If Wilson's principle (suitably weakened) is admitted, then all physically interesting operators must obey some equal time commutation relations, with well-known operators on the right-hand side, and presumably there are fairly small algebraic systems to which these equal time commutators belong. The dimensions of the operators constrain severely the nature of the algebra involved. For example, suppose $SU_3 \times SU_3$ is broken by a quantity u belonging to the representation $(3, \bar{3}) \oplus (\bar{3}, 3)$ and having a single dimension l_u. Then, if $l_u = -3$, we may well have the algebraic system proposed years ago by one of us (M.G.-M.) in which F_i, F_i^5, $\int u_i d^3x$, and $\int v_i d^3x$ obey the E.T.C. relations of U_6, as in the quark model. If $l_u = -2$, however, then we would have $\int u_i d^3x$ and $d/dt \int u_i d^3x$ commuting to give a set of quantities including $\int u_i d^3x$, and so forth.

We have described scaling in this section as if the dimensions l were closely related to the dimensions $\bar{l}$ obtained by equal time commutation with the dilation operator D in Section 2. Let us now demonstrate that this is so.

To take a simple case, suppose that in the light cone commutator of an operator $\mathcal{O}\cdots$ with itself, the same operator $\mathcal{O}\cdots$ occurs in the expansion on the right-hand side. Then we have a situation crudely described by the equation

$$[\mathcal{O}\cdots(z), \mathcal{O}\cdots(0)] \mathrel{\widehat{=}} \cdots + (z)^l\mathcal{O}\cdots(0) + \cdots, \tag{4.1}$$

where l is the principal dimension of $\mathcal{O}\cdots$. Here $(z)^l$ means any function of z with dimension l, and we must have that because of conservation of dimension. Now under equal time commutation with D, say $\mathcal{O}\cdots$ exhibits dimension $\bar{l}$. Let $z_0 \to 0$ and perform the equal time commutation, according to Eq. (2.3). We obtain

$$\begin{aligned}(iz \cdot \nabla - 2i\bar{l})[\mathcal{O}\cdots(z), \mathcal{O}\cdots(0)] &= -i\bar{l}(z)^l\mathcal{O}\cdots(0) \\ &= (il - 2i\bar{l})(z)^l\mathcal{O}\cdots(0)\end{aligned} \tag{4.2}$$

so that $l = \bar{l}$, as we would like.

Now to generalize the demonstration, we consider the infinite closed algebra of light cone commutators, construct commutators like (4.1) involving different operators, and from commutation with D as in (4.2) obtain equations

$$l_1 + l_2 - l_3 = \bar{l}_1 + \bar{l}_2 - \bar{l}_3, \tag{4.3}$$

where $\mathcal{O}_{\cdots}^{(1)}$ and $\mathcal{O}_{\cdots}^{(2)}$ are commuted and yield a term containing $\mathcal{O}_{\cdots}^{(3)}$ on the right. Chains of such relations can then be used to demonstrate finally that $l = \bar{l}$ for the various operators in which we are interested.

The subsidiary dimensions associated with symmetry breaking have not been treated here. They can be dealt with in part by isolating the expressions $\partial_\mu \mathcal{F}^5_{i\mu}$, $\theta_{\mu\mu}$, etc., that exhibit only the subsidiary dimensions and applying similar arguments to them. In that way we learn that also for subsidiary dimensions $\bar{l} = l$.

However, the subsidiary dimensions, while numerically equal for the two definitions of dimension, do not enter in the same way for the two definitions. The physical dimension l defined by light cone commutation always enters covariantly, while $\bar{l}$ is defined by equal time commutation with the quantity D and enters non-covariantly, as in the break-up of $\theta_{\mu\nu}$ into the leading term $\bar{\bar{\theta}}_{\mu\nu}$ of dimension -4 and the subsidiary ones of higher dimensions. If these others come from world scalars w_n of dimensions l_n, then we have

$$\theta_{\mu\nu} = \bar{\bar{\theta}}_{\mu\nu} + \sum_n \{(3 + l)\delta_{\mu\nu} + (4 + l)\delta_{\mu 0}\delta_{\nu 0}\} \frac{w_n}{3}, \tag{4.4}$$

so that we agree with the relations

$$\theta_{00} = \bar{\bar{\theta}}_{00} + \sum_n w_n, \tag{2.8}$$

$$-\theta_{\mu\mu} = \sum_n (l_n + 4) w_n. \tag{2.9}$$

Clearly, $\bar{\bar{\theta}}_{\mu\nu}$ is non-covariant.

To obtain the non-covariant formula from the covariant one, the best method is to write the light cone commutator of an operator with $\theta_{\mu\nu}$, involving physical dimensions l, and then construct $D = -\int x_\mu \theta_{\mu 0} d^3x$ out of $\theta_{\mu\nu}$ and allow the light cone commutator to approach an equal time commutator. The non-covariant formula involving $\bar{l}$ must then result.

As an example of non-covariant behavior of equal time commutation with D, consider such a commutator involving an arbitrary tensor operator $\mathcal{O}_{\rho\sigma}$ of dimension -4. We may pick up non-covariant contributions that arise from lower order terms near the light cone than those that give the dominant scaling behavior. We may have

$$[\theta_{\mu\nu}(x), \mathcal{O}_{\rho\sigma}(y)] = \text{leading term} + \partial_\mu\partial_\nu\partial_\rho\partial_\sigma\{\varepsilon(z_0)\delta(z^2)[\mathcal{O}(y) + \cdots]\} + \cdots$$

giving the result

$$\text{E.T.C.}\ [D, \mathcal{O}_{\rho\sigma}(0)] = 4i\theta_{\rho\sigma}(0) + \text{const.}\, \delta_{\rho 0}\delta_{\sigma 0}\mathcal{O}(0) + \cdots .$$

For commutation of D with a scalar operator, there is no analog of this situation.

5. BILOCAL OPERATORS

So far, in commuting two currents at points separated by a four-dimensional vector z_μ, we have expanded the right-hand side on the light cone in powers of z_μ. It is very convenient for many purposes to sum the series and obtain a single operator of low Lorentz tensor rank that is a function of z. In a barred commutator, it is a function of z only, but in an ordinary unbarred commutator, it is a function of z and $R \equiv (x+y)/2$, in other words, a function of x and y. We call such an operator a bilocal operator and write it as $\mathcal{O}\cdots(x, y)$ or, in barred form, $\bar{\mathcal{O}}\cdots(x, y)$.

We can, for example, write Eq. (3.3) in the form

$$\overline{[j_\mu(x), j_\nu(y)]} \mathrel{\hat{=}} t_{\mu\nu\rho\sigma}\{\varepsilon(z_0)\delta(z^2)\overline{\mathcal{O}_{\rho\sigma}(x, y)}\} + \text{longitudinal term}, \quad (5.1)$$

using the barred form of a bilocal operator $\mathcal{O}_{\rho\sigma}(x, y)$ that sums up all the tensors of higher and higher rank in Eq. (3.3).

Now in terms of bilocal operators we can formulate a much stronger hypothesis than the modified Wilson hypothesis mentioned in the last section. There we supposed that on the right-hand side of any light-cone commutators (unless the leading terms were forbidden for some reason) we would always have operators with $l = -J - 2$ and that for a given J and a given set of quantum numbers there would be very few of these, perhaps only one, and that the quantum numbers themselves would be greatly restricted (for example, to SU_3 octets and singlets). Here we can state the much stronger conjecture that for a given set of quantum numbers the bilocal operators appearing on the right are very few in number (and perhaps there is only one in each case), with the quantum numbers greatly restricted. That means that instead of an arbitrary series $\mathcal{O}_{\rho\sigma} + \text{const.}\ z_\lambda z_\mu \mathcal{O}_{\rho\sigma\lambda\mu} + \text{const.}' z_\lambda z_\mu z_\alpha z_\beta \mathcal{O}_{\alpha\beta\rho\sigma\lambda\mu} + \cdots$, we have a unique sum $\mathcal{O}_{\rho\sigma}(x, y)$ with all the constants determined. The same bilocal operator will appear in many commutators, then, and its matrix elements (for example, between proton and proton with no charge of momentum) will give universal deep inelastic form factors.

Let us express in terms of bilocal operators the idea mentioned in the last section that all tensor operators appearing on the right-hand side of the light cone current commutators may themselves be commuted according to conservation of dimension on the light cone, but lead to the same set of operators, giving a closed light cone algebra of an infinite number of local operators of all tensor ranks. We can sum up all these operators to make bilocal operators and commute those, obtaining, on the right-hand side according to the principle mentioned above, the same bilocal operators. Thus we obtain a light cone algebra generated by a small finite number of bilocal operators. These are the bilocal operators that give the most singular terms on the light cone in any commutator of local operators, the terms that give scaling behavior. (As we have said, in certain cases they may be forbidden to occur and positive powers of masses would then appear instead of dimensionless coefficients.)

This idea of a universal light cone algebra of bilocal operators with $l = -J - 2$ is a very elegant hypothesis, but one that goes far beyond present experimental evidence. We can hope to check it some day if we can find situations in which limiting cases of experiments involve the light cone commutators of light cone commutators. Attempts have been made to connect differential cross sections for the Compton effect with such mathematical quantities;[5] it will be interesting to see what comes of that and other such efforts.

A very important technical question arises in connection with the light cone algebra of bilocal operators. When we talk about the commutators of the individual local operators of all tensor ranks, we are dealing with just two points x and y and with the limit $(x-y)^2 \to 0$. But when we treat the commutator of bilocal operators $\mathcal{O}(x, u)$ and $\mathcal{O}(y, v)$, what are the space-time relationships of x, u, y, and v in the case to which the commutation relations apply? We must be careful, because if we give too liberal a prescription for these relationships we may be assuming more than could be true in any realistic picture of hadrons.

The bilocal operators arise originally in commutators of local operators on the light cone, and therefore we are interested in them when $(x-u)^2 \to 0$ and $(y-v)^2 \to 0$. In the light cone algebra of bilocal operators, we are interested in singularities that are picked up when $(x-y)^2$ or when $(u-v)^2 \to 0$ or when $(x-v)^2 \to 0$ or when $(u-y)^2 \to 0$. But do we have to have all six quantities simultaneously brought near to zero? That is not yet clear. In order to be safe, let us assume here that all six quantities do go to zero.

6. LIGHT CONE ALGEBRA ABSTRACTED FROM A QUARK PICTURE

Can we postulate a particular form for the light cone algebra of bilocal operators?

We have indicated above that if the Stanford experiments, when extended and refined, still suggest the absence of logarithmic terms the vanishing of the longitudinal cross section, and a difference between neutron and proton in the deep inelastic limit, then it looks as if in this limit Nature is following free field theory, or interacting field theory with naïve manipulation of operators, rather than what we know about the perturbation expansions of renormalised field theory. We might, therefore, look at a simple relativistic field theory model and abstract from it a light cone algebra that we could postulate as being true of the real system of hadrons. The simplest such model would be that of free quarks.

In the same way, the idea of an algebra of equal-time commutators of charges or charge densities was abstracted ten years ago from a relativistic Lagrangian model of a free spin 1/2 triplet, what would nowadays be called the quark triplet. The essential feature in this abstraction was the remark that turning on certain kinds of strong interaction in such a model would not affect the equal time commutation relations, even when all orders of perturbation theory were included; likewise, mass differences breaking the symmetry under SU_3 would not disturb the equal time commutation relations of SU_3.

We are faced, then, with the following question. Are there non-trivial field theory models of quarks with interactions such that the light cone algebra of free quarks remains undisturbed to all orders of naïve perturbation theory? Of course, the interactions will make great changes in the operator commutators inside the light cone; the question is whether the leading singularity on the light cone is unaffected. Let us assume, for purposes of our discussion, that the answer is affirmative. Then we can feel somewhat safe from absurdity in postulating for real hadrons the light cone algebras of free quarks, and indeed of massless free quarks (since the masses do not affect the light cone singularity).

Actually, it is easy to construct an example of an interacting field theory in which our condition seems to be fulfilled, namely a theory in which the quark field interacts with a neutral scalar or pseudoscalar "gluon" field ϕ. We note the fact that the only operator series in such a theory that fulfills $l = -J-2$ and contains $\phi(x)$ is the following: $\phi(x)\phi(x)$, $\phi(x)\partial_\mu\phi(x)\cdots$. But these operators do not seem to appear in light cone expansions of products of local operators consisting only of quark fields, like the currents. A different situation prevails in a theory in which the "gluon" is a vector meson, since in that case we can have the operator series $\bar{q}(x)\gamma_\mu B_\nu(x)q(x)$, $\bar{q}(x)\gamma_\mu B_\nu B_\rho q(x)$, $\cdots$, contributing to the Bjorken limit. The detailed behavior of the various "gluon" models is being studied by Llewellyn Smith.[18]

In the following, we consider the light cone algebra suggested by the quark model. We obtain for the commutator of two currents on the light cone (connected part only):

$$
\begin{aligned}
&[\mathscr{F}_{i\mu}(x), \mathscr{F}_{j\nu}(y)] \\
&\quad \hat{=} \frac{1}{4\pi}\partial_\rho[\varepsilon(z_o)\delta(z^2)]\{if_{ijk}[s_{\mu\nu\rho\sigma}(\mathscr{F}_{k\sigma}(x,y) + \mathscr{F}_{k\sigma}(y,x)) \\
&\qquad + i\varepsilon_{\mu\nu\rho\sigma}(\mathscr{F}^5_{k\sigma}(y,x) - \mathscr{F}^5_{k\sigma}(x,y))] + d_{ijk}[s_{\mu\nu\rho\sigma}(\mathscr{F}_{k\sigma}(x,y) \\
&\qquad - \mathscr{S}_{k\sigma}(y,x)) - i\varepsilon_{\mu\nu\rho\sigma}(\mathscr{F}^5_{k\sigma}(y,x) + \mathscr{F}^5_{k\sigma}(x,y))]\},
\end{aligned}
$$

$$
\begin{aligned}
&[\mathscr{F}^5_{i\mu}(x), \mathscr{F}_{j\nu}(y)] \\
&\quad \hat{=} \frac{1}{4\pi}\partial_\rho[\varepsilon(z_o)\delta(z^2)]\{if_{ijk}[s_{\mu\nu\rho\sigma}(\mathscr{F}^5_{k\sigma}(x,y) + \mathscr{F}^5_{k\sigma}(y,x)) \\
&\qquad + i\varepsilon_{\mu\nu\rho\sigma}(\mathscr{F}_{k\sigma}(y,x) - \mathscr{F}_{k\sigma}(x,y))] \\
&\qquad + d_{ijk}[s_{\mu\nu\rho\sigma}(\mathscr{F}^5_{k\sigma}(x,y) - \mathscr{F}^5_{k\sigma}(y,x)) \\
&\qquad - i\varepsilon_{\mu\nu\rho\sigma}(\mathscr{F}_{k\sigma}(y,x) + \mathscr{F}_{k\sigma}(x,y))]\}, \\
&[\mathscr{F}^5_{i\mu}(x), \mathscr{F}^5_{j\nu}(y)] = [\mathscr{F}_{i\mu}(x), \mathscr{F}_{j\nu}(y)], \\
&s_{\mu\nu\rho\sigma} = \delta_{\mu\rho}\delta_{\nu\sigma} + \delta_{\nu\rho}\delta_{\mu\sigma} - \delta_{\mu\nu}\delta_{\rho\sigma}, \qquad z = x - y.
\end{aligned} \tag{6.1}
$$

If we go to the equal time limit in (6.1) we pick up the current algebra relations for the currents; in fact we obtain, for the space integrals of all componets of nine vector and nine axial-vector currents, the algebra[19] of $U_6 \times U_6$.

Note that we can get similar relations for the current anti-commutators or for the products of currents on the light cone, just by replacing

$$
\frac{1}{4\pi}\partial_\rho[\varepsilon(z_0)\delta(z^2)] \text{ by } -\frac{i}{4\pi^2}\partial_\rho\frac{1}{z^2} \text{ or by } -\frac{i}{8\pi^2}\partial_\rho\frac{1}{z^2 + i\varepsilon z_0}
$$

respectively. Perhaps we can abstract these relations also and use them for hadron theory.

In (6.1) we have introduced bilocal generalizations of the vector and axial-vector currents, which in a quark model correspond to products of quark fields:

$$
\begin{aligned}
\mathscr{F}_{k\sigma}(x,y) &\sim \bar{q}(x)\frac{i}{2}\lambda_k\gamma_\sigma q(y), \\
\mathscr{F}^5_{k\sigma}(x,y) &\sim \bar{q}(x)\frac{i}{2}\lambda_k\gamma_\sigma\gamma_5 q(y).
\end{aligned} \tag{6.2}
$$

Note that the products in (6.2) have to be understood as "generalized Wick products". The c-number part in the product of two quark fields is already excluded, since it does not contribute to the connected current commutator. The c-number part is measured by vacuum processes like e^+e^- annihilation. Assuming that the disconnected part of the commutator on the light cone is also dictated by the quark model, we would obtain

$$\sigma_{\mathrm{tot}\,e^+e^-} \sim \text{const.}/s \text{ for } e^+e^- \text{ annihilation},$$

where s is as usually defined: $s = -(p_1 + p_2)^2$. In particular, we would get

$$\sigma_{\mathrm{tot}}(e^+ e^- \text{ into hadrons}) \to (\Sigma Q^2)\sigma_{\mathrm{tot}}(e^+ e^- \text{ into muons})$$

with $\Sigma Q^2 = (2/3)^2 + (1/^3)^2 + (1/3)^2 = 2/3$.

Now we go on to close the algebraic system of (6.1), where local currents occur on the left-hand side and bilocal ones on the right.

Let us assume that the bilocal generalizations of the vector and axial vector currents are the basic entities of the scheme. Again using the quark model as a guideline on the light cone, we obtain the following closed algebraic system for these bilocal operators:

$$[\mathscr{F}_{i\mu}(x,u), \mathscr{F}_{j\nu}(y,v)]$$

$$\hat{=} \frac{1}{4\pi}\partial_\rho\{\varepsilon(x_0 - v_0)\delta[(x-v)^2]\}(if_{ijk} - d_{ijk})(s_{\mu\nu\rho\sigma}\mathscr{F}_{k\sigma}(y,u)$$

$$+ i\varepsilon_{\mu\nu\rho\sigma}\mathscr{F}^5_{k\sigma}(y,u))$$

$$+ \frac{1}{4\pi}\partial_\rho\{\varepsilon(u_0 - y_0)\partial[(u-y)^2]\}(if_{ijk} + d_{ijk})$$

$$\cdot (s_{\mu\nu\rho\sigma}\mathscr{F}_{k\sigma}(x,v) - i\varepsilon_{\mu\nu\rho\sigma}\mathscr{F}^5_{k\sigma}(x,v)),$$

$$[\mathscr{F}^5_{i\mu}(x,u), \mathscr{F}_{j\nu}(y,v)] \tag{6.3}$$

$$\hat{=} \frac{1}{4\pi}\partial_\rho\{\varepsilon(x_0 - v_0)\delta[(x-v)^2]\}(if_{ijk} - d_{ijk})$$

$$(s_{\mu\nu\rho\sigma}\mathscr{F}^5_{k\sigma}(y,u) + i\varepsilon_{\mu\nu\rho\sigma}\mathscr{F}_{k\sigma}(y,u))$$

$$+ \frac{1}{4\pi}\partial_\rho\{\varepsilon(u_o - y_o)\delta[(u-y)^2]\}(if_{jik} + d_{ijk})$$

$$\cdot (s_{\mu\nu\rho\sigma}\mathscr{F}^5_{k\sigma}(x,v) - i\varepsilon_{\mu\nu\rho\sigma}\mathscr{F}_{k\sigma}(x,v)),$$

$$[\mathscr{F}^5_{i\mu}(x,u), \mathscr{F}^5_{j\nu}(y,v)] \hat{=} [\mathscr{F}_{i\mu}(x,u), \mathscr{F}_{j\nu}(y,v)]$$

Similar relations might be abstracted for the anticommutators and products of two bilocal currents near the light cone. The relations (6.3) are assumed to be true if

$$(x-u)^2 \approx 0, \qquad (u-y)^2 \approx 0,$$
$$(u-v)^2 \approx 0, \qquad (x-y)^2 \approx 0,$$
$$(x-v)^2 \approx 0, \qquad (u-v)^2 \approx 0,$$

This condition is obviously fulfilled if the four points x, u, y, v are distributed on a straight line on the light cone. The algebraic relations (6.3) can be used, for example, to determine the light cone commutator of two light cone commutators and relate this more complicated case to the simpler case of a light cone commutator. It would be interesting to propose experiments in order to test the relations (6.3).

7. LIGHT CONE ALGEBRA AND DEEP INELASTIC SCATTERING

In the last section we have emphasized that perhaps the light cone is a region of very high symmetry (scale and $SU_3 \times SU_3$ invariance). Furthermore, we have abstracted from the quark model certain algebraic properties that might be right on the light cone. Now we should like to mention some general relations that we can obtain using this light cone algebra. But let us first consider the weak interactions in the deep inelastic region.

We introduce the weak currents $J_\mu^+(x)$, $J_\nu^-(x)$ and consider the following expression:

$$W_{\mu\nu}(q) = \frac{1}{4\pi}\int d^4z e^{-iq\cdot z}\langle p|[J_\mu{}^+(z), J_\nu{}^-(0)]|p\rangle$$

$$= \left(\delta_{\mu\nu} - \frac{q_\mu q_\nu}{q^2}\right)\left(W_1{}^+ - \frac{(p\cdot q)^2}{q^2}W_2{}^+\right) - \frac{i}{2}\varepsilon_{\mu\nu\alpha\beta}p_\alpha q_\beta W_3{}^+$$

$$+ \frac{\delta_{\mu\nu}(p\cdot q)^2 + p_\mu p_\nu q^2 - (p_\mu q_\nu + p_\nu q_\mu)p\cdot q}{q^2} W_2{}^+ + q^\mu q^\nu W_4{}^+$$

$$+ (q_\mu p_\nu + q_\nu p_\mu)W_5{}^+ + i(q_\mu p_\nu - q_\nu p_\mu)W_6{}^+ . \tag{7.1}$$

In general, we have to describe the inelastic neutrino hadron processes by six structure functions. From naïve scaling arguments we would expect in the deep inelastic limit:

$$W_1{}^+ \to F_1(\xi), \qquad -q\cdot pW_2{}^+ \to F_2(\xi),$$
$$-q\cdot pW_3{}^+ \to F_3(\xi), \qquad -q\cdot pW_4{}^+ \to F_4(\xi), \tag{7.2}$$
$$-q\cdot pW_5{}^+ \to F_5(\xi), \qquad -q\cdot pW_6{}^+ \to F_6(\xi).$$

The formulae above have the most general form, valid for arbitrary vectors $J_\mu(x)$. We neglect the T-violating effects, which may in any case be 0 on the light cone: $F_6 = 0$. We have already stressed that the weak currents are conserved on the light cone, and we conclude:

$$F_4(\xi) = F_5(\xi) = 0. \tag{7.3}$$

Equation (7.3) is an experimental consequence of the $SU_3 \times SU_3$ symmetry on the light cone, which may be tested by experiment. In the deep inelastic limit we have only three non-vanishing structure functions, corresponding to a conserved current.

It is interesting to note that there is the possibility of testing the dimension l of the divergence of the axial vector current, if our scaling hypothesis is right. We write, for the weak axial vector current,

$$\partial_\mu \mathscr{F}^5_{\pm\mu} = c\cdot v_\pm(x) \tag{7.4}$$

where $v_\pm(x)$ is a local operator of dimension l, and c is a parameter with non-zero dimension.

According to our assumptions about symmetry breaking, c can be written as a positive power of a mass. Using (7.1), we obtain

$$\begin{aligned} q^\mu q^\nu W_{\mu\nu}{}^+(q) &= \frac{c^2}{4\pi}\int d^4z e^{-iq\cdot z}\langle p|[v_+(z), v_-(0)]|p\rangle \\ &= (q^2)^2 W_4{}^+ - 2q^2 q\cdot pW_5{}^+. \end{aligned} \tag{7.5}$$

We define:

$$D(q^2, q\cdot p) = \frac{1}{4\pi}\int d^4z e^{-iq\cdot z}\langle p|[v_+(z), v_-(0)]|p\rangle. \tag{7.6}$$

If we assume that D scales in the deep inelastic region according to the dimension l of $v_\pm(x)$, we obtain

$$\lim_{bj} (-p\cdot q)^{-l-3} D(q^2, q\cdot p) = \phi(\xi) \tag{7.7}$$

where $\phi(\xi)$ denotes the deep inealstic structure function for the matrix element (7.6).

Using (7.5) we obtain

$$\lim_{bj} (-p \cdot q)^{5+l} (\xi^2 W_4^+ - 2\xi W_5^+) = c^2 \phi(\xi). \qquad (7.8)$$

If we determine experimentally the scaling properties of W_4 and W_5, then we can deduce from (7.8) the dimension l of $v_{\pm}(x)$. This l is the same quantity as the dimension l_u discussed in Section 2, provided the $SU_3 \times SU_3$ violating term in the energy has a definite dimension.[20]

In order to apply the light cone algebra of Section 6, we have to relate the expectation values of the bilocal operators appearing there to the structure function in question. This is done in Appendix II, where we give this connection for arbitrary currents. We use Eqs. (A.12) and (A.13), where the functions $S^k(\xi)$, $A^k(\xi)$ are given by the expectation value of the symmetric and antisymmetric bilocal currents (Eq. (A.8)), and obtain:

(a) for deep inelastic electron-hadron scattering:

$$F_2^{ep}(\xi) = \xi\left(\frac{2}{3}\sqrt{\frac{2}{3}}A^0(\xi) + \frac{1}{3\sqrt{3}}A^8(\xi) + \frac{1}{3}A^3(\xi)\right) \qquad (7.9)$$

(b) for deep inelastic neutrino-hadron scattering:

$$F_2^{\nu p}(\xi) = \xi(2S^3(\xi) + 2\sqrt{\frac{2}{3}}A^0(\zeta) + \frac{2}{\sqrt{3}}A^8(\xi)) \qquad (7.10)$$

$$F_3^{\nu p}(\xi) = 2A^3(\xi) - 2\sqrt{\frac{2}{3}}S^0(\xi) - \frac{2}{\sqrt{3}}S^8(\xi). \qquad (7.11)$$

In (7.5) and (7.6) we have neglected the Cabibbo angle, since $\sin^2\theta_c = 0.05 \approx 0$.

Both in (7.4) and (7.6), $A^3(\xi)$ occurs as the only isospin dependent part, and we can simply derive relations between the structure functions of different members of an isospin multiplet, e.g., for neutron and proton:

$$6 \cdot (F_2^{en} - F_2^{ep}) = \xi \cdot (F_3^{\nu p} - F_3^{\nu n}). \qquad (7.12)$$

This relation was first obtained by C. H. Llewellyn Smith[7] within the "parton" model. One can derive similar relations for other isospin multiplets.

In the symmetric bilocal current appear certain operators that we know. The operator $j_\mu(x) = i\bar{q}(x)\gamma_\mu q(x)$ has to be identical with the hadron current (we suppress internal indices) in order to give current algebra. But we know their expectation values, which are given by the corresponding quantum number. In such a way we can derive a large set of sum rules relating certain moments of the structure functions to their well-known expectation values.

We give only the following two examples, which follow immediately from (7.9), (7.10), (7.11):

$$\int_{-1}^{1} \frac{d\xi}{\xi}(F_2^{\nu p}(\xi) - F_2^{\nu n}(\xi)) = \int_{-1}^{1} \frac{d\xi}{\xi}(F_2^{\nu p}(\xi) - F_2^{\nu p}(-\xi))$$
$$= 4s_1^3(p) = 4. \tag{7.13}$$

Here $s_1^3(p)$ means, as in Appendix II, the proton expectation value of $2F_3$. This is the Adler sum rule,[21] usually written as

$$\int_0^1 \frac{d\xi}{\xi}(F_2^{\nu p}(\xi) - F_2^{\nu n}(\xi)) = 2. \tag{7.14}$$

From (7.11) we obtain:

$$\int_1^1 (F_3^{\nu p} + F_3^{\nu n})d\xi = -2(2s_1^0(p) + s_1^8(p)) = -12 \tag{7.15}$$

or

$$\int_{-0}^{1} (F_3^{\nu p} + F_3^{\nu n})d\xi = -6, \tag{7.16}$$

which is the sum rule first derived by Gross and Llewellyn Smith.[22]

If we make the special assumption that we are abstracting our light cone relations from a pure quark model with no "gluon field" and non-derivative couplings, we can get a further set of relations. Of course, no such model is known to exist in four dimensions that is even renormalizable, much less super-renormalisable as we would prefer to fit in with the ideas presented here. Nevertheless, it may be worthwhile to examine sum rules that test whether Nature imitates the "pure quark" case.

The point is that when we expand the bilocal quantity $\mathcal{F}_{0\alpha}(x, y)$ to first order in $y - x$, we pick up a Lorentz tensor operator, a singlet under SU_3, that corresponds in the quark picture to the operator $1/2\{\bar{q}(x)\gamma_\mu\partial_\nu q(x) - \partial_\nu\bar{q}(x)\gamma_\mu q(x)\}$, which, if we symmetrize in μ and ν and ignore the trace, is the same as the stress-energy-momentum tensor $\theta_{\mu\nu}$ in the pure quark picture. But the expected value of $\theta_{\mu\nu}$ in any state of momentum p is just $2p_\mu p_\nu$ and so we obtain sum rules for the pure quark case.

We consider the isospin averaged expressions:

$$(F_2^{ep}(\xi) + F_2^{en}(\xi)) = 2\xi\left\{\frac{2}{3}\sqrt{\frac{2}{3}}A^0(\xi) + \frac{1}{3}\frac{1}{\sqrt{3}}A^8(\xi)\right\}$$

$$(F_2^{\nu p}(\xi) + F_2^{\nu n}(\xi)) = 2\xi\left\{2\sqrt{\frac{2}{3}}A^0(\xi) + \frac{2}{\sqrt{3}}A^8(\zeta)\right\}$$

and obtain

$$6(F_2^{ep} + F_2^{en}) - (F_2^{\nu p} + F_2^{\nu n}) = 4\sqrt{\frac{2}{3}}A^0(\xi)$$

$$= 4\sqrt{\frac{2}{3}}(a_1^0\delta(\xi) - \frac{1}{2!}a_3^0\delta''(\xi))\cdots)$$

In pure quark theories we have $a_1^0 = \sqrt{2/3}$ and we obtain

$$6\int_{-1}^{1}(F_2^{ep} + F_2^{en})d\xi - \int_{-1}^{1}(F_2^{\nu p}(\xi) + F_2^{\nu n}(\xi))d\xi = 8/3$$

or, for the physical region $0 \leq \xi \leq 1$:

$$6\int_0^1(F_2^{ep} + F_2^{en})d\xi - \int_0^1(F_2^{\nu p} + F_2^{\nu n})d\xi = 4/3. \quad (7.17)$$

The sum role (7.17) can be tested by experiment. This will test whether one can describe the real world of hadrons by a theory resembling one with only quarks, interacting in some unknown non-linear fashion.

The scaling behavior in the deep inelastic region may be described by the "parton model".[4,5] In the deep inelastic region, the electron is viewed as scattering in the impulse approximation off point-like constituents of the hadrons ("partons"). In this case the scaling function $F_2^e(\xi)$ can be written as

$$F_2^e(\xi) = \sum_N P(N) \ (\sum_i Q_i^2)_N \xi f_N(\xi) \tag{7.18}$$

where we sum up over all "partons" (Σ_i) and all the possibilities of having N partons (Σ_N). The momentum distribution function of the "partons" is denoted by $f_N(\xi)$, the charge of the i-th "parton" by Q_i. We compare (7.9) with (7.18):

$$\begin{aligned} F_2^e(\xi) &= \xi \ \frac{2}{3}A^0(\xi) + \frac{1}{6}A^8(\xi) + \frac{1}{3}A^3(\xi)) \\ &= \sum_N P(N) \left(\sum_i Q_i^2)_N \xi f_N(\xi).\right) \end{aligned} \tag{7.19}$$

As long as we do not specify the functions $f_N(\xi)$ and $P(N)$, the "parton model" gives us no more information than the generalization of current algebra to the light cone as described in the last sections. If one assumes special properties of these functions, one goes beyond the light cone algebra of the currents, that means beyond the properties of the operator products on the light cone. Such additional assumptions, e.g., statistical assumptions about the distributions of the "partons" in relativistic phase space, appear in the light cone algebra approach as specific assumptions about the matrix elements of the expansion operators on the light cone. These additional assumptions are seen, in our approach, to be model dependent and somewhat arbitrary, as compared to results of the light cone algebra. Our results can, of course, be obtained by "parton" methods and are mostly well-known in that connnection.

It is interesting to consider the different sum roles within the "parton model". The sum rules (7.14) and (7.16) are valid in any "quark-parton" model; so is the symmetry relation (7.12). The sum rule (7.17) is a specific property of a model consisting only of quarks. If there is a "gluon" present, we obtain a deviation from 4/3 on the right-hand side, which measues the "gluon" contribution to the energy-momentum tensor.

Our closed algebra of bilocal operators on the light cone has, of course, a parallel in the "parton" model. However, it is again much easier using our approach to disentangle what may be exactly true (formulae for light cone commutators of light cone commutators) from what depends on specific matrix elements and is therefore model dependent. It would be profitable to apply such an analysis to the work of Bjorken and Paschos, in the context of "partons", on scaling in the Compton effect on protons.

As an example of a "parton model" relation that mingles specific assumptions about matrix elements with more general ideas of light cone algebra and abstraction from a pure quark model, we may take the allegation that in the pure quark case we have $\int F_2^{en}(\xi)d\xi = 2/9$. Light cone algebra and the pure quark assumption do not imply this.

8. CONCLUDING REMARKS

There are many observations that we would like to make and many unanswered questions that we would like to raise about light cone algebra. But we shall content ourselves with just a few remarks.

First comes the question of whether we can distinguish in a well-defined mathematical way, using physical quantities, between a theory that makes use of SU_3 triplet representations locally and one that does not. If we can, we must then ask whether a theory that has triplets locally necessarily implies the existence of real triplets (say real quarks) asymptotically. Dashen (private communication) raises these two questions by constructing local charge operators $\int_V \mathscr{F}_{i0} d^3x$ over a finite volume. (This construction is somewhat illegitimate, since test functions in field theory have to be multiplied by δ functions in equal time charge density commutators and should therefore have all derivatives, not like the function that Dashen uses, which is unity inside V and zero outside.) If his quantities F_i^V make sense, they obey the commutation rules of SU_3 and we can ask whether for any V our states contain triplet (or other triality $\neq 0$) representations of this SU_3. Dashen then suggests that our bilocal algebra probably implies that local triplets in this sense are present; if the procedure and the conclusion are correct, we must ask whether real quarks are then implied.

The question of quark statistics is another interesting one. If quarks are real, then we cannot assign them para- Fermi statistics of rank 3, since that is said to violate the factoring of the S-matrix for distant subsystems. However, if somehow our quarks are permanently bound in oscillators (and our theory is thus perhaps equivalent to a bootstrap theory with no real quarks), then they could be parafermions of rank 3. They can be bosons, too, if they are not real, but only if there is a spinless fermion (the "soul" of a baryon) that accompanies the three quarks in each baryon.

Another topic is the algebra of $U_6 \times U_6 \times O_3$ that is implied at equal times for the integrals of the current component and the angular momentum.[19] Is that algebra really correct or is it too strong an assumption? Should it be replaced at $P_z = \infty$ by only the "good-good" part of the algebra?

If we do have the full algebra, then the quark kinetic part of the energy density is uniquely defined as the part behaving like **(35, 1)** and **(1, 35)** with $L = 1$, i.e., like $\alpha \cdot \nabla$.

If we abstract relations from a pure quark picture without gradient couplings, then this quark kinetic part of $\theta_{\mu\nu}$ is all there is apart from the trace contribution. In that case, we have the equal time commutation relation for the whole energy operator:

$$\sum_{r=1}^{3} \sum_{i=1}^{8} [\int \mathscr{F}_{ir} d^3x, [\int \mathscr{F}_{ir} d^3x, P_0]] = 16/3\, P_0 + \text{scale violating terms.}$$

This relation, in the pure quark case, can be looked at in another way. It is an equal time consequence of the relation

$$\theta_{\mu\nu} = \lim_{y \to x} \frac{3\pi^2}{32} \partial_\mu \partial_\nu \{(z^2)^2 \mathscr{F}_{i\alpha}(x) \mathscr{F}_{i\alpha}(y)\} + \text{scale violating terms}$$

that holds when the singlet tensor term in the light cone expansion of $\mathscr{F}_{i\mu}(x)\mathscr{F}_{j\nu}(y)$ is just proportional to $\theta_{\mu\nu}$, as in the pure quark case. This relation is what, in the pure quark version of the light cone algebra (extended to light cone products), replaces the Sugawara[23] model, in which $\theta_{\mu\nu}$ is proportional to $\mathscr{F}_{i\mu}\mathscr{F}_{i\nu}$, with dimension -6. Our expression is much more civilized, having $l = -4$ as it should. A more general equal time commutator than the one above, also implied by the pure quark case, is the following:

$$\sum_{r=1}^{3} [\mathscr{F}_{ir}(x), \partial_0 \mathscr{F}_{ir}(y)] = 16i/3\,\theta_{00}\delta(x-y) + \text{scale breaking terms.}$$

Another important point that should be emphasized is that the $U_6 \times U_6$ algebra requires the inclusion of a ninth vector current $\mathscr{F}_{0\alpha}$ and a ninth axial vector current $\mathscr{F}^5_{0\alpha}$, and that the Latin index for SU_3 representation components in Appendix II has to run from 0 to 8. Now if the term in the energy density that breaks $SU_3 \times SU_3$ follows our usual conjecture and behaves like $-u_0 - cu_8$ with c near $-\sqrt{2}$ and if the chiral symmetry preserving but scale breaking term δ is just a constant, then as $u \to 0$ scale invariance and chiral invariance become good, but the mass formula for the pseudoscalar mesons indicates that we do not want $\partial_\alpha \mathscr{F}^5_{0\alpha}$ to be zero in that limit.[10] Yet $\mathscr{F}^5_{0\alpha}$ is supposed to be conserved on the light cone. Does this raise a problem for the idea of $\delta =$ const. or does it really raise the whole question of the relation of the light cone limit and the formal limit $u \to 0$, $\delta \to 0$?

If there are dilations, with $m^2 \to 0$ in the limit of scale invariance while other masses stay finite, how does that jibe with the light cone limit in which all masses act as if they go to zero? Presumably there is no contradiction here, but the situation should be explored further.

Finally, let us recall that in the specific application of scaling to deep inelastic scattering, the functions $F(\xi)$ connect up with two important parts of particle physics. As $\xi \to 0$, if we can interchange this limit with the Bjorken limit, we are dealing with fixed q^2 and with $p \cdot q \to \infty$ and the behavior of the F's comes directly from the Regge behavior of the corresponding exchanged channel. If $\alpha_P(0) = 1$, then $F_2^{ep}(\xi) + F_2^{en}(\xi)$ goes like a constant at $\xi = 0$, i.e., $\xi^{1-\alpha_P(0)}$, while $F_2^{ep}(\xi) - F_2^{en}(\xi)$ goes like $^{1-\alpha_\rho(0)}$, etc.

As $\xi \to 1$, as emphasized by Drell and Yan[8], there seems to be a connection between the dependence of $F(\xi)$ on $1-\xi$ and the dependence of the elastic form factors of the nucleons on t at large t.

ACKNOWLEDGEMENTS

We would like to thank J. D. Bjorken, R. P. Feynman, and C. H. Llewellyn Smith for stimulating conversations about the relation of our work to previous work on "partons". One of us (H. F.) would like to express his gratitude to the DAAD, to SLAC, and to the AEC high energy physics group at Caltech for support.

The ninth Section, prepared for the Tel Aviv Conference, contains a number of points that have been elaborated between the Conference and the time of publication, especially matters concerned with "anomalies". For many enlightening discussions of these questions, we are deeply indebted to W. Baarden, and to the staff of the Theoretical Study Division of CERN.

REFERENCES

1. K. G. Wilson, *Phys. Rev.* **179**, 1499 (1969).
2. R. Brandt and G. Preparata, CERN preprint TH-1208.
3. Y. Frishman, Phys. Rev. Lett. **25**, 966 (1970).
4. R. P. Feynman, *Proceedings of Third High Energy Collision Conference at State University of New York*, Stony Brook, Gordon and Breach, 1970.
5. J. D. Bjorken and E. A. Paschos, Phys. Rev. **185**, 1975 (1969).
6. P. Landshoff and J. C. Polkinghorne, Cambridge University DAMTP preprints (1970).
7. C. H. Llewellyn Smith, Nucl. Phys. **B17**, 277 (1970).
8. S. D. Drell and T. Yan, Phys. Rev. Lett. **24**, 181 (1970).
9. Note we use the metric $\delta_{\mu\nu} = (1, 1, 1, -1)$ and the covariant state normalization $\langle p's' | ps \rangle = (2\pi)^3 \, 2p_0^3 \, (p-p') \, \delta_{s's}$.
10. M. Gell-Mann, *Proceedings of Third Hawaii Topical Conference on Particle Physics*, Western Periodicals Co., Los Angeles, 1969.
11. F. von Hippel and J. K. Kim, *Phys. Rev.* **D1**, 151 (1970).
12. J. Ellis, Physics Lett. **33B**, 591 (1970).
13. R. F. Dashen and T. P. Cheng, Institute for Advanced Study preprint (1970).
14. J. Ellis, P. Weisz, and B. Zumino, *Phys. Lett.* **34B**, 91 (1971).
15. E. D. Bloom, G. Buschhorn, R. L. Cottrell, D. H. Coward, H. DeStaebler, J. Drees, C. L. Jordan, G. Miller, L. Mo, H. Piel, R. E. Taylor, M. Breidenbach, W. R. Ditzler, J. I. Friedman, G. C. Hartmann, H. W. Kendall, and J. S. Poucher, Stanford Linear Accelerator Center preprint SLAC-PUB-796 (1970) (report presented at the XVth International Conference on High Energy Physics, Kiev, USSR, 1970).
16. H. Leutwyler and J. Stern, *Nucl. Phys.* **B20**, 77 (1970); R. Jackiw, R. Van Royen, and G. B. West, *Phys. Rev.* **D2**, 2473 (1970).
17. S. Ciccariello, R. Gatto, G. Sartori, and M. Tonin, *Phys. Lett.* **30B**, 546 (1969); G. Mack, *Phys. Rev. Lett.* **25**, 400 (1970). J. M. Cornwall and R. E. Norton, *Phys. Rev.* **177**, 2584 (1968) used a different approach to accomplish about the same result. Instead of expanding light cone commutators, they use equal time commutators with higher and higher time derivatives, sandwiched between states at infinite momentum. That amounts to roughly the same thing, and represents an alternative approach to light cone algebra.
18. C. H. Llewellyn Smith, to be published.
19. R. P. Feynman, M. Gell-Mann, and G. Zeieg, *Phys. Rev. Lett.* **13**, 678 (1964). The idea was applied to many important effects by J. D. Bjorken, Phys. Rev. **148**, 1467 (1966).
20. J. Mandula, A. Schwimmer, J. Weyers, and G. Zweig have proposed independently this test of the dimension l_μ and are publishing a full account of it.
21. S. Adler, *Phys. Rev.* **143**, 154 (1966).
22. D. J. Gross and C. H. Llewellyn Smith, *Nucl. Phys.* **B14**, 337 (1969).
23. H. Sugawara, *Phys. Rev.* **170**, 1659 (1968).
24. J. M. Cornwall and R. Jackiw, UCLA preprint (1971).
25. D. J. Gross and S. B. Treiman, Princeton University preprint (1971).
26. W. Bardeen, private communication.
27. M. Gell-Mann and F. E. Low *Phys. Rev.* **95**, 1300 (1954); M. Baker and K. Jonson, *Phys. Rev.* **183**, 1292 (1969).
28. A. H. Muller, *Phys. Rev.* **D2**, 2963 (1970).
29. J. D. Bjorken, Talk given at the same Conference.

CHAPTER 7

Light-Cone Current Algebra, π° Decay, and $e^{+}e^{-}$ Annihilation

W. A. Bardeen
H. Fritzsch
M. Gell-Mann

1. INTRODUCTION

The indication from deep inelastic electron scattering experiments at SLAC that Bjorken scaling may really hold has motivated an extension of the hypotheses of current algebra to what may be called light-cone current algebra.[1] As before, one starts from a field theoretical quark model (say one with neutral vector "gluons") and abstracts exact algebraic results, postulating their validity for the real world of hadrons. In light-cone algebra, we abstract the most singular term near the light cone in the commutator of two-vector or axial vector currents, which turns out to be given in terms of bilocal current operators that reduce to local currents when the two space-time points coincide. The algebraic properties of these bilocal operators, as abstracted from the model, give a number of predictions for the Bjorken functions in deep inelastic electron and neutrino experiments. None is in disagreement with experiment. These algebraic properties, by the way, are the same as in the free quark model.

Reprinted from *Scale and Conformal Symmetry in Hadron Physics*, ed. R. Gatto

From the mathematical point of view, the new abstractions differ from the older ones of current algebra (commutators of "good components" of current densities at equal times or on a light plane) in being true only formally in a model with interactions, while failing to each order of renormalized perturbation theory, like the scaling itself. Obviously it is hoped that, if the scaling works in the real world, so do the relations of light-cone current algebra, in spite of the lack of cooperation from renormalized perturbation theory in the model.

The applications to deep inelastic scattering involve assumptions only about the connected part of each current commutator. We may ask whether the disconnected part—for example, the vacuum expected value of the commutator of currents—also behaves in the light-cone limit as it does formally in the quark-gluon model, namely, the same as for a free quark model. Does the commutator of two currents, sandwiched between the hadron vacuum state and itself, act at high momenta exactly as it would for free quark theory? If so, then we can predict immediately and trivially the high-energy limit of the ratio

$$\sigma(e^+ + e^- \to \text{hadrons})/\sigma(e^+ + e^- \to \mu^+ + \mu^-)$$

for one-photon annihilation.

In contrast to the situation for the connected part and deep inelastic scattering, the annihilation results depend on the statistics of the quarks in the model. For three Fermi-Dirac quarks, the ratio would be $(\frac{2}{3})^2 + (-\frac{1}{3})^2 + (-\frac{1}{3})^2 = \frac{2}{3}$, but do we want Fermi-Dirac quarks? The relativistic "current quarks" in the model, which are essentially square roots of currents, are of course not identical with "constituent quarks" of the naive, approximate quark picture of baryon and meson spectra. Nevertheless, there should be a transformation, perhaps even a unitary transformation, linking constituent quarks and current quarks (in a more abstract language, a transformation connecting the symmetry group $[SU(3) \times SU(3)]_{W,\infty,\text{strong}}$ of the constituent quark picture of baryons and mesons, a subgroup of $[SU(6)]_{W,\infty,\text{strong}}$,[2] with the symmetry group $[SU(3) \times SU(3)]_{W,\infty,\text{currents}}$,[3] generated by the vector and axial vector charges). This transformation should certainly preserve quark statistics. Therefore the indications from the constituent quark picture that quarks obey peculiar statistics should suggest the same behavior for the current quarks in the underlying relativistic model from which we abstract the vacuum behavior of the light-cone current commutator.[4]

In the constituent quark picture of baryons,[5] the ground-state wave function is described by $(\mathbf{56}, \mathbf{1}), L=0^+$ with respect to [SU(6)×SU(6)× SU(3)] or $(\mathbf{56}, L_z=0)$ with respect to $[\mathrm{SU}(6)\times \mathrm{O}(2)]_W$. It is totally symmetric in spin and SU(3). In accordance with the simplicity of the picture, one might expect the space wave function of the ground state to be totally symmetric. The entire wave function is then symmetrical. Yet baryons are to be antisymmetrized with respect to one another, since they do obey the Pauli principle. Thus the peculiar statistics suggested for quarks has then symmetrized in sets of three and otherwise antisymmetrized. This can be described in various equivalent ways. One is to consider "para-Fermi statistics of rank 3"[6] and then to impose the restriction that all physical particles be fermions or bosons; the quarks are then fictitious (i.e., always bound) and all physical three-quark systems are totally symmetric overall. An equivalent description, easier to follow, involves introducing nine types of quarks, that is, the usual three types in each of three "colors," say red, white, and blue. The restriction is then imposed that all physical states and all observable quantities like the currents be singlets with respect to the SU(3) of color (i.e., the symmetry that manipulates the color index). Again, the quarks are fictitious. Let us refer to this type of statistics as "quark statistics."

If we take the quark statistics seriously and apply it to current quarks as well as constituent quarks, then the closed-loop processes in the models are multiplied by a factor of 3, and the asymptotic ratio $\sigma(e^+e^-\rightarrow \text{hadrons}) / \sigma(e^+e^-\rightarrow\mu^+\mu^-)$ becomes $3\cdot\frac{2}{3}=2$.

Experiments at present are too low in energy and not accurate enough to test this prediction, but in the next year or two the situation should change. Meanwhile, is there any supporting evidence? Assuming that the connected light-cone algebra is right, we should like to know whether we can abstract the disconnected part as well, and whether the statistics are right. In fact, there is evidence from the decay of the π^0 into 2γ. It is well known that in the partially conserved axial current (PCAC) limit, with $m_\pi^2\rightarrow 0$, Adler and others[7] have given an exact formula for the decay amplitude $\pi^0\rightarrow 2\gamma$ in a "quark-gluon" model theory. The amplitude is a known constant times $(\sum Q_{1/2}^2-\sum Q_{-1/2}^2)$, where the sum is over the types of quarks and the charges $Q_{1/2}$ are those of $I_z=\frac{1}{2}$ quarks, while the charges $Q_{-1/2}$ are those of $I_z=-\frac{1}{2}$ quarks. The amplitude agrees with experiment, within the errors, in both sign and magnitude if $\sum Q_{1/2}^2 - \sum Q_{-1/2}^2=1$.[8] If we had three Fermi-Dirac quarks, we would have $(\frac{2}{3})^2-(-\frac{1}{3})^2=\frac{1}{3}$, and the decay rate would be wrong by a factor of $\frac{1}{9}$. With "quark statistics," we get $\frac{1}{3}\cdot 3=1$ and everything is all right, assuming that PCAC is applicable.

There is, however, the problem of the derivation of the Adler formula. In the original derivation a renormalized perturbation expansion is applied to the "quark-gluon" model theory, and it is shown that only the lowest-order closed-loop diagram survives in the PCAC limit,[9] so that an exact expression can be given for the decay amplitude. Clearly this derivation does not directly suit our purposes, since our light-cone algebra is not obtainable by renormalized perturbation theory term by term. Of course, the situation might change if all orders are summed.

Recently it has become clear that the formula can be derived without direct reference to renormalized perturbation theory, from considerations of light-cone current algebra. Crewther has contributed greatly to clarifying this point,[10] using earlier work of Wilson[11] and Schreier.[12] Our objectives in this chapter are to call attention to Crewther's work, to sketch a derivation that is somewhat simpler than his, and to clarify the question of statistics.

We assume the connected light-cone algebra, and we make the further abstraction, from free quark theory or formal "quark-gluon" theory, of the principle that not only commutators but also products and physically ordered products of current operators obey scale invariance near the light cone, so that, apart from possible subtraction terms involving four-dimensional δ functions, current products near the light cone are given by the same formula as current commutators, with the singular functions changed from $\epsilon(z_0)\delta[(z^2)]$ to $(z^2-i\epsilon z_0)^{-1}$ for ordinary products or $(z^2-i\epsilon)^{-1}$ for ordered products.

Then it can be shown from consistency arguments that the only possible form for the disconnected parts (two-, three-, and four-point functions) is that given by free quark theory or formal "quark-gluon" theory, with only the coefficient needing to be determined by abstraction from a model. (In general, of course, the coefficient could be zero, thus changing the physics completely.) Then, from the light-cone behavior of current products, including connected and disconnected parts, the Adler formula for $\pi^0\to2\gamma$ in the PCAC limit can be derived in terms of that coefficient.

If we take the coefficient from the model with "quark statistics," predicting the asymptotic ratio of $\sigma(e^+e^-\to\text{hadrons})/\sigma(e^+e^-\to\mu^+\mu^-)$ to be 2 for one-photon annihilation, we obtain the correct value of the $\pi^0\to2\gamma$ decay amplitude, agreeing with experiment in magnitude and sign. Conversely, if for any reason we do not like to appeal to the model, we can take the coefficient from the observed $\pi^0\to2\gamma$ amplitude and predict in that way that the asymptotic value of $\sigma(e^+e^-\to\text{hadrons})/\sigma(e^+e^-\to\mu^+\mu^-)$ should be about 2.

Some more complicated and less attractive models that agree with the observed $\pi^0\to2\gamma$ amplitude are discussed in Section 3.

2. LIGHT-CONE ALGEBRA

The ideas of current algebra stem essentially from the attempt to abstract, from field theoretic quark models with interactions, certain algebraic relations obeyed by weak and electromagnetic currents to all orders in the strong interaction, and to postulate these relations for the system of real hadrons, while suggesting possible experimental tests of their validity. In four dimensions, with spinor fields involved, the only renormalizable models are ones that are barely renormalizable, such as a model of spinors coupled to a neutral vector "gluon" field. Until recently, the relations abstracted, such as the equal-time commutation relations of vector and axial charges or charge densities, were true in each order of renormalized perturbation theory in such a model. Now, however, one is considering the abstraction of results that are true only formally, with canonical manipulation of operators, and that fail, by powers of logarithmic factors, in each order of renormalized perturbation theory, in all barely renormalizable models (although they might be all right in a super-renormalizable model, if there were one).

The reason for the recent trend is, of course, the tendency of the deep inelastic electron scattering experiments at SLAC to encourage belief in Bjorken scaling, which fails to every order of renormalized perturbation theory in barely renormalizable models. There is also the availability of beautiful algebraic results, with Bjorken scaling as one of their predictions, if formal abstractions are accepted. The simplest such abstraction is that of the formula giving the leading singularity on the light cone of the connected part of the commutator of the vector or axial vector currents,[1] for example:

$$
\begin{aligned}
[\mathcal{F}_{i\mu}(x),\mathcal{F}_{j\nu}(y)] &\hat{=} [\mathcal{F}_{i\mu}{}^5(x),\mathcal{F}_{j\nu}{}^5(y)] \\
&\hat{=} \frac{1}{4\pi}\partial_\rho\{\epsilon(x_0-y_0)\delta[(x-y)^2]\} \\
&\quad\times\{(if_{ij\lambda}-d_{ijk})[s_{\mu\nu\rho\sigma}\mathcal{F}_{k\sigma}(y,x)+i\epsilon_{\mu\nu\rho\sigma}\mathcal{F}_{k\sigma}{}^5(y,x)] \\
&\quad+(if_{ijk}+d_{ijk})[s_{\mu\nu\rho\sigma}\mathcal{F}_{k\sigma}(x,y)-i\epsilon_{\mu\nu\rho\sigma}\mathcal{F}_{k\sigma}{}^5(x,y)]\}
\end{aligned}
\quad (1)
$$

On the right-hand side we have the connected parts of bilocal operators $\mathcal{F}_{i\mu}(x,y)$ and $\mathcal{F}_{i\mu}{}^5(x,y)$, which reduce to the local currents $\mathcal{F}_{i\mu}(x)$ and $\mathcal{F}_{i\mu}{}^5(x)$ as $x \to y$. The bilocal operators are defined as observable quantities only in the vicinity of the light-cone, $(x-y)^2=0$. Here

$$s_{\mu\nu\rho\sigma}=\delta_{\mu\rho}\delta_{\nu\sigma}+\delta_{\nu\rho}\delta_{\mu\sigma}-\delta_{\mu\nu}\delta_{\rho\sigma}.$$

Formula 1 gives Bjorken scaling by virtue of the finite matrix elements assumed for $\mathsf{F}_{i\mu}(x,y)$ and $\mathsf{F}_{i\mu}{}^5(x,y)$; in fact, the Fourier transform of the matrix element of $\mathsf{F}_{i\mu}(x,y)$ is just the Bjorken scaling function. The fact that all charged fields in the model have spin $\frac{1}{2}$ determines the algebraic structure of the formula and gives the prediction $(\sigma_L/\sigma_T)_{Bj}\to 0$ for deep inelastic electron scattering, not in contradiction with experiment. The electrical and weak charges of the quarks in the model determine the coefficients in the formula, and give rise to numerous sum rules and inequalities for the SLAC-MIT experiments in the Bjorken limit, again none in contradiction with experiment.

The formula for the leading light-cone singularity in the commutator contains, of course, the physical information that near the light cone we have full symmetry with respect to SU(3)×SU(3) and with respect to scale transformations in coordinate space. Thus there is conservation of dimension in the formula, with each current having $l=-3$ and the singular function $x-y$ also having $l=-3$.

A simple generalization of the abstraction that we have considered turns into a closed system, called the basic light-cone algebra. Here we commute the bilocal operators as well, for instance, $\mathsf{F}_{i\mu}(x,u)$ with $\mathsf{F}_{j\nu}(y,v)$, as all of the six intervals among the four space-time points approach 0, so that all four points tend to lie on a lightlike straight line in Minkowski space. Abstraction from the model gives us, on the right-hand side, a singular function of one coordinate difference, say $x-v$, times a bilocal current $\mathsf{F}_{i\alpha}$ or $\mathsf{F}_{i\alpha}{}^5$ at the other two points, say y and u, plus an expression with (x,v) and (y,u) interchanged, and the system closes algebraically. The formulas are just like Eq. 1. We shall assume here the validity of the basic light-cone algebraic system, and discuss the possible generalization to products and to disconnected parts. In Section 4, we conclude from the generalization to products that the form of an expression like $\langle \mathrm{vac}|\mathsf{F}_{i\alpha}(x)$ $\mathsf{F}_{j\beta}(y,z)|\mathrm{vac}\rangle$ for disconnected parts is uniquely determined from the consistency of the connected light-cone algebra to be a number N times the corresponding expression for three free Fermi-Dirac quarks, when $x,y,$ and z tend to lie on a straight lightlike line. The $\pi^0\to 2\gamma$ amplitude in the PCAC approximation is then calculated in terms of N and is proportional to it. Thus we do not want N to be zero.

The asymptotic ratio $\sigma(e^+e^-\to\text{hadrons})/\sigma(e^+e^-\to\mu^+\mu^-)$ from one-photon annihilation is also proportional to N. We may either determine N from the observed $\pi^0\to 2\gamma$ amplitude and then compute this asymptotic ratio approximately, or else appeal to a model and abstract the exact value of N, from which we calculate the amplitude of $\pi^0\to 2\gamma$. In a model, N depends on the statistics of the quarks, which we discuss in the next section.

3. STATISTICS AND ALTERNATIVE SCHEMES

As we remarked in Section 1, the presumably unwanted Fermi-Dirac statistics for the quarks, with $N=1$, would give $\sigma(e^+e^- \to \text{hadrons}) / \sigma(e^+e^- \to \mu^+\mu^-) \to 2/3$. (Such quarks could be real particles, if necessary.) Now let us consider the case of "quark statistics," equivalent to para-Fermi statistics of rank 3 with the restriction that all physical particles be bosons or fermions. (Quarks are then fictitious, permanently bound. Even if we applied the restriction only to baryons and mesons, quarks would still be fictitious, as we can see by applying the principle of cluster decomposition of the S-matrix.)

The quark field theory model or the "quark-gluon" model is set up with three fields, q_R, q_B, and q_W, each with three ordinary SU(3) components, making nine in all. Without loss of generality, they may be taken to anticommute with one another as well as with themselves. The currents all have the form $\bar{q}_R q_R + \bar{q}_B q_B + \bar{q}_W q_W$, and are singlets with respect to the SU(3) of color. The physical states too are restricted to be singlets under the color SU(3). For example, the $q\bar{q}$ configuration for mesons is only $\bar{q}_R q_R + \bar{q}_B q_B + \bar{q}_W q_W$, and the qqq configuration for baryons is only $q_R q_B q_W - q_B q_R q_W + q_W q_R q_B - q_R q_W q_B + q_B q_W q_R - q_W q_B q_R$. Likewise all the higher configurations for baryons and mesons are required to be color singlets.

We do not know how to incorporate such restrictions on physical states into the formalism of the "quark-gluon" field theory model. We assume without proof that the asymptotic light-cone results for current commutators and multiple commutators are not altered. Since the currents are all color singlets, there is no obvious contradiction.

The use of quark statistics then gives $N=3$ and $\sigma(e^+e^- \to \text{hadrons}) / \sigma(e^+e^- \to \mu^+\mu^-) \to 2$. This is the value that we predict.

We should, however, examine other possible schemes. First, we might treat actual para-Fermi statistics of rank 3 for the quarks without any further restriction on the physical states. In that case, there are excited baryons that are not fermions and are not totally symmetric in the $3q$ configuration; there are also excited mesons that are not bosons. Whether the quarks can be real in this case without violating the principle of "cluster decomposition" (factorizing of the S-matrix when a physical system is split into very distant subsystems) is a matter of controversy; probably they cannot. In this situation, N is presumably still 3.

Another situation with $N=3$ is that of a physical color SU(3) that can really be excited by the strong interaction. Excited baryons now exist that are in octets, decimets, and so on with respect to color, and mesons in octets and higher configurations. Many conserved quantum numbers exist, and new interactions may have to be introduced to violate them. This is a wildly speculative scheme. Here the nine quarks can be real if necessary, that is, capable of being produced singly or doubly at finite energies and identified in the laboratory.

We may consider a still more complicated situation in which the relationship of the physical currents to the current nonet in the connected algebra is somewhat modified, namely, the Han-Nambu scheme.[13] Here there are nine quarks, capable of being real, but they do not have the regular quark charges. Instead, the u quarks have charges $1, 1, 0$, averaging to $\frac{2}{3}$; the d quarks have charges $0, 0, -1$, averaging to $-\frac{1}{3}$; and the s quarks also have charges $0, 0, -1$, averaging to $-\frac{1}{3}$. In this scheme, not only can the analog of the color variable really be excited, but also it is excited even by the electromagnetic current, which is no longer a "color" singlet. Since the expressions for the electromagnetic current in terms of the current operators in the connected algebra are modified, this situation cannot be described by a value of N. It is clear, however, from the quark charges, that the asymptotic behavior of the disconnected part gives, in the Han-Nambu scheme, $\sigma(e^+e^-\to\text{hadrons})/\sigma(e^+e^-\to\mu^+\mu^-)\to 4$. Because the formulas for the physical currents are changed, numerical predictions for deep inelastic scattering are altered too. For example, instead of the inequality $\frac{1}{4}\leqslant[F^{en}(\xi)/F^{ep}(\xi)]\leqslant 4$ for deep inelastic scattering of electrons from neutrons and protons, we would have $\frac{1}{2}\leqslant[F^{en}(\xi)/F^{ep}(\xi)]\leqslant 2$. However, comparison of asymptotic values with experiment in this case may not be realistic at the energies now being explored. The electromagnetic current is not a color singlet; it directly excites the new quantum numbers, and presumably the asymptotic formulas do not become applicable until above the thresholds for the new kinds of particles. Thus, unless and until entirely new phenomena are detected, the Han-Nambu scheme really has little predictive power.

A final case to be mentioned is one in which we have ordinary "quark statistics" but the usual group SU(3) is enlarged to SU(4) to accomodate a "charmed" quark u' with charge $\frac{2}{3}$ which has no isotopic spin or ordinary strangeness but does have a nonzero value of a new conserved quantum number, charm, which would be violated by weak interactions (in such a way as to remove the strangeness-changing part from the commutator of the hadronic weak charge operator with its Hermitian conjugate). Again the expression for the physical currents in terms of our connected algebra is altered, and again the asymptotic value of $\sigma(e^+e^-\to\text{hadrons})/\sigma(e^+e^-\to\mu^+\mu^-)$ is changed, this time to $[(\frac{2}{3})^2+(-\frac{1}{3})^2+(-\frac{1}{3})^2+(\frac{2}{3})^2]\cdot 3=\frac{10}{3}$. Just as in the Han-Nambu scheme, the predictive power is very low here until the energy is above the threshold for making "charmed" particles.

We pointed out in Section 1 that for three Fermi-Dirac quarks the Adler amplitude is too small by a factor of 3. For all the other schemes quoted above, however, it comes out just right and the decay amplitude of $\pi^0\to 2\gamma$ in the PCAC limit agrees with experiment. One may verify that for all of these schemes $\sum Q_{1/2}{}^2-\sum Q_{-1/2}{}^2=1$.The various schemes are summarized in the following table.

Scheme	$\frac{(e^+e^-\to\text{hadrons})}{(e^+e^-\to\mu^+\mu^-)}$	Can quarks be real?
"Quark statistics"	2	No
Para-Fermi statistics rank 3	2	Probably not
Nine Fermi-Dirac quarks	2	Yes
Han-Nambu, Fermi-Dirac	4	Yes
Quark statistics + charm	10/3	No
Para-Fermi, rank 3 + charm	10/3	Probably not
Twelve Fermi-Dirac + charm	10/3	Yes

In what follows, we shall confine ourselves to the first scheme, as requiring the least change in the present experimental situation.

4. DERIVATION OF THE $\pi^0\to 2\gamma$ AMPLITUDE IN THE PCAC APPROXIMATION

In the derivation sketched here, we follow the general idea of Wilson's and Crewther's method. We lean more heavily on the connected light-cone current algebra, however, and we do not need to assume full conformal invariance of matrix elements for small values of the coordinate differences.

To discuss the $\pi^0\to 2\gamma$ decay in the PCAC approximation, we shall need an expression for

$$\langle \text{vac}|\mathcal{F}_{e\alpha}(x)\mathcal{F}_{e\beta}(y)\mathcal{F}_{3\gamma}{}^5(x)|\text{vac}\rangle$$

when $x \approx y \approx z$. (Here e is the direction in SU(3) space of the electric charge.) In fact, we shall consider general products of the form

$$\langle \text{vac}|F(x_1)F(x_2)\cdots F(x_n)|\text{vac}\rangle$$

where F's stand for components of any of our currents, and we shall examine the leading singularity when $x_1, x_2, \ldots, x_n$ tend to lie among a single lightlike line. (The case when they tend to coincide is then a specialization.)

We assume not only the validity of the connected light-cone algebra, which implies scale invariance for commutators near the light cone, but also scale invariance for products near the lightcone, with leading dimension $l = -3$ for all currents. There may be subtraction terms in the products, or at least in physical ordered products, for example, subtractions corresponding to four-dimensional δ functions in coordinate space; these are often determined by current consrvation. But apart from the subtraction terms the current products near the light cone have no choice, because of causality and their consequent analytic properties in coordinate space, but to obey the same formulas as the commutators, with $i\pi\epsilon(z_0)\delta(z^2)$ replaced by $\frac{1}{2}(z^2 - iz_0\epsilon)^{-1}$ for products and $\frac{1}{2}(z^2 - i\epsilon)^{-1}$ for physical ordered products.

Our general quantity $\langle \text{vac}|F(x_1)F(x_2)\cdots F(x_n)|\text{vac}\rangle$ may now be reduced, using successive applications of the product formulas near the light cone and ignoring possible subtraction terms, since all the intervals $(x_i - x_j)^2$ tend to zero, as they do when all the points x_i tend to lie on the same lightlike line.

A contraction between two currents $F(x_i), F(x_j)$ gives a singular function $S(x_i - x_j)$ times a bilocal $F(x_i x_j)$. If we now contract another local current with the bilocal, we obtain $S(x_i - x_j)S(x_k - x_j)F(x_i, x_k)$ and so on.

As long as we do not exhaust the currents, our intermediate states have particles in them and we are using the connected algebra generalized to products. Finally, we reach the stage where we have a string of singular functions multiplied by $\langle \text{vac}|F(x_i, x_j)F(x_k)|\text{vac}\rangle$, and the last contraction amounts to knowing the disconnected matrix element of a current product. However, the leading singularity structure of this matrix element can also be determined from the light-cone algebra by requiring consistent reductions of the three current amplitudes.

We can algebraically reduce a three-current amplitude in two possible ways. For each reduction, the algebra implies the existence of a known light-cone singularity. The reductions may also be carried out for an amplitude with a different ordering of the currents. One reduction of this amplitude yields the same two-point function as before, whereas the other reduction implies the existence of a second singularity in the two-point function. Hence we may conclude that the leading singularity of the two-point function when all points tend to a light line is given by the product of the two singularities identified by these reductions. Similarly, the leading singularity of the three-current amplitude is given by the product of the three singularities indicated by the different reductions. Since the connected light-cone algebra can be abstracted from the free quark model, the result of this analysis implies that the leading singularities of the two- and three-point functions are also given by the free quark model (say, with Fermi-Dirac quarks) and the only undetermined parameter is an overall factor, N, by which all vacuum amplitudes must be multiplied.

Since the singularity structure of the two-point function is determined, we can identify at least a part of the leading light line singularity of the n current amplitudes. Each different reduction of the n current amplitudes implies free quark singularities associated with this reduction. For two, three, and four current amplitudes, all of the singularities can be directly determined from the different reductions. For the five and higher-point functions not all of the singularities can be directly determined, but it is plausible that these others also have the free quark structure.

For the asymptotic value of $\sigma(e^+e^- \to \text{hadrons})/\sigma(e^+e^- \to \mu^+\mu^-)$, we are interested in the vacuum expected value of the commutator of two electromagnetic currents, and it comes out equal to N times a known quantity. Similarly, more complicated experiments testing products of four currents, for example, e^+e^- annihilation into hadrons and a massive muon pair or "γ" $-$ "γ" annihilation into hadrons, might be considered. Also these processes are, in the corresponding deep inelastic limit, completely determined by the number N.

Returning to $\pi^0 \to 2\gamma$ in the PCAC approximation, we have $\langle \text{vac}|\mathcal{F}_{e\alpha}(x)\, \mathcal{F}_{e\beta}(y)\mathcal{F}_{3\gamma}{}^5(z)|\text{vac}\rangle$ as the three space-time points approach a lightlike line, apart from subtraction terms, in terms of N times a known quantity. We now need only appeal to Wilson's argument (as elaborated by Crewther). The vacuum expected value of the physically ordered product $T(\mathcal{F}_{e\alpha}(x), \mathcal{F}_{e\beta}(y), \partial_\gamma \mathcal{F}_{3\gamma}{}^5(z))$, taken at low frequencies, is what we need for the $\pi^0 \to 2\gamma$ decay with PCAC, and the Wilson-Crewther argument shows that it is determined from the small-distance behavior of $\langle \text{vac}|\mathcal{F}_{e\alpha}(x)\mathcal{F}_{e\beta}(y)\mathcal{F}_{3\gamma}{}^5(z) |\text{vac}\rangle$, with the subtraction terms (which are calculable from current conservation in this case) playing no rôle. This remarkable superconvergence result, that the low-frequency matrix element can be calculated from a surface integral around the leading short-distance singularity (which is the same as the singularity if all three points tend to a lightlike line), makes possible the derivation of $\pi^0 \to 2\gamma$ in the PCAC approximation from the light-cone current algebra. We come out with the Adler result (i.e., the result for three Fermi-Dirac quarks) multiplied by N.

Thus the connected light-cone algebra provides a link between the $\pi^0 \to 2\gamma$ decay and the asymptotic ratio $\sigma(e^+e^- \to \text{hadrons})/\sigma(e^+e^- \to \mu^+\mu^-)$. Of course, one might doubt the applicability of PCAC to π^0 decay, or to any process in which other currents are present in addition to the axial vector current connected to the pion by PCAC. If the connected algebra is right, including products, then failure of the asymptotic ratio of the e^+e^- cross sections to approach the value 2 would be attributed either to such a failure of PCAC when other currents are present or else to the need for an alternative model such as we discussed in Section 3.

As a final remark, let us mention the "finite theory approach," as discussed in ref. 4 in connection with the light-cone current algebra. Here the idea is to abstract results not from the formal "quark-gluon" field theory model, but rather from the sum of all orders of perturbation theory (insofar as that can be studied) under two special assumptions. The assumptions are that the equation for the renormalized coupling constant that allows for a finite coupling constant renormalization has a root and that the value of the renormalized coupling constant is that root. Under these conditions, the vacuum expected values of at least some current products are less singular than in the free theory. Since the Adler result still holds in the "finite theory case," the connected light-cone algebra would have to break down. In particular, the axial vector current appearing in the commutator of certain vector currents is multiplied by an infinite constant. There are at present two alternative possibilities for such a "finite theory":

1. Only vacuum expected values of products of singlet currents are less singular than in the free theory;[14] only the parts of the algebra that involve singlet currents are wrong (e.g., the bilocal singlet axial vector current is infinite); the e^+e^- annihilation cross section would still behave scale invariantly.

2. All vacuum expected values of current products are less singular than in the free theory; the number N is zero; all bilocal axial vector currents are infinite; the e^+e^- annhilation cross section would decrease more sharply at high energies than in the case of scale invariance.

1. ACKNOWLEDGMENTS

For discussions, we are indebted to D. Maison, B. Zumino, and other members of the staff of the Theoretical Study Division of CERN. We are pleased to acknowledge also the hospitality of the Theoretical Study Division.

REFERENCES

1. H. Fritzsch and M. Gell-Mann, "Proceedings of the Coral Gables Conference on Fundamental Interactions at High Energies, January 1971," in *Scale Invariance and the Light Cone,* Gordon and Breach, New York, (1971).
2. H.J. Lipkin and S. Meshkov, *Phys. Rev. Letters,* **14**, 670 (1965).
3. R. Dashen and M. Gell-Mann, *Phys. Letters,* **17**, 142 (1965).
4. H. Fritzsch and M. Gell-Mann, *Proceedings of the International Conference on Duality and Symmetry in Hadron Physics,* Weizmann Science Press of Israel, Jerusalem, 1971.
5. G. Zweig, CERN Preprints TH. 401 and 412 (1964).
6. See, for example, O. W. Greenberg, *Phys. Rev. Letters,* **13**, 598 (1964).
7. S. L. Adler, *Phys. Rev.,* **177**, 2426 (1969); J. S. Bell and R. Jackiw, *Nuovo Cimento,* **60A**, 47 (1969).
8. S. L. Adler, in *Lectures on Elementary Particles and Fields* (1970 Brandeis University Summer Institute), MIT Press, Cambridge, Mass., 1971, and references quoted therein.
9. S. L. Adler and W. A. Bardeen, *Phys. Rev.,* **182**, 1517 (1969).
10. R. J. Crewther, Cornell preprint (1972).
11. K. G. Wilson, *Phys. Rev.,* **179**, 1499 (1969).
12. E. J. Schreier, *Phys. Rev. D,* **3**, 982 (1971).
13. M. Han and Y. Nambu, *Phys. Rev.,* **139**, 1006 (1965).
14. See also B. Schroer, Chapter 3 in this volume (p. 42).

Acta Physica Austriaca, Suppl. IX, 733–761 (1972)

QUARKS *

BY

M. GELL-MANN
CERN - Geneva+

In these lectures I want to speak about at least two interpretations of the concept of quarks for hadrons and the possible relations between them.

First I want to talk about quarks as "constituent quarks". These were used especially by G. Zweig (1964) who referred to them as aces. One has a sort of a simple model by which one gets elementary results about the low-lying bound and resonant states of mesons and baryons, and certain crude symmetry properties of these states, by saying that the hadrons act as if they were made up of subunits, the constituent quarks q. These quarks are arranged in an isotopic spin doublet u, d and an isotopic spin singlet s, which has the same charge as d and acts as if it had a slightly higher mass.

---

*Lecture given at XI. Internationale Universitätswochen für Kernphysik, Schladming, February 21 - March 4, 1972.

+On leave from CALTECH, Pasadena. John Simon Guggenheim Memorial Fellow.

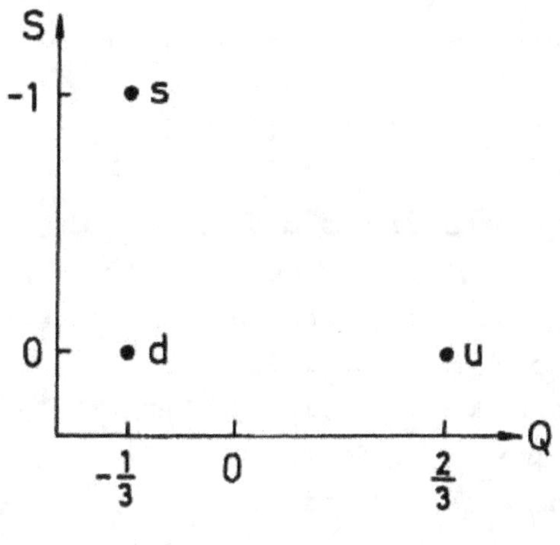

The antiquarks $\bar{q}$ of course, have the opposite behaviour. The low-lying bound and resonant states of baryons act like qqq and those of the mesons like $q\bar{q}$. Other configurations, e.g., $q\bar{q}q\bar{q}$, $qqqq\bar{q}$, etc., are called exotic, but they certainly exist in the continuum and may have resonances corresponding to them.

In this way one builds up the low-lying meson and baryon states and it is frequently useful to classify them in terms of an extremely crude symmetry group $U_6 \times U_6 \times O_3$, where one U_6 is for the quarks (three states of charge and two spin states) and one for the antiquarks, whereas O_3 represents a sort of angular momentum between them. This symmetry is however badly violated in the lack of degeneracy of the spectrum. The mesons then have as the lowest representation

$$(\underline{6}, \underline{\bar{6}}), \qquad L^P = 0^-$$

which gives the pseudoscalar and vector mesons, nine of each, just the ones which have been observed. (L=0 would normally have parity plus, but since we have a q and a $\bar{q}$

the intrinsic parity is minus.) The next pattern would be

$$(\underline{6},\ \bar{\underline{6}}), \qquad L^P = 1^+$$

and this gives us the tensor mesons, axial vector mesons, another kind of axial vector mesons with opposite charge conjugation, and scalar mesons. All of these kinds have been seen, although not yet quite nine in every case. Then one can go up with L=2,3,... where there is just scattered experimental information. But whatever experimental information exists is at least compatible with this trivial picture. As is well known this group is not very well conserved in the spectrum and the states are badly split, e.g., m_π=140 MeV and $m_{\eta'}$=960 MeV.

For the baryons the lowest configuration is assigned to

$$(\underline{56},\ \underline{1}), \qquad L^P = 0^+$$

that is three quarks in a totally symmetric state of spin, isospin, etc., without antiquarks; the parity is plus by definition. This gives the baryon octet with spin $\frac{1}{2}$ and just above it the decimet with spin $\frac{3}{2}$, which agree with the lowest-lying and best-known states of the baryons. The next thing one expects would be an excitation of one unit of angular momentum which changes the symmetry to the mixed symmetry under permutations:

$$(\underline{70},\ \underline{1}), \qquad L^P = 1^-$$

and that seems to be a reasonable description of the low lying states of reversed parity. If one goes on to higher configurations things become more uncertain both experimentally and theoretically; presumably

$$(\underline{56}, \underline{1}), \quad L^P = 2^+$$

exists and contains the Regge excitations of the corresponding ground state, likewise

$$(\underline{70}, \underline{1}), \ L^P = 3^- \quad \text{and so on.}$$

In doing this we have to assume something peculiar about the statistics obeyed by these particles, if we want the model to be simple. One expects the ground state to be totally symmetric in space. If the quarks obeyed the usual Fermi-Dirac statistics for spin $\frac{1}{2}$ particles, then there would be an over-all antisymmetry and one would obtain a totally antisymmetric wave function in spin, isospin and strangeness, whereas ($\underline{56}$, $\underline{1}$) is the totally symmetric configuration in these quantum numbers. What most people have assumed therefore from the beginning (1963) is that the quarks obey some unusual kind of statistics in which every set of three has to be symmetrized but all other bonds have to be made antisymmetric, so that, e.g., two baryons are antisymmetric with respect to each other. One version of this came up under the name of parastatistics, precisely "para-Fermi statistics of rank three", which gives a generalization of the result I just described. I will discuss it in a slightly different way, which is equivalent to para-Fermi statistics of rank three with the restriction that physical baryon states are all fermions and physical meson states are all bosons.

We take three different kinds of quarks, that is nine altogether, and call the new variable distinguishing the sets "color", for example red, white and blue (R-W-B).

The nine kinds of quarks are then individually Fermi-Dirac particles, but we require that all physical baryon and meson states be singlets under the SU_3 of "color". This means that for the meson $q\bar{q}$ configuration we now have

$$q_R\bar{q}_R + q_W\bar{q}_W + q_B\bar{q}_B$$

and for a baryon qqq we have

$$q_Rq_Wq_B - q_Wq_Rq_B + q_Bq_Rq_W - q_Rq_Bq_W + q_Wq_Bq_R - q_Bq_Wq_R \quad ,$$

which is totally antisymmetric in color and permits the baryon to be totally symmetric in the other variables space, spin, isospin and strangeness. This restriction to color singlet states for real physical situations gives back exactly the sort of statistics we want.

Now if this restriction is applied to all real baryons and mesons, then the quarks presumably cannot be real particles. Nowhere have I said up to now that quarks have to be real particles. There might be real quarks, but nowhere in the theoretical ideas that we are going to discuss is there any insistence that they be real. The whole idea is that hadrons act as if they are made up of quarks, but the quarks do not have to be real.

If we use the quark statistics described above, we see that it would be hard to make the quarks real, since the singlet restriction is not one that can be easily applied to real underlying objects; it is not one that factors: a singlet can be made up of two octets and these can be removed very far from each other such that the system over-all still is a singlet, but then we see the two pieces as octets because of the factoring property of the S matrix. If we adopt this point of view we are then faced with two alternatives: one is that there are three quarks, fictitious and obeying funny statistics; the other is that there are actually three triplets of real quarks, which is possible but unpleasant. In the latter case we would replace the singlet restriction with the assumption that the low lying states are singlets and one has to pay a large price in energy to get the colored SU_3 excited. I would prefer to adopt the first point of view, at least for these lectures.

Various crude symmetries and other related methods have been applied to these constituent quarks. First of all there is the famous subgroup of the classifying $U_6 \times U_6 \times U_3$, namely $[U_6]_w \times [O_2]_w$ which is applied to processes involving only one direction in space, like a vertex or forward and backward scattering (in general, collinear processes). $[O_2]_w$ has the generator L_z (assuming z is the chosen direction) and $[U_6]_w$ consists of the generators

$$\frac{1}{2}(\sum_i \lambda_i + \sum_j \lambda'_j),$$

$$\frac{1}{2}(\sum_i \lambda_i \sigma_{iz} + \sum_j \lambda'_j \sigma'_{jz}),$$

$$\frac{1}{2}(\sum_i \lambda_i \sigma_{ix} - \sum_j \lambda'_j \sigma'_{jx}),$$

$$\frac{1}{2}(\sum_i \lambda_i \sigma_{iy} - \sum_j \lambda'_j \sigma'_{jy}),$$

where the sum over i extends over the constituent quarks and the primed sum over j extends over the constituent antiquarks; we have 36 operations. There is a very crude symmetry of collinear processes under this group.

Another thing that has been done is to draw simple diagrams following quark lines through the vertices and the scattering. These have been recently used by Harari and Rosner, who called them "twig" diagrams after Zweig, who introduced them in 1964.

The twig diagram, e.g., for a meson-meson-meson vertex, looks like

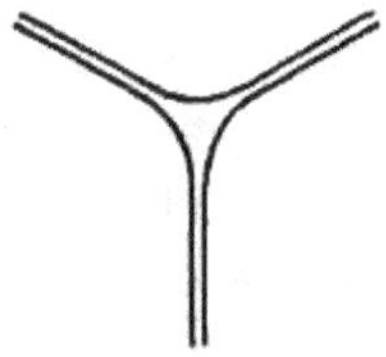

But another form

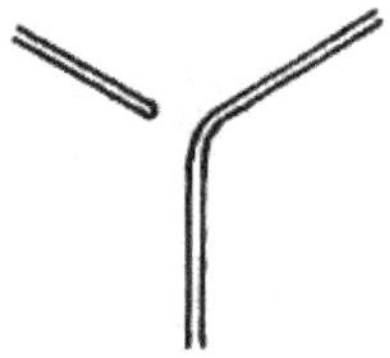

is forbidden by "Zweig's rule". This rule then leads to important experimental results, especially that the ϕ cannot decay appreciably into a ρ and π ($\phi \not\to \rho + \pi$), since the ϕ is composed of strange and antistrange quarks whereas ρ and π have only ordinary up and down quarks, and, therefore, the decay could take place only via the forbidden diagram. Similarly, we have the baryon-baryon-meson vertex

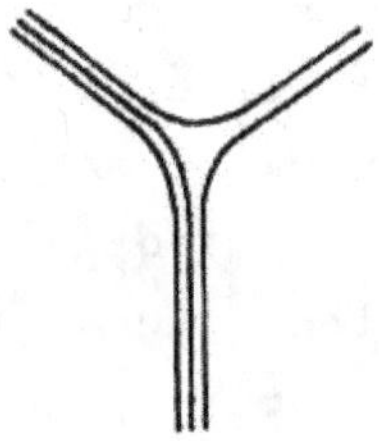

One can extend this concept to scattering processes and get a graphical picture of the so-called duality approach to scattering, e.g., for meson-meson scattering one can introduce the following diagram:

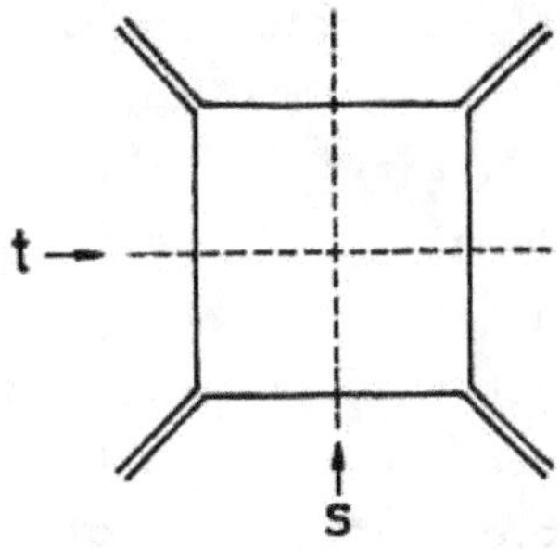

If we cut the diagram in the s and t channels we get $q\bar{q}$ in both cases: therefore, in meson-meson scattering we have ordinary non-exotic mesons in the intermediate states and exchange non-exotic mesons. We run into something of a trap, though, if we try to apply this to baryon-antibaryon scattering, because then we have a situation like

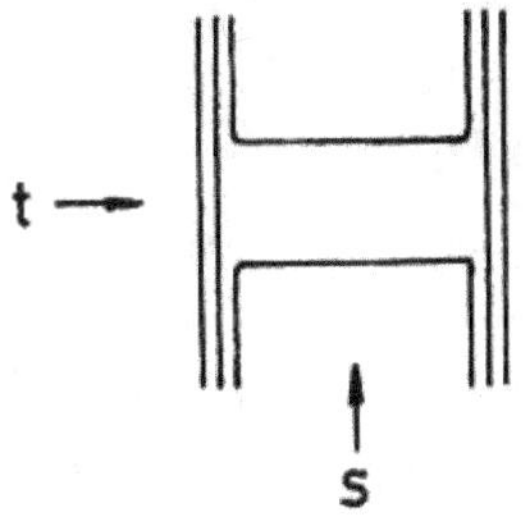

where the intermediate state is $q\bar{q}q\bar{q}$, which includes exotic configurations. In order to interpret this inconsistency different people have done different things.

The diagrams have been used in two different ways: one involves saying what the diagram means mathematically, and the other one involves not saying what it means mathematically. This is possible since here we do not have a priori a definite mathematical rule for computing the diagram, in contrast to a Feynman diagram for which a specific integral always exists. But we can obtain some results by never giving such a rule, only by noticing that we have zero when there is no diagram. Those so-called "null-relations" have been used by Schmid, by Harari and Rosner, by Zweig, Weyers and Mandula and by others for establishing a number of extremely useful sum rules. They give a correspondence between lack of resonances in the s channel, in places where the resonances would have to be exotic, and exchange degeneracy in the t channel. Exchange degeneracy is a noticeable feature of low energy hadron physics and a number of cases of agreement with experiments have been obtained.

All I want to say about the null-relation approach is that from the point of view of constituent quarks we are dealing here with a non-exotic approximation, because we are leaving out exotic exchanges, and that cannot be expected to be completely right. The simple null-equation duality approach is just another feature of the same kind of approximation we were talking about before, i.e., the classification under $U_6 \times U_6 \times O_3$ and the rough symmetry of collinear processes under $[U_6]_W \times [O_2]_W$, and when it fails that resembles a failure of such an approximation.

Another school of people consists of those who do the Veneziano duality kind of work and actually attempt to assign mathematical meaning to these diagrams. They go very far and construct almost complete theories of hadron scattering by means of extending these simple diagrams to ones with any number of quark pairs, but they run into trouble with negative probabilities or negative mass squares and the difficulty of introducing quark spin. There are also some difficulties with high energy diffraction scattering, etc. So that approach is not yet fully successful, while the much more modest null-relation approach has borne some fruit. However, if they overcome their difficulties, the members of the other school will have produced a full-blown hadron theory and advanced physics by a huge step.

There is a second use of quarks, as so-called "current quarks", which is quite different from their use as constituent quarks; we have to distinguish carefully between the two types in order to think about quarks in a reasonable manner. Unfortunately many authors including, I regret to say, me, have in past years written things that tended to confuse the two. In the following discussion of current quarks we attempt to write down properties that may be exact, at least to all orders in the strong interaction, with the weak, electromagnetic and gravitational interactions treated as perturbations. (It is necessary always to include gravity because the first order coupling to gravity is the stress-energy-momentum tensor and the integral over this tensor gives us the energy and momentum which we have to work with.)

When I say we attempt exact statements I do not mean that they are automatically true - there is also the incidental matter that they have to be confirmed by experiment, but the statements have a chance to be exact. Such statements which are supposed to be exact at least in certain limits or in certain well-defined approximations, or even generally exact, are to be contrasted with statements which are made in an ill-defined approximation or a special model whose domain of validity is not clearly specified. One frequently sees allegedly exact statements mixed up with these vague model statements and when experiments confirm or fail to confirm them it does not mean anything. Of course, we all have to work occasionally with these vague models because they give us some insight into the problem but we should carefully distinguish highly model-dependent statements from statements that have the possibility of being true either exactly or in a well defined limit.

The use of current quarks now is the following:we say that currents act as if they were bilinear forms in a relativistic quark field.We introduce a quark field,presumably one for the red,white and blue quarks and then we have for the vector currents in weak and electromagnetic interaction

$$F_{i\mu} \sim i\bar{q}_R \gamma_\mu \frac{\lambda_i}{2} q_R + i\bar{q}_W \gamma_\mu \frac{\lambda_i}{2} q_W + i\bar{q}_B \gamma_\mu \frac{\lambda_i}{2} q_B \quad .$$

The symbol $\sim$ means the vector current "acts like" this bilinear combination. Likewise the axial vector current acts like

$$F^5_{i\mu} \sim i\bar{q}_R \gamma_\mu\gamma_5 \frac{\lambda_i}{2} q_R + i\bar{q}_W \gamma_\mu\gamma_5 \frac{\lambda_i}{2} q_W + i\bar{q}_B \gamma_\mu\gamma_5 \frac{\lambda_i}{2} q_B \quad .$$

The reason why I want all these colors at this stage is that I would like to carry over the funny statistics for the current quarks, and eventually would like to suggest a transformation which takes one into the other, conserving the statistics but changing a lot of other things. An important feature of this discussion will be the following: is there any evidence for the current quarks that they obey the funny statistics? the answer is yes, and the evidence depends on a theoretical result due to many people but principally S.Adler.

The result is that in the PCAC limit one can compute exactly the decay rate of $\pi^0 \rightarrow 2\gamma$. The basis on which Adler derived it was a relativistic renormalized quark-gluon field theory treated in renormalized perturbation theory order by order, and there the lowest order triangle diagram gives the only surviving result in the PCAC limit:

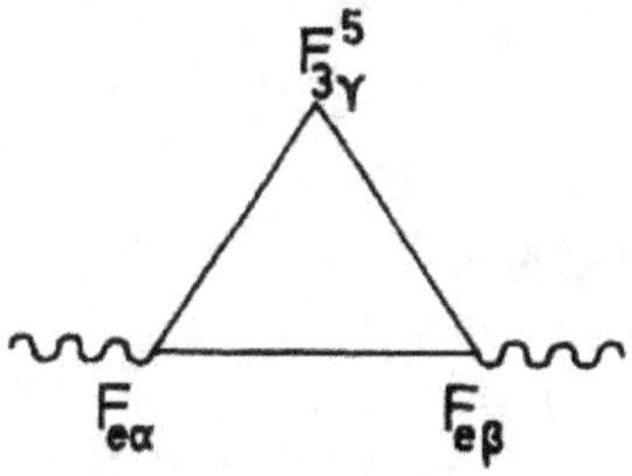

Here F_e are the electromagnetic currents, F^5_3 the third component of the axial current that is converted into a π^0 through PCAC. We reject this derivation because order by order evaluation of a renormalizable quark-gluon field theory does not lead to scaling in the deep inelastic limit. Experiments at SLAC up to the present time are incapable of proving or disproving such a thing as scaling in the Bjorken limit but they are certainly suggestive and we would like to accept the Bjorken scaling. So we must reject the basis of Adler's derivation but we can derive this result in other ways, consistent with scaling, as I shall describe briefly later on.

What is sometimes said about Adler's computation is that this result completely contradicts the quark model. What is true is that it completely contradicts a hypothetical quark model that practically nobody wants, with Fermi-Dirac statistics. It agrees beautifully on the other hand with the ancient model that nobody would conceivably believe today in which things are made up of neutrons and protons. If you take basic neutrons and protons you get for an over-all coefficient of that diagram the following: we have charges squared multiplied by the third component of I, which means +1 for up isotopic spin and -1 down isotopic spin, so for protons and neutrons as basic constituents we get

$$(p, n): +1(1)^2 -1(0)^2 = 1$$

and with this Adler obtained exactly (within experimental errors) the right experimental decay rate for π^0 and even the right sign for the decay amplitude. If we take quarks u, d, s we get $(u, d, s): +1(2/3)^2 - 1(-1/3)^2 = 1/3$. So we obtain a decay rate which is wrong by a factor of 9. However, if we have the funny statistics - say in the easiest way with the red, white and blue color - we should put in

$$\begin{array}{lcl} u_R, d_R, s_R & & \\ u_W, d_W, s_W & : & 3[+1(2/3)^2 - 1(-1/3)^2] = 1 \quad , \\ u_B, d_B, s_B & & \end{array}$$

remembering that the current is a singlet in R-W-B, but in the summation we obtain a loop for each color. So we get back the correct result. Thus, there is this, in my mind, very convincing piece of evidence from the current quarks too for the funny statistics such as the constituent quarks seem to obey. The transformation between them should preserve statistics and so the picture seems to be a consistent one.

The relation "acts like" (∿) which we used to define the current quarks can be strengthened as we introduce more and more properties of the currents which are supposed to be like the properties of these expressions. In other words there will be a hierarchy of strength of abstraction from such a field theory to the properties that we suggest are the exact characteristics of the vector and axial vector currents. We have to be very careful then to abstract as much as we can so as to learn as much as we can from the current quark picture, but not to abstract too much, otherwise first of all experiments may prove us wrong, and secondly that it may involve us with the existence of actual quarks, maybe even free quarks - and that, of course, would be a disaster.

If quarks are only fictitious there are certain defects and virtues. The main defect would be that we will never experimentally discover real ones and thus will never have a quarkonics industry. The virtue is that then there are no basic constituents for hadrons - hadrons act as if they were made up of quarks but no quarks exist - and, therefore, there is no reason for a distinction between the quark and bootstrap picture: they can be just two different descriptions of the same system, like wave mechanics and matrix mechanics. In one case you talk about the bootstrap and when you solve the equations you get something that looks like a quark picture; in the other case you start out with quarks and discover that the dynamics is given by bootstrap dynamics. An important question about the bootstrap approach is whether the bootstrap gives everything or whether there are some symmetry considerations to be supplied aside from the bootstrap. I do not know the answer, but some people claim to have direct information from heaven about it.

Let us go back to the current quarks. Besides the V and A currents we might have well defined tensor (T), scalar (S) and pseudoscalar (P) currents, which would act like

$$T_{i\mu\nu} \sim \bar{q}\, \lambda_i\, \sigma_{\mu\nu}\, q$$

$$S_i \sim \bar{q}\, \lambda_i\, q$$

$$P_i \sim i\bar{q}\, \lambda_i\, \gamma_5\, q \quad .$$

I think these currents all can be physically defined: the S and P currents would be related to the divergences of the V and A currents and the tensor currents would arise when you commute the currents with their divergences.

The first of the most elementary abstractions was the introduction of the $SU_3 \times SU_3$ charges, that is

$$\int F_{io}\, d^3x = F_i$$

$$\int F^5_{io}\, d^3x = F^5_i$$

with their equal time commutators. We do not have very good direct evidence that this is true, but the best evidence comes from the Adler-Weisberger relation which is in two forms: first the pure one, namely just these commutators

$$[F^5_i, F^5_j] = i\, f_{ijk}\, F_k$$

which give sum rules for neutrino reactions, and a second form which involves the use of PCAC giving sum rules for pion reactions, and those have been verified. So an optimist would say that the commutation relations are okay and PCAC is okay; the pessimist might say that they are both wrong but they compensate. That will be checked relatively soon as neutrino experiments get sophisticated enough to test the pure form. In the meantime I will assume that the equal-time commutators (ETC) and PCAC are okay. In order to get the Adler-Weisberger relation it is necessary to apply the ETC of light-like charges for the special kinematic condition of infinite momentum (in the z direction). We make the additional assumption that between finite mass states we can saturate the sum over intermediate states by finite mass states. In the language of dispersion theory this amounts to an assumption of unsubtractedness; in the language of light-cone theory it amounts to smoothness on the light cone. In this way Adler and Weisberger derived their simple sum rule.

We are considering here the space integrals of the time components of the V and A currents, but not those of the x or y components at $P_z=\infty$. We are restricting ourselves, in order to have saturation by finite mass intermediate states, to "good" components of the currents, those with finite matrix elements at $P_z=\infty$. These are $F_{io} \approx F_{iz}$ and $F^5_{io} \approx F^5_{iz}$ at $P_z=\infty$. The "bad" components F_{ix}, F_{iy}, F^5_{ix}, F^5_{iy} have matrix elements going like P_z^{-1} at infinite P_z. The components $F_{iz}-F_{io}$ and $F^5_{iz}-F^5_{io}$ have matrix elements going like P_z^{-2} and are "terrible".

One generalization that we can make of the algebra of $SU_3 \times SU_3$ charges is to introduce the tensor currents $T_{i\mu\nu}$. In the case of the tensor currents the good components are

$$T_{ixo} \cong T_{ixz}$$
$$T_{iyo} \cong T_{iyz}$$

from which we construct the charges

$$"T_{ix}" = \int T_{ixo}\, d^3x$$

$$"T_{iy}" = \int T_{iyo}\, d^3x$$

at $P_z=\infty$ and adjoin them to F_i and F_i^5. Thus we get a system of 36 charges and that just gives us a $[U_6]_W$. One reason why I introduced these tensor currents is that it is simpler to work with $[U_6]_W$, which we have met before, rather than with one of its subgroups $SU_3\times SU_3$. These charges then form generators of an algebra which we call

$$[U_6]_{W,\infty,\text{currents}}\ ;$$

to be contrasted with

$$[U_6]_{W,\infty,\text{strong}}$$

which had to do with the constituent quarks. The contrast is between the $[U_6]_{W,\infty,\text{strong}}$ which is essentially approximate in its applications (collinear processes), and the $[U_6]_{W,\infty,\text{currents}}$ drawn from presumably exact commutation rules of the currents and having to do with current quarks. Although they are isomorphic they are not equal.

For those who cannot stand the idea of introducing the tensor currents, we can just reduce both groups to their subgroups $SU_3\times SU_3$. In that case for $[U_6]_{W,\infty,\text{currents}}$ we are discussing only the vector and axial vector charges and for $[U_6]_{W,\infty,\text{strong}}$ only the so-called coplanar subgroup. Then we can make the same remark that these two are not equal, they are mathematically similar but their matrix elements are totally different. So one of them is the transform of the other in some sense.

The transformation between current and constituent quarks is then phrased here in a way which does not involve quarks; we discuss it as a transformation between $[U_6]_{w,\infty,\text{currents}}$ and $[U_6]_{w,\infty,\text{strong}}$ (or their respective subgroups). What would happen if they were equal? We know that for $[U_6]_{w,\infty,\text{strong}}$ the low-lying baryon and meson states belong approximately to simple irreducible representations $\underline{35}$, $\underline{1}$, $\underline{56}$, etc. If this were true also of $[U_6]_{w,\infty,\text{currents}}$ then we would have the following results:

$$-\frac{G_A}{G_V} \cong \frac{5}{3}$$

which we know is more like $5/3\sqrt{2}$; the anomalous magnetic moments of neutron and proton would be approximately zero, while they are certainly far from zero, and so on.

Many authors have in fact investigated the mixing under this group, and they found that there is an enormous amount of mixing, e.g., the baryon is partly $\underline{56}$, $L_z=0$ and partly $\underline{70}$, $L_z=\pm1$ and the admixture is of the order of 50%. There are higher configurations, too.

So these two algebras are not closely equal although they have the same algebraic structure. And there is some sort of a relation between them, which might be a unitary one, but we cannot prove that since they do not cover a complete set of quantum numbers. But we can certainly look for a unitary transformation connecting the two algebras and my student J. Melosh is pursuing that problem. He has found this transformation for free quarks where it is simple and leads to a conserved $[U_6]_{w,\text{strong}}$. But, of course, we are not dealing with free quarks and have to look at a more complicated situation. What I want to emphasize is that here we have the definition of the search; the search is for a transformation connecting the two algebras. In popular language we can refer to it as a search for the transformation connecting constituent quarks and current quarks.[+]

Let me mention here the work of another student of mine, Ken Young, who has cleaned up this past year at Caltech work that Dashen and I began about 7 years ago, and which we continued sporadically ever since. That is the attempt to represent the charges and also the transverse Fourier components of the charge densities at infinite momentum completely with non-exotic states, so as to make a non-exotic relativistic quark model as a representation of charge density algebra. We ran into all kinds of troubles, particularly with the existence of states with negative mass squares and the failure of the operators of different quarks to commute with each other. Young seems to have shown that these difficulties are a property of trying to represent the charge algebra at infinite momentum with non-exotics alone. Therefore, the transformation which connects the two algebras does not just mix up non-exotic states but also brings in higher representations that contain exotics. In simple lay language the transformation must bring in quark pair contributions and the constituent quark looks then like a current quark dressed up with current quark pairs.

We therefore must reject all the extensive literature, which I am proud not to have contributed to over the last few years, in which the constituent quarks are treated as current quarks and the electromagnetic current is made to interact through a simple current operator $F_{e\mu}$ with what

[+]Buccella, Kleinert and Savoy have suggested a simple phenomenological form of such a transformation.

are essentially constituent quarks. That is certainly wrong.

Another way of describing the infinite momentum and the smoothness assumption is to perform an alibi instead of an alias transformation, i.e., instead of letting everything go by you at infinite momentum you leave it alone at finite momentum and you run by it. These two are practically equivalent. In that case one is not talking about infinite momentum but about the behaviour in co-ordinate space as we go to a light-like plane and about the commutators of light-like charges which are integrated over a light-like plane instead of an equal-time plane. Leutwyler, Stern and a number of other people have especially emphasized this approach. From that again one can get the Adler-Weisberger relation, and so forth.

On the light-like plane (say z+t=0) we have the commutation rules not only for the charges, but also for the local densities of the good components of the currents, namely $F_{io}+F_{iz}$ and $F^5_{io}+F^5_{iz}$ for V and A, with the possible adjunction of the good components $T_{ixo}+T_{ixz}$ and $T_{iyo}+T_{iyz}$ of the tensor currents. Especially useful is the algebra of these quantities integrated over the variable z-t and Fourier-transformed with respect to the variables x and y. We obtain the operators $F_i(\vec{k}_\perp)$, $F^5_i(\vec{k}_\perp)$, $T_{ix}(\vec{k}_\perp)$, and $T_{iy}(\vec{k}_\perp)$ and they have commutation relations like

$$[F_i(\vec{k}_\perp),\ F_j(\vec{k}'_\perp)] = i\ f_{ijk}\ F_k(\vec{k}_\perp+\vec{k}'_\perp) \quad .$$

By the way, if we take this equation to first order in $\vec{k}_\perp$ and $\vec{k}'_\perp$ we get the so-called Cabibbo-Radicati sum rule for photon-nucleon collisions, which seems to work quite well.[+]

[+]Strictly, the operators $F_i(\vec{k}_\perp)$ are not exactly the Fourier transforms of integrals of current densities but rather these Fourier transforms multiplied by

$$\exp\{i[k_x(\Lambda_x+J_y)+k_y(\Lambda_y-J_x)][P_o+P_z]^{-1}\}\ ,$$

So far we have abstracted from quark field theory relations that were true not only in a free quark field model but also to all orders of strong interactions in a field model with interactions, say through a neutral "gluon" field coupled to the quarks. (We stick to a vector "gluon" picture so that we can use the scalar and pseudoscalar densities S_i and P_i to describe the divergences of the vector and axial vector currents respectively.) We shall continue to make use of abstractions limited in this way, since if we took all relations true in a free quark model we would soon be in trouble: we would be predicting free quarks! However, in what follows we consider the abstraction of propositions true only formally in the vector "gluon" model with interactions, not order by order in renormalized perturbation theory. We do this in order to get Bjorken scaling, which fails to each order of renormalized perturbation theory in a barely renormalizable model like the quark-vector "gluon" model but which, as mentioned before, we would like to assume true.

That leads us to light-cone current algebra, on which I have worked together with Harald Fritzsch. It has been studied by many other people as well, including Brandt and Preparata, Leutwyler et al., Stern et al., Frishman, and, of course, Wilson, who pioneered in this field although he disagrees with what we do nowadays. Those whose work is most similar to ours are Cornwall and Jackiw and Llewellyn-Smith.

The first assumption in light-cone current algebra is to abstract from free quark theory or from formal vector "gluon" theory the leading singularity on the light cone $(x-y)^2 \approx 0$ of the connected part of the commutator of two currents at space-time points x and y. For V and A currents we find

---

where $\vec{K}$ is the Lorentz-boost operator, $\vec{J}$ is the total angular momentum, and the component $P_o + P_z$ of the energy-momentum-four-vector is conserved by all the operators $F_i(\vec{k}_\perp)$.

$$[F_{i\mu}(x),\ F_{j\nu}(y)] \hat{=} [F^5_{i\mu}(x),\ F^5_{j\nu}(y)] \hat{=}$$

$$\frac{1}{4\pi}\,\partial_\rho\{\varepsilon(x_o-y_o)\delta((x-y)^2)\}\{(i\ f_{ijk}-d_{ijk})(s_{\mu\nu\rho\sigma}F_{k\sigma}(y,x) + $$
$$+\ i\varepsilon_{\mu\nu\rho\sigma}F^5_{k\sigma}(y,x)) +$$

$$+\ (i\ f_{ijk}+d_{ijk})(s_{\mu\nu\rho\sigma}F_{k\sigma}(x,y)-i\ \varepsilon_{\mu\nu\rho\sigma}F^5_{k\sigma}(x,y))\}\ ,$$

$$[F_{i\mu}(x),F^5_{j\nu}(y)] \hat{=}$$

$$\frac{1}{4\pi}\,\partial_\rho\{\varepsilon(x_o-y_o)\delta((x-y)^2)\}\{(i\ f_{ijk}-d_{ijk})(s_{\mu\nu\rho\sigma}F^5_{k\sigma}(y,x) +$$
$$+\ i\varepsilon_{\mu\nu\rho\sigma}F_{k\sigma}(y,x)) +$$

$$+\ (i\ f_{ijk}+d_{ijk})(s_{\mu\nu\rho\sigma}F^5_{k\sigma}(x,y)-i\ \varepsilon_{\mu\nu\rho\sigma}F_{k\sigma}(x,y))\}\ .$$

On the right-hand side we have the connected parts of bilocal operators $F_{k\sigma}(x,y)$ and $F^5_{k\sigma}(x,y)$ that reduce to the local currents $F_{k\sigma}(x)$ and $F^5_{k\sigma}(x)$ as $y\to x$. The bilocal operators are defined only in the vicinity of $(x-y)^2=0$. Here $s_{\mu\nu\rho\sigma}=\delta_{\mu\rho}\delta_{\nu\sigma}+\delta_{\nu\rho}\delta_{\mu\sigma}-\delta_{\mu\nu}\delta_{\rho\sigma}$.

The formulae give Bjorken scaling by virtue of the finite matrix elements assumed for $F_{k\sigma}(x,y)$ and $F^5_{k\sigma}(x,y)$; in fact the Fourier transform of the matrix element of $F_{k\sigma}(x,y)$ is just the Bjorken form factor. The fact that all charged fields in the model have spin 1/2 determines the algebraic structure of the formula and gives the prediction $\sigma_L/\sigma_T \underset{bj}{\to} 0$ for deep inelastic electron scattering, not in contradiction with experiment. The electrical and weak charges of the quarks in the model determine the coefficients in this formula, and give rise to numerous sum rules and inequalities for the SLAC-MIT experiments and for corresponding neutrino and antineutrino experiments in the Bjorken limit, none in contradiction with experiment, although the inequality $1/4 \leq F^{en}(\xi)/F^{ep}(\xi) \leq 4$ appears to be tested fairly severely at the lower end near $\xi=1$.

The formula for the leading light-cone singularity in the commutator contains, of course, the physical information that near the light-cone we have full symmetry with respect to $SU_3 \times SU_3$ and with respect to scale transformations in co-ordinate space. Thus there is conservation of dimension in the formula, with each current having $\ell=-3$ and the singular function of x-y also having $\ell=-3$.

A simple generalization of the abstraction we have considered turns it into a closed system, called the basic light-cone algebra. Here we commute the bilocal operators as well, for instance $F_{i\mu}(x,u)$ with $F_{j\nu}(y,v)$, as all the six intervals among the four space-time points approach 0, so that all four points tend to lie on a light-like straight line in Minkowski space. Abstraction from the model gives us on the right-hand side a singular function of one coordinate difference, say x-v, times a bilocal current $F_{k\sigma}$ or $F^5_{k\sigma}$ at the other two points, say y and u, plus an expression with (x,v) and (y,u) interchanged, and the system closes algebraically. The formulae are just like the ones for local currents.

We shall assume here the validity of the basic lightcone algebraic system, and discuss possible applications and generalizations.

First of all, we may consider what happens when the points x and u lie on a light-like plane with one value of z+t and y and v lie on another light-like plane with a slightly different value of z+t, and we let these values approach each other.

For commutators of good components of currents, the limit is finite, and we get a generalization of the light-plane algebra of commutators of good densities $F_{io}+F_{iz}$ and $F^5_{io}+F^5_{iz}$. There is now a fourth argument in each density, namely the internal co-ordinate η, which runs only in the light-like direction z-t. As before, we get the most useful results by integrating over the average z-t and Fourier-transforming with respect to the transverse average co-ordinates x and y, obtaining operators $F_i(\vec{k}_\perp, \eta)$, $F^5_i(\vec{k}_\perp, \eta)$, with commutation relations like

$$[F_i(\vec{k}_\perp, \eta), F_j(\vec{k}'_\perp, \eta')] = i\, f_{ijk}\, F_k(\vec{k}_\perp + \vec{k}'_\perp, \eta + \eta') \quad .$$

Remember that $\vec{k}_\perp$ is in momentum space and η in (relative) co-ordinate space.

The non-local operators $F_i(\vec{k}, \eta)$ acting on the vacuum create strings of mesons with all values of the meson spin angular momentum J. In fact, a power series expansion in η of $F_i(\vec{k}_\perp, \eta)$ is just an expansion in η^{J-1}. At large η, we can reggeize and obtain a dominant term in $\eta^{\alpha(-k^2_\perp)-1}$, where $\alpha(-k^2_\perp)$ is the leading Regge trajectory in the relevant meson channel, for instance P or ρ. (In the work on these questions, Fritzsch has played a particularly important role.) The couplings of the Regge poles in the bilocals are proportional to the hadronic couplings of meson Regge poles.

If we commute bad components with bad components as the two light planes approach each other, then the leading singularity on the light-cone leads to a singular term that goes like a δ-function of the difference in coordinates $z+t$. This singular term, which is multiplied by a good component of a bilocal current on the right-hand side, gives the Bjorken scaling in deep inelastic scattering. Unlike the good-good commutators on the light plane, it involves a commutator of local quantities on the left giving a bilocal on the right, a bilocal of which the matrix elements give the Fourier transforms of the Bjorken scaling functions $F(\xi)$.

We may now generalize, if we keep abstracting from the vector gluon model, to a connected light-cone algebra involving V, S, T, A and P densities, where the divergences of the V and A currents are proportional to the S and P currents respectively, with coefficients that correspond in the model to the three bare quark masses, forming a diagonal 3×3 matrix M. The divergences of the axial vector currents, for example, are given by masses M multiplied by normalized pseudoscalar densities P_i.

The scalar and pseudoscalar densities are all bad, and do not contribute to the good-good algebra on the light plane, but we can commute two of these densities as the light planes approach each other and obtain the singular Bjorken term. In fact, the leading light-cone singularity in the commutator of two pseudoscalars or two scalars just involves the same vector bilocal densities as the leading singularity in the commutator of two vector densities

$$[P_i(x),\ P_j(y)] \hat{=} [S_i(x),\ S_j(y)] \hat{=} [F_{i\mu}(x),\ F_{j\mu}(y)]$$

so that the P's and S's give Bjorken functions that are not only finite but known from deep inelastic electron and neutrino experiments. The Bjorken limit of the commutator of two divergences of vector or axial vector currents is also measurable in deep inelastic neutrino experiments, albeit very difficult ones, since they involve polarization and also involve amplitudes that vanish when the lepton masses vanish. The important thing is that the shapes of the form factors in such experiments are predictable from known Bjorken functions and the overall strength is given by the "bare quark mass" matrix M, which is thus perfectly measurable, according to our ideas, even though the quarks themselves are presumably fictitious and have no real masses.

The next generalization we may consider is to abstract the behaviour of current products as well as commutators near the light-cone. Here we need only abstract the principle that scale invariance near the light-cone applies to products as well as commutators. The result is that products of operators, and even physical ordered products, are given, apart from subtraction terms that act like four-dimensional δ functions, by the same expressions as commutators, with $\varepsilon(x_o-y_o)\delta((x-y)^2)$ replaced by $1/[(x-y)^2-i(x_o-y_o)\varepsilon]$ for ordinary products and by $1/[(x-y)^2-i\varepsilon]$ for physical ordered products. The subtraction terms can often be determined from current conservation; sometimes they are zero and sometimes, for certain processes, they do not matter even when they are non-zero.

Using the current products, one can design experiments to test the bilocal-bilocal commutators, for example fourth order cross-sections like those for $e^-+p\rightarrow e^-+X+\mu^++\mu^-$, where X is any hadronic state, summed over X.

Using products, and employing consistency arguments, we can determine the form of the leading light-cone singularity in the disconnected part of the current commutator, i.e., the vacuum expected value of a current commutator, and it turns out to be the same as in free quark theory or formal quark "gluon" theory. The constant in front is not determined in this way, and we must abstract it from the model. It depends on the statistics. With our funny "quark statistics" or with nine real quarks, the constant is three times as large as for three Fermi-Dirac quarks.

We can then predict the asymptotic cross-section for $e^{+}+e^{-}\to$hadrons using single photon annihilation, namely

$$\frac{\sigma(e^{+}+e^{-}\to \text{hadrons})}{\sigma(e^{+}+e^{-}\to\mu^{+}+\mu^{-})} \to 3\left[\left(\frac{2}{3}\right)^{2}+\left(-\frac{1}{3}\right)^{2}+\left(-\frac{1}{3}\right)^{2}\right] = 2$$

where we would have obtained 2/3 with three Fermi-Dirac quarks.

We are now in a position to go back and rederive the Adler result for the rate of $\pi^{0}\to 2\gamma$ in the PCAC approximation. Following the lead of Crewther, who first showed how such an alternative derivation could be given, Bardeen, Fritzsch and I use the connected light-cone algebra and the disconnected result just given to obtain the Adler result without invoking renormalized perturbation theory. The answer, as we indicated earlier, agrees with the experimental $\pi^{0}\to 2\gamma$ amplitude in both sign and magnitude.

A final generalization, about which Fritzsch and I are not so convinced as we are of the others, involves a change in our approach from considering only quantities based on currents that couple to electromagnetism and the weak interaction to including quantities that are not physically determinable in that way. I have mentioned that bilocals like $F_{i\mu}(x,y)$, which are analogous to quantities in the model that involve one quark operator and one anti-quark operator, can be applied to the vacuum to create Regge sequences of non-exotic meson states. It might also be useful to define trilocals $B_{\alpha\beta\gamma,abc}(x,y,z)$ that are analogous to operators in the model involving three quark operators at x, y and z, when these points lie on the same straight light-like line, and to abstract their algebraic properties from the model, so that sequences of baryon states could be produced from the vacuum. We could, in fact, construct operators that would, between the vacuum and hadron states, give a partial Fock space for hadrons with any number of quarks and antiquarks lying on a straight light-like line. Whether this makes sense, and how many properties of hadrons we can calculate from such a partial Fock space of "wave functions" we do not know.

If we go too far in this direction, and try to construct a complete Fock space for quarks and antiquarks on a light-like plane, abstracting the algebraic properties from free quark theory, we are in danger of ending up with real quarks, and perhaps even with free real quarks, as mentioned before. In our work, we are always between Scylla and Charybdis; we may fail to abstract enough, and miss important physics, or we may abstract too much and end up with fictitious objects in our models turning into real monsters that devour us.

In connection with the written version of these lectures, I should like to thank Dr. Heimo Latal and his collaborators for the excellent lecture notes that they provided me. I should also like to thank Dr. Oscar Koralnik of Geneva for providing the beautiful table on which most of my writing was done. I acknowledge with thanks the hospitality of the Theoretical Study Division of CERN.

Current Algebra: Quarks and What Else?

Harald Fritzsch*†

and

Murray Gell–Mann**†

CERN, Geneva, Switzerland

Proceedings of the XVI International Conference on High Energy Physics, Chicago, 1972. Volume 2, p. 135 (J. D. Jackson, A. Roberts, eds.)

Abstract

After receiving many requests for reprints of this article, describing the original ideas on the quark gluon gauge theory, which we later named QCD, we decided to place the article in the e–Print archive.

*On leave from the Max–Planck–Institut für Physik und Astrophysik. München, Germany.

†Present address: Lauritsen Laboratory of High Energy Physics, California, Institut of Technology, Pasadena, California.

**John Simon Guggenheim Memorial Foundation Fellow.

I. Introduction

For more than a decade, we particle theorists have been squeezing predictions out of a mathematical field theory model of the hadrons that we don't fully believe – a model containing a triple of spin 1/2 fields coupled universally to a neutral spin 1 field, that of the "gluon". In recent years, the triplet is usually taken to be the quark triplet, and it is supposed that there is a transformation, presumably unitary, that effectively converts the current quarks of the relativistic model into the constituent quarks of the naive quark model of baryon and meson spectrum and couplings.

We abstract results that are true in the model to all orders of the gluon coupling and postulate that they are really true of the electromagnetic and weak currents of hadrons to all orders of the strong interaction. In this way we build up a system of algebraic relations, so–called current algebra, and this algebraic system gets larger and larger as we abstract more and more properties of the model.

In section III, we review briefly the various stages in the history of current algebra. The older abstractions are correct to each order of renormalized perturbation theory in the model[1)], while the more recent ones, those of light cone current algebra, are true to all orders only formally[3)]. We describe the results of current algebra[2)] in terms of commutators on or near a null plane, say $x_3 + x_0 = 0$.

In section IV, we attempt to describe, in a little more detail, using null plane language, the system of commutation relations valid in a free quark model that are known to remain unchanged (at least formally) when the coupling to a vector "gluon" is turned on. These equations give us a formidable body of information about the hadrons and their currents, supposedly exact as far as the strong interaction is concerned, for comparison with experiment. However, they by no means exhaust the degrees of freedom present in the model; they do not yield an algebraic system large enough to contain a complete description of the hadrons. In an Appendix, the equations of Section IV are related to form factor algebra.

In Section V, we discuss how further commutation of the physical quantities arising from light cone algebra leads, in the model field theory, to results dependent on the coupling constant, to formulae in which gluon field strength operators occur in bilocal current operators proliferate. Only when these relations are included do we finally get an algebraic system that contains nearly all the degrees of freedom of the model. We may well ask, however, whether it is the right algebraic system. We discuss briefly how the complete description of the hadrons involves the specification and slight enlargement of this algebraic system, the choice of representation of the algebra that corresponds to the complete set of hadron states, and the form of the mass or the energy operator, which must be expressible in terms of the algebra when it is complete. The choice of representation may be dictated by the algebra, and if so that would justify the use of a quark and gluon Fock space by some "parton" theorists.

Finally, in Section VI, it is suggested that perhaps there are alternatives to the vector gluon model as sources of information or as clues for the construction of the true hadron theory. Assuming we have described the quark part of the model correctly, can we replace the gluons by something else? The "string" or "rubber band" formulation, in ordinary coordinate space, of the zeroth approximation to the dual resonance model, is suggested as an interesting example.

Before embarking on our discussion of current algebra, we discuss in Section II the crucial point that quarks are probably not real particles and probably obey special statistics, along with related matters concerning the gluons of the field theory model.

II. FICTITIOUS QUARKS AND "GLUONS" AND THEIR STATISTICS

We assume here that quarks do not have real counterparts that are detectable in isolation in the laboratory – they are supposed to be permanently bound inside the mesons and baryons. In particular, we assume that they obey the special quark statistics, equivalent to "para–Fermi statistics of rank three" plus the requirement that mesons always be bosons and baryons fermions. The simplest description of quark statistics involves starting with three triplets of quarks, called red, white, and blue, distinguished only by the parameter referred to as color. These nine mathematical entities all obey Fermi–Dirac statistics, but real particles are required to be singlets with respect to the SU_3 of color, that is to say combinations acting like

$$\bar{q}_R q_R + \bar{q}_B q_B + \bar{q}_W q_W \text{ or } q_R q_B q_W - q_B q_R q_W - q_R q_W q_B - q_W q_B q_R + q_W q_R q_B + q_B q_W q_R \,. \quad (1)$$

The assumption of quark statistics has been common for many years, although not necessarily described in quite this way, and it has always had the following advantage: The constituent quarks as well as current quarks would obey quark statistics, since the transformation between them would not affect statistics, and the constituent quark model would then assign the lowest-lying baryon states (56 representation) to a symmetrical spatial configuration, as befits a very simple model.

Nowadays there is a further advantage. Using the algebraic relations abstracted formally from the quark–gluon model, one obtains a formula for the π^0 decay amplitude in the PCAC approximation, one that works beautifully for quark statistics but would fail by a factor 3 for a single Fermi–Dirac triplet[4)].

We have the option, no matter how far we go in abstracting results from a field theory model, of treating only color singlet operators. All the currents, as well as the stress–energy–momentum tensor $\Theta_{\mu\nu}$ that couples to gravity and defines the theory, are color singlets. We may, if we like, go further and abstract operators with three quark fields, or four quark fields and an antiquark field, and so forth, in order to connect the vacuum with baryon states, but we still need select only those that are color singlets in order to connect all physical hadron states with one another.

It might be a convenience to abstract quark operators themselves, or other non–singlets with respect to color, along with fictitious sectors of Hilbert space with triality non–zero, but it is not a necessity. It may not even be much of a convenience, since we would then, in describing the spatial and temporal variation of these fields, be discussing a fictitious spectrum for each fictitious sector of Hilbert space, and we probably don't want to load ourselves with so much spurious information.

We might eventually abstract from the quark–vector–gluon field theory model enough algebraic information about the color singlet operators in the model to describe all the degrees of freedom that are present.

For the real world of baryons and mesons, there must be a similar algebraic system, which may differ in some respects from that of the model, but which is in principle knowable. The operator $\Theta_{\mu\nu}$ could then be expressed in terms of this system, and the complete Hilbert space of baryons and mesons would be a representation of it. We would have a complete theory of the hadrons and their currents, and we need never mention any operators other than color singlets.

Now the interesting question has been raised lately whether we should regard the gluons as well as the quarks as being non–singlets with respect to color[5)]. For example, they could form a color octet of neutral vector fields obeying the Yang–Mills equations. (We must, of course, consider whether it is practical to add a common mass term for the gluon in that case – such a mass term would show up physically as a term in $\Theta_{\mu\nu}$ other than the quark bare mass term. In the past, we have referred to such an additional term that violates scale invariance, but does not violate $SU_3 \times SU_3$ as δ and its dimension as l_δ. Nowadays, ways of detecting expected values of δ are emerging.)[6)].

If the gluons of the model are to be turned into color octets, then an annoying asymmetry between quarks and gluons is removed, namely that there is no physical channel with quark quantum numbers, while gluons communicate freely with the channel containing the ω and ϕ mesons. (In fact, this communication of an elementary gluon potential with the real current of baryon number makes it very difficult to believe that all the formal relations of light cone current algebra could be true even in a "finite" version of singlet neutral vector gluon field theory.)

If the gluons become a color octet, then we do not have to deal with a gluon field strength standing alone, only with its square, summed over the octet, and with quantities like $\bar{q}\,(\partial_\mu - ig\sigma_A B_{A\mu})\,q$, where the σ's are the eight 3×3 color matrices for the quark and the B's are the eight gluon potentials.

Now, suppose we look at such a model field theory, with colored quarks and colored gluons, including the stress–energy–momentum tensor. Basically the questions we are asking are the following:

1. Up to what point does the algebraic system of the color singlet operators for the real hadrons resemble that in the model? What is it in fact?

2. Up to what point does the representation of the algebraic system by the Hilbert space of physical hadron states resemble that in the model? What is it in fact?

3. Up to what point does $\Theta_{\mu\nu}$, expressed in term of the algebraic system, resemble that in the model? What is it in fact?

The measure of our ignorance is that for all we know, the algebra of color singlet operators, the representation, and even the form of $\Theta_{\mu\nu}$ could be exactly as in the model! We don't yet know how to extract enough consequences of the model to have a decisive confrontation with experiment, nor can we solve the formal equations for large g.
If we were solving the equations of a model, the first question we would ask is: Are the quarks really kept inside or do they escape to infinity? By restricting physical states and interesting operators to color singlets only, we have to some extent begged that question. But it re-emerges in the following form:
With a given algebraic system for the color singlet operators, can we find a locally causal $\Theta_{\mu\nu}$ that yields a spectrum corresponding to mesons and baryons and antibaryons and combinations thereof, or do we find a spectrum (in the color singlet states) that looks like combinations of free quarks and antiquarks and gluons?
In the next three Sections we shall usually treat the vector gluon, for convenience, as a color singlet.

III. REVIEW OF CURRENT ALGEBRA

In this section we sketch the gradual extension of algebraic results abstracted from free quark theory that remain true, either in renormalized perturbation theory or else only formally, when the coupling to a neutral vector gluon field is turned on.
The earlier abstractions were of equal–time commutation relations of current components. It was soon found that useful sum rules could best be derived from these by taking matrix elements between hadron states of equal P_3 as $P_3 \to \infty$, selecting the "gluon" components of the currents (those with matrix elements finite in this limit rather than tending to zero), and adding the postulate that, in the sum over intermediate states in the commutator, only states of finite mass need be considered. Thus formulae like the Adler–Weisberger and Cabibbo–Radicati sum rules were obtained and roughly verified by experiment.
Nowadays, the same procedure is usually accomplished in a slightly different way that is a bit cleaner – the hadron momenta are left finite instead of being boosted by a limit of Lorentz transformations, and the equal time surface is transformed by a corresponding limit of Lorentz transformations into a null plane, with $x_3 + x_0 =$ constant, say zero. The hypothesis of saturation by finite mass intermediate states is replaced by the hypothesis that the commutation rules of good components can be abstracted from the model not only on an equal time plane, but on a null plane as well[7,8].
In the last few years, the process of abstraction has been extended to a large class of algebraic relations (those of "light cone current algebra") that are true only formally in the model, but fail to each order of renormalized perturbation theory - they would be true to each order if the model were super–renormalizable. The motivation has been supplied by the compatibility of the deep inelastic electron scattering experiments performed at SLAC with the scaling predictions of Bjorken, which is the most basic feature of "light

cone current algebra". The Bjorken scaling limit $q^2 \to \infty, 2p \cdot q \to \infty, \xi \equiv q^2/(-2p \cdot q)$ finite) corresponds in coordinate space to the singularity on the light cone $(x-y)^2 = 0$ of the current commutator $[j(x), j(y)]$, and the relations of light cone current algebra are obtained by abstracting the leading singularity on the light cone from the field theory model. The singular function of $x - y$ is multiplied by a bilocal current operator $\Theta(x, y)$ that reduces to a familiar local current as $x - y \to 0$. The Bjorken scaling functions $F(\xi)$ are Fourier transforms of the expected values of the bilocal operators. Numerous predictions emerge from the relations abstracted from the quark–gluon model for deep inelastic and neutrino cross–sections. For example, the spin 1/2 character for the quanta bearing the charge in the model is reflected in the prediction $\sigma_L/\sigma_T \to 0$, while the charges of the quarks are reflected in the inequalities $1/4sF^{\mathrm{en}}(\xi)/F^{\mathrm{ep}}(\xi) \leq 4$. So far there is no clear sign of my contradiction between the formulae and the experimental results.
We may go further and abstract from the model also the light–cone commutators of bilocal currents, in the limit in which all the intervals among the four points approach zero, that is to say, when all four points tend to lie on a light–like line. The same bilocal operators then recur as coefficients of the singularity, and the algebraic system closes.
The light cone results can be reformulated in terms of the null plane. We consider a commutator of local currents at two points x and y and allow the two points to approach the same null plane, say

$$x_+ \equiv x_3 + x_0 = 0, y_+ \equiv y_3 + y_0 = 0 \tag{2}$$

As mentioned above, when both current components are "good", we obtain a local commutation relation on the null plane, yielding another good component, or else zero. But when neither component is good, there is a singularity of the form

$$\delta(x_+ - y_+) \tag{3}$$

and the coefficient is a bilocal current on the null plane. It is this singularity, arising from the light–cone singularity, that gives the Bjorken scaling.
On the null plane, with $x_+ = 0$, the three coordinates are the transverse spacelike coordinates x_1 and x_2 (called $x_\perp$) and the lightlike coordinate $x_- \equiv x_3 - x_0$. Our bilocal currents $O(u, y)$ on the nullplane are functions of four coordinates: x_-, y_- and $x_\perp = y_\perp$, since the interval between x and y is lightlike.
We may now consider the commutator of two bilocal currents defined on neighboring null planes (in each case with a lightlike interval between the two arguments of the bilocal current). Again, when neither current component is good, there is a δ–function singularity of the spacing between the two null planes and the coefficient is a bilocal current defined on the common limiting null plane. In this language, as before in the light cone language, the system of bilocal currents closes.
We may commute two good components of bilocal currents on the same null plane, and,

as for local currents, we obtain a good component on the right–hand side, without any δ–function singularity at coincidence of the two null planes. Thus the good components of the bilocal currents $O(u, y)$ form a Lie algebra on the null plane, a generalization of the old Lie algebra of local good components on the null plane (recovered by putting $x_- = y_-$).

Now, how far can we generalize this new Lie algebra on the null plane and still obtain exact formulae, formally true to all orders in the coupling constant, but independent of it, so that free quark formulae apply?

In the next section, we take up that question, but first we summarize the situation of current algebra on and near the null plane.

IV. SUMMARY OF LIGHT CONE AND NULL PLANE RESULTS

Let us now be a little more explicit. We are dealing with 144 bilocal quantities $\mathcal{F}_{j\alpha}, \mathcal{F}_{j\alpha}, S_j, F$ and $T_{j\alpha\beta}$ all functions of $x - y$ with $(x-y)^2 \to 0$. Let us select the 3–direction for our null planes. Then in the model we can set $B_+ \equiv B_3 + B_0 = 0$ for the gluon potential by a choice of gauge. The gauge–invariance factor $\exp ig \int_y^x B \cdot dl$ for a straight line path on a null plane is just $\exp\left[i\frac{g}{2}B_+(x_- - y_-)\right] = 1$. Thus we have simple correspondences between our quantities and operators in the model:

$$\mathcal{F}_{j\alpha}(x, y) \sim \frac{i}{2}\bar{q}(x)\lambda_j\gamma_\alpha q(y), \text{ etc.}$$

and we have introduced the notation $\mathcal{D}\left(x, y, \frac{i}{2}\lambda_j\gamma_\alpha\right)$, etc., where

$$\mathcal{D}(x, y, G) \sim \bar{q}(x)Gq(y) \sim q^+(x)(\beta G)q(y)\,. \tag{4}$$

We are dealing with $\mathcal{D}(x, y, G)$ for every (12×12) matrix G, with

$$\mathcal{F}^5_{j\alpha}(x, y) = \mathcal{D}\left(x, y, \frac{i}{2}\lambda_j\gamma_\alpha, \gamma_5\right) S_j(x, y) = \mathcal{D}\left(x, y, \frac{1}{2}\lambda_j\right), \tag{5}$$

$$P_j(x, y) = \mathcal{D}\left(x, y, \frac{i}{2}\lambda_j\gamma_5\right), \text{ and } T_{j\alpha\beta}(x, y) = \mathcal{D}\left(x, y, \frac{i}{2}\lambda_j\sigma_{\alpha\beta}\right). \tag{6}$$

The good components, in the old equal–time $P_3 \to \infty$ language, were those with finite matrix elements between states of finite mass and $P_3 \to \infty$. By contrast, bad components were those with matrix elements going like P_3^{-1} and terrible components those with matrix elements going like P_3^{-2}.

In the null plane language, good components are those for which βG is proportional to $1+\alpha_3$; thus the 36 good components are $\mathcal{F}_{j+}, \mathcal{F}^5_{j+}, \mathcal{T}_{j1+}, \mathcal{T}_{j2+}$ for $j = 0\ldots 8$. The terrible components are those for which βG is proportional to $1-\alpha_3$, hence $\mathcal{F}_{j-}, \mathcal{F}^5_{j-}, \mathcal{T}_{j1}$, and $\mathcal{T}_{j2-}$. The rest are bad; they have βG anticommuting with α_3 so that α_3 is -1 on the left and +1 on the right or vice versa.
Now the leading light cone singularity in the commutator of two bilocals is just given by the formula

$$[(\mathcal{D}(x,y,G), \mathcal{D}(u,v,G'))] \hat{=} \mathcal{D}(x,v,iG\gamma_\mu G')\,\partial_\mu\Delta(y-u) - \mathcal{D}(u,y,iG'\gamma_\mu G)\,\partial_\mu\Delta(v-x), \tag{7}$$

with $\Delta(z) = (2\pi)^{-1}\,\varepsilon(z_0)\,\delta(z^2)$.

When we commute two operators with coordinates lying on neighboring null planes with separation Δx_+, a singularity of the type $\delta(\Delta x_+)$ appears (as we have mentioned in Section III) multiplied by a bilocal operator, with coordinates lying in the common null plane as $\Delta x_+ \to 0$, and it is this term that gives rise to Bjorken scaling. The term in question comes from the component $\frac{\partial}{\partial z_+}\Delta(z)$ in $\partial_\mu\Delta(z)$, and is thus multiplied by $\mathcal{D}(x,\nu,iG\gamma_+G')$ and $\mathcal{D}(u,y,iG'\gamma_+G)$. Now $\beta(iG\gamma_+G') = (\beta G)(1-\alpha_3)(\beta G')$, so it is clear that the singular Bjorken scaling term vanishes for good–good and good–bad commutators. In the case of the other components, we have, schematically, [bad, bad] $\to$ good, [bad, terrible] $\to$ bad, and [terrible, terrible] $\to$ terrible for the Bjorken singularity.
The vector and axial vector local currents $\mathcal{F}_{j\alpha}(x,x)$ and $\mathcal{F}^5_{j\alpha}(x,x)$ occur, of course, in the electromagnetic and weak interactions. The local scalar and pseudoscalar currents occur in the divergences of the non–conserved vector and the axial vector currents, with coefficients that are linear combinations of the bare quark masses, m_u, m_d and m_s, treated as a diagonal matrix. (Here m_u would equal m_d if isotopic spin conservation were perfect, while the departure of m_s from the common value of m_u and m_d is what gives rise to SU_3 splitting; the non–vanishing of m is what breaks $SU_3 \times SU_3$).
We see that all the 144 bilocals are physically interesting, including the tensor currents, because they all occur in the commutators of these local V, A, S, and P densities as coefficients of the $\delta(\Delta x_+)$ singularity. Commuting a local scalar with itself or a local pseudoscalar with itself leads to the same bilocal as commuting a transverse component of a vector with itself, and thus the light cone commutator of current divergences is predicted to lead to Bjorken scaling functions that are proportional to those observed in the light cone commutation of currents, while the coefficients permit the experimental determination of the squares of the quark bare masses. Unfortunately, the relevant experiments are difficult. (The finiteness of the bare masses, as compared with the divergences encountered term in renormalized perturbation theory in a gluon model, presumably has the same origin as the scaling, which also fails term by term in renormalized perturbation theory.)

As we have outlined in Section III, we begin the construction of the algebraic system on the null plane by commuting the good bilocals with one another. The leading singularity on the light cone (Eq.(4.1)) gives rise to the simple closed algebra we have mentioned, but we need also the additional assumption that lower singularities on the light cone give no contribution to the good–good commutators on the null plane. This additional assumption can be squeezed out of the model in various ways. The simplest, however, is to use canonical quantization of the quark–gluon model on the null plane.
In the model, the quark field q is written as $q_+ + q_-$, where $q_\pm = \frac{1}{2}\left(1 \pm \alpha_3\right) q$. Then q_+ obeys the canonical rules $\{q_{+\alpha}(x), q_{+\beta}(y)\} = 0, \left\{q_{+\alpha}(x).q^+_{+\beta}(y)\right\} = \delta^{(3)}(x-y)\frac{1}{2}\left(l+a_3\right)_{\alpha\beta}$ on the null plane, where $\delta^{(3)}(x-y) = \delta\left(x_\perp - y_\perp\right)\delta\left(x_- - y_-\right)$. Thus for any good matrices βA_{++} and (βB_{++}), we have on the null plane

$$\left[\mathcal{D}\left(x, y, \beta A_{++}\right), \mathcal{D}\left(u, v, \beta B_{++}\right)\right] =$$

$$\mathcal{D}\left(x, v\beta A_{++}B_{++}\right)\delta^{(3)}(y-u) - \mathcal{D}\left(u, y, \beta B_{++}A_{++}\right)\delta^{(3)}(v-x),$$

which is just what we would get from (4.1) with no additional contribution from lower light cone singularities.
The good–good commutation relations (4.2) on the null plane, together with the equations (4.1) for the leading light cone singularity in the commutator of two bilocal currents, illustrate how far we can go with abstracting free quark formulae that are formally unchanged in the model when the gluon coupling is turned on.
One may go further in certain directions. For example, the formulae for the leading light cone singularity presumably apply to disconnected as well as connected parts of matrix elements, and thus the question of the vacuum expected value of a bilocal operator arises. In the model, the coefficient of the leading singularity as $(x-y)^2 \to 0$ of such an expected value is formally independent of the coupling constant, and we abstract that as well – the answer here is dependent on statistics, however, and we assume the validity of quark statistics. Thus we obtain predictions like the following:

$$\sigma\left(e^+ + e^- \to \ \text{hadrons}\right)/\sigma\left(e^+ + e^- \to \mu^+ + \mu^-\right) \to 2 \tag{8}$$

at high energy to lowest order in the fine structure constant.

The leading light cone singularity of an operator product, or of a physical order (T^*) product, may also be abstracted from the model, except for certain subtraction terms (often calculable and / or unimportant) that behave like four-dimensional δ-functions in coordinate space. To go from a commutator formula to a physical ordered product formula, we simply perform the substitutions

$$(2\pi)^{-1}\varepsilon(z)\delta\left(z^2\right) \to \left(4\pi^2 i\right)^{-1}\left(z^2 - iz_0\varepsilon\right)^{-1} \to \left(4\pi^2 i\right)^{-1}\left(z^2 - i\varepsilon\right)^{-1}. \tag{9}$$

With the aid of the product formulae and the vacuum expected values, we obtain the PCAC value of the $\pi^0 \to 2\gamma$ decay amplitude.

Other exact abstractions from the vector gluon model that do not contain g are divergence and curl relations for local V and A currents:

$$\begin{aligned} \frac{\partial}{\partial x_\mu}\mathcal{D}\left(x, x, \frac{i}{2}\lambda_i\,\gamma_\mu\right) &= \mathcal{D}\left(x, x, \frac{i}{2}\left[m, \lambda_i\right]\right), \\ \frac{\partial}{\partial x_\mu}\mathcal{D}\left(x, x, \frac{i}{2}\lambda_i\gamma_\mu\gamma_5\right) &= \mathcal{D}\left(x, x, \frac{i}{2}\left\{m, \lambda_i\right\}\gamma_5\right), \end{aligned} \tag{10}$$

but we also have, as presented elsewhere[2)],

$$\begin{aligned} \frac{\partial}{\partial x_\nu}\mathcal{D}\left(x, x, \frac{1}{2}\lambda_i\,\sigma_{\mu\nu}\right) &= -\mathcal{D}\left(x, x, \frac{i}{2}\left\{m, \lambda_i\right\}\gamma_\nu\right) \\ &\quad + \left[\left(\frac{\partial}{\partial x_\nu} - \frac{\partial}{\partial y_\nu}\right)\mathcal{D}\left(x, y, \frac{i}{2}\lambda_i\right)\right]_{x=y} \end{aligned} \tag{11}$$

$$\begin{aligned} \frac{\partial}{\partial x_\nu}\mathcal{D}\left(x, x, \frac{1}{2}\lambda_i\sigma_{\mu\nu}\gamma_5\right) &= -\mathcal{D}\left(x, x, \frac{i}{2}\right)\left[m, \lambda_i\right]\gamma_\nu\gamma_5 \\ &\quad + \left[\left(\frac{\partial}{\partial x_\nu} - \frac{\partial}{\partial y_\nu}\right)\mathcal{D}\left(x, y, \frac{i}{2}\lambda_i\gamma_5\right)\right] \end{aligned} \tag{12}$$

and a number of other formulae, including the following:

$$\left[\left(\frac{\partial}{\partial x_\nu} - \frac{\partial}{\partial y_\nu}\right)\mathcal{D}\left(x, y, \frac{i}{2}\lambda_i\gamma_\nu\right)\right]_{x=y} = \mathcal{D}\left(x, x, \frac{i}{2}\left\{\lambda_i, m\right\}\right) \tag{13}$$

In the last three formulae, it must be pointed out that for a general direction of $x - y$ we have the gauge–invariant correspondence

$$\mathcal{D}\left(x, y, G\right) \sim \bar{q}(x) G q(y) \exp \mathrm{ig} \int_y^x B \cdot dl, \tag{14}$$

which is independent of the path from y to x when the coordinate difference and the path are taken as first order infinitesimals. The first internal derivative

$$\left[\left(\frac{\partial}{\partial x_\mu} - \frac{\partial}{\partial y_\mu}\right)\mathcal{D}(x, y, G)\right]_{x=y} \tag{15}$$

is physically interesting for all directions μ (and not just the $-$ direction), as a result of Lorentz covariance.
In Eqs. (4.5–4.7), we have for the moment thrown off the restriction to a single null plane. In the next Section, we return to the consideration of the algebra on the null plane, and we see how further extensions give a much wider algebra, in which departures from free quark relations begin to appear.

V. THE FURTHER EXTENSION OF NULL PLANE ALGEBRA

We now look beyond the commutation relations of good bilocals on the null plane. In the model, then, we have to examine operators containing q_- or q_-^+ or both. The Dirac equation in the gauge we are using ($B_+ = 0$ on the null plane) tells us that we have

$$-2i\frac{\partial q_-}{\partial x_-} = (\alpha_\perp \cdot (-i\nabla_\perp - gB_\perp) + \beta m)\, q_+. \tag{16}$$

In terms of Eq. (5.1), we can review the various anticommutators on the null plane. We have already discussed the trivial one,

$$\left(q_+(x), q_+^\dagger(y)\right) = \delta(x_- - y_-) . \frac{1}{2}(1+\alpha_3)\,\delta(x_\perp - y_\perp) . \tag{17}$$

Using (5.1), (5.2), the fact that $B_\perp$ commutes with q_+ on the null plane, and the equal–time anticommutator $\left\{q_-, q_+^\dagger\right\} = 0$, we obtain well–known result

$$\left\{q_-(x), q_+^\dagger(y)\right\} = \frac{i}{4}\varepsilon(x_- - y_-)\left[\alpha_\perp \cdot \left(i\nabla_\perp^{(y)} - gB_\perp(y)\right) + \beta m\right]\frac{1}{2}(1+\alpha_3)\,\delta(x_\perp - y_\perp) . \tag{18}$$

Using the same method a second time, one finds, for $y_- > x_-$,

$$\begin{aligned}\left\{q_-(x), q_-^\dagger(y)\right\} = &-\frac{1}{8}\int_{x_-}^{y_-} dr_-\left[\alpha_\perp\left(-i\nabla_\perp^{(x)} - gB_\perp(x_\perp, r_-)\right) + \beta m\right]^2\left(\frac{1-\alpha_3}{2}\right)\delta(x_\perp - y_\perp)\\ &+i\frac{g^2}{32}\int_{x_-}^{y_-} dy'_-\int_{x_-}^{y'_-} dx'_-\left[\alpha_\perp q_+\left(x_\perp, x'_-\right); q_+\left(y_\perp, y'_-\right)\alpha_\perp\right]\delta(x_\perp - y_\perp)\\ &+\delta(x_+ - y_+)\left(\frac{1-\alpha_3}{2}\right)\delta(x_\perp - y_\perp),\end{aligned} \tag{19}$$

where the singularity at the coincidence of the two null planes appears as an unpleasant integration constant. This singularity is, of course, responsible in the model for the Bjorken singularity in the commutator of two bad or terrible operators.
Because of the singularity, it is clumsy to construct the wider algebra by commuting all

our bilocals with one another. Instead, we adopt the following procedure. Whenever a bilocal operator corresponds to one in the model containing $q_-^+(x)$, we differentiate with respect to x_-; whenever it corresponds to one in the model containing $q_{(y)}$, we differentiate with respect to y_-. Thus we "promote" all our bilocals to good operators. We construct the wider algebra by starting with the original good bilocals and these promoted bad and terrible bilocals. We commute all of these, commute their commutators, and so forth, until the algebra closes. Then, later on, if we want to commute an unpromoted operator, we use the information contained in equations of the model like (5.1) - (5.3) to integrate over x_- or y_- or both and undo the promotion. (A similar situation obtains for operators corresponding to those in the model containing the longitudinal gluon potential B_-.)
Now let us classify the matrices βG into four categories:
the good ones, $\beta G = A_{++}$, with $\alpha_3 = 1$ on both sides;
the bad ones $\beta G = A_{+-}$ that have $\alpha_3 = 1$ on the left and -1 on the right;
the bad ones $\beta G = A_{-+}$ that have $\alpha_3 = -1$ on the left and $+1$ on the right;
and the terrible ones $\beta G = A_{--}$, with $\alpha_3 = -1$ on both sides.

Then, wherever q_- or q_-^+ appears, we promote the operator by differentiating q_- or q_-^+ with respect to its argument in the – direction. We obtain, then:

$$\mathcal{D}\,(x, y, \beta A_{++})\,,$$

the good operators, unchanged;

$\frac{\partial}{\partial x_-}\mathcal{D}\,(x, y, \beta A_{-+})$ and $\frac{\partial}{\partial y_-}\,(x, y, \beta, A_{+-})$ promoted bad operators:

and

$\frac{\partial}{\partial x_-}\frac{\partial}{\partial y_-}\mathcal{D}\,(x, y, \beta A_{--})$, promoted terrible operators.

All 144 of these operators now are given, in the model, in terms of q_+ and q_+^+, but the promoted bad and terrible operators involve the expressions $(\nabla_\perp - igB_\perp)\,q_+$ and $(\nabla_\perp + igB_\perp)\,q_+^+$. In fact, substituting the Dirac equation for $\frac{\partial q_-}{\partial x_-}$ into the definitions of the promoted bad and terrible operators, we see that we obtain good operators (with coefficients depending on bare quark masses) and also good matrices sandwiched between $(\nabla_\perp + igB_\perp)\,q_+^+$ and q_+ or between q_+^+ and $(\nabla_\perp - igB_\perp)\,q_+$ or between $(\nabla_\perp + igB_\perp)\,q_+^+$ and $(\nabla_\perp - igB_\perp)\,q_+$.
The null plane commutators of all these operators with one another are finite, well–defined, and physically meaningful, but they lead to an enormous Lie algebra that is not identical with the one for free quarks, but instead contains nearly all the degrees of freedom of the model.
Let us first ignore any lack of commutation of the B's with one another. We keep commuting the operators in question with one another. When $\nabla_\perp \pm igB_\perp$ appears acting on a $\delta^{(3)}$

function, we can always perform an integration and fold it over onto an operator. Thus the number of applications of $\nabla_\perp \pm igB_\perp$ grows without limit. Since these gauge derivatives do not commute with one another, but give field strengths as commutators, it can easily be seen that we end up with all possible operators corresponding to $\bar{q}_+(x)Gq_+(y)$ acted on by any gauge invariant combination of transverse gradients and potentials. We have to put it differently, the operators corresponding to $\bar{q}_+(x)Gq_+(y) \exp ig \int_P B \cdot dl$ for any pair of points x and y on the null plane connected by any path P lying in the null plane. We could think of these as operators $\mathcal{D}(x, y, G, P)$ depending on the path P, with $\beta G = A_{++}$.

In fact the B's do not commute with another in the model, and so we get an even more complicated result. We have

$$[B_{\perp i}(x), B_{\perp j}(y)] \sim \varepsilon\,(x_- - y_-)\,\delta\,(x_\perp - y_\perp)\,\delta_{ij} \tag{20}$$

on the null plane, and the commutation of promoted bad and terrible bilocals with one another leads to operators corresponding to $\bar{q}_+(x)Gq_+(y)\bar{q}_+(a)G'q_+(b)$. Further commutation then introduces an unlimited number of sideways gradients, gluon field strengths, and additional quark pairs, until we end up with all possible operators of the model that can be constructed from equal numbers of $\bar{q}_+$'s and q_+'s at any points on the null plane and from exponentials of $ig \int B \cdot dl$ for any paths connecting these points.

If we keep track of color, we note that only color singlets are generated. If the gluons are a color octet Yang–Mills field, we must make suitable changes in the formalism but again we find that only color singlets are generated. The coupling constant g that occurs is, of course, the bare coupling constant. If may not be intrinsic to the algebraic system (equivalent to that of quarks and gluons) on the null plane, but it certainly enters importantly into the way we reach the system starting from well–known operators.

A troublesome feature of the extended null plane algebra is the apparent absence of operators corresponding to those in the model that contain only gluon field strengths and no quark operators; for a color singlet gluon, the field strength itself would be such an operator, while for a color octet gluon we could begin with bilinear forms in the field strength in order to obtain color singlet operators. Can we obtain these quark–free operators by investigating discontinuities at the coincidence of coordinates characterizing quark and antiquark fields in the model? At any rate, we certainly want these quarkfree operators included in the extended algebra.

Now when our algebra has been extended to include the analogs of all relevant operators of the model on the null plane that are color singlets and have baryon number $A = 0$, then the Hilbert space of all physical hadron states with $A = 0$ is an irreducible representation of the algebra.

If we wish, we might as well extend the algebra further by including the analogs of color singlet operators of the model (on the null plane) that would change the number of baryons. In that case, the entire Hilbert space of all hadron states is an irreducible representation of the complete algebra. From now on, let us suppose that we are always dealing with the complete color singlet algebra (whether the one abstracted from the quark–gluon model or some other) and with the complete Hilbert space, which is an irreducible representation of it.

The representation may be determined by the algebra and the uniqueness of the physical vacuum. We note that we are dealing with arbitrarily multilocal operators, functions of any number of points on the null plane. We can Fourier transform with respect to all these variables and obtain Fourier variables $(k_+, k_\perp)$ in place of the space coordinates. Since $B_+ = 0$, there is no formal obstacle to thinking of each k_+ as being like the contribution of the individual quark, antiquark or gluon to the total $P_+ = \sum k_+$. Now $P_+ = 0$ for the vacuum, and for any other state we can get $P_+ = 0$ only by taking $P_z \to -\infty$. The same kind of smoothness assumption that allows scaling can allow us to forget about matrix elements to such infinite momentum states. In that case, we have the unique vacuum state of hadrons as the only state of $P_+ = 0$, while all others have $P_+ > 0$. All Fourier components of multilocal operators for which $\sum k_+ < 0$ annihilate the physical vacuum. (Note in the null plane formalism we do not have to deal with a fictitious "free vacuum" as in the equal–time formalism.) The Fourier components of multilocal operator with $\sum k_+ > 0$ act on the vacuum to create physical states, and the orthogonality properties of these states and the matrix elements of our operators sandwiched between them are determined largely or wholly by the algebra. The details have to be studied further to see to what extent the representation is really determined. (The vacuum expected values contain one adjustable parameter in the case of free quarks, namely the number of colors.) Once we have the representation of the complete color singlet algebra on the null plane, as well as the algebra itself, then the physical states of hadrons can all be written as linear combinations of the normalized basis states of the representation. These coefficients represent a normalized set of Fock space wave functions for each physical hadron state, with orthogonality relations for orthogonal physical states. Since the matrix elements of all null plane operators between basis states are known, the matrix elements between physical states of bilocal currents or other operators of interest are all calculable in terms of the Fock space wave functions[9)].

This situation is evidently the one contemplated by "parton" theorists such as Feynman and Bjorken; they suppose that we know the complete algebra, that it comes out to be a quark–gluon algebra, and that the representation is the familiar one, so that there is a simple Fock space of quark, antiquark, and gluon coordinates. In the Fourier transform, negative values of each k_+ correspond to destruction and positive values to creation.

Now the listing of hadron states by quark and gluon momenta is a long way from listing by meson and baryon moments. However, as long as we stick to color singlets, there is not necessarily any obstacle to getting one from the other by taking linear combinations. The operator $M^2 = -P^2 - P_+P_-$ has to be such that its eigenvalues correspond to meson and baryon configurations, and not to a continuum of quarks, antiquarks and gluons.

The important physical questions are whether we have the correct complete algebra and

representation, and what the correct form of $\Theta_{\mu\nu}$ or P_μ or M^2 is, expressed in terms of that algebra.
In the quark–gluon model we have $\Theta_{\mu\nu} = \Theta_{\mu\nu}^{\text{quark}} + \Theta_{\mu\nu}^{\text{glue}}$, where

$$\begin{aligned}\Theta_{\mu\nu}^{\text{quark}} &= \frac{1}{4}\bar{q}\gamma_\mu\left(\partial_\nu - igB_\nu\right)q + \ldots q + \frac{1}{4}\bar{q}\gamma_\nu\left(\partial_\mu - igB_\mu\right)q \\ &\quad -\frac{1}{4}\left(\partial_\mu + igB_\mu\right)\bar{q}\gamma_\nu q - \frac{1}{4}\left(\partial_\nu + igB_\nu\right)\bar{q}\gamma_\mu q\,, \end{aligned} \tag{21}$$

and $\Theta_{\mu\nu}^{\text{glue}}$ does not involve the quark variables at all. The term $\Theta_{\mu\nu}^{\text{quark}}$, by itself, has the wrong commutation rules to be a true $\Theta_{\mu\nu}$ (unless $g = 0$). For example, $\left(P_1^{\text{quark}}, P_2^{\text{quark}}\right) \neq 0$. The correct commutation rules are restored when we add the contribution from $\Theta_{\mu\nu}^{\text{glue}}$.
We can abstract from the quark–gluon model some or all the properties of $\Theta_{\mu\nu}$, in terms of the null plane algebra. We see that in the model we have

$$\Theta_{++}^{\text{quark}} = \left[\left(\frac{\partial}{\partial y_-} - \frac{\partial}{\partial x_-}\right)\mathcal{D}\left(x, y, \frac{1}{2}\gamma_+\right)\right]_{x=y} \tag{22}$$

and, as is well–known, the expected value of the right–hand side in the proton state can be measured by deep inelastic experiments with electrons and neutrinos. All indications are that it is not equal to the expected value of Θ_{++}, but rather around half of that, so that half is attributable to gluons, or whatever replaces them in the real theory.

In general, using the gauge–invariant definition of $\mathcal{D}$, we have in the model

$$\Theta_{\mu\nu}^{\text{quark}} = \left[\left(\frac{\partial}{\partial y_\nu} - \frac{\partial}{\partial x_\nu}\right)\mathcal{D}\left(x, y, \frac{1}{4}\gamma_\mu\right) + \left(\frac{\partial}{\partial y_\mu} - \frac{\partial}{\partial x_\mu}\right)\mathcal{D}\left(x, y, \frac{1}{4}\gamma_\nu\right)\right]_{x=y} \tag{23}$$

and Eq. (4.7) then gives us the obvious result

$$-\,\Theta_{\mu\nu}^{\text{quark}} = \mathcal{D}\left(x, x, m\right)\,. \tag{24}$$

Whereas in (5.5) we are dealing with an operator that belongs to the null plane algebra generated by good, promoted bad, and promoted terrible bilocal currents, other components of $\Theta_{\mu\nu}^{\text{quark}}$ are not directly contained in the algebra, neither are the bad and terrible local currents, nor their internal derivatives in directions other than $-$. In order to obtain the commutation properties of all these operators with those actually in the algebra, we must, as we mentioned above, undo the promotions by abstracting the sort of information contained in (5.3) and (5.4). Thus we are really dealing with a wider mathematical system than the closed Lie algebra abstracted from that of operators in the model containing $q_+^\dagger, q_+$ and $B_\perp$ only.

We shall assume that the true algebraic system of hadrons resembles that of the quark–gluon model at least to the following extent:

1) The null plane algebra of good components (4.2) and the leading light cone singularities (4.1) are unchanged.

2) The system acts as if the quarks had vectorial coupling in the sense that the divergence equation (4.3) and (4.4) are unchanged.

3) There is a gauge derivative of some kind, with path–dependent bilocals that for an infinitesimal interval become path–independent. Eqs. (4.5) - (4.7) are then defined and we assume they also are unchanged.

4) The expression (5.6) for $\Theta_{\mu\nu}^{\text{quark}}$ is also defined and we assume it, too, is unchanged, along with its corollary (5.7).

About the details of the form of the path–dependent null plane algebra arising from the successive application of gauge derivatives, we are much less confident, and correspondingly we are also less confident of the nature of the gluons, even assuming that we can decide whether to use a color singlet or a color octet. What we do assert is that there is some algebraic structure analogous to that in quark–gluon theory and that it is in principle knowable.

One fascinating problem, of course, is to understand the conditions under which we can have an algebra resembling that for quarks and gluons and yet escape having real quarks and gluons. Under what conditions do the bilocals act as if they were the products of local operators without, in fact, being seen. We seek answers to this and other questions by asking "Are there models other than the quark–gluon field theory from which we can abstract results? Can we replace $\Theta_{\mu\nu}^{\text{glue}}$ by something different and the gauge–derivative by a different gauge–derivative?"

VI. ARE THERE ALTERNATIVE MODELS?

In the search for alternatives to gluons, one case worth investigating is that of the simple dual resonance model. It can be considered in three stages: first, the theory of a huge infinity of free mesons of all spins; next, tree diagrams involving the interaction of these mesons; and finally loop diagrams. The theory is always treated as though referring to real mesons, and an S–matrix formulation is employed in which each meson is always on the mass shell.

Now the free stage of the model can easily be reformulated as a field theory in ordinary coordinate space, based on a field operator Φ that is a function not of one point in space, but of a whole path – it is infinitely multilocal. The free approximation to the dual resonance model is then essentially the quantum theory of a relativistic string or linear rubber band in ordinary space.

The coupling that leads, on the mass shell, to the tree diagrams of the dual resonance model has not so far been successfully reformulated as a field theory coupling but we shall assume that this can be done. Then the whole model theory, including the loops, would be a theory of a large infinity of local meson fields, all described simultaneously by a grand infinitely multilocal field Φ, couples to themselves and one another. The mesons, in the free approximation, lie on straight parallel Regge trajectories with a universal slope α'.

In the simplest form of such a theory, the grand field Φ (path) can be resolved into local fields $\phi(R), \Phi_{n\mu}(R), \Phi_{n\mu,n'\mu'}(R), \ldots$. There is a single scalar, a single infinity of vectors, a double infinity of tensors and scalars, and so forth. The matrices $a_{n\mu}$ and $a^+_{n\mu}$ of the dual theory connect these components of Φ with one another.

Perhaps the model theory of a gluon field can be replaced by a field theory version of a dual resonance model; the properties of operators, including $\Theta_{\mu\nu}$, would be abstracted from the new model instead of the old one. With $\alpha' \neq 0$, a term δ would naturally appear that violates scale invariance and is not related to the bare quark masses. (Probably $l_\delta = 0$ here rather than -2 as in the case of a gluon mass.) The gauge derivative in the other portion of $\Theta_{\mu\nu}$, referring to the quarks, would then involve a special linear combination of the $\Phi_{n\nu}(R)$ instead of the gluon potential $B_\mu(R)$.

An amusing point is that in the limit of a dual resonance theory as $\alpha' \to 0$ (so that the trajectories become flat), with attention concentrated on the value $\alpha = 1$, if the mathematics of a Lie group is built into the model, then the mass shell predictions become those of the corresponding massless Yang–Mills theory[10)]. That suggests that one might even try a dual resonance model as a replacement of a color octet Yang–Mills gluon model, with abstraction of the properties of color singlet operators.

We are not at all sure that what we are discussing here is a practical scheme, and if it is, we do not know how the resulting algebraic system differs from that of gluons. We put it forward merely in order to stimulate thinking about whether or not here are candidates for the algebra, the representation, and the form of $\Theta_{\mu\nu}$ other than those suggested by the gluon model.

Our attempt to use the dual model to construct a field theory has no bearing on whether the mass–shell dual model can lead to a complete S–matrix theory of hadrons; our suggestion resembles the use of limits of dual theories to obtain unified theories of weak and electromagnetic interactions or the theory of gravity.

One interesting speculation that is independent of what model we use for the stuff to which quarks are coupled is that perhaps when we perform the mathematical transformation from current quarks to constituent quarks and obtain the crude naive quark model of meson and baryon spectra and couplings, the gluons or whatever they are will also be approximately transformed into fictitious constituents, so that meson states would appear that act as if they were made of gluons rather than $q\bar{q}$ pairs. If there are indeed ten low–lying scalar mesons rather than nine, then we might interpret the tenth one (roughly speaking, the ε° meson) as the beginning of such a sequence of extra Su_3 singlet meson states. (A related question, much debated by specialists in the usual, mass–shell dual models, is whether the infinite sequence of meson and baryon Regge trajectories, all rising indefinitely and straight and parallel in zeroth approximation, should be extended to exotic channels, i. e., those with quantum numbers characteristic of $qqqq\bar{q}, q\bar{q}q\bar{q}$ etc.).

Let us end by emphasizing our main point, that it may well be possible to construct an explicit theory of hadrons, based on quarks and some kind of glue, treated as fictitious, but with enough physical properties abstracted and applied to real hadrons to constitute a complete theory. Since the entities we start with are fictitious, there is no need for any conflict with the bootstrap or conventional dual model point of view.

APPENDIX – BILOCAL FORM FACTOR ALGEBRA

We have described in Section III and IV a Lie algebra of good components of bilocal operators on a null plane. The generators are 36 functions of x_-, y_- and $x_\perp = y_\perp$, namely $\mathcal{F}_{j+}, \mathcal{F}^5_{j+}, \mathcal{T}_{jl+}$, and $\mathcal{T}_{j2+}$. We define $R \equiv 1/2(x+y)$ and $z \equiv x-y$; then we have functions of $R_\perp, R_-$, and z_-.
With z_- set equal to zero, we have just the usual good local operators on the null plane, related to the corresponding good local operators at equal times with $P_3 \to \infty$. We recall that in the early work using $P_3 \to \infty$ the most useful applications (fixed virtual mass sum rules) involved matrix elements with no change of longitudinal momentum, i. e., transverse Fourier components of the operators. Dashen and Gell–Mann[11] studied these operators and found that between finite mass states their matrix elements do not depend separately on the transverse momenta of the initial and final states, but only on the difference, which is the Fourier variable $k_\perp$. Thus they obtained a "form factor algebra" generated by operators $F_i(k_\perp)$ and $F_i^5(k_\perp)$, to which, of course, one may adjoin $T_{il}(k_\perp)$ and $T_{i2}(k_\perp)$.
We may consider the analogous quantities using the null plane method and generating to bilocals:

$$F_i(k_\perp, z_-) \equiv$$

$$\int d^4R\delta(R_+)\mathcal{F}_{i+}(R, z_-) \ \exp\ ik_1\left[R_1 + P_+^{-1}(\Lambda_1 + J_2)\right] \ \exp\ ik_2\left[R_2 + P_+^{-1}(\Lambda_2 - J_1)\right] \tag{25}$$

and so forth. Here the integration over R_- assures us that $P_+ \equiv P_0 + P_3$ is conserved by the operator. (We note that Minkowski[12] and others have studied the interesting problem of extracting useful sum rules from operators unintegrated over R_-, but we do not discuss that here.) The quantities $P_+^{-1}(\Lambda_1 + J_2)$ and $P_+^{-1}(\Lambda_2 - J_1)$ act like negatives of center–of–mass coordinates, $-\bar{R}_1$ and $-\bar{R}_2$, since on the null plane $x_+ = 0$ we have $\Lambda_1 + J_2 = -\int R_1\Theta_{++}d^4R\delta(R_+)$ and $\Lambda_1 + J_1 = -\int R_2\Theta_{++}d^4R\delta(R_+)$, while $P_+ = \int \Theta_{++}d^4R(R_+)$.
Our bilocal form factor algebra has the commutation rules

$$\left[F_i(k_\perp, z_-), F_j\left(k'_\perp, z'_-\right)\right] = i\, f_{ijk}F_k\left(k_\perp + k'_\perp, z_- + z'_-\right), \tag{26}$$

etc., where the structure constants in general are those of $[U_6]_w$. Putting $z_- = z_-' = 0$, we obtain exactly the form factor algebra of Dashen and Gell–Mann. If we specialize further to $k_\perp = k'_\perp = 0$, we obtain the algebra $[U_6]_{w,\infty,\text{ currents}}$, of vector, axial vector, and tensor charges. It is not, of course, identical to the approximate symmetry algebra $[U_6]_{w,\infty \text{ strong}}$, for baryon and meson spectra and vertices, but is related to it by a transformation, probably unitary. That is the transformation which we have described crudely as connecting current quarks and constituent quarks.
The behavior of the operators $F_i(k_\perp)$, etc., with respect to angular momentum in the

s–channel is complicated and spectrum–dependent; it was described by Dashen and Gell–Mann in their angular condition[10)]. There is a similar angular condition for the bilocal generalizations $F_i\,(k_\perp, z_-)$, etc.

The behavior of $F_i\,(k_\perp, z_-)$ and the other bilocals with respect to angular momentum in the cross–channel is, in contrast, extremely simple. If we expand $F_i\,(k_\perp, z_-)$ or $F_i^5\,(k_\perp, z_-)$ in powers of z_-, each power z_-^n corresponds to a single angular momentum, namely $J = n + 1$.

As we expand $F_i\,(k_\perp, z_-)$, etc., in power series in z_-, we note that each term, in ${z_-}^{J-1}$, has a pole in $k_\perp^2$ at $k_\perp^2 + M^2 = 0$, where M is the mass of any meson of spin J. By an extension of the Regge procedure, we can keep $k_\perp^2$ fixed and let the angular momentum vary by looking at the asymptotic behavior of matrix elements of $F_i\,(k_\perp, z_-)$, etc., at large z_-. A Regge pole in the cross channel gives a contribution $z_-^{\alpha\left(-k_\perp^2\right)}\beta\,(k_\perp^2)\,\left[\sin\,\pi\alpha\,(-k_\perp^2)\right]^{-1}$ and a cut gives a corresponding integral over α. Thus the bilocal form factor $F_i\,(k_\perp, z_-)$ couples to each Reggeon in the non–exotic meson system in the same way that $\mathcal{F}_i\,(k_\perp)$ couples to each vector meson. The contribution of each Regge pole to the asymptotic matrix element of $F_i\,(k_\perp, z_-)$ between hadron states A and B is given by the coupling of $\mathcal{F}_i\,(k_\perp, z_-)$ to that Reggeon multiplied by the strong coupling constant of the Reggeon to A and B.

It would be nice to substitute the Regge asymptotic behavior of $F_i\,(k_\perp, x_-)$ etc., into the commutation rules and obtain algebraic relations among the Regge residues. Unfortunately, the asymptotic limit is not approached uniformly in the different matrix elements, and the asymptotic Regge formulae cannot, therefore, be used for the operators everywhere in the equations (A.2); only partial results can be extracted.

References

1. M. Gell–Mann, Phys. Rev. 125, 1067 (1962) and Physics 1, 63 (1964).

2. H. Fritzsch, M. Gell–Mann, Proceedings of the Coral Gables Conference on Fundamental Interactions at High Energies, January 1971, in "Scale Invariance and the Light Cone", Gordon and Breach Ed. (1971), and Proceedings of the International Conference on Duality and Symmetry in Hadron Physics, Weizmann Science Press (1971).
   J. M. Cornwall, R. Jackiw, Phys. Rev. D4, 367, (1971).
   C.H. Llewellyn Smith, Phys. Ref. D4, 2392, (1971).

3. D. J. Gross, S. B. Treiman, Phys. Rev. D4, 1059, (1971).

4. M. Gell–Mann, Schladming Lectures 1972, CERN–preprint TH 1543.
   W. A. Bardeen, H. Fritzsch, M. Gell–Mann, Proceedings of the Topical Meeting on Conformal Invariance in Hadron Physics, Frascati, May 1972.

5. J. Wess (Private communication to B. Zumino).

6. H. Fritzsch, M. Gell–Mann and A. Schwimmer, to be published.
   D. J. Broadhurst and R. Jaffe, to be published.

7. H. Leutwyler, J. Stern, Nuclear Physics B20, 77 (1970).

8. R. Jackiw, DESY Summer School Lectures 1971, preprint MIT–CTP 236.

9. G. Domokos, S. Kövesi–Domokos, John Hopkins University preprint C00–3285–22, 1972.

10. A. Neveu, J. Scherk, Nuclear Physics B36, 155, 1972.

11. R. Dashen, M. Gell–Mann, Phys. Rev. Letters 17, 340 (1966).
    M. Gell–Mann, Erice Lecture 1967, in: Hadrons and their Interactions, Academic Press, New York–London, 1968.
    S.–J. Chang, R. Dashen, L. O'Raifeartaigh, Phys. Rev. 182, 1805 (1969).

12. P. Minkowski, unpublished (private communication).

Volume 47B, number 4 PHYSICS LETTERS 26 November 1973

ADVANTAGES OF THE COLOR OCTET GLUON PICTURE ☆

H. FRITZSCH*, M. GELL-MANN and H. LEUTWYLER**
California Institute of Technology, Pasadena, Calif. 91109, USA

Received 1 October 1973

It is pointed out that there are several advantages in abstracting properties of hadrons and their currents from a Yang–Mills gauge model based on colored quarks and color octet gluons.

In the discussion of hadrons, and especially of their electromagnetic and weak currents, a great deal of use has been made of a Lagrangian field theory model in which quark fields are coupled symmetrically to a neutral vector "gluon" field. Properties of the model are abstracted and assumed to be true for the real hadron system. In the last few years, theorists have abstracted not only properties true to each order of the coupling constant (such as the charge algebra $SU_3 \times SU_3$ and the manner in which its conservation is violated) but also properties that would be true to each order only if there were an effective cutoff in transverse momentum (for example, Bjorken scaling, V-A light cone algebra, extended V-A-S-T-P light cone algebra with finite quark bare masses, etc.).

We suppose that the hadron system can be described by a theory that resembles such a Lagrangian model. If we accept the stronger abstractions like exact asymptotic Bjorken scaling, we may have to assume that the propagation of gluons is somehow modified at high frequencies to give the transverse momentum cutoff. Likewise a modification at low frequencies may be necessary so as to confine the quarks and antiquarks permanently inside the hadrons.

The resulting picture could be equivalent to that emerging from the bootstrap-duality approach (in which quarks and gluons are not mentioned initially), provided the baryons and mesons then turn out to behave as if they were composed of quarks and gluons.

We assume here the validity of quark statistics (equivalent to para-Fermi statistics of rank three, but with restriction of baryons to fermions and mesons to bosons). The quarks come in three "colors", but all physical states and interactions are supposed to be singlets with respect to the SU_3 of color. Thus, we do not accept theories in which quarks are real, observable particles; nor do we allow any scheme in which the color non-singlet degrees of freedom can be excited. Color is a perfect symmetry. (We should mention that even if there is a fourth "charmed" quark u′ in addition to the usual u, d, and s, there are still three colors and the principal conclusions set forth here are unaffected.)

For a long time, the quark-gluon field theory model used for abstraction was the one with the Lagrangian density

$$L = -\bar{q}\,[\gamma_\alpha(\partial_\alpha - igB_\alpha\lambda_0) + M]q + L_B. \qquad (1)$$

Here M is the diagonal mechanical mass matrix of the quarks and L_B is the Lagrangian density of the free neutral vector field B_α, which is a color singlet. Recently, it has been suggested [1] that a different model be used, in which the neutral vector field $B_{A\alpha}$ is a color octet ($A = 1 \ldots 8$) and we have

$$L = -\bar{q}\,[\gamma_\alpha(\partial_\alpha - igB_{A\alpha}\chi_A) + M]q + L_B \text{ (Yang–Mills)}, \qquad (2)$$

where χ_A is the color SU_3 analog λ_i. In this communication we discuss the advantages of abstracting properties of hadrons from (2) rather than (1).

We remember, of course, that the real description of hadrons may involve a mysterious alteration of L_B to $\hat{L}_B$ or of L_B(Y-M) to $\hat{L}_B$(Y-M), where the new

☆ Work supported in part by the U.S. Atomic Energy Commission. Prepared under Contract AT(11-1)-68 for the San Francisco Operations Office, U.S. Atomic Energy Commission. Work supported in part by a grant from the Alfred P. Sloan Foundation.

* On leave from Max-Planck-Institut für Physik und Astrophysik, München, Germany.

** On leave from Institute for Theoretical Physics, Bern, Switzerland.

Lagrangian has the needed properties at high and low frequencies to give scaling and confinement respectively. No convincing example of such a situation has ever been given. In ref. [1], it was suggested the required new gluon propagation might be supplied in a model where $B_{A\mu}$ appears as one mode of a quantized string in a multilocal field theory version of a dual picture for the glue. (The mass-shell version of such a dual scheme, for particles treated as real, is known to reduce to a Yang–Mills theory as the slope parameter α' for Regge trajectories tends toward zero.) Another suggestion [2] is that somehow the free gluon propagator contains, instead of the factor $1/q^2$, a factor μ^2/q^4, where μ is some mass. All such suggestions are, for the moment, mere speculations.

It may be, of course, that there is no modification at high frequencies, in which case we would probably not have exact asymptotic Bjorken scaling. Also, modification at low frequencies may not be necessary for confinement.

A modified theory would clearly have an operator term δ in the energy density that violates scale invariance but not $SU_3 \times SU_3$, while the unmodified one would either lack δ or generate it spontaneously. A theory with $\delta = 0$ would have a massless scalar dilaton as $M \to 0$.

The simplest and most obvious advantage of (2) over (1) is that the gluons are now just as fictitious as the quarks. The color octet gluon field $B_{A\alpha}$ does not communicate with any physical channel, since the physical states are all color singlets; in contrast, the color singlet gluon field B_α would have the same quantum numbers as the baryon current, the ϕ meson, and so forth. Since in (2) the gluon is unphysical, we have no objection to the occurrence of long-range forces in its fictitious channel, produced either by massless gluons in the unmodified version or by the noncanonical glue propagation in the modified version. These fictitious long-range forces and the associated infrared divergencies could provide a mechanism for confining all color nonsinglets permanently. They would not be present in physical hadronic interactions, where long-range forces are know to be absent.

The second advantage is that we can see in (2) a *hint* as to why Nature selects color singlets. Looking at the crudest nonrelativistic, weak-coupling approximation to (2), we find a potential

$$g^2 (2\pi)^{-1} \sum_{i \neq j} r_{ij}^{-1} C_{iA} C_{jA} \ ,$$

where the C_{iA} are the color octet SU_3 charges of the various quarks, antiquarks, and gluons. Then it is easy to envisage a situation in which the only states with deep attraction would be the color singlets. (We suppose that in the true theory the other states become completely unphysical.)

Recently, this point has been given publicity by Lipkin [3], who treats, however, a Han–Nambu picture in which color nonsinglets can be physically excited by electromagnetism and in which there are three triplets of real quarks with integral charges that average to 2/3, −1/3, and −1/3. We have rejected such a picture. In fact, a serious argument against it is the clash between the color octet Yang–Mills gauge on the one hand and the electromagnetic gauge or the Yang–Mills gauge of unified weak and electromagnetic interactions on the other. Since, in our work, the weak and electromagnetic currents form color singlets, we encounter no such difficulty.

A third and very important advantage of the color octet gluon scheme has been pointed out by L.B. Okun in a private communication to H. Pagels. Okun's point is that in (1) there is no distinction between ordinary SU_3 and the SU_3 of color in the limit $m_u = m_d = m_s$, and thus we would have the symmetry of SU_9 (or of SU_6 for $m_u = m_d$) where these groups combine color SU_3 and ordinary SU_3. No evidence of such extended symmetries exists. In (2), of course, these annoying symmetries are not present.

A fourth apparent advantage of the color octet gluon scheme has recently been demonstrated [4] using the asymptotic perturbation theory method of Gell-Mann and Low. Assuming that the method is valid (sum of asymptotic forms of orders of perturbation theory equaling asymptotic form of sum), one can have a situation in which the bare coupling constant is zero, there are no anomalous dimensions for color singlet quantities, and the behavior of light cone commutators comes closer to scaling behavior than in the color singlet vector gluon case (1). However, actual Bjorken scaling does not occur; instead, each moment $\int F_2(\xi)\xi^n \, d\xi$ of the Bjorken scaling function appears multiplied, in the Bjorken limit, by a distinct power $(\ln q^2)^{p_n}$, where $-q^2$ is the virtual photon mass squared.

Volume 47B, number 4 PHYSICS LETTERS 26 November 1973

That sort of violation of Bjorken scaling is not contradicted by present experiments. Furthermore, many sum rules and symmetry principles of light cone current algebra would be preserved.

For us, the result that the color octet field theory model comes closer to asymptotic scaling than the color singlet model is interesting, but not necessarily conclusive, since we conjecture that there may be a modification at high frequencies that produces true asymptotic scaling.

There is one more advantage of the color octet gluon scheme over the color singlet scheme, and it is the main point we wish to stress in this communication. In either scheme, there is an anomalous divergence of the axial vector baryon current $F^5_{i\alpha}$. While, for the other eigth axial vector currents $F^5_{i\alpha}$ $(i = 1 \ldots 8)$, we have simply

$$\partial_\alpha F^5_{i\alpha}(x) = \mathcal{D}(x, x, \mathrm{i}\gamma_5 \{\tfrac{1}{2}\lambda_i, M\}), \tag{3}$$

the divergence equation for $F^5_{0\alpha}$ is [5]

$$\partial_\alpha F^5_{0\alpha} = \mathcal{D}(x, x, \mathrm{i}\sqrt{\tfrac{2}{3}} M\gamma_5) + \sqrt{6}\, g^2 (8\pi^2)^{-1} G_{\mu\nu} G^*_{\mu\nu}, \tag{4}$$

where $\mathcal{D}(x, y, G)$ is the physical operator that corresponds in a free quark model to $\bar{q}(x) G q(y)$, and $G_{\mu\nu} = \partial_\mu B_\gamma - \partial_\nu B_\mu$ for the color singlet case, while $G_{A\mu\nu} = \partial_\mu B_{A\nu} - \partial_\nu B_{A\mu} + g f_{ABC} B_{B\mu} B_{C\nu}$ for the color octet case.

Here the extra term in (4) arises from a several-gluon effect in the strong interaction analogous to the two-photon effect in the familiar electromagnetic triangle anomaly [6], which contributes a term $e^2(16\pi^2)^{-1} F_{\mu\nu} F^*_{\mu\nu}$ to the divergence of F^5_3.

It was shown [6] that in renormalizable gluon models the anomalous divergence arises essentially from the lowest order triangle diagram.

Wilson has demonstrated [7] that the anomaly is the consequence of a singularity in coordinate space. In field theory models this singularity comes from low order quark loop diagrams, since higher order corrections are less singular and do not contribute. Therefore, in a theory in which the gluon propagation is less singular at small distances than in the canonical one, the anomaly coefficient will be unchanged, since the quark propagation is left canonical.

In the color singlet gluon picture, the anomalous divergence term in (4) is necessarily associated [5] with an anomalous singularity in the bilocal current $F^5_{0\alpha}(x,y)$ as $z^2 = (x-y)^2$ tends to zero:

$$F^5_{0\alpha}(x,y) \mathrel{\hat{=}} 3\mathrm{i}(2\pi^2)^{-1} g G^*_{\alpha\beta} z_\beta (z^2)^{-1}. \tag{5}$$

The existence of such a term, while not contradicted by experiment so far, would destroy the light cone algebra *as a system* since one of the bilocal currents arising from commutation of two physical currents would be infinite on the light cone. In any case, we have assumed that the full light cone algebra is correct or at most violated by powers of logarithms, and we therefore cannot tolerate the term (5).

In ref. [5], this situation was posed as a puzzle: how to get rid of the anomalous singularity in $F^5_{0\alpha}(x,y)$, while retaining the anomalous divergence term for $\partial_\alpha F^5_{0\alpha}(x)$ given by triangle diagram.

The color octet gluon scheme solves the puzzle. The anomalous divergence term in $\partial_\alpha F^5_{0\alpha}(x)$ is unchanged, except for replacing $G_{\mu\nu} G^*_{\mu\nu}$ by $G_{A\mu\nu} G^*_{A\mu\nu}$, but it is now associated with a singularity as $z^2 \to 0$ not in $F^5_{0\alpha}(x,y)$, but in a different formal quantity, the corresponding color octet operator, which we may call $F^5_{0A\alpha}(x,y)$:

$$F^5_{0A\alpha}(x,y) \mathrel{\hat{=}} 3\mathrm{i}(2\pi^2)^{-1} g G^*_{A\alpha\beta} z_\beta (z^2)^{-1}. \tag{6}$$

Since $F^5_{0A\alpha}(x,y)$ is not a physical operator, being a color octet, we can have no objection to its being singular on the light cone.

To summarize, then, the fifth advantage of the color octet gluon scheme is that we get rid of the unacceptable anomalous singularity (5) in $F^5_{0\alpha}(x,y)$.

Now we can believe and make use of the anomalous divergence term in (4). This term looks as if it could be very useful in connection with the PCAC idea. Let us assume that the strong form of PCAC is correct [8]. Formally, we mean by this that as the bare quark masses tend to zero and the generators of $SU_3 \times SU_3$ become conserved, the conservation occurs according to the Nambu–Goldstone pattern, with eight massless pseudoscalar mesons. Physically, we mean that the real world of hadrons is not terribly far from such a situation, and not far at all from a situation with $SU_2 \times SU_2$ conserved and three massless pions. The bare quark masses are such that $m_\mathrm{u} \approx m_\mathrm{d} \ll m_\mathrm{s}$ and the ratios $M^2_\pi : M^2_\mathrm{K} : M^2_\eta$ are not very different from 0:1:4/3.

It has always been a great mystery why, if we abstract relations from a field theory model like (1) or (2), we do not have in the limit $M \to 0$ the conservation

Volume 47B, number 4 PHYSICS LETTERS 26 November 1973

of nine axial vector currents and the existence of nine massless pseudoscalar mesons. Turning on the quark bare masses, with $m_u \approx m_d \ll m_s$, we would have four nearly massless pseudoscalar mesons instead of three, in bad disagreement with observation. To put in another way, as m_u and m_d tend to zero, we would have $U_2 \times U_2$ conservation and four massless pseudoscalar mesons.

The mystery might appear to be resolved, since the anomalous term in (4) breaks the conservation of $F^5_{0\alpha}$ even in the limit $M \to 0$ and so in that limit it looks as if there need not be a ninth massless pseudoscalar meson ‡, and in the limit $m_u \to 0$, $m_d \to 0$ it looks as if there need not be a fourht one.

Unfortunately, the extra term in (4) is itself a divergence of another (non-gauge invariant) pseudovector, and thus as $M \to 0$ we still have the conservation of a modified axial vector baryon charge; we must still explain why this new ninth charge seems to correspond neither to a parity degeneracy of levels nor to a massless Nambu–Goldstone boson as $M \to 0$.

It is important to find the explanation ‡. Assuming that strong PCAC does not fail, we conjecture that the question is closely related to the question of whether there are modifications of Yang–Mills gluon propagation and, if so, what is the nature of those modifications.

Two of us (H.F. and M.G-M.) would like to thank S. Adler, W.A. Bardeen, R. Crewther, H. Pagels and A. Zee for useful conversations and the Aspen Center for Physics for making those conversations possible.

‡ In ref. [5], the authors, appalled at the anomalous singularity that accompanied the anomalous divergence in the color singlet gluon case, discussed the possibility of somehow getting rid of the anomalous divergence and finding a different explanation of the absence of a ninth pseudoscalar meson as $M \to 0$. The alternative explanation tentatively offered was that F^5_0, cummuting with $SU_3 \times SU_3$, could vanish in the limit $M \to 0$ when applied to "single particle states" instead of giving either parity doubling or a ninth massless pseudoscalar meson. However, using the full group $(SU_6)_{W,\text{currents}}$ of the light-like vector, axial vector, and tensor charges, we wee that F^5_0 fails to commute with the tensor charges T_{ix} and T_{iy}, and all matrix elements of those charges would have to vanish between "single particle states". The same is true of the modified F^5_0 that includes the effect of the anomalous divergence. It seems unlikely that all "single particle" matrix elements of T_{ix} and T_{iy} vanis as $M \to 0$.

References

[1] H. Fritzsch and M. Gell-Mann, Proc. XVI Intern. Conf. on High energy physics, Chicago, 1972, Vol. 2, p. 135.

[2] K. Kaufmann, private communication.

[3] H. Lipkin (Weizmann Institute) preprint 1973.

[4] H.D. Politzer (Harvard) preprint 1973; D. Gross and F. Wilczek (Princeton) preprint 1973.

[5] H. Fritzsch and M. Gell-Mann, Proc. Intern. Conf. on Duality and symmetry in hadron physics (Weizmann Science Press, 1971).

[6] J. Schwinger, Phys. Rev. 82 (1951) 664. S.L. Adler, Phys. Rev. 177 (1969) 2426. J.S. Bell and R. Jackiw, Nuovo Cimento 60A (1969) 47. S.L. Adler and W.A. Bardeen, Phys. Rev. 182 (1969) 1517.

[7] K. Wilson, Phys. Rev. 179 (1969) 1499.

[8] The weight of evidence is now in favor of strong PCAC. See H. Fritzch, M. Gell-Mann and H. Leutwyler, in preparation.

Supergravity
P. van Nieuwenhuizen & D.Z. Freedman (eds.)

COMPLEX SPINORS AND UNIFIED THEORIES

Murray Gell-Mann*†
CERN
Geneva, Switzerland

Pierre Ramond†
452-48 California Institute of Technology
Pasadena, California 91125, U.S.A.

and

Richard Slansky
Theoretical Division, University of California
Los Alamos Scientific Laboratory
Los Alamos, NM 87545 U.S.A.

We were told by Frank Yang in his welcoming speech that supergravity is a phenomenon of theoretical physics. Why, at this time, is it not more than that? Self-coupled extended supergravity, especially for $N = 8$, seems very close to the overall unified theory for which all of us have yearned since the time of Einstein. There are no quanta of spin >2; there is just one graviton of spin 2; there are N gravitini of spin 3/2, just right for eating the N Goldstone fermions of spin 1/2 that are needed if N-fold supersymmetry is to be violated spontaneously; there are $N(N-1)/2$ spin 1 bosons, perfectly suited to be the gauge bosons for SO_N in the theory with self-coupling. There are $N(N-1)(N-2)/6$ spin 1/2 Majorana particles, and with the simplest assignments of charge and colour they include isotopic doublets of quarks and leptons. The theory is highly non-singular in perturbation theory, and the threatened divergence at the level of three loops has not even been demonstrated. The apparently arbitrary cancellation of huge contributions of opposite sign to the cosmological constant (from self-coupling on the one hand and from spontaneous violation of supersymmetry on the other) has been phrased in such an elegant way that it may be acceptable. (Of course, if we follow Hawking et al., we may not even need to cancel out the cosmological constant!)

What is wrong then? Of course the spontaneous violation of SO_N and of supersymmetry is not known to happen in the supergravity theory. But what seems much worse, the spectrum of elementary particles includes too few spin 1 and spin 1/2 objects to agree with the list that we would like to see on the basis of our experience at energies $\lesssim 50$ GeV. Of course, looking up at the Planck mass of $\sim 2 \times 10^{19}$ GeV, we are in a position of greater inferiority than an ant staring up at a skyscraper (facing a factor of only 10^6 or so) and it may not be reasonable to expect that what looks elementary to us should be elementary on a grand scale.

---

* Permanent address: 452-48 California Institute of Technology, Pasadena, California 91125 U.S.A. Work supported in part by a grant from the Alfred P. Sloan Foundation.

† Work supported in part by the U.S. Department of Energy under contract No. DE-AC-03-79ER0068.

Nevertheless, we make the comparison and we find that $SO_8 \supset SU_3^c \times U_1 \times U_1$ but $SO_8 \not\supset SU_3^c \times SU_2 \times U_1$, so that there is no room for the $X^\pm$ intermediate bosons of the charged current weak interaction. Among the spin 1/2 particles, we have room for at most two flavours of lepton (say e and ν_e) and four flavours of quark (say u, d, c and s). Even these numbers may be reduced if we try to locate the Goldstone fermions among the elementary spin 1/2 particles or if we use the generator of the second U_1 in a restrictive way.

We would then be forced to regard all or most of the known quarks and leptons as non-elementary, as well as at least two of the intermediate bosons of the weak interaction. The broken Yang-Mills theory of the weak interaction would be only an effective gauge theory, not a fundamental one. All this may prove to be the case, and we will then have to understand the rather complicated relation existing between the elementary particles of the theory and the elementary particles as we perceive them today.

Various investigators have looked into superconformal supergravity, in which one tries to use the full SU_N as a gauge group; such a theory is plagued with particles appearing as multiple poles in propagators, involving difficulties with negative probabilities or lack of causality. Ignoring these serious difficulties, we may ask about the algebraic description of the spin 1/2 fermions in such a theory Apparently they are again connected with third-rank tensor representations, forming part of $(\underset{\sim}{N} + \bar{\underset{\sim}{N}})^3_A$ of SU_N (where A means totally antisymmetrized) instead of being assigned to $(\underset{\sim}{N})^3_A$ of SO_N.

This tendency to assign the spin 1/2 fermions to a tensor representation, probably a third rank tensor, of SO_N or SU_N, exists even in theories having nothing to do with supergravity. We may, for example, consider a composite model of quarks and leptons, in which they are made up of N kinds of fermionic sub-units. We may think of such a scheme in algebraic terms as assigning these sub-units to the representation N of SO_N or SU_N and the quarks and leptons to tensors that are part of $(N)^3$ of SO_N or part of $(\underset{\sim}{N} + \bar{\underset{\sim}{N}})^3$ of SU_N, provided each known particle is made up of three sub-units. (Of course one might use five or a higher odd number and obtain fifth rank tensors and so forth, but three is much simpler.)

Now what indications come from the attempts to construct a unified Yang-Mills theory? Do they also point to such a third-rank tensor for the spin 1/2 fermions?

We turn, then, to the program of formulating a broken Yang-Mills theory of strong and weak interactions, with an effective energy of unification between 10^{14} GeV and the Planck mass. This program is only slightly less immodest in conception than the overall unification program of self-coupled supergravity. For the sake of expressing all the Yang-Mills coupling constants in terms of a single one, a simple group G is employed. (Actually one could use G × G, G × G × G, etc. with discrete symmetries connecting the factors, but we shall treat here only the case of a single G.)

The smallest G that has been used is SU_5; the known left-handed spin 1/2 fermions are then assigned to three families, each belonging to the reducible representation $\bar{5} + 10$ of SU_5, where $\bar{5}$ contains $\bar{d}$, e^-, and ν_e for the lowest family, while 10 consists of d, u, $\bar{u}$ and e^+. The combination $\bar{5} + 10$ is anomaly-free. The violation of symmetry takes place in two stages. First the symmetry SU_5 is broken down to $SU_3^c \times SU_2 \times U_1$ by means of a non-zero vacuum expected value of an operator transforming like the adjoint representation 24, with no direct effect on the fermion masses, and then $SU_2 \times U_1$ is broken down to U_1^{e-m} by means of operators transforming like $5 + \bar{5}$, with perhaps an admixture of 45, giving the masses of the quarks and leptons. The detailed work is done using explicit spinless Higgs boson fields, with various constants for mass, for self-coupling and for coupling to fermions, constants that must be delicately adjusted to make the masses of the fermions and of the intermediate boson for weak interactions tiny with respect to the unification mass. A quantity of roughly similar magnitude, the renormalization-group-invariant mass Λ of QCD, is tiny with respect to the unification mass for a totally different reason, namely the smallness of the unified coupling constant, which is $\sim 10^{-2}$ near the unification mass, corresponding to the fine structure constant at low energies, and is proportional to the reciprocal of the logarithm of $(10^{14}\ \text{GeV})/\Lambda$. Despite some successes, which we mention below, the SU_5 scheme seems to us a temporary expedient rather than a final theory, because of the arbitrariness associated with the Higgs bosons and also because the particles and antiparticles among the left-handed spin 1/2 fermions have no relation to each other (i.e., there is no C or P operator for the theory).

The SU_5 scheme has at least two successes: a roughly correct prediction of the weak angle θ_W and the prediction that after allowing for renormalization $m_b \approx m_\tau$, which works quite well. The violation of $SU_2 \times U_1$ by $5 + \bar{5}$ of SU_5 would also give $m_s \approx m_\mu$ and $m_d \approx m_e$ after renormalization. The second of these relations does not work but might be subject to large corrections because the quantities are so small; the first might work if the usual estimates of m_s are in error - otherwise some admixture of a 45 of SU_5 has been suggested, along with $5 + \bar{5}$, but affecting mainly the two lower families.

We have studied various complex spinor schemes that reduce to the SU_5 system after some symmetry violation. Work on such schemes has also been done by Georgi et al. at Harvard, Susskind and collaborators at Stanford, Wilczek and Zee, Gürsey et al. in the case of E_6, and no doubt by many others. Early investigations of complex spinor assignments were carried out by Fritzsch and Minkowski.

First, let us restrict our attention to a single family, say the third one, assuming that the t quark exists and that $\nu_{\tau L}$ is nearly massless. We note that the reason that $\bar{5} + 10$ of SU_5 is anomaly-free is that the complex spinor representation 16 of SO_{10} breaks up into $1 + \bar{5} + 10$ of SU_5, where the singlet can give rise to no SU_5 anomaly, and all representations of $SO_n (n \neq 6)$ are anomaly-free.

The 16-dimensional spinor possesses a C symmetry to start with, connecting $\tau_L^- \rightleftarrows \tau_L^+$, $b_L \rightleftarrows (\bar{b})_L$, etc., and the 16th particle is just the missing $(\bar{\nu}_\tau)_L$. Symmetry violations giving fermion masses must correspond to representations contained in the symmetrized square of the fermion representation. We note that in SO_{10} we have $(16)^2_S = 10 + 126$ and that with respect to SU_5 we have $10 \to 5 + \bar{5}$ and $126 \to 1 + 45 + 10 + \bar{15} + \bar{5} + 50$. An operator transforming like the SU_5 singlet piece of 126 would break the SO_{10} symmetry down to SU_5 and would give a Majorana mass term of the form $(\bar{\nu}_\tau)^2_L + \nu^2_{\tau R}$ to the unobserved neutrino, one that had better be very large if the scheme is to work.

Meanwhile, the 10 of SO_{10} would give rise to equal Dirac masses for b and τ (apart from renormalization) and also to equal Dirac masses for t and ν_τ. The Dirac mass for the neutrino leads directly to a small effective mass $m(\nu_{\tau L}) \sim m^2_{Dirac}/m_{Majorana}$. If $m(\nu_{\tau L}) \approx 1$ eV, then neutrinos account for a modest fraction of the missing matter in the universe and give a moderate contribution to the gravitational closure of galaxies and clusters of galaxies. Putting $m_{Dirac} \approx m_t \approx 30$ GeV at a guess, the corresponding value of $m_{Majorana}$ would be $\approx 10^{12}$ Gev. If $m_{Majorana}$ is very much smaller than that, the cosmological effects become too large; if $m_{Majorana}$ is much larger, that is harmless, but the cosmological effects become negligible.

We can examine SO_{10} in a different way by using the decomposition $SO_{10} \supset SO_6 \times SO_4$ where algebraically SO_6 is equivalent to SU_4 and SO_4 to $SU_2 \times SU_2$. We have, then, effectively $SO_{10} \supset SU_2 \times SU_2 \times SU_4$, where the first SU_2 is that of the weak interactions, the second one the corresponding SU_2 for left-handed antiparticles or right-handed particles and SU_4 is the generalization of SU_3^c introduced in a different connection by Pati and Salam, in which leptons appear as having a fourth colour. The representations of SO_{10} then decompose as follows:

$$\underline{16} \rightarrow (\underline{2},\underline{1},\underline{4}) + (\underline{1},\underline{2},\underline{\bar{4}}), \quad \underline{\overline{16}} \rightarrow (\underline{1},\underline{2},\underline{4}) + (\underline{2},\underline{1},\underline{\bar{4}}), \quad \underline{10} \rightarrow (\underline{2},\underline{2},\underline{1}) + (\underline{1},\underline{1},\underline{6}),$$
$$\underline{126} \rightarrow (\underline{2},\underline{2},\underline{15}) + (\underline{1},\underline{1},\underline{6}) + (\underline{3},\underline{1},\underline{10}) + (\underline{1},\underline{3},\underline{\overline{10}}).$$

The representations $\underline{1}$, $\underline{4}$, $\underline{4}$, $\underline{15}$, $\underline{10}$ and $\underline{\overline{10}}$ of SU_4 each contain one colour singlet. We see that the Dirac mass term coming from $\underline{10}$ is just of the form $(\underline{2},\underline{2},\underline{1})$, while the Majorana mass term for $(\bar{\nu}_e)_L$ or ν_R coming from $\underline{126}$ is a component of $(\underline{1},\underline{3},\underline{\overline{10}})$. We must not use $(\underline{3},\underline{1},\underline{10})$, which would introduce an unwanted triplet violation of SU_2^{weak} and would give a mass directly to the left-handed neutrino. A possible danger is that radiative corrections might give rise to a large or uncontrollable term of that kind anyway, in addition to the term $m^2_{Dirac}/m_{Majorana}$, since the left-handed neutrino Majorana mass is not prohibited by a selection rule.

Such a selection rule exists in the SU_5 scheme, where an ungauged quantity that distinguishes $\underline{\bar{5}}$ from $\underline{10}$ and a gauged generator of SU_5 are simultaneously violated, preserving a linear combination, which is the baryon number minus the lepton number. The conservation of this quantity prohibits neutrino mass altogether. Here an unwanted massless spinless Goldstone boson is fed to an unwanted massless spin 1 gauge boson to give a massive spin 1 boson. This trick, which we have studied in connection with conserving baryon number (perhaps an obsolete idea now) can be applied whenever there is a reducible representation of G for the fermions (or even for spinless elementary particles if there are some).

A further generalization of SU_5 for one family might make use of the lowest complex spinor representation $\underline{27}$ of E_6, which breaks down to $\underline{16} + \underline{10} + \underline{1}$ of SO_{10}. Here one would have to marry the new SO_{10}-singlet neutrino to the unwanted $(\bar{\nu})_L$ of $\underline{16}$, allowing them to share a huge Dirac mass, and one would have to do it in such a way as to leave the ν_L of $\underline{16}$ with a small mass or none at all. At the same time one would have to assign high masses to the members of the $\underline{10}$ of SO_{10} in the $\underline{27}$ of E_6, in order to get them out of the way, leaving just the fifteen fermions of the SU_5 scheme.

What we have seen from the example of one family is that a complex spinor representation, while it involves us in delicate questions of neutrino mass, does permit the description of left-handed fermions by a single irreducible representation of G and in such a way that the asymmetry between the SU_2^{weak} assignments of particle and antiparticle is rather natural, while the whole system possesses an initial symmetry C between left-handed particles and left-handed antiparticles, a symmetry that interchanges SU_2^{weak} and another SU_2.

It is also clear that in such a scheme the dimensions of the representations that violate the symmetry, for example in the generation of fermion masses, tend to be large and that the arbitrary character of the violation scheme employing elementary Higgs bosons is strongly emphasized. It seems to us that one must hope for a situation in which, somehow, spontaneous symmetry violation is achieved dynamically

Although we do not, of course, exclude the existence of some spinless elementary fields, provided they are not the arbitrary ones of the elementary Higgs boson method, we may look as an example at a theory with just gauge bosons and elementary spin 1/2 fields and imagine what hypothetical dynamical spontaneous symmetry breaking would be like.

We would like to point out first that if the (say) left-handed fermion representation in such a theory is reducible, then ungauged quantum numbers arise that commute with the gauge group. When these are violated spontaneously, that necessarily leads to unwanted massless Goldstone bosons unless the trick described above is used and global conservation laws result. If all the irreducible representations are inequivalent, then such globally conserved quantities are Abelian and tolerable, but if there are equivalences among representations, as in the case of several families transforming alike, then an ungauged non-Abelian family group arises and that would have to be matched with an isomorphic subgroup of G with resulting global conservation of a third isomorphic non-Abelian group relating the families. That would not agree at all with observation, and we conclude therefore that having united each family in an irreducible representation of SO_{10} or E_6 we had better consider all the fermions as belonging to a single irreducible representation of the gauge group G.

This can be done in two different ways. Either we go to a higher-dimensional representation of the same group that we used for one family or else we enlarge the group and assign the fermions to a relatively low-lying representation of the bigger group. In the case of complex spinor representations, we could try, as an example of the first approach, the $\underset{\sim}{1728}$ of E_6, contained in $\underset{\sim}{27} \times \underset{\sim}{78}$. As examples of the second approach, we can take the lowest-dimensional complex spinors of larger groups, and the only larger groups possessing such spinors are SO_{14} (lowest dimensional spinor $\underset{\sim}{64}$), SO_{18} (lowest-dimensional spinor $\underset{\sim}{256}$), SO_{22} (lowest-dimensional spinor $\underset{\sim}{1024}$), etc. We have studied both possibilities, but we shall describe here the case of the lowest spinors of SO_{4n+2}

A great deal of thought has been devoted to the question of what dynamical spontaneous symmetry breaking would be like for a theory containing elementary fields for gauge bosons and fermions only. Weinberg, Dimopoulos and Susskind, and various other theorists have drawn some important conclusions, including the following, which we specialize to the case of an irreducible fermion representation.

Symmetry reduction occurs through "condensations", that is non-zero vacuum expected values of operators that break symmetries. If the symmetry group of the kinetic energy is H and if $G_1 \subset G$ and $H_1 \subset H$ are the subgroups left invariant by these condensations, then the generators of G_1 correspond to exact conservation laws and massless gauge bosons, those of G/G_1 to massive gauge bosons, those of H_1/G_1 to modified Goldstone bosons that acquire mass a a result of the gauge coupling and those of $(H/G)/(H_1/G_1)$ to approximate conservation laws, broken by the gauge coupling . The flavour-non-singlet pseudoscalar mesons would be modified Goldstone bosons, and the PCAC condensation $\langle\bar{q}q\rangle_{vac} \sim \Lambda^3$ presumably occurring in QCD would contribute only $\sim e\Lambda$ to the masses of the weak intermediate bosons. If one or more additional factors of the exactly conserved strong colour group exist (we prefer to call them primed colour, etc.), then these could have higher renormalization-group-invariant masses Λ', etc., and a primed colour group with $\Lambda' \sim 10^3$ GeV could give a condensation of fermions possessing primed colour that would account for the weak intermediate boson masses. Some of the corresponding pseudoscalar primed mesons would serve as effective Higgs bosons to be eaten by these intermediate gauge bosons. The mixing between these primed mesons and ordinary pseudoscalar mesons would be rather small. There would be no real ultra-violet fermion masses, but only medium-frequency or infra-red masses of order Λ for quarks, Λ' for fermions possessing primed colour, and so forth, and then masses obtained by sharing these medium-frequency masses through radiative corrections - these last would simulate ultra-violet masses up to fairly high energies.

A great deal of the algebraic behaviour of such symmetry-breaking schemes should be simulated by generalized non-linear σ-models. If those are embedded in linearized σ-models, then one has some connection with the algebraic properties of explicit Higgs boson theories.

An important question is whether the many condensations required for symmetry breaking in a unified theory can be explained by the strong long-range interactions that appear in the same theory. This is a problem, for example, in connection with any condensation leading to Majorana masses for the unwanted neutrinos.

Now let us return to the notion that G might be SO_{4n+2} with the left-handed fermions placed in the 2^{2n}-dimensional complex spinor representation. Let us consider the example of SO_{18}, which evidently contains $SO_8 \times SO_{10}$. We can decompose the 256 of SO_{18} as $(8_{sp},\overline{16}) + (8'_{sp},16)$ of $SO_8 \times SO_{10}$, where 8_{sp} and $8'_{sp}$ are the two real inequivalent spinors of SO_8. We can now write $SO_8 \supset Sp_4 \times SU_2$, where the vectorial octet 8_v of SO_8 can be made to give $(4,2)$ of $Sp_4 \times SU_2$ and $8'_{sp}$ of SO_8 likewise, while 8_{sp} of SO_8 gives $(1,3) + (5,1)$ of $Sp_4 \times SU_2$. The 256 of SO_{18} then becomes $(1,3,\overline{16}) + (5,1,16) + (4,2,\overline{16})$ of $Sp_4 \times SU_2 \times SO_{10}$. If we interpret Sp_4 as a supplementary factor of the exactly conserved colour group, which becomes $SU_3^c \times Sp_4^{c'}$, and SU_2 as a gauged family subgroup of SO_{18}, we see that the only fundamental left-handed fermions without primed colour are three families of 16-dimensional spinors of SO_{10}, and we glimpse a possible agreement with experiment.

We note that $Sp_4^{c'}$, if there were no fermions would have the same renormalization-grou behaviour as SU_3^c in lowest order, and we would need a special explanation for its reaching the strong-coupling regime at a much higher mass than SU_3^c. The differing fermion corrections might make a difference; so might the possibility that as we come down in mass from the unification region SO_8 remains undivided over a considerable interval before splitting into $Sp_4 \times SU_2$.

In the same way, $SO_{14} \supset SO_4 \times SO_{10}$ and SO_4 is actually algebraically equivalent to $Sp_2 \times SU_2$; the 64 of SO_{14} decomposes into $(1,2,16) + (2,1,\overline{16})$ of $Sp_2 \times SU_2 \times SO_{10}$ and we would have two families lacking primed colour. Similarly, $SO_{22} \supset SO_{12} \times SO_{10}$ and $SO_{12} \supset Sp_6 \times SU_2$; the 1024 of SO_{22} decomposes into $(1,4,16) + (14,2,16)$ $+ (6,3,\overline{16}) + (14',1,\overline{16})$ and we would have four families lacking primed colour.

As far as representations giving fermion mass are concerned, we have the following situation:

SO_{10}: $(16)^2_S = (10)^5_A(\text{self-dual}) + (10)^1 = 126 + 10;$

SO_{14}: $(64)^2_S = (14)^7_A(\text{self-dual}) + (14)^3_A = 1716 + 364;$

SO_{18}: $(256)^2_S = (18)^9_A(\text{self-dual}) + (18)^5_A + (18)^1 = 24310 + 8568 + 18;$

SO_{22}: $(1024)^2_S = (22)^{11}_A(\text{self-dual}) + (22)^7_A + (22)^3_A = 352{,}716 + 170{,}544 + 1540;$

and so forth. It looks in each case as if the Majorana mass term comes from the highest-dimensional representation and the Dirac masses of the familiar fermions from the next-highest-dimensional one, if such a scheme is to work. The Dirac masses then obey an important constraint, which equates a function of the charged lepton mass matrix with the same function of the $Q = -1/3$ quark mass matrix. Since in each case the matrix is dominated, according to experience, by the highest mass, these two masses must be roughly equal, and for three families that explains the relation $m_b \sim m_\tau$ after renormalization.

The question is, of course, left open as to <u>why</u> the mass matrix for three families is so close to

$$\begin{pmatrix} C & 0 & 0 \\ 0 & 0 & 0 \\ 0 & 0 & 0 \end{pmatrix}$$

for each kind of particle. With the families described as a triplet of SU_2 rather than a triplet of SU_3, that is rather mysterious, since it corresponds to a miraculous compensation of a scalar and a quadrupole term under SU_2. Under SU_3, of course, they would combine to form a 6 of SU_3 and the approximate matrix above would correspond to the intervention of the component of 6 invariant under the maximal little group SU_2. Unfortunately we are not dealing here with a family SU_3.

The Dirac masses of neutrinos and of $Q = +2/3$ quarks would obey the same relation as that for the charged leptons and $Q = -1/3$ quarks. The Majorana masses of the neutrinos are also subject to a constraint if they come from the highest-dimensional representation for the fermion mass. Of course, the mass matrix for neutrinos is not easy to detect and at best requires delicate experiments that we shall describe elsewhere.

In summary, the idea of assigning left-handed spin 1/2 fermions to a complex spinor representation of a gauge group SO_{4n+2} (or conceivably E_6) has a number of attractive features, although some difficulties as well. As alternatives for an irreducible representation, we have, of course, the possibility of a real or pseudoreal representation, giving a vector-like theory in which all the known fermions must be accompanied by heavy partners that have weak interactions of opposite handedness; or a complex representation of a unitary group, which when irreducible generally leads to anomalies and thus to divergences, and is also rather hard to reconcile with observation.

If we suppose that the familiar quarks and leptons are really to be assigned to a complex spinor representation of a group SO_{4n+2} or E_6, can we reconcile that idea with the notion that there is some truth in extended supergravity where the spin 1/2 fermions are placed in a third rank antisymmetric tensor representation of SO_N?

We have looked, in collaboration with Jon Rosner, for an analogue of supersymmetry that might lead to a theory with assignments like $\underset{\sim}{1}$ of E_6 for $J_z = 2$, $\underset{\sim}{27}$ of E_6 for $J_z = 3/2$, adjoint $\underset{\sim}{78}$ of E_6 for $J_z = 1$, and for $J_z = 1/2$ some representation contained in $\underset{\sim}{27}\times\underset{\sim}{78}$, like $\underset{\sim}{1728}$ of E_6. We have searched for the same kind of scheme using SO_{10}, and we have even tried non-associative systems in an effort to find something that would work. So far we have had no success.

It seems likely anyway that if supergravity or some similar future theory is correct, then there must be only an indirect relation between the elementary fields of the theory and the particles that appear to us today to be elementary. If the known fermions behave, for a given handedness, like a complex spinor representation of SO_{4n+2} or E_6, then the relation is not even that of a composite model. All or most of the familiar particles would have to correspond to particle-like solutions of the fundamental equations, with a different algebraic behavior from that of the fundamental fields.

In this talk we have only sketched the subject of complex spinor representations and related topics. Elsewhere we present a proper account of our own work and adequate references to the work of others.

We have also taken a rather schizophrenic approach, shuttling back and forth between extended supergravity on the one hand and a particular kind of unified Yang-Mills theory on the other. The ideas underlying the two approaches have to be compared more carefully.

20

PARTICLE THEORY FROM S-MATRIX TO QUARKS

Murray Gell-Mann
California Institute of Technology

CONTENTS

Reprinted from *Symmetries in Physics* (1600–1980), eds. M. G. Doncel, A. Hermann, L. Michel and A. Pais (Universitat Autònoma de Barcelona), pp. 474–497.

It is a great pleasure to be here in Catalunya and to participate in this meeting on the important subject of the history of physical ideas. I am particularly pleased that this meeting is in a sense dedicated to my old friend Louis Michel. (It has been predicted by Professor Telegdi that some day in the distant future there will be a «station de métro» and even a «boucherie chevaline» named after our colleague, in the tradition of Louis Arago and others. For the moment it is wonderful to have this memorial to Professor Michel while he is still alive.)

This is called the first «trobada» on the history of scientific ideas. But there was another one on the history of elementary particle physics last summer, in Paris, presumably the zeroth. I was fortunate enough to be invited to give a talk there on some very early work of mine, on strangeness and related matters. For that talk as for this one, I do not pretend to have done any significant amount of historical research. I am far too lazy to have done more than look up a few papers here and there and glance at them. These talks should be considered as raw material for historical research, like reminiscences of some aged farmer discussing the rural landscape of his youth.

I had hoped to make the written report more systematic than the somewhat confused oral presentation, but that hope has not been fulfilled. However, the confusion has the merit that it reflects well the situation during the entire period that I am going to describe.

I find these reminiscences bittersweet, and the preparation of the talk has been somewhat painful for me. Sometimes, of couse, thinking about the evolution of my research in physics is amusing; and there is occassionally the delight of remembering past triumphs. But there is also the recollection of lost opportunities to see clearly what was going on.

There used to be a jingle posted on the wall of certain doughnut shops in the U.S. that ran:

As you ramble on through life, Brother,
Whatever be your goal,
Kepp your eye upon the doughnut,
And not upon the hole.

I am afraid that I am one of those that do not follow the injunction in the jingle; I tend to keep my eye on the hole in the toroidal doughnut. As Harald Fritzsch remarked to me, that means that you brood about having invested in stocks at the wrong time, you regret having taken the wrong job, and so forth. And so in physics I regret not having taken certain ideas seriously enough at certain times and being unnecessarily confused on many occasions when it would have been perfectly possible to think straight.

I have decided that in this talk I shall emphasize some of the confusions and hesitations in the progress of my own ideas and sometimes those of others.

At this point I should refer to a disagreement between two different schools of theoretical physics about how to report theoretical ideas. I belonged to one of these schools, not as a result of careful choice, but simply through personal predilection.

And Yang and Lee, in the middle fifties, exemplified for me the other school. I thought that it was not fair for a theorist to propose several contradictory theories at one time, that the theorist should save his money and then bet on one idea that he really thought was right. And I believed also that proposing a wrong theory counted as having written a wrong paper. The other school argued, at least at one time, many years ago, that this was not so, that a paper was wrong only if the author made a mathematical mistake, that the point of a theoretical paper was to demonstrate the consequences of a particular set of assumptions. The assumptions did not need to be correct in nature, and one could in different papers try different contradictory assumptions without disgracing one's self. My feeling was that a theorist should be judged by the correctness of his guesses about nature, that his reputation should be gauged by the number right minus the number wrong, or even the number right minus twice the number wrong.

I remember talking about somewhat the same methods of scoring with Henry Kissinger, when he started to work for Nixon. We discussed Dean Rusk, who is said to have distinguished himself during the war in Korea in the following way: Allen Whiting has discovered, through intelligence methods, that a Chinese army group was missing from Manchuria in September of 1950. He was told to go to our Assistant Secretary of State for Far Eastern Affairs and tell him that the Chinese army group was missing and was most likely in North Korea, in the path of General MacArthur. Whiting did as he was advised, and the Assistant Secretary for Far Eastern Affairs is said to have replied, speaking of the Chinese leaders: «They would not dare.» Was it on the strength of that brilliant prediction he was made Secretary of State a few years later? Kissinger agreed with me in 1968 that people in such positions should be scored by number right minus twice number wrong. I do not know whether he would still agree.

The reason I have brought up my reluctance to discuss several competing theories at once is that it explains why I rarely shared my agonies of choice with the readers of my articles. If I had felt differently I could in many cases have saved myself a good deal of hesitation and anguish. I could have explained in print the confusions that were bothering me and the choices that I believed we faced.

During this period, I enjoyed agreeable and productive collaborations with a number of distinguished colleagues. I have no wish to slight their contributions, which were of the greatest importance, but in these reminiscences I shall emphasize mainly the evolution of my own thinking, with all its fits and starts. In the same spirit, I shall not always distinguish sharply between my ideas that were published fairly promptly and those that were delayed a long time in submission for publication or never printed at all. Much of my communication with other physicists was in the form of lectures, seminars, and conversations.

1. The renormalization group and the possible failure of old-fashioned field theory

Let me begin my reminiscences with the very hot summer of 1953, thirty years ago. I talked about some events of that summer at the Paris meeting, but now I shall discuss others. I spent part of the summer sweltering in Urbana, where the University of Illinois is located, working with my good friend Francis Low on what is now called the renormalization group, what we called quantum electrodynamics at small distances. We thought of ourselves as working on a very far-out problem, because we were discussing phenomena that would occur at enormous energies like exp 137 times the electron mass. Nowadays, everybody talks about such energies, but at that time they seemed ridiculously high. Of course we hoped that in meson theories and so forth, the relevant energy would be very much lower, because the coupling constant would be larger, and so the results might have some immediate relevance, but we had justified doubts about the correctness of those theories.

Petermann and Stueckelberg, I understand, worked on something similar at around the same time. And I think they used the phrase «renormalization group», which subsequently became standard.

Also writing on this subject were Bogolyubov and Shirkov. Their work consisted, as far as I can tell, largely in copying our analysis. But they accused us of having used the wrong gauge, and said that therefore our work was wrong and needed to be replaced by theirs. Now, in a gauge invariant theory, I do not know why using a different gauge is a crime, but for many years people referred to Bogolyubov and Shirkov rather than to us. Even my student Ken Wilson learned about the renormalization group from their paper rather than ours! The situation was rectified only in the book by Bjorken and Drell, who by some historical research, discovered the actual situation. Now, I mention this not in order to complain but because it comes up importantly in a moment.

In simple glorified perturbation theory, where denominators are expanded in perturbation series, the vacuum polarization in quantum electrodynamics has this form:

$$d(k^2) = \frac{D(k^2)}{D_{free}(k^2)} \simeq \frac{1}{1 - \dfrac{\alpha}{3\pi} \ln \dfrac{k^2}{m^2}}, \qquad (1)$$

where $D(k^2)$ is the photon propagator function. This formula is very well known; of course, if it were exact, it would contradict the positivity of the theory, because $d(k^2)$, which has to be positive, as we and others showed, would go to infinity and start again from minus infinity. The theory would therefore be self-inconsistent, and α would have to be zero.

Low and I found the results:

$$d(k^2) = \frac{e_0^2(k^2)}{e_1^2}, \qquad (2)$$

and

$$\int_{e_1^2+O(e_1^4)}^{e_0^2(\Lambda)} \frac{dx}{\psi(x)} = \ln \frac{\Lambda^2}{m^2}, \tag{3}$$

where e_1^2 is the renormalized coupling constant, e_0^2 is the bare coupling constant as a function of the cut-off Λ, and ψ is the renormalization-group function. (For some mysterious reason, $\psi(x)$ has been re-christened $\sqrt{x}\ \beta(\sqrt{x})$ by some people.)

We noticed that there are two possibilities if the theory makes sense. One is for the integral to diverge at infinity, in which case $e_0^2(\infty)$ is infinity, and

$Z_3 \equiv \dfrac{e_1^2}{e_0^2(\infty)}$ is zero.

The other possibility is to have the integral diverging before x reaches infinity in which case the upper limit is finite for $\Lambda=\infty$, there is a finite unrenormalized charge in QED, and Z_3 is different from zero.

We did not list the remaining possibility, that the integral converges, in which case the theory is inconsistent. However, shortly afterward, Källén suggested that this kind of inconsistency might occur, that a formula with the properties of eq.(1) might actually hold in QED. Källén referred to this situation as a theory with a «ghost». Pauli, too, discussed such a possibility. And Landau and his collaborators in the Soviet Union claimed to prove it, calling it the «zeroness of the charge», because for consistency one would have to put $\alpha = \dfrac{e_1^2}{4\pi}$ equal to zero.

This question is still unresolved after thirty years, and is the subject of research and violent controversy right at this very moment. A lot of «pure people» (pure mathematical physics types, of whose work I cannot understand a single word) have been studying the matter, and indicate that they are «close to a proof» that this kind of thing actually happens, at least in $\lambda\varphi^4$ theory. Now how does a mathematically pure person get close to a proof? I do not know. «Computer people», theorists working with a salt crystal lattice for space-time, have also worked on the problem, and they claim too that they are close to showing the zeroness of the charge, as $\Lambda\to\infty$, at least for $\lambda\varphi^4$ theory. Nick Khuri has analyzed all these claims and says that he does not believe there are any good arguments yet. He finds that so far the pure people have proved a theorem that is inadequate, because it is based on a hypothesis that may not be true and that he believes may well be false. He also finds that the computer people have not yet computed just the right quantities, and that their evidence is therefore not useful either. So, if he is right, the question is still completely open and the arguments so far would appear to be no more conclusive than those given in the 1950's by Landau and collaborators. I am, of course, not saying that the conjecture is false, only that it is still unproved.

The question is still of importance today, because if $\lambda\varphi^4$ theory makes no sense by itself and QED makes no sense by itself, and if the standard model is embedded in a larger, presumably unified theory (say a Yang-Mills theory with formal scale invariance), then that unified theory would have to have very special properties in order to make sense. The restriction would be severe, probably a restriction to the case of asymptotic freedom, which itself may be related to the restrictions imposed by supersymmetry. Howard Georgi and I have done some thinking on this subject recently.

In 1956 I was with the first substantial group of Western physicists that visited Moscow after the war. Some of the people in the group had been there twenty years before, like Weisskopf. For the younger ones, like me, it was, of course, our first visit, and it was something like visiting another planet. We had inadequate information about what was going on there, and there were many surprises. Khrushchev had just delivered his speech about Stalin, millions of people had been released from captivity, and the atmosphere was extraordinary. Our plane from Copenhagen landed in the middle of the night at Vnukovo and we were met on the tarmac by Tamm, Landau, Pomeranchuk, and many other leaders of the physics section of the Academy of Sciences. Some of the oldtimers in our group knew them from twenty years before. The rest of us became acquainted with them instantly. Landau was a very interesting, exciting person but very full of his own ideas; it was difficult, in fact almost impossible, to argue with him. In the course of our discussions over the next few days, he maintained, along with his entire crew (Pomeranchuk, Abrikosov, Yoffe, and so on) that they had proved the zeroness of the charge, arguing from leading logs in perturbation theory.

I explained that $\psi(x)$ needed to be known for large values of x in order to answer the question, and that such information could not be obtained from leading logs in perturbation theory. But I made no impression, because they thought of our work as Bogolyubov's and Bogolyubov was the enemy! They would not talk about $\psi(x)$ and the discussion ended.

2. Dispersion relations and the «S-matrix» program

Now let us return to 1953. I was working not only on the renormalization group, but also on strange particles, as I described at the Paris meeting, and on dispersion relations, with Murph Goldberger. Murph and I were engaged in a program of extracting as many general results as possible from local field theory, by proof if possible but otherwise by finding rules that held to every order of perturbation theory. Thus we discovered (or perhaps re-discovered) the crossing relations among amplitudes on or off the mass-shell.

Around the same time, we found in the literature the dispersion relation for the forward scattering without spin-flip of a massless particle like the photon, and we set about generalizing that dispersion relation as much as possible. First, Murph and I found the spin-flip dispersion relation for photon scattering in the forward direction, and also the low energy limit of the real part of the same amplitude; together with the oddness of the amplitude under crossing, we had all the elements of a useful formula. Then, after I left for Caltech, Goldberger and some other collaborators found the forward dispersion relations for non-zero mass. Relations that had been thought to be good only for the photon now worked also for mesons, provided one extrapolated the mass-shell amplitudes to imaginary momenta over a small part of the range of integration. Then there were the non-forward dispersion relations, which were found by many different groups: by Goldberger and collaborators, by Capps and Takeda, by Polkinghorne and me at Caltech, and by many others in the course of 1955 and 1956.

The discovery of the non-forward relations made possible the observation that dispersion relations, together with other known properties of amplitudes, were almost enough to specify a field theory. I described that idea at the Rochester meeting in 1956. I showed that to each order of perturbation theory, if one uses crossing relations, unitarity (thus picking up lower orders of perturbation theory in the calculation of the absorptive part of each amplitude), and then dispersion relations (to calculate the dispersive part from the absorptive part), one can generate each order of perturbation theory from the lower orders, provided there are suitable boundary conditions in momentum space, especially at infinite momenta. Crossing, analyticity, and unitarity, together with such boundary conditions, would give all the scattering amplitudes. Furthermore, the whole procedure would be carried out on the mass shell, provided one generalized the mass shell to include imaginary momenta.

I then mentioned casually that the program I was outlining, if treated non-perturbatively, was reminiscent of Heisenberg's hope of writing down the S-matrix directly instead of calculating it from field theory. I indicated that I was talking about a mass-shell determination of the scattering amplitudes in a particular field theory, using certain conditions in order to specify the theory. (The part about the conditions did not get into the written report, which was in the third person, but I remember saying it.)

Goldberger and I tried to teach these notions to Geoffrey Chew. It was very difficult, because he resisted furiously. Among other things, he disliked the idea that there were mysterious boundary conditions (and perhaps other conditions) that would distinguish one theory from another and complete the information necessary to give the whole S-matrix.

In 1958 in Geneva, Geoffrey Chew quoted my Rochester remarks of 1956; apparently by then he had started to believe them. In the meantime, in the Soviet Union, Landau had caught on to these ideas. In his group they were now busily studying analyticity properties of the S-matrix, trying to obtain them from perturbation theory. Jon Mathews at Caltech was doing much the same sort of thing, but he did not reach as wide an audience as Landau. Landau, however, went on to enunciate a special dogma, starting from his impression that field theory was no damned good because all the coupling constants were zero. He proposed that while field theory was wrong, the program of using dispersion relations, crossing relations, and unitarity to calculate the S-matrix was right and could be used as a substitute for field theory. Whereas I had suggested the program as another way of dealing with field theory, Landau now proposed that it was correct and field theory incorrect. I could never understand that point of view and I still cannot.

Of course, the general program could allow for field theories (such as Yang-Mills theories, which were known at the time but not known to be renormalizable or asymptotically free) that one had overlooked and that might have better properties, but that is not the way the dogma was expressed. Presumably, in today's language, Landau wanted to say that non-asymptotically-free field theory was no good, but that was the only kind we knew about. In any case, we still do not know, as I mentioned earlier, whether non-asymptotically-free field theory has to have zero charges.

In Kiev in 1959, I had terrible arguments with Landau on the subject of condemning field theory and welcoming dispersion theory. As usual, it was impossible, really, to argue. He did not give an inch. Then Geoffrey Chew adopted the same sort of point of view in La Jolla in 1961.

I had suggested the La Jolla meeting and then turned it over to Keith Brueckner to organize as a way of celebrating the beginning of U.C.S.D. He had just moved to La Jolla, and it seemed to be a good way to put his department on the map. He did a beautiful job of organizing the meeting. Robert Oppenheimer attended, and I remember that Keith and I entertained him at the Hotel del Coronado, where there are (so to speak) heads of stuffed admirals on the walls. There was an aged, uniformed waiter who took our order. As a compliment to Robert because he fancied himself as a connoisseur of wine, I asked him to select the wine for lunch. He looked at the wine list, which did not give much information, and asked the waiter the year of a particular Bordeaux. The waiter replied: «Sir, all our wines are at least five years old.»

There were no Proceedings of that conference. That is a pity, because we heard Goldberger and Blankenbecler on Regge poles and Mandelstam on the Mandelstam representation, I talked about the current algebra in the Eightfold Way, and so forth and so on. There were many very interesting first reports of discoveries, all of which are lost to historians. But the most dramatic talk was that of Geoffrey Chew, who said, in effect: «I have always been a simple worker in physics: I do my calculations, I maintain a low profile, and I do not make grand pronouncements. But this time I have something important to say. Field theory is no damned good; instead we must use the S-matrix theory.» He did not say dispersion relations, crossing, unitarity and so on, but «S-matrix theory». He allowed that electromagnetism and gravity might be different. But for the strong interaction he insisted that we abandon field theory and go over to «S-matrix theory».

As I mentioned just now, I have never succeeded in understanding that point of view. While before 1961 one of the great pleasures of working in theoretical physics was the possibility of discussing theories with Geoffrey Chew, after 1961 it became very difficult.

His collaborator Stanley Mandelstam did not agree with the dogma. He did very important work on the «S-matrix» program, but he did not subscribe to the Chuvian religion. He thought, like me, that if the nonlinear system of unitarity and dispersion relations on the mass shell possessed a solution, then the set of linear equations that take us off the mass shell could well have a solution too, so there would be some kind of field theory for each «S-matrix» theory.

Associated with the dispersion theory or «S-matrix» approach to the hadrons was the «bootstrap» picture of the hadrons, as discussed by Chew and Frautschi; and about that I was quite enthusiastic. Using suitably smooth boundary conditions at infinite momenta in the system of unitarity, dispersion, and crossing relations, one was to calculate the properties of the hadrons in principle without introducing any fundamental objects. None of the observable hadrons would be any more fundamental than any other – we called that principle «nuclear democracy». Perhaps we should have called it «hadronic egalitarianism». Anyway, it was a good idea. In practice, one studied two colliding hadrons and demanded that the set of exchanged hadrons in the «t and u channels» be the same as the set of resonant and bound hadrons formed in the «s channel». I liked the approach, but I complained during the middle sixties that Geoffrey Chew and his group kept beating to death one state, the rho meson, and one kind of system, with two pions, rather than making approximate calculations with an infinite number of intermediate states in each channel and a correspondingly infinite number of systems. The approximation could involve assuming narrow resonances only. Some Caltech postdocs, Dolen, Horn, and Schmid, took this up in 1967 in a seminal paper on «duality», in which they suggested that one could in fact approximate the intermediate states by an infinite set of resonances in the s, t, and u channels and that these resonances could be the same as the external particles. This observation led to the Veneziano dual-resonance model, and later to the Neveu-Schwarz model. Remarkably, a recent variant of the Neveu-Schwarz model in ten dimensions is now being proposed as a possible universal theory of all the forces and all the elementary particles of nature, with ten-dimensional supergravity and ten-dimensional unified Yang-Mills theory as an approximation. Of course, the slope of the trajectory has changed a little bit, by a factor of 10^{38} or so. But apart from that trivial modification, the dual-resonance or string theories are again at the forefront of research.

In 1963, when I developed the quark proposal, with confined quarks, I realized that the bootstrap idea for hadrons and the quark idea with confined quarks can be compatible with each other, and that both proposals result in «nuclear democracy», with no observable hadron being any more fundamental than any other. This point is emphasized in my lecture to the Royal Institution in London in 1966.

What worried me about the bootstrap was that it distinguished hadrons sharply from leptons, while the weak and electromagnetic interactions treated them nearly alike. A feature that I liked about the quarks (the current quarks), when I found them, was that they presented an analogy with the leptons, in electromagnetic and weak interactions.

3. Hadron approximate symmetries and Yang-Mills theories for the electro-weak and strong interactions

In 1957 I tried an approximate symmetry for the hadrons, the «global symmetry», based on SO(4), which has the same algebra as SU(2)×SU(2). I shall not discuss it at length here, except to say that it was wrong, and that the mass formula for the N, Ξ, Λ, and Σ had the factors one and three in the wrong places. «Global Symmetry» gave the approximate mass formula:

$$\frac{1m_{\Lambda} + 3m_{\Sigma}}{4} \approx \frac{m_{N} + m_{\Xi}}{2}. \tag{4}$$

If we interchange the 1 and the 3, we get a much better formula, as I realized at that time. Nambu, I remember, noticed the same thing. But we did not know at that time what kind of theory would interchange them. A few years later, the *Eightfold Way* scheme gave the same approximate formula with the 1 and the 3 interchanged and was successful.

I shall not dwell on the complications of the discovery of the weak interaction theory in 1957. But it is well known that it required the rejection of a considerable number of experimental results. Here I shall begin to refer to Table 1, where I have listed a large number of confusions of the period, my confusions anyway, some of them shared by other people. These confusions are of several kinds. Some are mathematical questions about field theory (part A). Others are theoretical questions about how to explain crucial experimental results (parts C and D). Only a few involve confusions arising from wrong experimental results or wrong preliminary reports of experiments (part B). In the weak interaction domain there was a series of wrong experimental results, such as that of Rustad and Ruby on a scalar and tensor interaction in β-decay, the reported failure of $\pi^{\pm}$ to decay into electron and neutrino, and so forth and so on. (An example of a wrong result that held sway for a time in another part of particle physics is the report that the Σ-Λ relative parity was minus instead of plus.)

I was writing a review article on weak interaction with Arthur Rosenfeld during the spring and summer of 1957, when I came to the conclusion that there was a chance that the universal Fermi interaction could still be right, and have the form V−A, provided that we could ignore various experimental results. We included the idea in the review article, where we called it the «last stand» of the universal Fermi interaction.

Meanwhile, George Sudarshan, working with Bob Marshak, was making a similar suggestion, but in a more confident manner. I recall a summit meeting with them on the weak interaction at the RAND Corporation (not usually considered a place for that kind of activity). There was an exchange of views, and we mentioned our section on the «last stand», while they told us of their plans to write an article. Felix Boehm was there, too, describing his recent experimental results, which indicated V−A.

I then went on vacation, after further discussions of the «last stand» with Felix Boehm, who described it to Feynman when he returned from Brazil in my absence. Feynman got tremendously excited, expanded the idea somewhat, and wrote a long paper on it. When I returned we decided to modify it and sign it together, but it bore Feynman's stamp, including a notation that I would normally not have used. However, we made some useful proposals in that paper. We also did research that was not included and that Feynman and I presented in successive talks at the American Physical Society meeting at Stanford around Christmas of 1957.

That was the time when Okun' was visiting from Moscow. We taught him to play Monopoly, and sent him back to the Soviet Union with a Monopoly game. That may account for some of his difficulties in attending subsequent meetings. He was, by the way, a very good Monopoly player.

At that meeting I was chosen to talk about the intermediate boson, which we called X^+ and X^- (and which I still call X^+ and X^-!), and about the different possibilities for explaining the absence of $\mu \to e + \gamma$ decay. On behalf of Feynman and me, I described our calculation of $\mu \to e + \gamma$, using the intermediate boson, with electromagnetic properties derived from what is essentially the modern theory, of Yang-Mills type. We had found a finite result. The same result was obtained by Gerald Feinberg, who published it a little later. We showed that $\mu \to e + \gamma$ would go very quickly if there were a serious intermediate boson. Therefore there were only two possibilities, both of which I mentioned. One was that there was a cut-off below the mass of the intermediate boson or there was no intermediate boson at all. The other was that there were two kinds of neutrinos: «red» and «blue», for electrons and muons respectively. I wrote this speech up for publication, but we never submitted it, and Feynman gradually lost his enthusiasm for the red and blue neutrinos. When Bludman presented his work on two kinds of neutrinos the following year at Gatlinburg, Feynman did not like them any more. My worry about the red and blue neutrinos was not so much about whether they were a good idea as about how they could be verified. It did not occur to me, as it did later to Pontecorvo and to Lee and Yang, that it would soon be possible to do accelerator experiments with high energy neutrinos that could actually verify the existence of the two kinds of neutrinos.

During 1958, Feynman and I, together with the postdocs Richard Norton and Keith Watson, worked on the notion of a charged intermediate boson with dimensionless coupling constant approximately equal to e, and we saw that the mass of the intermediate boson had to be around 100 GeV. That was a very high energy, and we did not see any immediate possibility for verifying it. Also we were terribly concerned about the absence of the decay $\mu \to e + \gamma$ unless there were two different neutrinos. And of course, in connection with a neutral intermediate boson, we were worried about how to get a neutral current without a strangeness-changing term, the usual difficulty that was fixed years later by means of charm.

I played briefly with the idea of having two neutral intermediate bosons, as in the model that was published later by Lee and Yang, who called them schizons, because they behaved like $I=1$ for strangeness-preserving couplings and like $I=\frac{1}{2}$ for strangeness-changing couplings. Unfortunately, this model seemed to offer no explanation for the absence of a strangeness-changing neutral weak current, coupled to leptons, although it could explain the absence of a weak non-leptonic interaction with $|\Delta S|=2$. Later, Glashow and I were able to show that it did not fit in with Yang-Mills theory; there was no Lie algebra of the right kind to make it work.

In my list of confusions in Table 1, I should like to point out particularly the one (C 1.) referring to the electron with its neutrino and the muon with a different neutrino, making four leptons. (We did not, of course, know about the third family at that time.) In that case the intermediate boson would be O.K., and $\mu \rightarrow e+\gamma$ would not create any problems. And in that case the universality of the weak interactions required that the hadron analog of these doublets be something like proton and «neutron times cos θ plus Λ times sin θ» for a small angle θ. In 1959 Lévy and I worked out the size of this angle, and compared it successfully with a small discrepancy in the usual formulation of the universality of the Fermi constants. Our value of around 15 degrees has survived to this day. Later on, for quarks, the doublet became u and «d cos θ + s sin θ». But there was always lurking in the background the other possibility that there was only one neutrino, coupled to e + μ. That, of course, gave difficulties with the charged intermediate boson, because of the decay $\mu \rightarrow e+\gamma$, but we knew there were troubles with the neutral intermediate boson anyway, because of the strangeness-changing neutral weak current and $\Delta S=2$ problems. With a single neutrino, the hadron analog of ν and e+μ would be something like p and n+Λ, with the p Λ coupling conjectured to be much reduced by renormalization. I hesitated between these two possibilities for years, although I preferred the case of two neutrinos. In September of 1960 at Rochester, I actually explained the two alternatives carefully, with their consequences for the weak couplings of hadrons.

In our paper on the weak interaction, written in 1957, Feynman and I included the conserved vector current idea. That idea had actually been suggested some years before by Gershtein and Zeldovich in the Soviet Union, but we did not know that. The fact that the strangeness-preserving vector current for the charge-exchange weak interaction in the case of hadrons was the current for the plus and minus components of isotopic spin indicated to me that the entire charge-exchange weak current, including strangeness-changing and leptonic terms, must be the current of some weak charge operator and its Hermitian conjugate, with interesting algebraic properties. Not only that, but the hadronic part, when broken up into vector and axial vector charges and strangeness-changing and strangeness-preserving charges, would have to generate a larger algebra. I became very interested in the commutation relations and in how these commutation relations would close in each case, for the total weak charge and, in the case of hadrons, for the individual pieces.

I did not call the resulting algebra in either case a Lie algebra, because I had forgotten what a Lie algebra was. I had studied Lie algebras when I was an undergraduate, taking advanced mathematics courses at Yale. But you know what

Table 1. Some theoretical and experimental confusions of the late fifties and early sixties

(A) *Some mathematical questions:*

1. Is Yang-Mills theory generalizable beyond products of SU(2) and U(1) factors for the gauge group?
2. Is Yang-Mills theory with zero mass renormalizable?
3. Is Yang-Mills theory with hard masses, equal or unequal, renormalizable?
4. Is there a soft mass mechanism?
5. Are old-fashioned field theories consistent, or must their coupling constants be zero?
6. Is there another kind of field theory? (As it turned out, asymptotically free.)
7. Is the «S-matrix» approach different from field theory?

(B) *Some examples of confusion resulting from experimental error:*

1. Beta-decay weak interaction scalar and tensor?
2. $\pi \not\to e\nu$?
3. Σ-Λ parity odd?

(C) *Some physical confusions connected with missing charm:*

1. $\nu_e\ e;\quad \nu_\mu\ \mu$	$X^\pm$ OK	$p,\ \frac{n+\Lambda\varepsilon}{\sqrt{1+\varepsilon^2}}$
	versus	
$\nu,\ e+\mu$	$X^\pm$ difficult	$p,\ n+\Lambda$

with lack of fourth hadron flavor to correspond to four leptons in first alternative above.

2. Strangeness-changing neutral weak current in SU(2)×U(1) scheme in absence of fourth hadron flavor, and resulting problem of $\Delta S = 2$ transitions if there is only one such current.
3. Fourth hadron flavor versus TrQ=0 for quarks alone, and TrQ=0 for quarks alone versus TrQ=0 for quarks and leptons together.

(D) *Some physical confusions connected with missing quarks and color:*

1. Appeal of Yang-Mills theories for both strong and electro-weak interactions versus clashing of the two in chiral flavor space.
2. Flavor algebra for hadrons, besides being approximate symmetry, also generating gauge group for Yang-Mills theory of strong interaction?
3. Algebra generated by pieces of weak charge same as chiral flavor algebra for hadrons? (By the way, is the electro-weak gauge group just SU(2)×U(1) or is it some larger subgroup of the chiral flavor group?)
4. Appeal of hadron scheme with fundamental entities like p, n, Λ coupled to neutral vector boson(s) and with electro-weak couplings showing a parallel between these entities and leptons

   versus

   «nuclear democracy» (no observable hadrons distinguished as more fundamental than others) and appeal of not having to put Σ and Ξ into separate 15 representation if they are in an octet with the nucleon.

(E) *Some points not clearly understood about quark statistics:*

1. Confinement of «mathematical quarks» connected with suppression of para-baryons and para-mesons.
2. Parastatistics with suppression of observable para-hadrons equivalent to color with suppression of colored hadrons.

mathematics education is like, rarely emphasizing the intuitive and computational, concentrating on general results with theorems and proofs. While I followed the courses and did well in them, I never really understood what was going on. Mathematicians usually give only trivial examples; in the case of a Lie group, they might discuss U(1). A generous mathematician might include SU(2), but not SU(3) or E_6. Reduction of representations as one goes from a group to a subgroup is discussed, but again non-trivial examples are unlikely to be presented.

As a graduate student in physics, I struggled manfully to read the work of Professor Wigner published in 1936-1937: the Harvard tercentenary lecture on the nucleus and then his 1937 *Physical Review* paper on SU(4). I read his papers over and over again, I consulted books on group theory, but I still did not properly understand what it was all about.

In 1951 I attended the beautiful lectures by Giulio Racáh or Rácah. (He used the former pronunciation in Italian and the latter in Hebrew; if you were drinking the Italian wine with his name on the label, you would presumably put the stress on the second syllable.) I listened to his lectures, and received the notes, taken by my friends Park and Merzbacher, but I did not really understand the material. The reason was not that the lectures were not elegant, or that they were not explicit. The problem was his accent. Of course I understood the words; I have no trouble following English spoken with a foreign accent. But his accent was so remarkable that I could not hear the substance. Every English word was pronunced with a perfect Florentine accent. For example he would say: «Tay vah-loo-ay eess toh eeg». (The value is too high). So I never really learned about Lie algebras, and I had to rediscover them.

In any case, from 1957 through 1960 I worried about the commutation relations of the weak charge and their possible relation to a theory of the weak and electromagnetic interactions of the Yang-Mills type, and at the same time I worried about the presumably larger algebra generated by the separate pieces of the weak charge. I knew that universality of strength of the weak interaction had to depend on the commutation relations of the weak charges, which are nonlinear, and that universality of at least differences of electric charge would be similarly controlled since the weak charge operator always changes electric charge by one unit. But my thinking on the whole subject was hampered by my failure to realize that the possible algebraic systems were known and tabulated.

I spent the academic year 1959-60 in Paris, at the Collège de France and at other Parisian institutions. There I continued, with Lévy, work that I had started with Feynman at Caltech on the derivation of the Goldberger-Treiman relation. This led to the idea that the divergence of the axial vector current was proportional to a pion field. Then, with Lévy, Bernstein, Fubini, and Thirring, I progressed to a more sophisticated idea. After all, who knows what a pion field is? The main thing is that the divergence should be a «soft» operator, so that its matrix elements would be dominated at low frequencies by the contribution from the pion intermediate state. This was «PCAC», the principle of the partially conserved axial vector current.

According to PCAC, as the divergence of the current goes to zero, m_π tends to zero rather than m_N. In the limit, there is a realization of the «Nambu-Goldstone» mechanism, which those authors were developing independently around the same time.[1]

During 1959-60, Shelly Glashow came to visit me in Paris, and he explained his theory of SU(2)×U(1) Yang-Mills vector bosons, with mixing of the weak and electromagnetic interactions. We did not call it SU(2)×U(1), but that was what it was. Applied to leptons, it was like the present electro-weak theory except for the Higgs mechanism added by Weinberg in 1967 and discussed also by Salam. I gave the first easily intelligible presentation of this theory in September, 1960, attributing it of course to Glashow, at the Rochester meeting. The abstract was sent in longhand from Kisoro, Uganda, where I was looking at wild gorillas, with apologies for the lack of secretarial help.

All of this was very nice, provided the strangeness-changing contribution to the weak charge was omitted. But if it was included, then Shelly's theory seemed to be in trouble, because there would be a strangeness-changing neutral weak current and serious problems with $\Delta S=2$ transitions. Charm was, of course, the way out, but we did not use it, and consequently the puzzle remained unresolved.

I interested myself in the algebra that would be generated by the various pieces of the weak charge, those connected with the vector and axial vector, strangeness-preserving and strangeness-changing currents. If I had understood Lie algebras in 1959 and early 1960, I would have realized that (without charm) the system of commutation relations could close on the chiral algebra of SU(3)×SU(3). But I did not. The notion that all the relevant algebraic systems were tabulated and that I could look them up in a book was still strange to me.

In Table 1 (part A), we see some more confusions of that time. The Yang-Mills theory with zero vector boson mass, was it renormalizable or not? And what about the renormalizability with masses? For a while, Glashow claimed wrongly that a Yang-Mills theory was completely renormalizable, even with masses, even with different masses for different vector bosons.

Salam and Kumar were proving, in London during 1959-60, that this was false, but that the unbroken theory was renormalizable. Later on, in my *Eightfold Way* report, I advertised for a «soft mass» for vector bosons in Yang-Mills theory. Also, in the course of our PCAC work in Paris, Lévy and I had played with the sigma model, both linear and nonlinear (which had also been studied by Schwinger and by Gürsey, respectively, in different connections). By January 1961, I had all the materials necessary for inventing Higgs bosons as a mechanism for providing the «soft mass» needed for the renormalizability of a Yang-Mills theory with broken

1. Thus the algebra that we would now call the SU(2)×SU(2) chiral flavor algebra is realized as an approximate symmetry by having the charges of the vector currents (isotopic spin components) give approximate hadron degeneracies and having those of the axial vector currents give nearly massless pions. I was to extend this idea later to flavor SU(3)×SU(3). Even on this major point I wavered briefly in early 1964, when I tried using the charges of axial vector currents as very approximate degeneracy symmetries.

symmetry. However, I failed to see the relevance of the sigma model and I did not solve the problem.

I had continued to play with Yang-Mills theory for the weak interaction all during 1958 and 1959. Early in 1959 I decided that Yang-Mills theory must also be relevant to the strong interaction.[2] But it was very difficult to understand how the two could be related. Late in 1959, in Paris, I decided to find out what where the possible generalizations of the Yang-Mills theory —those two authors had given only the case of isotopic spin, what we would nowadays call SU(2). Quantum electrodynamics was, of course, an example of a gauge theory using U(1). We could readily conceive, therefore, of a gauge theory involving what we would call today a product of SU(2) factors and U(1) factors. When Shelly visited and presented his scheme, that fitted in easily because it corresponded to one factor of each type. The question was whether any other generalization existed. I worked and worked in my office in the Collège de France and finally I wrote down, as the necessary and sufficient conditions, the canonical relations:

$$[\mathbf{F}_i, \mathbf{F}_j] = ic_{ijk}\mathbf{F}_k\,, \tag{5}$$

where c_{ijk} is real and totally antisymmetric and the $\mathbf{F}_i$ are Hermitian charge operators. I had no idea what were the possible realizations of this formula.

Every day I would have lunch with my French friends and drink wine, and afterwards I would come back and struggle with drowsiness in my office. I worked through the cases of three operators, four operators, five operators, six operators, and seven operators, trying to find algebras that did not correspond to what we would now call products of SU(2) factors and U(1) factors. I got all the way up to seven dimensions and found none, of course. At that point I said: «That's enough!» I did not have the strength after drinking all that wine to try eight dimensions. Unfortunately I did not pay sufficient attention to the identity of one of my regular companions at lunch. It was Professor Serre, one of the world's greatest experts on Lie algebras. I knew, of course, that Serre was a famous pure mathematician, but I did not know what his specialty was. It never occurred to me to ask him about my equations, and I doubt whether he would have given me the answer, if I had asked him. Probably this canonical form for the commutation relations was far too explicit for Serre. He would have preferred something like:

$$T : s \longrightarrow u\,. \tag{6}$$

At least that is my impression of the kind of formula that mathematicians like to write.

2. I had always thought that understanding classification and symmetry would lay the groundwork for discovering the dynamics of the weak (or electro-weak) interaction and of the strong interaction, but in Yang-Mills theory it was clear that the determination of the dynamics involved merely choosing the symmetries to be gauged. Thus the study of symmetry was more important than ever.

It was not until December 1960 at Caltech that a young mathematician called Block told me: «What you have there is a canonical form for the commutation relations of a Lie algebra generating a product of U(1) factors and simple compact Lie groups in a unitary representation.» I did not know that, of course; I had been reinventing those Lie algebras.

By this time Glashow was spending a year with me at Caltech, and we had been discussing how to generalize the Yang-Mills trick and how to apply it to weak and strong interactions. While he was away in Massachusetts, I received enlightenment about Lie algebras and immediately thought up the *Eightfold Way.* But on his return we resumed our collaboration.

Two pieces of work now emerged. Glashow and I wrote up our research on generalizing Yang-Mills theory from the group SU(2) to any product of simple compact Lie groups and U(1) factors; and we discussed our attempts to construct a Yang-Mills theory for the strong interaction and one for the weak and electromagnetic ones (such as Glashow's SU(2)×U(1) theory, which has since proved to be essentially correct). I wrote my Caltech Report on *The Eightfold Way,* in which I suggested flavor SU(3) as an approximate symmetry of the hadrons, with symmetry violation coming from the eighth component of an octet and leading to the baryon and meson mass formulae. The baryon octet mass formula was just the one, eq. (4) with the 1 and 3 interchanged, that I had wanted. In the Report I also proposed a Yang-Mills theory for the strong interaction based on flavor SU(3).

In both papers there was a fundamental difficulty, namely the absence of color space that permits us today to have the Yang-Mills gauge groups for strong and for electro-weak interactions operating on different sets of coordinates. Shelly and I had to cram our strong and electro-weak Yang-Mills theories into chiral flavor space, where they clashed in an intolerable way. While our work on generalizing the Yang-Mills idea was elegant, our attempts to construct both strong and electro-weak theories without color were necessarily clumsy.

I therefore had to face the question of whether to maintain or abandon my particular theory of the strong interaction based on flavor SU(3). I decided to abandon it as I converted my Caltech Report into an article for the *Physical Review;* the successful features of the theory could be replaced by the assumption of «vector dominance». Thus, the SU(3)×SU(3) chiral flavor algebra was to be used as an approximate symmetry, with the charges of the vector currents giving approximate degeneracy of hadron states and those of the axial vector currents giving approximately massless pseudoscalar mesons (the PCAC idea). At the same time, these currents would obey local current algebra relations, which are a kind of pre-condition for Yang-Mills theory, and thus they could include electromagnetic and weak currents that would be suitable sources for a Yang-Mills electro-weak theory. The pieces of the weak and electromagnetic charges would generate the whole chiral flavor algebra, while the total weak and electromagnetic charge operators would generate the electro-weak subalgebra.

I put off for a long time the construction of a Yang-Mills theory for the strong interaction. The abstract approach fitted in nicely with the assignment of the $J=\frac{1}{2}^{+}$ baryons to an octet. But then I wavered for a while on both the approach and the assignment.

Back in 1958 and 1959, I had played with the idea, which Sakata was to publish and elaborate, of treating the baryons p, n, and Λ as fundamental and then trying to explain others as composites of two «fundamental» baryons and one «fundamental» anti-baryon. Since at that time I did not understand SU(3), I used the permutation group S_3 to analyze the symmetry properties of the model. In this indirect manner I found the relation that in the language of SU(3) representations could be given as:

$$3 \times 3 \times \bar{3} = 3 + 3 + \bar{6} + 15 . \tag{7}$$

The Σ and Ξ hyperons, along with some other hypothetical ones, would have to be put into the 15. At the time I did not find this possibility at all attractive, but Sakata adopted it.

After I wrote my Caltech Report on *The Eightfold Way* in January 1961, I became worried by the reports that the Σ and Λ hyperons had opposite parity. Also, my work on current algebra, involving currents for flavor SU(3)×SU(3), persuaded me that there should in some sense be three fundamental spin $\frac{1}{2}$ hadrons. Thus for a while I embraced what was by then the Sakata model, and submitted to the *Physical Review* a version of my work on flavor SU(3) in which the octet assignment of baryons was almost entirely suppressed.

During the summer of 1961, though, my faith in the octet assignment of baryons was restored; the odd parity rumor was receding and the beauty of the scheme continued to impress me. I added Section VIII to my *Physical Review* paper, in which I treated the octet assignment in detail. Since the early part of the paper used the Sakata model for explaining flavor SU(3) and SU(3)×SU(3), it was easy to emphasize the notion of abstracting these symmetries from a concrete model based on three fundamental objects. My agonizing over the correctness of the octet assignment for baryons thus led to a useful presentation of the subject. A very natural groundwork was being laid for the quark model of 1963.

In 1961, however, I had to cope with an additional confusion, the continuing puzzle about two neutrinos versus one and the small weak angle θ versus equal bare strength for strangeness-preserving and strangeness-changing weak currents, with the latter reduced by renormalization. In my early work on the *Eightfold Way,* I leaned away from the correct hypothesis of two neutrinos and a small weak angle and toward the wrong one, because of the parallel between the triplet substrate of SU(3) on the one hand and three leptons on the other. Of course, if we had suggested charm at that time, this confusion would have disappeared, along with the problem of the neutral strangeness-changing current; four flavors for leptons and four for hadrons would have been so neat. (I noted in my Paris talk of 1982 that Pais and I had presented a model at the Glasgow meeting in 1954 that contained what was essentially charm, but of course the «charmed» hadrons in the scheme were not found during the fifties and sixties, and I did not revive the idea.)

I should mention that there was a subtle reason why I stayed away from charm, even after it had been suggested by Bjorken and Glashow in 1964. Here is yet another source of confusion. In the *Eightfold Way* scheme, electric charge is a generator of the flavor SU(3) group, and correspondingly when we go over to quarks the sum of the quark charges $\left(\frac{2}{3} - \frac{1}{3} - \frac{1}{3}\right)$ is zero. If we bring in charm, then electric charge is no longer a generator of the resulting flavor SU(4), and the sum of the quark charges $\left(\frac{2}{3} + \frac{2}{3} - \frac{1}{3} - \frac{1}{3}\right)$ is no longer zero; the electric charge operator contains a piece that is outside the SU(4) algebra and commutes with it. I wanted to avoid that situation and to have the electric charge operator for quarks as a generator of the relevant algebra. I should have realized that such a narrow requirement was foolish since for the leptons the trace of electric charge is not zero either. In fact, in a modern unified theory of strong, electromagnetic, and weak interactions, we must consider each whole family of quarks and leptons together, including the three quark colors, for the trace to come out $(0 - 1) + 3 \times \left(\frac{2}{3} - \frac{1}{3}\right) = 0$, so that the electric charge can be a generator of a hypothetical unified Yang-Mills gauge group without a U(1) factor, such as SU(5) or SO(10). I did not appreciate this point fully until about 1973.

To return to the winter of 1960-61, Yuval Ne'eman invented the *Eightfold Way* independently and published it without much hesitation. David Speiser also thought up the SU(3) octet assignment for baryons, I understand, but was discouraged from publishing it. They will no doubt tell about their adventures.

In 1962, at the International Conference in Geneva, I heard a presentation of the experimental discovery of a baryon resonance that was an excited Ξ. It occurred to me immediately that along with the $J = \frac{3}{2}^+$, $I = \frac{3}{2}$ excited nucleon and a recently discovered excited Σ, it could form part of a decimet with $J = \frac{3}{2}^+$. I wrote down the mass formula and realized that in the case of the decimet the formula gave equal spacing. The three known masses were around 1240 MeV, 1385 MeV, and 1530 MeV, with the last mass being the one just reported. I became very excited and jumped up to address the meeting. I pointed out the likelihood that we were dealing with a $J = \frac{3}{2}^+$ decimet in the *Eightfold Way* scheme, and predicted the existence of Ω^- at around 1675 MeV, a metastable baryon of spin $\frac{3}{2}$ that could decay spectacularly into $\pi^- + \Xi^0$ or $K^- + \Lambda^0$. (I was so excited that at first I called it Ω^0 and had to correct myself.) At lunch, at the request of Leitner and Samios of Brookhaven, I wrote a note on a paper napkin to Maurice Goldhaber, the Director of Brookhaven, asking him to let them search for Ω^-. After two years and after scanning two million feet of film, they finally found, in 1964, the two events (one $\pi^- \Xi^0$ decay and one $K^- \Lambda^0$ decay) that confirmed the *Eightfold Way* scheme with flavor SU(3) as an excellent approximate symmetry.

4. Quarks

Now we get to quarks. In early 1963, lecturing at M.I.T., on leave from Caltech, I tried to work out for my lectures the minimal set of fundamental hadronic entities. I found various sets of four objects if I insisted on integral charges; no scheme of that kind looked particularly attractive. But then in March I went to Columbia on a visit, and at the Faculty Club there Bob Serber asked me why it was not possible to use my formula:

$$3 \times 3 \times 3 = 1 + 8 + 8 + 10\,, \tag{8}$$

to obtain the baryons. I replied that I had tried to get the baryons that way, but that the fundamental entities would then have fractional charges —I showed him (again on a paper napkin) the fractional charges: $\frac{2}{3}$ and $-\frac{1}{3}$. He said in effect: «Oh well, I see why you do not do it then.» But thinking about the matter afterwards, that day and the next morning, it occurred to me that if the bootstrap approach were correct, then any fundamental hadrons would have to be unobservable, incapable of coming out of the baryons and mesons to be seen individually, and that if they were unobservable, they might as well have fractional charge.

Later on, after I had named the quarks,[3] I referred to these trapped fundamental entities as «mathematical quarks» —by which I always meant that they were permanently stuck inside the baryons and mesons. (Conceivably, they were almost trapped and very rarely, with great difficulty, they could come out and be seen in very sensitive experiments —then I would call them «real quarks».) But usually I thought of quarks as being permanently trapped, right from the first day.

Except for the bootstrap idea, the notion of «nuclear democracy», I do not know if I would have come so readily to the notion that fundamental hadrons ought to be unobservable and that therefore it was all right for them to have fractional charge. Of course, it was known that fractionally charged particles were absent or at least very rare in nature, and so it was particularly appropriate to think of the quarks as being unobservable; an alternative hypothetical set of integrally charged fundamental hadrons would not have this additional, experimental reason for being unobservable in isolation. That is what I meant, in one of my early papers on quarks, when I mentioned a less attractive model with a larger number of integrally charged fundamental objects and wrote that such objects «would be more likely to be real than quarks». I was not by any means rejecting the quark idea, merely remarking that fractionally charged quarks had an empirical reason for not being «real» (i.e., directly observable) that integrally charged counterparts would not have. At my lecture to the Royal Institution of London in 1966, I emphasized that «mathematical» (that is, confined) quarks would be compatible with the bootstrap idea. Later

3. By the way, in referring to the quotation about «three quarks» in *Finnegans Wake*, I was thinking of the three quarks in a baryon, corresponding, as we learned later, to the three colors. Thus the quotation is still apposite.

that same year, in my introductory review lecture at the International Conference in Berkeley, I said: «One may think of mathematical quarks as the limit of real light quarks confined by a barrier, as the barrier goes to an infinitely high one.» I was still using the terms «mathematical» and «real» to denote what today might be called «confined» and «incompletely confined» quarks. Now why did I use that language? Probably because I dreaded philosophical discussions about whether particles could be considered real if they were permanently confined. While a colleague of mine falsely claims to have a doctor's prescription forbidding him to engage in philosophical debates, I really do have one, given to me by a physician who was a student in one of my extension courses at U.C.L.A..

Even in my original letter on quarks, published early in 1964, I stressed their compatibility with the dispersion theory program and the likelihood that the quarks would not emerge. At that time I referred to the «mathematical» nature of the quarks as resulting from the limit of infinite mass rather than an infinite barrier.

The notion of confined quarks fitted in well with the abstract approach that I had taken in 1961 with respect to the octet assignment of baryons. In neither case did I insist on observable fundamental hadrons like p, n, and Λ. In 1961, except for a period of wavering, I took the position that SU(3) of flavor could be abstracted from something like the p, n, and Λ model and used as an approximate symmetry with whatever representations were needed; and in 1963 I maintained that quarks that were probably not «real» could be the substrate of the operation of SU(3).

On my first visit to Japan in the spring of 1964 I encountered with something of a shock the exactly contrary attitude of the group of Marxist theoretical physicists that included Sakata. Yukawa, who did not entirely share their views, had nevertheless helped to place many of these theorists in academic positions in the Kansai region, in the vicinity of his Institute in Kyoto. I was visiting Yukawa's Institute, and a meeting was arranged in my office with Taketani, Ohnuki, Maki, and some others. Sakata, who was in the building at the time, did not come. All of these men were strongly opposed to my abstract approach, which they condemned as smacking of «bourgeois or revisionist idealism». They had missed the octet assignment of baryons in flavor SU(3) on account of their *a priori* requirement of concreteness, and now they insisted that if there were basic hadrons they must be integrally charged and observable. Presumably, they wanted concrete basic objects that could be explained to the masses. It was interesting to see these very intelligent theoretical physicists, working on exactly the right problems with suitable mathematical methods, missing right answers because of their fixed philosophical positions. If I had designed a lesson in the virtues of pragmatism, I could not have found a better one, and I described that lesson in a public lecture in Tokyo sponsored by the newspaper *Yomiuri Shimbun*.

George Zweig arrived at the idea of quarks independently and somewhat later than I did, and his approach was significantly different. He had been a student of mine, and by 1963 he was on the faculty of Caltech. I was on leave from January to September of 1963, and so I did not have occasion to discuss my ideas about quarks with George after I first began to work seriously on them in March. When I returned to Pasadena in the fall, he had just left to spend the academic year 1963-64 at CERN in Geneva, and it was there that he started to think about quarks, under the name of «aces».

My only attempt to convey the notion of quarks to CERN during that fall was in a telephone call to my old teacher, Viki Weisskopf, who was the Director General at that time. After I had explained to him that I had been working for a few months on an exciting idea that each baryon was made of three fractionally charged particles, he said: «Please, Murray, let us be serious; this is an international call.» It is no wonder that the news of the idea never trickled down to George.

During that fall I wrote my first letter about the quarks. After several miserable experiences with *Physical Review Letters,* I had decided to use *Physics Letters* instead and they printed my communication with no difficulty. I shudder to think of the trouble I would have had if I had sent it to *Physical Review Letters.*

Meanwhile, George started to work on «aces». He never thought of them as confined, and to this day he is studying the theoretical chemistry of unconfined quarks and their possible role in catalyzing controlled thermonuclear reactions. Of course, by now we know that quarks must be pretty well confined, but there is still the possibility of a slight leak somewhere in the physics that could lead to the existence of a few isolated quarks in the environment; George is still betting on such a leak. It would be fascinating if he were right; a flourishing quarkonics industry would grow up, and money for particle physics would be easy to find.

Another difference between George Zweig's work and mine in the early days was the following: I particularly emphasized the «current quarks», the basic quark fields that enter into the weak and electromagnetic currents and present a close analogy with lepton fields. He concentrated more on «constituent quarks», the transformed objects out of which the hadrons are effectively constructed and on which the spectra of baryons and mesons depend. Another amusing difference is that I had sent my results voluntarily to a European journal, while George got into a furious argument with the head of the CERN Theory Division over a rule that required him to use a European journal, and ended up not publishing his work. Some of it appeared much later in the Proceedings of the 1964 Erice Summer School.

Now what about quark statistics? In September, 1963, I investigated whether para-Fermi statistics with index three would explain that fact that the three quarks in a baryon appeared to be in a symmetrical state of orbital, spin, and flavor variables. What I found was that the resulting baryons would include not only a normal one obeying Fermi-Dirac statistics but also some «para-baryons», which I was unwilling to buy. I therefore refused to commit myself on the matter of statistics. Greenberg later published the idea of parastatistics for the quarks, but as far as I know he did not resolve the problem of the para-hadrons.

In 1966, in my Royal Institution lecture, I alluded to the fact that «mathematical», i.e. confined quarks would make it easier to understand the puzzle of the statistics, but I did not clearly formulate the principle that confinement of paraquarks might involve the confinement of the parastatistics and result in hadrons that would be fermions and bosons. It was not until 1971, in collaboration with Harald Fritzsch, that I understood that color, with prohibition of the emergence of colored objects, is equivalent to parastatistics with prohibition of the emergence of para-objects. I wish that I had understood that much earlier, but it was only in 1971 that Fritzsch and I saw it clearly and proposed the confinement of color.

When, in the fall of 1963, I showed an early version of my first letter about quarks to Stanley Mandelstam, he exclaimed that I was trying to start a counter-revolution. I explained that if the quarks were not directly observable there would be no contradiction of the bootstrap approach to the scattering amplitudes («S-matrix») on the mass shell, and I believe Mandelstam was satisfied. Geoffrey Chew, however, was not, and he never approved of my work on the quarks and related subjects. There was a fundamental asymmetry in the situation, because I approved of his work on the «S-matrix» program and indeed I participated actively in that program (which I had proposed in 1956), doing research on Regge poles and other S-matrix singularities. In 1964, despite our disagreements, he and Arthur Rosenfeld and I wrote a joint article for the *Scientific American* on hadrons. We mentioned the bootstrap idea and «nuclear democracy», Regge poles, the classification of hadron states, and so forth. Flavor SU(3) was included as an approximate symmetry, but as I recall there was not a word about quarks.

As far as I know, something like the bootstrap program might still qualify as a valid description of hadrons, if the right boundary conditions at infinite momenta are imposed on the theory; but it has not proved as useful as the quark description, which led to the development of QCD.

Before discussing QCD, let me relate an anecdote that bears on the significance of the discovery of quarks. I gave a series of lectures on quarks in Cambridge, England in 1966. Dirac attended them all. As usual, he fell asleep shortly after each lecture began, awoke at the end, and then asked penetrating questions about the subject of each lecture. How he manages to do that I have never discovered. Anyway, his questions revealed that he was fond of the quark scheme, and one day I asked him why he liked quarks when most of my colleagues thought they were crazy. He replied by asking: «They do have spin one-half, don't they?» I am sure that he meant: «They do obey my equation, don't they?» He had grasped the essential point that the quarks, like the leptons, are Dirac particles that are coupled calculably, and in some sense weakly, to various fields. In other words, in discovering quarks we have found hadron building blocks that are just as elementary as the leptons. At very small distances, of course, there might turn out to be constituents for both, but surely the quarks and leptons are equally elementary or equally composite.

In 1966, in *Preludes to Theoretical Physics,* Nambu wrote an article in which some features of QCD were anticipated, especially the exchange of colored gluons. He did not, however, actually employ a Yang-Mills theory. Moreover, he used the Han-Nambu scheme, in which electric charge varied with color in such a way that the quarks were given integral charges, and color was not confined —the quarks could emerge singly. If I had known about his article during the next six years, especially after Fritzsch and I had started to collaborate, I would probably have proposed the relevant modifications: Yang-Mills color theory, no dependence of charge on color, fractional charges, and color confinement leading to quark and gluon confinement. Unfortunately, I never saw his article or heard of it. The volume in which it appeared was a *Festschrift* in honor of the sixtieth birthday of Viki Weisskopf, and I was so ashamed of not having managed to contribute to it that I never looked at it.

I did not learn about Nambu's article until after Fritzsch and I worked on QCD in 1971 and 1972. At that time, we had to overcome confusion about yet another matter, whether the fundamental hadronic theory, apart from quarks, would be a theory of colored strings or a local Yang-Mills field theory of gluons, with strings constituting only an approximation. We settled on the latter, in other words QCD, but only after passing through a period of uncertainty, principally caused by the properties of QCD in the limit of vanishing quark masses. We did not fully understand the subtle processes by which scale invariance and conservation of total quark helicity are violated in that limit. In both cases, anomalies explain the difficulties that bothered us.

Painful as it is to describe, this story of confusion, ignorance, mistakes, difficulties, and vacillations may be the most interesting aspect of our particular corner of the history of science. This mode of description is related to the approach through hindsight, which is condemned by many historians who work in the history of science, but is embraced by many scientists working in the field. To me it is important not only to situate scientific ideas in the context of their time but also to try to figure out what relation the scientists' thinking bore to what we now know to be correct, how close they came to the right answers, or why they missed them. I hope that my reflections, based as they are on looking less at the doughnut than at the hole, have been useful.

DOE RESEARCH AND
DEVELOPMENT REPORT

Remarks given at the Celebration of Victor Weisskopf's 80th Birthday, CERN, Geneva, September 20, 1988

Murray Gell-Mann

California Institute of Technology, Pasadena, CA 91125

Forty years ago, I arrived at M.I.T. as a graduate student. I was discouraged at having been rejected by Princeton and granted insufficient financial aid by Harvard. The only really friendly letter that I received from a graduate school in physics was one from M.I.T. welcoming me as a potential student and as a research assistant in theoretical physics to a certain Professor Weisskopf, of whom I had never heard, but who added a personal letter of invitation of his own. I have described elsewhere how that letter arrived as I was contemplating suicide, as befits someone rejected by the Ivy League. It occurred to me however, (and it is an interesting example of non-commutation of operators) that I could try M.I.T. first and kill myself later, while the reverse order of events was impossible.

To my great surprise, the theoretical physics group at the dreaded M.I.T. was fun. Thoughts of suicide quickly disappeared. I became a member of a warm family when I entered the world of Viki, his postdoctoral fellows, his graduate students, and his charming secretary, Inge Reethof (whom I remember as frequently racing dangerously in her high heels to answer the telephone).

I shared a large office at various times with fellow students like David Jackson (who is here today) and Larry Biedenharn, and with postdoctoral fellows such as Murph Goldberger, John Blatt, and Satio Hayakawa. Francis Friedman, who died very young a few years later, was a postdoctoral fellow or assistant professor, lazy but bright, who often joined us for discussions and contributed a great deal.

Quite frequently, the whole group would go out to dinner together, sometimes to a vaguely French-style restaurant in Boston that served bear meat (quite tough, by the way, though sweet). I shall return in a little while to our discussions at the restaurant.

After dinner, we would go back to Viki's office and try to concentrate on physics, discussing the interpretation of recent experimental results in nuclear or elementary particle physics, theoretical puzzles, and, of course, those elementary models of physical phenomena that are so dear to Viki's heart. (Sometimes we would be invited to Viki's home, where we encountered his wonderful wife Ellen. The first time I met her, I was delighted that she was doing 360° turns on a gymnasium bar; obviously she was no more stuffy than Viki!)

Several speakers have alluded to Viki's emphasis on substantive progress in understanding rather than high-flown language or elegant mathematics. That was a very salutary antidote to the natural tendency of a youthful theoretician with formal talent to be dazzled by mathematical elegance. But the lesson goes much further than the mere advice (so useful to a social scientist, for example) to use no more mathematics than is warranted by the depth of theoretical understanding and the richness of the data. Let me explain what I mean by saying that it goes further.

At that time, theoretical science and the core of pure mathematics, where geometry, algebra, and analysis come together, had been drifting apart for a generation or so. None of us foresaw that a generation later the frontier of theoretical physics would be reunited with that core of mathematics, a dramatic development in the history of science, nowadays beautifully illustrated by the example of superstring theory or even that of Yang-Mills theory. Despite that development, the basic message of Viki is still applicable, because even the advanced mathematics that is now indispensable to the proper understanding of much recent progress in scientific theory can be explained in a relatively straightforward way à la Viki and need not be conveyed, à la Bourbaki, in a manner that obfuscates the rôle of intuition and example.

In such a discussion, it should be remarked, it is a little clumsy to keep using phrases like "à la Viki." A great man deserves an adjective, like Darwinian or Newtonian or even Freudian. Of course, "Weisskopfian" is not particularly euphonious. We could invoke the Greek language and use an adjective like "leukocephalic," but that sounds too academic and reminds us besides of medical terminology that most of us would rather forget. "Victorious" has a nice ring to it, but it already means something else. So does "Victorian." Perhaps we should settle for "Vikiesque." As to orthography, Maurice Jacob points out that the task of deciding on *k* versus another *qu* can safely be left to the French Academy.)

Now too much has already been said about the Vikiesque style of doing calculations and of giving class lectures, with factors of 2 and π and i flying in all directions. (It wasn't until I met first Goldberger and then his teacher Enrico Fermi that I realized that it was possible to get such factors right.) I have found that in many cases professors who are famous for their smooth, beautifully prepared lectures have their notes transformed into students' notes with very little interaction on the way with the students' brains; and such professors sometimes gloss over conceptual difficulties or important points of principle. If they use beautiful tricks, like Enrico, the student may be at a loss when he has to solve his own problems and the corresponding tricks don't come to mind. Of course none of that happened in Viki's case, and the occasional arithmetical Schlamperei usually inspired the students to rush home and get everything right. (Dave Jackson claims that he was able to do it on the spot!) Those students were not likely to forget what they had learned in that way. *And the scientific principles always came across.*

It was in that spirit that Viki's quotation from Niels Bohr was so apt, that an expert is someone who has made all possible mistakes.

Now, to return to the dinners. We talked about so many subjects: world politics, including Viki's experiences in the Soviet Union in the mid-1930's; Zipf's law, which I shall describe; questions about genetics and biological evolution; problems of creativity in art and science; and many others. Do many young physicists today have such a wide spectrum of interests as we did? I'm not so sure. That was just the time when physics was turning from a calling into a career, from an obscure cultural subject into a conspicuous technical one.

Next, Zipf's law! (The pronunciation is not Tsipf, but Zipf, as the man in question was a somewhat eccentric American with a link of some sort to Harvard.) It is still not properly understood, but at that very time it was helping to propel Benoit Mandelbrot into the study of scaling behavior in Nature. It is a crude law that applies to the frequency distribution of words in a text, in any human language; to the populations of cities; to the money values of different exports from a given country in a given year; and to many other sets of figures such as are given in the World Almanac. Plot the value against the ordinal number n (the largest, the next largest, the next-to-next largest, etc.). Roughly, the curve will be proportional to $1/n$. That means the initial digit will be a one very roughly forty-five times as often as it will be a nine. If you are not familiar with the effect, look up some statistics in the World Almanac and try it, and then attempt to explain the law! It may well be related, by the way, to the widespread phenomenon of "$1/f$ noise," proportional to the reciprocal of frequency, in physical science, another effect that is poorly understood. Zipf's law came up in a discussion of the statistical distribution of widths of nuclear levels. I give it as a striking example of the way our conversation flowed from one topic to another, without the respectful attention to disciplinary boundaries that is such a stultifying feature of much of our intellectual life.

These days, a number of us have begun to create, at the Santa Fe Institute in New Mexico, in a convent rented from the Catholic Church, the hub of a world network of researchers, with the highest qualifications, from the physical sciences (physicists, mathematicians, chemists, computer scientists), the biological sciences, the behavioral and the social sciences (linguists, psychologists, economists, archaeologists), and even humanistic subjects like history, to study, mainly from the theoretical point of view, the concept of complexity and the behavior of complex systems. We are concerned not only with physical and chemical systems, but also with complex adaptive systems, as represented by the behavior of the immune system in mammals (theoretical immunology); by learning and thinking; by the global economy as an evolving complex system; by biological evolution in general; by prebiotic chemical evolution (of which biological evolution is only a prolongation, making use of organisms like us); by the generation of new strategies for game-playing by means of artificial evolution in computers; by the evolution of human languages; by the rise and collapse of cultures; and by the study of quantum and classical information, entropy, and complexity in the universe. Here it is necessary to find those people who are good at perceiving and gradually understanding some of the features that these diverse complex systems have in common, and also those who are responsibly familiar with the details of the different fields, and especially subfields, that are being unified.

We hope not only to make progress in these difficult scientific and scholarly endeavors, but also to challenge the tyranny of the disciplines. In order to maintain standards of excellence, our culture has established a whole apparatus of university departments, professional societies, professional journals, and sections of funding agencies, all contributing to the preservation of barriers between the disciplines.

We must find new ways to establish standards of excellence that transcend the increasingly obsolescent categories of the traditional disciplines, what de Groot might call promotion standards for encyclopedists.

In helping to found and guide this ambitious institute, I am inspired to a considerable degree by the experience of those bear-meat dinners of forty years ago in Boston. If we are successful, and the Santa Fe Institute is known to history as a place where great advances are made in the understanding of Nature, including human affairs, and in the unification of human culture, you, Viki, deserve a good deal of the credit. If, as you probably think right now, it will go down in history as a crank idea, we will try to shield you from any blame.

In closing, let me say to you that we look forward to your autobiography, which I have urged you to write for so long, as well as to the various volumes of which Leon Lederman showed us the title pages.

Let me say also, at this birthday celebration, to you and to Ellen, on behalf of all of us, that we wish you both many more years of life and happiness.

Proc. 3rd Int. Symp. Foundations of
Quantum Mechanics, Tokyo, 1989, pp. 321–343

Quantum Mechanics in the Light of Quantum Cosmology[§]

Murray GELL-MANN and James B. HARTLE[†]

California Institute of Technology, Pasadena, CA 91125, USA
[†]Department of Physics, University of California, Santa Barbara, CA 93106, USA

We sketch a quantum-mechanical framework for the universe as a whole. Within that framework we propose a program for describing the ultimate origin in quantum cosmology of the "quasiclassical domain" of familiar experience and for characterizing the process of measurement. Predictions in quantum mechanics are made from probabilities for sets of alternative histories. Probabilities (approximately obeying the rules of probability theory) can be assigned only to sets of histories that approximately decohere. Decoherence is defined and the mechanism of decoherence is reviewed. Decoherence requires a sufficiently coarse-grained description of alternative histories of the universe. A quasiclassical domain consists of a branching set of alternative decohering histories, described by a coarse graining that is, in an appropriate sense, maximally refined consistent with decoherence, with individual branches that exhibit a high level of classical correlation in time. We pose the problem of making these notions precise and quantitative. A quasiclassical domain is emergent in the universe as a consequence of the initial condition and the action function of the elementary particles. It is an important question whether all the quasiclassical domains are roughly equivalent or whether there are various essentially inequivalent ones. A measurement is a correlation with variables in a quasiclassical domain. An "observer" (or information gathering and utilizing system) is a complex adaptive system that has evolved to exploit the relative predictability of a quasiclassical domain, or rather a set of such domains among which it cannot discriminate because of its own very coarse graining. We suggest that resolution of many of the problems of interpretation presented by quantum mechanics is to be accomplished, not by further scrutiny of the subject as it applies to reproducible laboratory situations, but rather by an examination of alternative histories of the universe, stemming from its initial condition, and a study of the problem of quasiclassical domains.

§1. Quantum Cosmology

If quantum mechanics is the underlying framework of the laws of physics, then there must be a description of the universe as a whole and everything in it in quantum-mechanical terms. In such a description, three forms of information are needed to make predictions about the universe. These are the action function of the elementary particles, the initial quantum state of the universe, and, since quantum mechanics is an inherently probabilistic theory, the information available about our specific history. These are sufficient for every prediction in science, and there are no predictions that do not, at a fundamental level, involve all three forms of information.

A unified theory of the dynamics of the basic fields has long been a goal of elementary particle physics and may now be within reach. The equally fundamental, equally necessary search for a theory of the initial state of the universe is the objective of the discipline of quantum cosmology. These may even be related goals; a single action function may describe both the hamiltonian and the initial state.[a)]

There has recently been much promising progress in the search for a theory of the quantum initial condition of the universe.[3)] Such diverse observations as the large scale homogeneity and isotropy of the universe, its approximate spatial flatness, the spectrum of density fluctuations from which the galaxies grew, the thermodynamic arrow of time, and the existence of classical spacetime may find a

a) As in the "no boundary" and the "tunneling from nothing proposals" where the wave function of the universe is constructed from the action by a Euclidean functional integral in the first case or by boundary conditions on the implied Wheeler-DeWitt equation in the second. See, *e.g.*, refs. 1 and 2.

unified, compressed explanation in a particular simple law of the initial condition.

The regularities exploited by the environmental sciences such as astronomy, geology, and biology must ultimately be traceable to the simplicity of the initial condition. Those regularities concern specific individual objects and not just reproducible situations involving identical particles, atoms, etc. The fact that the discovery of a bird in the forest or a fossil in a cliff or a coin in a ruin implies the likelihood of discovering another similar bird or fossil or coin cannot be derivable from the laws of elementary particle physics alone; it must involve correlations that stem from the initial condition.

The environmental sciences are not only strongly affected by the initial condition but are also heavily dependent on the outcomes of quantum-probabilistic events during the history of the universe. The statistical results of, say, proton-proton scattering in the laboratory are much less dependent on such outcomes. However, during the last few years there has been increasing speculation that, even in a unified fundamental theory, free of dimensionless parameters, some of the observable characteristics of the elementary particle system may be quantum-probabilistic, with a probability distribution that can depend on the initial condition.[4]

It is not our purpose in this article to review all these developments in quantum cosmology.[3] Rather, we will discuss the implications of quantum cosmology for one of the subjects of this conference—the interpretation of quantum mechanics.

§2. Probability

Even apart from quantum mechanics, there is no certainty in this world; therefore physics deals in probabilities. In classical physics probabilities result from ignorance; in quantum mechanics they are fundamental as well. In the last analysis, even when treating ensembles statistically, we are concerned with the probabilities of particular events. We then deal with the probabilities of deviations from the expected behavior of the ensemble caused by fluctuations.

When the probabilities of particular events are sufficiently close to 0 or 1, we make a definite prediction. The criterion for "sufficiently close to 0 or 1" depends on the use to which the probabilities are put. Consider, for example, a prediction on the basis of present astronomical observations that the sun will come up tomorrow at 5:59 AM ± 1 min. Of course, there is no certainty that the sun will come up at this time. There might have been a significant error in the astronomical observations or the subsequent calculations using them; there might be a non-classical fluctuation in the earth's rotation rate or there might be a collision with a neutron star now racing across the galaxy at near light speed. The prediction is the same as estimating the probabilities of these alternatives as low. How low do they have to be before one sleeps peacefully tonight rather than anxiously awaiting the dawn? The probabilities predicted by the laws of physics and the statistics of errors are generally agreed to be low enough!

All predictions in science are, most honestly and most generally, the probabilistic predictions of the *time histories* of particular events in the universe. In cosmology we are necessarily concerned with probabilities for the single system that is the universe as a whole. Where the universe presents us effectively with an ensemble of identical subsystems, as in experimental situations common in physics and chemistry, the probabilities for the ensemble as a whole yield definite predictions for the statistics of identical observations. Thus, statistical probabilities can be derived, in appropriate situations, from probabilities for the universe as a whole.[5]

Probabilities for histories need be assigned by physical theory only to the accuracy to which they are used. Thus, it is the same to us for all practical purposes whether physics claims the probability of the sun not coming up tomorrow is $10^{-10^{57}}$ or $10^{-10^{27}}$, as long as it is very small. We can therefore conveniently consider *approximate probabilities*, which need obey the rules of the probability calculus only up to some standard of accuracy sufficient for all practical purposes. In quantum mechanics, as we shall see, it is likely that only by this means can probabilities be assigned to interesting histories at all.

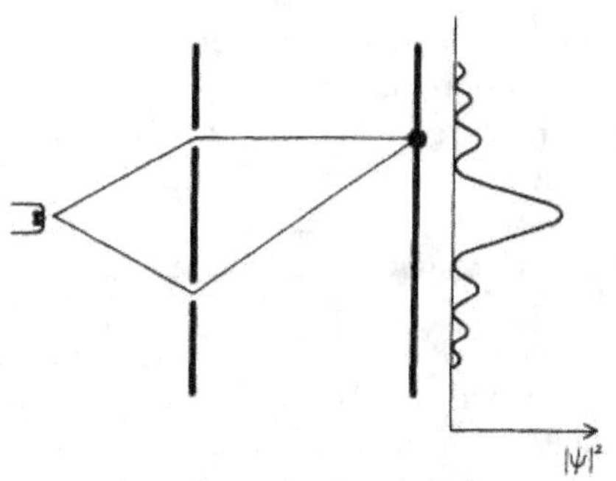

Fig. 1. The two-slit experiment. An electron gun at left emits an electron traveling towards a screen with two slits, its progress in space recapitulating its evolution in time. When precise detections are made of an ensemble of such electrons at the screen, it is not possible, because of interference, to assign a probability to the alternatives of whether an individual electron went through the upper slit or the lower slit. However, if the electron interacts with apparatus which measures which slit it passed through, then these alternatives decohere and probabilities can be assigned.

§3. Historical Remarks

In quantum mechanics not every history can be assigned a probability. Nowhere is this more clearly illustrated than in the two-slit experiment (Fig. 1). In the usual discussion, if we have not measured which slit the electron passed through on its way to being detected at the screen, then we are not permitted to assign probabilities to these alternative histories. It would be inconsistent to do so since the correct probability sum rules would not be satisfied. Because of interference, the probability to arrive at y is not the sum of the probabilities to arrive at y going through the upper and the lower slit:

$$p(y) \neq p_U(y) + p_L(y) \tag{1}$$

because

$$|\psi_L(y) + \psi_U(y)|^2 \neq |\psi_L(y)|^2 + |\psi_U(y)|^2. \tag{2}$$

If we *have* measured which slit the electron went through, then the interference is destroyed, the sum rule obeyed, and we *can* meaningfully assign probabilities to these alternative histories.

It is a general feature of quantum mechanics that one needs a rule to determine which histories can be assigned probabilities. The familiar rule of the "Copenhagen" interpretations described above is external to the framework of wave function and Schrödinger equation. Characteristically these interpretations, in one way or another, assumed as fundamental the existence of the classical domain we see all about us. Bohr spoke of phenomena that could be described in terms of classical language.[6] Landau and Lifshitz formulated quantum mechanics in terms of a separate classical physics.[7] Heisenberg and others stressed the central role of an external, essentially classical observer.[8] A measurement occurred through contact with this classical domain. Measurements determined what could be spoken about.

Such interpretations are inadequate for cosmology. In a theory of the whole thing there can be no fundamental division into observer and observed. Measurements and observers cannot be fundamental notions in a theory that seeks to discuss the early universe when neither existed. There is no reason in general for a classical domain to be fundamental or external in a basic formulation of quantum mechanics.

It was Everett who in 1957 first suggested how to generalize the Copenhagen framework so as to apply quantum mechanics to cosmology.[b] His idea was to take quantum mechanics seriously and apply it to the universe as a whole. He showed how an observer could be considered part of this system and how its activities—measuring, recording, and calculating probabilities—could be described in quantum mechanics.

Yet the Everett analysis was not complete. It did not adequately explain the origin of the classical domain or the meaning of the "branching" that replaced the notion of measurement. It was a theory of "many worlds" (what we would rather call "many histories"), but it did not sufficiently explain how these were defined or how they arose. Also, Everett's discussion suggests that a probability formula is somehow not needed in quantum

b) The original paper is by Everett.[9] The idea was developed by many, among them Wheeler,[10] DeWitt,[11] Geroch,[12] and Mukhanov[13] and independently arrived at by others, *e.g.*, Gell-Mann[14] and Cooper and Van Vechten.[15] There is a useful collection of early papers on the subject in ref. 16.

mechanics, even though a "measure" is introduced that, in the end, amounts to the same thing.

Here we shall briefly sketch a program aiming at a coherent formulation of quantum mechanics for science as a whole, including cosmology as well as the environmental sciences.[17] It is an attempt at extension, clarification, and completion of the Everett interpretation. It builds on many aspects of the post-Everett developments, especially the work of Zeh,[18] Żurek,[19] and Joos and Zeh.[20] In the discussion of history and at other points it is consistent with the insightful work (independent of ours) of Griffiths[21] and Omnès.[22] Our research is not complete, but we sketch, in this report on its status, how it might become so.

§4. Decoherent Sets of Histories

(a) *A Caveat*

We shall now describe the rules that specify which histories may be assigned probabilities and what these probabilities are. To keep the discussion manageable we make one important simplifying approximation. We neglect gross quantum variations in the structure of spacetime. This approximation, excellent for times later than 10^{-43} sec after the beginning, permits us to use any of the familiar formulations of quantum mechanics with a preferred time. Since histories are our concern, we shall often use Feynman's sum-over-histories formulation of quantum mechanics with histories specified as functions of this time. Since the hamiltonian formulation of quantum mechanics is in some ways more flexible, we shall use it also, with its apparatus of Hilbert space, states, hamiltonian, and other operators. We shall indicate the equivalence between the two, always possible in this approximation.

The approximation of a fixed background spacetime breaks down in the early universe. There, a yet more fundamental sum-over histories framework of quantum mechanics may be necessary.[23] In such a framework the notions of state, operators, and hamiltonian may be approximate features appropriate to the universe after the Planck era, for particular initial conditions that imply an approximately fixed background spacetime there. A discussion of quantum spacetime is essential for any detailed theory of the initial condition, but when, as here, this condition is not spelled out in detail and we are treating events after the Planck era, the familiar formulation of quantum mechanics is an adequate approximation.

The interpretation of quantum mechanics that we shall describe in connection with cosmology can, of course, also apply to any strictly closed sub-system of the universe provided its initial density matrix is known. However, strictly closed sub-systems of any size are not easily realized in the universe. Even slight interactions, such as those of a planet with the cosmic background radiation, can be important for the quantum mechanics of a system, as we shall see. Further, it would be extraordinarily difficult to prepare precisely the initial density matrix of any sizeable system so as to get rid of the dependence on the density matrix of the universe. In fact, even those large systems that are approximately isolated today inherit many important features of their effective density matrix from the initial condition of the universe.

(b) *Histories*

The three forms of information necessary for prediction in quantum cosmology are represented in the Heisenberg picture as follows:[24] The quantum state of the universe is described by a density matrix ρ. Observables describing specific information are represented by operators $\mathcal{O}(t)$. For simplicity, but without loss of generality, we shall focus on non-"fuzzy", "yes-no" observables. These are represented in the Heisenberg picture by projection operators $P(t)$. The hamiltonian, which is the remaining form of information, describes evolution by relating the operators corresponding to the same question at different times through

$$P(t)=e^{iHt/\hbar}P(0)\,e^{-iHt/\hbar}. \tag{3}$$

An exhaustive set of "yes-no" alternatives at one time is represented in the Heisenberg picture by *sets* of projection operators $(P_1^k(t), P_2^k(t), \cdots)$. In $P_\alpha^k(t)$, k labels the set, α the particular alternative, and t its time. An exhaustive set of exclusive alternatives satisfies

$$\sum_\alpha P^k_\alpha(t)=1, \quad P^k_\alpha(t)P^k_\beta(t)=\delta_{\alpha\beta}P^k_\alpha(t). \quad (4)$$

For example, one such exhaustive set would specify whether a field at a point on a surface of constant t is in one or another of a set of ranges exhausting all possible values. The projections are simply the projections onto eigenstates of the field at that point with values in these ranges. We should emphasize that an exhaustive set of projections need not involve a *complete* set of variables for the universe (one-dimensional projections)—in fact, the projections we deal with as observers of the universe typically involve only an infinitesimal fraction of a complete set.

Sets of alternative histories consist of *time sequences* of exhaustive sets of alternatives. A *history* is a particular sequence of alternatives, abbreviated $[P_\alpha]=(P^1_{\alpha_1}(t_1), P^2_{\alpha_2}(t_2), \cdots, P^n_{\alpha_n}(t_n))$. A *completely fine-grained* history is specified by giving the values of a complete set of operators at all times. One history is a *coarse graining* of another if the set $[P_\alpha]$ of the first history consists of sums of the $[P_\alpha]$ of the second history. The inverse relation is fine graining. The completely coarse-grained history is one with no projections whatever, just the unit operator!

The reciprocal relationships of coarse and fine graining evidently constitute only a partial ordering of sets of alternative histories. The arbitrary sets need not be related to each other by coarse/fine graining. The partial ordering is represented schematically in Fig. 2, where each point stands for a set of alternative histories.

Feynman's sum-over-histories formulation of quantum mechanics begins by specifying the amplitude for a completely fine-grained history in a particular basis of generalized coordinates $Q^i(t)$, say all fundamental field variables at all points in space. This amplitude is proportional to

$$\exp\,(iS[Q^i(t)]/\hbar), \quad (5)$$

where S is the action functional that yields the hamiltonian, H. When we employ this formulation of quantum mechanics, we shall introduce the simplification of ignoring fields with spins higher than zero, so as to avoid the complications of gauge groups and of fermion fields (for which it is inappropriate to discuss eigenstates of the field variables.) The operators $Q^i(t)$ are thus various scalar fields at different points of space.

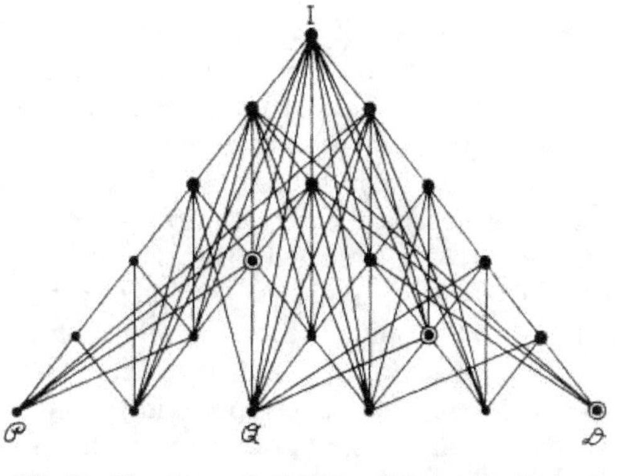

Fig. 2. The schematic structure of the space of *sets* of possible histories for the universe. Each dot in this diagram represents an exhaustive *set* of alternative histories for the universe. (This is not a picture of the branches defined by a given set!) Such sets, denoted by $\{[P_\alpha]\}$ in the text, correspond in the Heisenberg picture to time sequences $(P^1_{\alpha_1}(t_1), P^2_{\alpha_2}(t_2), \cdots P^n_{\alpha_n}(t_n))$ of sets of projection operators, such that at each time t_k the alternatives α_k are an orthogonal and exhaustive set of possibilities for the universe. At the bottom of the diagram are the completely fine-grained sets of histories each arising from taking projections onto eigenstates of a *complete set* of observables for the universe at *every time*. For example, the set $\mathcal{Q}$ is the set in which all field variables at all points of space are specified at every time. This set is the starting point for Feynman's sum-over-histories formulation of quantum mechanics. $\mathcal{P}$ might be the completely fine-grained set in which all field momenta are specified at each time. $\mathcal{D}$ might be a degenerate set of the kind discussed in §7 in which the *same* complete set of *operators* occurs at every time. But there are many other completely fine-grained sets of histories corresponding to all possible combinations of complete sets of observables that can be taken at every time.

The dots above the bottom row are coarse-grained sets of alternative histories. If two dots are connected by a path, the one above is a coarse graining of the one below—that is, the projections in the set above are *sums* of those in the set below. A line, therefore, corresponds to an operation of coarse graining. At the very top is the degenerate case in which complete sums are taken at every time, yielding no projections at all other than the unit operator! The space of sets of alternative histories is thus partially ordered by the operation of coarse graining.

The heavy dots denote the decoherent sets of alternative histories. Coarse grainings of decoherent sets remain decoherent. Maximal sets, the heavy dots surrounded by circles, are those decohering sets for which there is no finer-grained decoherent set.

Let us now specialize our discussion of histories to the generalized coordinate bases $Q^i(t)$ of the Feynman approach. Later we shall discuss the necessary generalization to the case of an arbitrary basis at each time t, utilizing quantum-mechanical tranformation theory.

Completely fine-grained histories in the coordinate basis cannot be assigned probabilities; only suitable coarse-grained histories can. There are at least three common types of coarse graining: (1) specifying observables not at all times, but only at some times: (2) specifying at any one time not a complete set of observables, but only some of them: (3) specifying for these observables not precise values, but only ranges of values. To illustrate all three, let us divide the Q^i up into variables x^i and X^i and consider only sets of ranges $\{\Delta^k_\alpha\}$ of x^i at times t_k, $k=1,\cdots,n$. A set of alternatives at any one time consists of ranges Δ^k_α, which exhaust the possible values of x^i as α ranges over all integers. An individual history is specified by particular Δ_α's at the times $t_1,\cdots t_n$. We write $[\Delta_\alpha]=(\Delta^1_{\alpha_1},\cdots\Delta^n_{\alpha_n})$ for a particular history. A *set* of alternative histories is obtained by letting $\alpha_1\cdots\alpha_n$ range over all values.

Let us use the same notation $[\Delta_\alpha]$ for the most general history that is a coarse graining of the completely fine-grained history in the coordinate basis, specified by ranges of the Q^i at each time, including the possibility of full ranges at certain times, which eliminate those times from consideration.

(c) *Decohering Histories*

The important theoretical construct for giving the rule that determines whether probabilities may be assigned to a given set of alternative histories, and what these probabilities are, is the decoherence functional D [(history)$'$, (history)]. This is a complex functional on any pair of histories in the set. It is most transparently defined in the sum-over-histories framework for completely fine-grained history segments between an initial time t_0 and a final time t_f, as follows:

$$D[Q'^i(t), Q^i(t)]=\delta(Q'^i_f-Q^i_f)\exp\{i(S[Q'^i(t)]-S[Q^i(t)])/\hbar\}\rho(Q'^i_0, Q^i_0). \tag{6}$$

Here ρ is the initial density matrix of the universe in the Q^i representation, Q'^i_0 and Q^i_0 are the initial values of the complete set of variables, and Q'^i_f and Q^i_f are the final values. The decoherence functional for coarse-grained histories is obtained from (6) according to the principle of superposition by summing over all that is not specified by the coarse graining. Thus,

$$D([\Delta_{\alpha'}], [\Delta_\alpha])=\int_{[\Delta_{\alpha'}]}\delta Q'\int_{[\Delta_\alpha]}\delta Q\,\delta(Q'^i_f-Q^i_f)\times e^{i\{(S[Q'^i]-S[Q^i])/\hbar\}}\rho(Q'^i_0, Q^i_0). \tag{7}$$

More precisely, the integral is as follows (Fig. 3): It is over all histories $Q'^i(t)$, $Q^i(t)$ that begin at Q'^i_0, Q^i_0 respectively, pass through the ranges $[\Delta_{\alpha'}]$ and $[\Delta_\alpha]$ respectively, and wind up at a common point Q^i_f at any time $t_f>t_n$. It is completed by integrating over Q'^i_0, Q^i_0, and Q^i_f.

The connection between coarse-grained histories and completely fine-grained ones is transparent in the sum-over-histories formulation of quantum mechanics. However, the sum-over-histories formulation does not allow us to consider directly histories of the most general type. For the most general histories one needs to exploit directly the transformation theory of quantum mechanics and for this the Heisenberg picture is convenient. In the Heisenberg picture D can be written

$$D([P_{\alpha'}], [P_\alpha])=\mathrm{Tr}\,[P^n_{\alpha'_n}(t_n)\cdots P^1_{\alpha'_1}(t_1)\rho P^1_{\alpha_1}(t_1)\cdots P^n_{\alpha_n}(t_n)]. \tag{8}$$

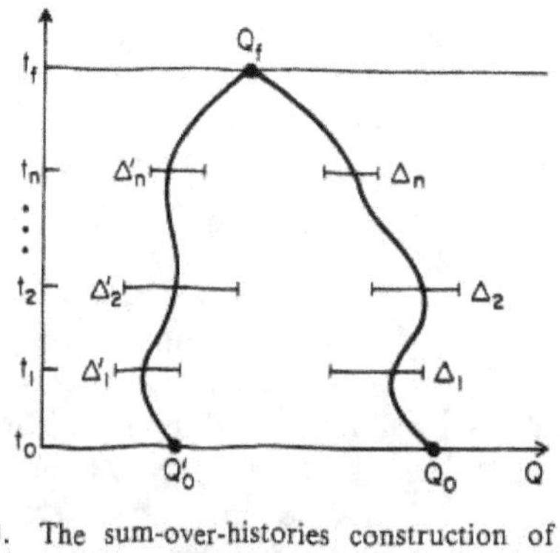

Fig. 3. The sum-over-histories construction of the decoherence functional.

The projections in (8) are time ordered with the earliest on the inside. When the P's are projections onto ranges Δ_α^k of values of the Q's, expressions (7) and (8) agree. From the cyclic property of the trace it follows that D is always diagonal in the final indices α_n and α_n'. (We assume throughout that the P's are bounded operators in Hilbert space dealing, for example, with projections onto ranges of the Q's and not onto definite values of the Q's). Decoherence is thus an interesting notion only for strings of P's that involve more than one time. Decoherence is automatic for "histories" that consist of alternatives at but one time.

Progressive coarse graining may be seen in the sum-over-histories picture as summing over those parts of the fine-grained histories not specified in the coarse-grained one, according to the principle of superposition. In the Heisenberg picture, eq. (8), the three common forms of coarse graining discussed above can be represented as follows: Summing on both sides of D over all P's at a given time and using (4) eliminates those P's completely. Summing over all possibilities for certain variables at one time amounts to factoring the P's and eliminating one of the factors by summing over it. Summing over ranges of values of a given variable at a given time corresponds to replacing the P's for the partial ranges by one for the total range. Thus, if $[\overline{P_\beta}]$ is a coarse graining of the set of histories $\{[P_\alpha]\}$, we write

$$D([\overline{P_{\beta'}}],[\overline{P_\beta}]) = \sum_{\substack{\text{all } P'_\alpha \\ \text{not fixed by } [\overline{P_{\beta'}}]}} \sum_{\substack{\text{all } P_\alpha \\ \text{not fixed by } [\overline{P_\beta}]}} D([P_{\alpha'}],[P_\alpha]). \quad (9)$$

In the most general case, we may think of the completely fine-grained limit as obtained from the coordinate representation by arbitrary unitary transformations at all times. All histories can be obtained by coarse-graining the various completely fine-grained ones, and coarse graining in its most general form involves taking arbitrary sums of P's, as discussed earlier. We may use (9) in the most general case where $[\overline{P_\beta}]$ is a coarse graining of $[P_\alpha]$.

A set of coarse-grained alternative histories is said to *decohere* when the off-diagonal elements of D are sufficiently small:

$$D([P_{\alpha'}],[P_\alpha]) \approx 0, \quad \text{for any } \alpha_k' \neq \alpha_k. \quad (10)$$

This is a generalization of the condition for the absence of interference in the two-slit experiment (approximate equality of the two sides of (2)). It is a sufficient (although not a necessary) condition for the validity of the purely diagonal formula

$$D([\overline{P_\beta}],[\overline{P_\beta}]) \approx \sum_{\substack{\text{all } P_\alpha \text{ not} \\ \text{fixed by } [\overline{P_\beta}]}} D([P_\alpha],[P_\alpha]). \quad (11)$$

The rule for when probabilities can be assigned to histories of the universe is then this: To the extent that a *set* of alternative histories decoheres, probabilities can be assigned to its individual members. The probabilities are the *diagonal* elements of D. Thus,

$$\begin{aligned} p([P_\alpha]) &= D([P_\alpha],[P_\alpha]) \\ &= \mathrm{Tr}\,[P^n_{\alpha_n}(t_n)\cdots P^1_{\alpha_1}(t_1)\rho P^1_{\alpha_1}(t_1)\cdots P^n_{\alpha_n}(t_n)], \end{aligned} \quad (12)$$

when the set decoheres. We will frequently write $p(\alpha_n t_n, \cdots \alpha_1 t_1)$ for these probabilities, suppressing the labels of the sets.

The probabilities defined by (11) obey the rules of probability theory as a consequence of decoherence. The principal requirement is that the probabilities be additive on "disjoint sets of the sample space". For histories this gives the sum rule

$$p([\overline{P_\beta}]) \approx \sum_{\substack{\text{all } P_\alpha \text{ not} \\ \text{fixed by } [\overline{P_\beta}]}} p([P_\alpha]) \quad (13)$$

These relate the probabilities for a set of histories to the probabilities for *all* coarser grained sets that can be constructed from it. For example, the sum rule eliminating all projections at only one time is

$$\begin{aligned} &\sum_{\alpha_k} p(\alpha_n t_n, \cdots \alpha_{k+1} t_{k+1}, \alpha_k t_k, \alpha_{k-1} t_{k-1}, \cdots, \alpha_1 t_1) \\ &\approx p(\alpha_n t_n, \cdots \alpha_{k+1} t_{k+1}, \alpha_{k-1} t_{k-1}, \cdots, \alpha_1 t_1). \end{aligned} \quad (14)$$

These rules follow trivially from (11) and (12). The other requirements from probability theory are that the probability of the whole sample space be unity, an easy consequence of (11) when complete coarse graining is performed, and that the probability for an empty set be zero, which means simply that the probability of any sequence containing a projection $P=0$ must vanish, as it does.

The $p([P_\alpha])$ are *approximate* probabilities for histories, in the sense of §2, up to the standard set by decoherence. Conversely, if a given standard for the probabilities is required by their use, it can be met by coarse graining until (10) and (13) are satisfied at the requisite level.

Further coarse graining of a decoherent set of alternative histories produces another set of decoherent histories since the probability sum rules continue to be obeyed. That is illustrated in Fig. 2, which makes it clear that in a progression from the trivial completely coarse graining to a completely fine graining, there are sets of histories where further fine graining always results in loss of decoherence. These are the *maximal* sets of alternative decohering histories.

These rules for probability exhibit another important feature: The operators in (12) are time-ordered. Were they not time-ordered (zig-zags) we could have assigned non-zero probabilities to conflicting alternatives at the same time. The time ordering thus expresses causality in quantum mechanics, a notion that is appropriate here because of the approximation of fixed background spacetime. The time ordering is related as well to the "arrow of time" in quantum mechanics, which we discuss below.

Given this discussion, the *fundamental formula* of quantum mechanics may be reasonably taken to be

$$D([P_{\alpha'}], [P_\alpha]) \approx \delta_{\alpha_1'\alpha_1} \cdots \delta_{\alpha_n'\alpha_n} p([P_\alpha]) \qquad (15)$$

for all $[P_\alpha]$ in a set of alternative histories. Vanishing of the off-diagonal elements of D gives the rule for when probabilities may be consistently assigned. The diagonal elements give their values.

We could have used a weaker condition than (10) as the definition of decoherence, namely the necessary condition for the validity of the sum rules (11) of probability theory:

$$D([P_\alpha], [P_{\alpha'}]) + D([P_{\alpha'}], [P_\alpha]) \approx 0 \qquad (16)$$

for any $\alpha_k' \neq \alpha_k$, or equivalently

$$\mathrm{Re}\,\{D([P_\alpha], [P_{\alpha'}])\} \approx 0. \qquad (17)$$

This is the condition used by Griffiths[21] as the requirement for "consistent histories". However, while, as we shall see, it is easy to identify physical situations in which the off-diagonal elements of D approximately vanish as the result of coarse graining, it is hard to think of a general mechanism that suppresses only their real parts. In the usual analysis of measurement, the off-diagonal parts of D approximately vanish. We shall, therefore, explore the stronger condition (10) in what follows. That difference should not obscure the fact that in this part of our work we have reproduced what is essentially the approach of Griffiths,[21] extended by Omnès.[22]

(d) *Prediction and Retrodiction*

Decoherent sets of histories are what we may discuss in quantum mechanics, for they may be assigned probabilities. Decoherence thus generalizes and replaces the notion of "measurement", which served this role in the Copenhagen interpretations. Decoherence is a more precise, more objective, more observer-independent idea. For example, if their associated histories decohere, we may assign probabilities to various values of reasonable scale density fluctuations in the early universe whether or not anything like a "measurement" was carried out on them and certainly whether or not there was an "observer" to do it. We shall return to a specific discussion of typical measurement situations in §11.

The joint probabilities $p(\alpha_n t_n, \cdots, \alpha_1 t_1)$ for the individual histories in a decohering set are the raw material for prediction and retrodiction in quantum cosmology. From them, the relevant conditional probabilities may be computed. The conditional probability of, one subset $\{\alpha_i t_i\}$, given the rest $\overline{\{\alpha_i t_i\}}$, is

$$p(\{\alpha_i t_i\} \mid \overline{\{\alpha_i t_i\}}) = \frac{p(\alpha_n t_n, \cdots, \alpha_1 t_1)}{p(\overline{\{\alpha_i t_i\}})}. \qquad (18)$$

For example, the probability for *predicting* alternatives $\alpha_{k+1}, \cdots \alpha_n$, given that the alternatives $\alpha_1 \cdots \alpha_k$ have already happened, is

$$p(\alpha_n t_n, \cdots \alpha_{k+1} t_{k+1} \mid \alpha_k t_k, \cdots, \alpha_1 t_1) = \frac{p(\alpha_n t_n, \cdots, \alpha_1 t_1)}{p(\alpha_k t_k, \cdots \alpha_1 t_1)}. \qquad (19)$$

The probability that $\alpha_{n-1}, \cdots \alpha_1$ happened in the *past*, given present data summarized by an alternative α_n at the present time t_n, is

$$p(\alpha_{n-1}t_{n-1},\cdots\alpha_1 t_1|\alpha_n t_n)=\frac{p(\alpha_n t_n,\cdots,\alpha_1 t_1)}{p(\alpha_n t_n)}. \quad (20)$$

Decoherence ensures that the probabilities defined by (18)-(20) will approximately add to unity when summed over all remaining alternatives, because of (14).

Despite the similarity between (19) and (20), there are differences between prediction and retrodiction. Future predictions can all be obtained from an effective density matrix summarizing information about what has happened. If ρ_{eff} is defined by

$$\rho_{\mathrm{eff}}=\frac{P^k_{\alpha_k}(t_k)\cdots P^1_{\alpha_1}(t_1)\rho P^1_{\alpha_1}(t_1)\cdots P^k_{\alpha_k}(t_k)}{\mathrm{Tr}\,[P^k_{\alpha_k}(t_k)\cdots P^1_{\alpha_1}(t_1)\rho P^1{}_{\alpha_1}(t_1)\cdots P^k_{\alpha_k}(t_k)]}, \quad (21)$$

then

$$p(\alpha_n t_n,\cdots\alpha_{k+1}t_{k+1}|\alpha_k t_k,\cdots,\alpha_1 t_1)$$
$$=\mathrm{Tr}\,[P^n_{\alpha_n}(t_n)\cdots P^{k+1}_{\alpha_{k+1}}(t_{k+1})\rho_{\mathrm{eff}}$$
$$\times P^{k+1}_{\alpha_{k+1}}(t_{k+1})\cdots P^n_{\alpha_n}(t_n)]. \quad (22)$$

By contrast, there is no effective density matrix representing present information from which probabilities for the past can be derived. As (20) shows, history requires knowledge of both present data *and* the initial condition of the universe.

Prediction and retrodiction differ in another way. Because of the cyclic property of the trace in (8), *any* final alternative decoheres and a probability can be predicted for it. By contrast we expect only certain variables to decohere in the past, appropriate to present data and the initial ρ. As the alternative histories of the electron in the two-slit experiment illustrate, there are many kinds of alternatives in the past for which the assignment of probabilities is prohibited in quantum mechanics. For those sets of alternatives that do decohere, the decoherence and the assigned probabilities typically will be approximate in the sense of §2. It is unlikely, for example, that the initial state of the universe is such that the interference is exactly zero between two past positions of the sun in the sky.

These differences between prediction and retrodiction are aspects of the arrow of time in quantum mechanics. Mathematically they are consequences of the time ordering in (8) or (12). This time ordering does not mean that quantum mechanics singles out an absolute direction in time. Field theory is invariant under CPT. Performing a CPT transformation on (8) or (12) results in an equivalent expression in which the CPT-transformed ρ is assigned to the far future and the CPT-transformed projections are anti-time-ordered. Either time ordering can, therefore, be used;[c)] the important point is that there is a knowable Heisenberg ρ from which probabilities can be predicted. It is by convention that we think of it as an "initial condition", with the projections in increasing time order from the inside out in (8) and (12).

While the formalism of quantum mechanics allows the universe to be discussed with either time ordering, the physics of the universe is time asymmetric, with a simple condition in what we call "the past." For example, the indicated present homogeneity of the thermodynamic arrow of time can be traced to the near homogeneity of the "early" universe implied by ρ and the implication that the progenitors of approximately isolated subsystems started out far from equilibrium at "early" times.

Much has been made of the updating of the fundamental probability formula in (19) and in (21) and (22). By utilizing (21) the process of prediction may be organized so that for each time there is a ρ_{eff} from which probabilities for the future may be calculated. The action of each projection, P, on both sides of ρ in (21) along with the division by the appropriate normalizing factor is then sometimes called the "reduction of the wave packet." But this updating of probabilities is no different from the classical reassessment of

c) It has been suggested[23)] that, for application to highly quantum-mechanical spacetime, as in the very early universe, quantum mechanics should be generalized to yield a framework in which both time orderings are treated simultaneously in the sum-over-histories approach. This involves including both exp (iS) and exp $(-iS)$ for each history and has as a consequence an evolution equation (the Wheeler-DeWitt equation) that is second order in the time variable. The suggestion is that the two time orderings decohere when the universe is large and spacetime classical, so that the usual framework with just one ordering is recovered.

probabilities that occurs after new information is obtained. In a sequence of horse races, the joint probability for the winners of eight races is converted, after the winners of the first three are known, into a reassessed probability for the remaining five races by exactly this process. The main thing is that, because of decoherence, the sum rules for probabilities are obeyed; once that is true, reassessment of probabilities is trivial.

The only non-trivial aspect of the situation is the choice of the string of P's in (8) giving a decoherent set of histories.

(e) *Branches (Illustrated by a Pure ρ)*

Decohering sets of alternative histories give a definite meaning to Everett's "branches". For a given such set of histories, the exhaustive set of $P^k_{\alpha_k}$ at each time t_k corresponds to a branching.

To illustrate this even more explicitly, consider an initial density matrix that is a pure state, as in typical proposals for the wave function of the universe:

$$\rho = |\Psi\rangle\langle\Psi| . \tag{23}$$

The initial state may be decomposed according to the projection operators that define the set of alternative histories

$$|\Psi\rangle = \sum_{\alpha_1\cdots\alpha_n} P^n_{\alpha_n}(t_n)\cdots P^1_{\alpha_1}(t_1)|\Psi\rangle$$
$$\equiv \sum_{\alpha_1\cdots\alpha_n} |[P_\alpha], \Psi\rangle . \tag{24}$$

The states $|[P_\alpha], \Psi\rangle$ are approximately orthogonal as a consequence of their decoherence

$$\langle[P_{\alpha'}], \Psi|[P_\alpha], \Psi\rangle \approx 0, \quad \text{for any } \alpha'_k \neq \alpha_k . \tag{25}$$

Equation (25) is just a reëxpression of (10), given (23).

When the initial density matrix is pure, it is easily seen that some coarse graining in the present is always needed to achieve decoherence in the past. If the $P^n_{\alpha_n}(t_n)$ for the last time t_n in (8) were all projections onto pure states, D would factor for a pure ρ and could never satisfy (10), except for certain special kinds of histories described near the end of §7, in which decoherence is automatic, independent of ρ. Similarly, it is not difficult to show that some coarse graining is required at any time in order to have decoherence of previous alternatives, with the same set of exceptions.

After normalization, the states $|[P_\alpha], \Psi\rangle$ represent the individual histories or individual branches in the decohering set. We may, as for the effective density matrix of (d), summarize present information for prediction just by giving one of these states, with projections up to the present.

(f) *Sets of Histories with the Same Probabilities*

If the projections P are not restricted to a particular class (such as projections onto ranges of Q^i variables), so that coarse-grained histories consist of arbitrary exhaustive families of projections operators, then the problem of exhibiting the decohering sets of strings of projections arising from a given ρ is a purely algebraic one. Assume, for example, that the initial condition is known to be a pure state as in (23). The problem of finding ordered strings of exhaustive sets of projections $[P_\alpha]$ so that the histories $P^n_{\alpha_n}\cdots P^1_{\alpha_1}|\Psi\rangle$ decohere according to (25) is purely algebraic and involves just subspaces of Hilbert space. The problem is the same for one vector $|\Psi\rangle$ as for any other. Indeed, using subspaces that are *exactly* orthogonal, we may identify sequences that *exactly* decohere.

However, it is clear that the solution of the mathematical problem of enumerating the sets of decohering histories of a given Hilbert space has no physical content by itself. No description of the histories has been given. No reference has been made to a theory of the fundamental interactions. No distinction has been made between one vector in Hilbert space as a theory of the initial condition and any other. The resulting probabilities, which can be calculated, are merely abstract numbers.

We obtain a description of the sets of alternative histories of the universe when the operators corresponding to the fundamental fields are identified. We make contact with the theory of the fundamental interactions if the evolution of these fields is given by a fundamental hamiltonian. Different initial vectors in Hilbert space will then give rise to decohering sets having different descriptions in terms of the fundamental fields. The probabilities acquire physical meaning.

Two different simple operations allow us to construct from one set of histories another set with a *different description* but the *same probabilities*. First consider unitary transformations of the P's that are constant in time and leave the initial ρ fixed:

$$\rho = U\rho U^{-1}, \tag{26}$$

$$\tilde{P}^k_\alpha(t) = UP^k_\alpha(t)U^{-1}. \tag{27}$$

If ρ is pure there will be very many such transformations; the Hilbert space is large and only a single vector is fixed. The sets of histories made up from the $\{\tilde{P}^k_\alpha\}$ will have an identical decoherence functional to the sets constructed from the corresponding $\{P^k_\alpha\}$. If one set decoheres, the other will and the probabilities for the individual histories will be the same.

In a similar way, decoherence and probabilities are invariant under arbitrary reassignments of the times in a string of P's (as long as they continue to be ordered), with the projection operators at the altered times unchanged as operators in Hilbert space. This is because in the Heisenberg picture every projection is at *any* time a projection operator for *some* quantity.

The histories arising from constant unitary transformations or from reassignment of times of a given set of P's will, in general, have very different descriptions in terms of fundamental fields from that of the original set. We are considering transformations such as (27) in an active (or alibi) sense so that the field operators and hamiltonian are unchanged. (The passive (or alias) transformations, in which these are transformed, are easily understood.) A set of projections onto the ranges of field values in a spatial region is generally transformed by (27) or by any reassignment of the times into an extraordinarily complicated combination of all fields and all momenta at all positions in the universe! Histories consisting of projections onto values of similar quantities at different times can thus become histories of very different quantities at various other times.

In ordinary presentations of quantum mechanics, two histories with different descriptions can correspond to physically distinct situations because it is presumed that various different hermitian combinations of field operators are potentially measurable by different kinds of external apparatus. In quantum cosmology, however, apparatus and system are considered together and the notion of physically distinct situations may have a different character.

§5. The Origins of Decoherence

What are the features of coarse-grained sets of histories that decohere, given the ρ and H of the universe? In seeking to answer this question it is important to keep in mind the basic aspects of the theoretical framework on which decoherence depends. Decoherence of a set of alternative histories is not a property of their operators *alone*. It depends on the relations of those operators to the density matrix ρ. Given ρ, we could, in principle, *compute* which sets of alternative histories decohere.

We are not likely to carry out a computation of all decohering sets of alternative histories for the universe, described in terms of the fundamental fields, anytime in the near future, if ever. However, if we focus attention on coarse grainings of particular variables, we can exhibit widely occurring mechanisms by which they decohere in the presence of the actual ρ of the universe. We have mentioned in §4(c) that decoherence is automatic if the projection operators P refer only to one time; the same would be true even for different times if all the P's commuted with one another. Of course, in cases of interest, each P typically factors into commuting projection operators, and the factors of P's for different times often fail to commute with one another, for example factors that are projections onto related ranges of values of the same Heisenberg operator at different times. However, these non-commuting factors may be correlated, given ρ, with other projection factors that do commute or, at least, effectively commute inside the trace with the density matrix ρ in eq. (8) for the decoherence functional. In fact, these other projection factors may commute with all the subsequent P's and thus allow themselves to be moved to the outside of the trace formula. When all the non-commuting factors are correlated in this manner with effectively commuting ones, then the off-diagonal terms in the decoherence functional vanish, in other words, decoherence results. Of course, all this behavior may be approximate, resulting in approximate decoherence.

This type of situation is fundamental in the interpretation of quantum mechanics. Non-commuting quantities, say at different times, may be correlated with commuting or effectively commuting quantities because of the character of ρ and H, and thus produce decoherence of strings of P's despite their non-commutation. For a pure ρ, for example, the behavior of the effectively commuting variables leads to the orthogonality of the branches of the state $|\Psi\rangle$, as defined in (24). We shall see that correlations of this character are central to understanding historical records (§10) and measurement situations (§11).

As an example of decoherence produced by this mechanism, consider a coarse-grained set of histories defined by time sequences of alternative approximate localizations of the center of mass of a massive body such as a planet or even a typical interstellar dust grain. As shown by Joos and Zeh,[20] even if the successive localizations are spaced as closely as a nanosecond, such histories decohere as a consequence of scattering by the 3° cosmic background radiation (if for no other reason). Different positions become correlated with nearly orthogonal states of the photons. More importantly, each alternative sequence of positions becomes correlated with a different orthogonal state of the photons at the final time. This accomplishes the decoherence and we may loosely say that such histories of the position of a massive body are "decohered" by interaction with the photons of the background radiation.

Other specific models of decoherence have been discussed by many authors, among them Joos and Zeh,[20] Caldeira and Leggett,[25] and Żurek.[26] Typically these discussions have focussed on a coarse graining that involves only certain variables analogous to the position variables above. Thus the emphasis is on particular non-commuting factors of the projection operators and not on correlated operators that may be accomplishing the approximate decoherence. Such coarse grainings do not, in general, yield the most refined approximately decohering sets of histories, since one could include projections onto ranges of values of the correlated operators without losing the decoherence.

The simplest model consists of a single oscillator interacting bilinearly with a large number of others, the coordinates of which are integrated over. Let x be the coordinate of the special oscillator, M its mass, ω_R its frequency renormalized by its interactions with the others, and $S_{\rm free}$ its free action. Consider the special case where the density matrix of the whole system, referred to an initial time, factors into the product of a density matrix $\bar\rho(x', x)$ of the distinguished oscillator and another for the rest. Then, generalizing slightly a treatment of Feynman and Vernon,[27] we can write D defined by (7) as

$$D([\Delta_{\alpha'}],[\Delta_\alpha])=\int_{[\Delta_{\alpha'}]}\delta x'(t)\int_{[\Delta_\alpha]}\delta x(t)\delta(x'_f-x_f)\times\exp\{i(S_{\rm free}[x'(t)]-S_{\rm free}[x(t)]+W[x'(t),x(t)])/\hbar\}\bar\rho(x'_0,x_0), \tag{28}$$

the intervals $[\Delta_\alpha]$ referring here only to the variables of the distinguished oscillator. The sum over the rest of the oscillators has been carried out and is summarized by the Feynman-Vernon influence functional $\exp(iW[x'(t),x(t)])$. The remaining sum over $x'(t)$ and $x(t)$ is as in (7).

The case when the other oscillators are in an initial thermal distribution has been extensively investigated by Caldeira and Leggett.[25] In the simple limit of a uniform continuum of oscillators cut off at frequency Ω and in the Fokker-Planck limit of $kT\gg\hbar\Omega\gg\hbar\omega_R$, they find

$$W[x'(t),x(t)]=-M\gamma\int dt[x'\dot x'-x\dot x+x'\dot x-x\dot x']+i\frac{2M\gamma kT}{\hbar}\int dt[x'(t)-x(t)]^2, \tag{29}$$

where γ summarizes the interaction strengths of the distinguished oscillator with its environment. The real part of W contributes dissipation to the equations of motion. The imaginary part squeezes the trajectories $x(t)$ and $x'(t)$ together, thereby providing approximate decoherence. Very roughly, primed and unprimed position intervals differing by distances d on opposite sides of the trace in (28) will decohere when spaced in time by intervals

$$t \gtrsim \frac{1}{\gamma}\left[\left(\frac{\hbar}{\sqrt{2MkT}}\right)\cdot\left(\frac{1}{d}\right)\right]^2. \tag{30}$$

As stressed by Żurek,[26] for typical macroscopic parameters this minimum time for decoherence can be many orders of magnitude smaller than a characteristic dynamical time, say the damping time $1/\gamma$. (The ratio is around 10^{-40} for $M \sim$ gm, $T \sim 300°$K, $d \sim$ cm!)

The behavior of a coarse-grained set of alternative histories based on projections, at times spaced far enough apart for decoherence, onto ranges of values of x alone, is then roughly classical in that the successive ranges of positions follow roughly classical orbits, but with the pattern of classical correlation disturbed by various effects, especially (a) the effect of quantum spreading of the x-coordinate, (b) the effect of quantum fluctuations of the other oscillators, and (c) classical statistical fluctuations, which are lumped with (b) when we use the fundamental formula. We see that the larger the mass M, the shorter the decoherence time and the more the x-coordinate resists the various challenges to its classical behavior.

What the above models convincingly show is that decoherence will be widespread in the universe for certain familiar "classical" variables. The answer to Fermi's question to one of us of why we don't see Mars spread out in a quantum superposition of different positions in its orbit is that such a superposition would rapidly decohere. We now proceed to a more detailed discussion of such decoherence.

§6. Quasiclassical Domains

As observers of the universe, we deal with coarse grainings that are appropriate to our limited sensory perceptions, extended by instruments, communication, and records, but in the end characterized by a great amount of ignorance. Yet we have the impression that the universe exhibits a finer-grained set of decohering histories, independent of us, defining a sort of "classical domain", governed largely by classical laws, to which our senses are adapted while dealing with only a small part of it. No such coarse graining is determined by pure quantum theory alone. Rather, like decoherence, the existence of a quasiclassical domain in the universe must be a consequence of its initial condition and the hamiltonian describing its evolution.

Roughly speaking, a quasiclassical domain should be a set of alternative decohering histories, maximally refined consistent with decoherence, with its individual histories exhibiting as much as possible patterns of classical correlation in time. Such histories cannot be *exactly* correlated in time according to classical laws because sometimes their classical evolution is disturbed by quantum events. There are no classical domains, only quasiclassical ones.

We wish to make the question of the existence of one or more quasiclassical domains into a *calculable* question in quantum cosmology and for this we need criteria to measure how close a set of histories comes to constituting a "classical domain". We have not solved this problem to our satisfaction, but, in the next few sections, we discuss some ideas that may contribute toward its solution.

§7. Maximal Sets of Decohering Histories

Decoherence results from coarse graining. As described in §4(b) and Fig. 2, coarse grainings can be put into a partial ordering with one another. A set of alternative histories is a coarse graining of a finer set if all the exhaustive sets of projections $\{P^k_\alpha\}$ making up the coarser set of histories are obtained by partial sums over the projections making up the finer set of histories.

Maximal sets of alternative decohering histories are those for which there are no finer-grained sets that are decoherent. It is desirable to work with maximal sets of decohering alternative histories because they are not limited by the sensory capacity of any set of observers — they can cover phenomena in all parts of the universe and at all epochs that could be observed, whether or not any observer was present. Maximal sets are the most refined descriptions of the universe that may be assigned probabilities in quantum mechanics.

The class of maximal sets possible for the universe depends, of course, on the completely fine-grained histories that are presented by

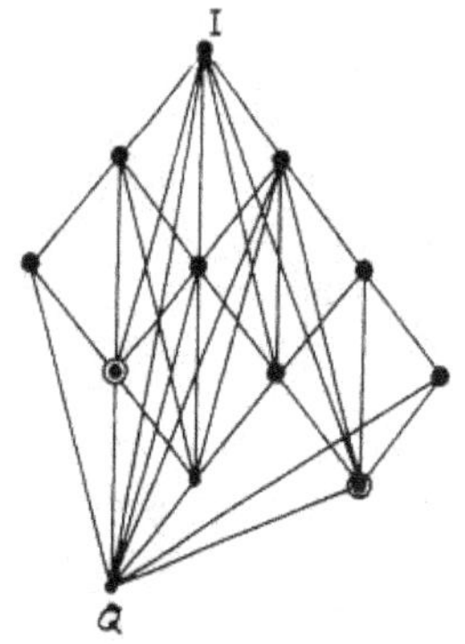

Fig. 4. If the completely fine grained histories arise from a single complete set of observables, say the set $\mathcal{Q}$ of field variables Q^i at each point in space and every time, then the possible coarse-grained histories will be a subset of those illustrated in Fig. 2. Maximal sets can still be defined but will, in general, differ from those of Fig. 2.

the actual quantum theory of the universe. If we utilize to the full, at each moment of time, all the projections permitted by transformation theory, which gives quantum mechanics its protean character, then there is an infinite variety of completely fine-grained sets, as illustrated in Fig. 2. However, were there some fundamental reason to restrict the completely fine grained sets, as would be the case if sum-over-histories quantum mechanics were fundamental, then the class of maximal sets would be smaller as illustrated in Fig. 4. We shall proceed as if all fine grainings are allowed.

If a full correlation exists between a projection in a coarse graining and another projection not included, then the finer graining including both still defines a decoherent set of histories. In a maximal set of decoherent histories, both correlated projections must be included if either one is included. Thus, in the mechanism of decoherence discussed in §5, projections onto the correlated orthogonal states of the 3°K photons are included in the maximal set of decohering histories along with the positions of the massive bodies. Any projections defining historical records such as we shall describe in §10, or values of measured quantities such as we shall describe in §11, must similarly be included in a maximal set.

More information about the initial ρ and H is contained in the probabilities of a finer-grained set of histories than in those of a coarser grained set. It would be desirable to have a quantitative measure of *how much* more information is obtained in a further fine graining of a coarse-grained set of alternative histories. Such a quantity would then measure how much closer a decoherent fine graining comes to maximality in a physically relevant sense.

We shall duscuss a quantity that, while not really a measure of maximality is useful in exploring some aspects of it. In order to construct this quantity the usual entropy formula is applied to sets of alternative decohering *histories* of the universe, rather than, as more usually, alternatives at a single time. We make use of the coarse-grained density matrix $\tilde{\rho}$ defined using the methods of Jaynes,[28] *but generalized to take account of the density matrix of the universe and applied to the probabilities for histories.* The density matrix $\tilde{\rho}$ is constructed by maximizing the entropy functional

$$S(\tilde{\rho}) = -\mathrm{Tr}\,(\tilde{\rho}\log\tilde{\rho}) \tag{31}$$

over all density matrices $\tilde{\rho}$ that satisfy the constraints ensuring that each

$$\mathrm{Tr}\,[P^n_{\alpha'_n}(t_n)\cdots P^1_{\alpha'_1}(t_1)\tilde{\rho}P^1_{\alpha_1}(t_1)\cdots P^n_{\alpha_n}(t_n)] \tag{32}$$

has the same value it would have had when computed with the density matrix of the universe, ρ, for a given set of coarse-grained histories. The density matrix $\tilde{\rho}$ thus reproduces the decoherence functional for this set of histories, and in particular their probabilities, but possesses as little information as possible beyond those properties.

A fine graining of a set of alternative histories leads to *more* conditions on $\tilde{\rho}$ of the form (32) than in the coarser-grained set. In non-trivial cases $S(\tilde{\rho})$ is, therefore, lowered and $\tilde{\rho}$ becomes closer to ρ.

If the insertion of apparently new P's into a chain is redundant, then $S(\tilde{\rho})$ will not be lowered. A simple example will help to illustrate this: Consider the set of histories consisting of projections $P^m_{\alpha_m}(t_m)$ which project onto an orthonormal basis for Hilbert space at one time, t_m. Trivial further decoherent fine grainings can be constructed as follows: At each other time t_k introduce a set of projections $P^k_{\alpha_k}(t_k)$ that, through the equations of motion, are identical operators in Hilbert space to the set $P^m_{\alpha_m}(t_m)$. In this way, even though we are going through the motions of introducing a completely fine-grained set of histories covering all the times, we are really just repeating the projections $P^m_{\alpha_m}(t_m)$ over and over again. We thus have a completely fine-grained set of histories that, in fact, consists of just one fine-grained set of projections and decoheres exactly because there is only one such set. Indeed, in terms of $S(\tilde{p})$ it is no closer to maximality than the set consisting of $P^m_{\alpha_m}(t_m)$ at one time. The quantity $S(\tilde{p})$ thus serves to identify such trivial refinements which amount to redundancy in the conditions (32).

We can generalize the example in an interesting way by constructing the special kinds of histories mentioned after eq. (25). We take t_m to be the final time and then adjoin, at earlier and earlier times, a succession of progressive coarse grainings of the set $\{P^m_{\alpha_m}(t_m)\}$. Thus, as time moves forward, the only projections are finer and finer grainings terminating in the one-dimensional $P^m_{\alpha_m}(t_m)$. We thus have again a set of histories in which decoherence is automatic, independent of the character of ρ, and for which $S(\tilde{p})$ has the same value it would have had if only conditions at the final time were considered. (In some related problems, a quantity like S that keeps decreasing as a result of adjoining projections at later times can be converted into an increasing quantity by adding an algorithmic complexity term.[29])

In a certain sense, $S(\tilde{p})$ for histories can be regarded as decreasing with time. If we consider $S(\tilde{p})$ for a string of alternative projections up to a certain time t_n, as in (32), and then adjoin an additional set of projections for a later time, the number of conditions on $\tilde{p}$ is increased and thus the value of $S(\tilde{p})$ is decreased (or, in trivial cases, unchanged). That is natural, since $S(\tilde{p})$ is connected with the lack of information contained in a set of histories and that information increases with non-trivial fine graining of the histories, no matter what the times for which the new P's are introduced.

The measure $S(\tilde{p})$ is closely related to other fundamental quantities in physics. One can, for example, show that when *used with the $\rho_{\rm eff}$ representing present data and with alternatives at a single time*, these techniques give a unified and generalized treatment of the variety of coarse grainings commonly introduced in statistical mechanics; and, as Jaynes and others have pointed out, the resulting $S(\tilde{p})$'s are the physical entropies of statistical mechanics. Here, however, these techniques are applied to time *histories* and the initial condition is utilized. The quantity $S(\tilde{p})$ is also related to the notion of thermodynamic depth currently being investigated by Lloyd.[30]

§8. Classicity

Some maximal sets will be more nearly classical than others. The more nearly classical sets of histories will contain projections (onto related ranges of values) of operators, for different times, that are connected to one another by the unitary transformations $e^{-iH(\Delta t)}$ and that are correlated for the most part along classical paths, with probabilities near zero and one for the successive projections. This pattern of classical correlation may be disturbed by the inclusion, in the maximal set of projection operators, of other variables, which do not behave in this way (as in measurement situations to be described later). The pattern may also be disturbed by quantum spreading and by quantum and classical fluctuations, as described in connection with the oscillator example treated in §5. Thus we can, at best, deal with *quasiclassical* maximal sets of alternative decohering histories, with trajectories that split and fan out at a result of the processes that make the decoherence possible. As we stressed earlier, there are no classical domains, only quasiclassical ones.

The impression that there is something like a classical domain suggests that we try to define quasiclassical domains precisely by searching for a measure of classicity for each of the maximal sets of alternative decohering histories and concentrating on the one (or ones) with maximal classicity. Such a measure would be applied to the elements of D and the corresponding coarse graining. It should favor predictability, involving patterns of classical correlation as described above. It should also favor maximal sets of alternative decohering histories that are relatively fine-grained as opposed to those which had to be carried to very coarse graining before they would give decoherence. We are searching for such a measure. It should provide a precise and quantitative meaning to the notion of quasiclassical domain.

§9. Quasiclassical Operators

What are the projection operators that specify the coarse graining of a maximal set of alternative histories with high classicity, which defines a quasiclassical domain? They will include, as mentioned above, projections onto comparable ranges of values of certain operators at sequences of times, obeying roughly classical equations of motion, subject to fluctuations that cause their trajectories to fan out from time to time. We can refer to these operators, which habitually decohere, as "quasiclassical operators". What these quasiclassical operators are, and how many of them there are, depends not only on H and ρ, but also on the epoch, on the spatial region, and on previous branchings.

We can understand the origin of at least some quasiclassical operators in reasonably general terms as follows: In the earliest instants of the universe the operators defining spacetime on scales will above the Planck scale emerge from the quantum fog as quasiclassical.[31] Any theory of the initial condition that does not imply this is simply inconsistent with observation in a manifest way. The background spacetime thus defined obeys the Einstein equation. Then, where there are suitable conditions of low temperature, etc., various sorts of hydrodynamic variables may emerge as quasiclassical operators. These are integrals over suitable small volumes of densities of conserved or nearly conserved quantities. Examples are densities of energy, momentum, baryon number, and, in later epochs, nuclei, and even chemical species. The sizes of the volumes are limited above by maximality and are limited below by classicity because they require sufficient "inertia" to enable them to resist deviations from predictability caused by their interactions with one another, by quantum spreading, and by the quantum and statistical fluctuations resulting from interactions with the rest of the universe. Suitable integrals of densities of approximately conserved quantities are thus candidates for habitually decohering quasiclassical operators. Field theory is local, and it is an interesting question whether that locality somehow picks out local densities as the source of habitually decohering quantities. It is hardly necessary to note that such hydrodynamic variables are among the principal variables of classical physics.[32]

In the case of densities of conserved quantities, the integrals would not change at all if the volumes were infinite. For smaller volumes we expect approximate persistence. When, as in hydrodynamics, the rates of change of the integrals form part of an approximately closed system of equations of motion, the resulting evolution is just as classical as in the case of persistence.

§10. Branch Dependence

As the discussion in §5 and §9 shows, physically interesting mechanisms for decoherence will operate differently in different alternative histories for the universe. For example, hydrodynamic variables defined by a relatively small set of volumes may decohere at certain locations in spacetime in those branches where a gravitationally condensed body (*e.g.*, the earth) actually exists, and may not decohere in other branches where no such condensed body exists at that location. In the latter branch there simply may be not enough "inertia" for densities defined with too small volumes to resist deviations from predictability. Similarly, alternative spin directions associated with Stern-Gerlach beams may decohere for those branches on which a photographic plate detects their beams and not in a branch where they recombine coherently instead. There are no variables that are expected to decohere universally. Even the mechanisms causing spacetime geometry at a given location to decohere on scales far above the Planck length cannot necessarily be expected to operate in the same way on a branch where the location is the center of a black hole as on those branches where there is no black hole nearby.

How is such "branch dependence" described in the formalism we have elaborated? It is not described by considering histories where the *set* of alternatives at one time (the k in a set of P_α^k) depends on *specific* alternatives (the α's) of sets of earlier times. Such dependence would destroy the derivation of the probability sum rules from the fundamental formula. However, there is no such obstacle to the set of alternatives at one time depending on the *sets* of alternatives at all previous times. It is by exploiting this possibility, together with the possibility of present records of past events, that we can correctly describe the sense in which there is branch dependence of decoherence, as we shall now discuss.

A record is a present alternative that is, with high probability, correlated with an alternative in the past. The construction of the relevant probabilities was discussed in §4, including their dependence on the initial condition of the universe (or at least on information that effectively bears on that initial condition). The subject of history is most honestly described as the construction of probabilities for the past, given such records. Even non-commuting alternatives such as a position and its momentum at different, even nearby times may be stored in presently commuting record variables.

The branch dependence of histories becomes explicit when sets of alternatives are considered that include records of specific events in the past. To illustrate this, consider the example above, where different sorts of hydrodynamic variables might decohere or not depending on whether there was a gravitational condensation. The set of alternatives that decohere must refer both to the records of the condensation *and* to hydrodynamic variables. Hydrodynamic variables with smaller volumes would be part of the subset with the record that the condensation took place and vice versa.

The branch dependence of decoherence provides the most direct argument against the position that a classical domain should simply be *defined* in terms of a certain set of variables (*e.g.*, values of spacetime averages of the fields in the classical action). There are unlikely to be any physically interesting variables that decohere independent of circumstance.

§11. Measurement Situations

When a correlation exists between the ranges of values of two operators of a quasiclassical domain, there is a *measurement situation*. From a knowledge of the value of one, the value of the other can be deduced because they are correlated with probability near unity. Any such correlation exists in some branches of the universe and not in others; for example, measurements in a laboratory exist only in those branches where the laboratory was actually constructed!

We use the term "measurement situation" rather than "measurement" for such correlations to stress that nothing as sophisticated as an "observer" need be present for them to exist. If there are many significantly different quasiclassical domains, different measurement situations may be exhibited by each one.

When the correlation we are discussing is between the ranges of values of two quasiclassical operators that *habitually* decohere, as discussed in §9, we have a measurement situation of a familiar classical kind. However, besides the quasiclassical operators, the highly classical maximal sets of alternative histories of a quasiclassical domain may include *other* operators having ranges of values strongly correlated with the quasiclassical ones at particular times. Such operators, not normally decohering, are, in fact, included among the decohering set only by virtue of their correlation with a habitually decohering one. In this case we have a measurement situation of the kind usually discussed in quantum mechanics. Suppose, for example, in the inevitable Stern-Gerlach experiment, that σ_z of a spin-1/2 particle is correlated with the orbit of an atom in an inhomogeneous magnetic field. If the two orbits decohere because of interaction with something else (the atomic excitations in a photographic plate for example), then the spin direction will be included in the maximal set of decoherent histories, fully correlated with the decohering orbital directions. The spin direction is thus measured.

The recovery of the Copenhagen rule for when probabilities may be assigned is immediate. Measured quantities are correlated with decohering histories. Decohering histories can be assigned probabilities. Thus in the two-slit experiment (Fig. 1), when the electron interacts with an apparatus that determines which slit it passed through, it is the decoherence of the alternative configurations of the apparatus that enables probabilities to be assigned for the electron.

Correlation between the ranges of values of operators of a quasiclassical domain is the *only* defining property of a measurement situation. Conventionally, measurements have been characterized in other ways. Essential features have been seen to be irreversibility, amplification beyond a certain level of signal-to-noise, association with a macroscopic variable, the possibility of further association with a long chain of such variables, and the formation of enduring records. Efforts have been made to attach some degree of precision to words like "irreversible", "macroscopic", and "record", and to discuss what level of "amplification" needs to be achieved.[33] While such characterizations of measurement are difficult to define precisely,[d] some can be seen in a rough way to be consequences of the definition that we are attempting to introduce here, as follows:

Correlation of a variable with the quasiclassical domain (actually, inclusion in its set of histories) accomplishes the amplification beyond noise and the association with a macroscopic variable that can be extended to an indefinitely long chain of such variables. The relative predictability of the classical world is a generalized form of record. The approximate constancy of, say, a mark in a notebook is just a special case; persistence in a classical orbit is just as good.

Irreversibility is more subtle. One measure of it is the cost (in energy, money, etc.) of tracking down the phases specifying coherence and restoring them. This is intuitively large in many typical measurement situations. Another, related measure is the negative of the logarithm of the probability of doing so. If the probability of restoring the phases in any particular measurement situation were significant, then we would not have the necessary amount of decoherence. The correlation could not be inside the set of decohering histories. Thus, this measure of irreversibility is large. Indeed, in many circumstances where the phases are carried off to infinity or lost in photons impossible to retrieve, the probability of recovering them is truly zero and the situation perfectly irreversible—infinitely costly to reverse and with zero probability for reversal!

Defining a measurement situation solely as the existence of correlations in a quasiclassical domain, if suitable general definitions of maximality and classicity can be found, would have the advantages of clarity, economy, and generality. Measurement situations occur throughout the universe and without the necessary intervention of anything as sophisticated as an "observer". Thus, by this definition, the production of fission tracks in mica deep in the earth by the decay of a uranium nucleus leads to a measurement situation in a quasiclassical domain in which the tracks directions decohere, whether or not these tracks are ever registered by an "observer".

§12. Complex Adaptive Systems

Our picture is of a universe that, as a consequence of a particular initial condition and of the underlying hamiltonian, exhibits at least one quasiclassical domain made up of suitably defined maximal sets of alternative histories with as much classicity as possible. The quasiclassical domains would then be a consequence of the theory and its boundary condition, not an artifact of our construction. How do we then characterize our place as a collectivity of observers in the universe?

Both singly and collectively we are examples of the general class of complex adaptive systems. When they are considered within quantum mechanics as portions of the universe, making observations, we refer to such complex adaptive systems as information gathering and utilizing systems (IGUSes). The general characterization of complex adaptive systems is the subject of much ongoing research, which we cannot discuss here. From a quantum-mechanical point of view the foremost characteristic of an IGUS is that, in

d) An example of this occurs in the case of "null measurements" discussed by Renninger,[34] Dicke,[35] and others. An atom decays at the center of a spherical cavity. A detector which covers all but a small opening in the sphere does *not* register. We conclude that we have measured the direction of the decay photon to an accuracy set by the solid angle subtended by the opening. Cartainly there is an interaction of the electromagnetic field with the detector, but did the escaping photon suffer an "irreversible act of amplification"? The point in the present approach is that the *set* of alternatives, detected and not detected, exhibits decoherence because of the place of the detector in the universe.

some form of approximation, however crude or classical, it employs the fundamental formula, with what amounts to a rudimentary theory of ρ, H, and quantum mechanics. Probabilities of interest to the IGUS include those for correlations between its memory and the external world. (Typically these are assumed perfect; not always such a good approximation!) An approximate fundamental formula is used to compute probabilities on the basis of present data, make predictions, control future perceptions on the basis of these predictions (i.e., exhibit behavior), acquire further data, make further predictions, and so on.

To carry on in this way, an IGUS uses probabilities for histories referring both to the future and the past. An IGUS uses decohering sets of alternative histories and therefore performs further coarse graining on a quasiclassical domain. Naturally, its coarse graining is very much coarser than that of the quasiclassical domain since it utilizes only a few of the variables in the universe.

The reason such systems as IGUSes exist, functioning in such a fashion, is to be sought in their evolution within the universe. It seems likely that they evolved to make predictions because it is adaptive to do so.[e] The reason, therefore, for their focus on decohering variables is that these are the *only* variables for which predictions can be made. The reason for their focus on the histories of a quasiclassical domain is that these present enough regularity over time to permit the generation of models (schemata) with significant predictive power.

If there is essentially only one quasiclassical domain, then naturally the IGUS utilizes further coarse grainings of it. If there are many essentially inequivalent quasiclassical domains, then we could adopt a subjective point of view, as in some traditional discussions of quantum mechanics, and say that the IGUS "chooses" its coarse graining of histories and, therefore, "chooses" a particular quasiclassical domain, or a subset of such domains, for further coarse graining. It would be better, however, to say that the IGUS evolves to exploit a particular quasiclassical domain or set of such domains. Then IGUSes, including human beings, occupy no special place and play no preferred role in the laws of physics. They merely utilize the probabilities presented by quantum mechanics in the context of a quasiclassical domain.

e) Perhaps as W. Unruh has suggested, there are complex adaptive systems, making no use of prediction, that can function in a highly quantum-mechanical way. If this is the case, they are very different from anything we know or understand.

§13. Conclusions

We have sketched a program for understanding the quantum mechanics of the universe and the quantum mechanics of the laboratory, in which the notion of quasiclassical domains plays a central role. To carry out that program, it is important to complete the definition of a quasiclassical domain by finding the general definition for classicity. Once that is accomplished, the question of how many and what kinds of essentially inequivalent quasiclassical domains follow from ρ and H becomes a topic for serious theoretical research. So is the question of what are the general properties of IGUSes that can exist in the universe exploiting various quasiclassical domains, or the unique one if there is essentially only one.

It would be a striking and deeply important fact of the universe if, among its maximal sets of decohering histories, there were one roughly equivalent group with much higher classicities than all the others. That would then be *the* quasiclassical domain, completely independent of any subjective criterion, and realized within quantum mechanics by utilizing only the initial condition of the universe and the hamiltonian of the elementary particles.

Whether the universe exhibits one or many maximal sets of branching alternative histories with high classicities, those quasiclassical domains are the possible arenas of prediction in quantum mechanics.

It might seem at first sight that in such a picture the complementarity of quantum mechanics would be lost; in a given situation, for example, *either* a momentum *or* a coördinate could be measured, leading to different kinds of histories. We believe that impression is illusory. The histories in which an observer, as part of the universe, measures p and the histories in which that observer measures x are decohering alternatives. The important point is that the decoherent histories of a quasiclassical domain contain all possible choices that might be made by all possible observers that might exist, now, in the past, or in the future for that domain.

The EPR or EPRB situation is no more mysterious. There, a choice of measurements, say, σ_x or σ_y for a given electron, is correlated with the behavior of σ_x or σ_y for another electron because the two together are in a singlet spin state even though widely separated. Again, the two measurement situations (for σ_x and σ_y) decohere from each other. But here, in each, there is also a correlation between the information obtained about one spin and the information that can be obtained about the other. This behavior, although unfortunately called "non-local" by some authors, involves no non-locality in the ordinary sense of quantum field theory and no possibility of signaling outside the light cone. The problem with the "local realism" that Einstein would have liked is not the locality but the *realism*. Quantum mechanics describes *alternative* decohering histories and one cannot assign "reality" simultaneously to different alternatives because they are contradictory. Everett[9] and others[11] have described this situation, not incorrectly, but in a way that has confused some, by saying that the histories are all "equally real" (meaning only that quantum mechanics prefers none over another except via probabilities) and by referring to "many worlds" instead of "many histories".

We conclude that resolution of the problems of interpretation presented by quantum mechanics is not to be accomplished by further intense scrutiny of the subject as it applies to reproducible laboratory situations, but rather through an examination of the origin of the universe and its subsequent history. Quantum mechanics is best and most fundamentally understood in the context of quantum cosmology. The founders of quantum mechanics were right in pointing out that something external to the framework of wave function and Schrödinger equation *is* needed to interpret the theory. But it is not a postulated classical world to which quantum mechanics does not apply. Rather it is the initial condition of the universe that, together with the action function of the elementary particles and the throws of quantum dice since the beginning, explains the origin of quasiclassical domain(s) within quantum theory itself.

Acknowledgements

One of us, MG-M, would like to acknowledge the great value of conversations about the meaning of quantum mechanics with Felix Villars and Richard Feynman in 1963–64 and again with Richard Feynman in 1987–88. He is also very grateful to Valentine Telegdi for discussions during 1985–86, which persuaded him to take up the subject again after twenty years. Both of us are indebted to Telegdi for further interesting conversations since 1987. We would also like to thank R. Griffiths for a useful communication and a critical reading of the manuscript and R. Penrose for a helpful discussion.

Part of this work was carried out at various times at the Institute for Theoretical Physics, Santa Barbara, the Aspen Center for Physics, the Santa Fe Institute, and the Department of Applied Mathematics and Theoretical Physics, University of Cambridge. We are grateful for the hospitality of these institutions. The work of JBH was supported in part by NSF grant PHY85-06686 and by a John Simon Guggenheim Fellowship. The work of MG-M was supported in part by the U.S. Department of Energy under contract DE-AC-03-81ER40050 and by the Alfred P. Sloan Foundation.

References

For a subject as large as this one it would be an enormous task to cite the literature in any historically complete way. We have attempted to cite only papers that we feel will be *directly* useful to the points raised in the text. These are not always the earliest nor are they always the latest. In particular we have not attempted to review or to cite papers where similar problems are discussed from different points of view.

1) J. B. Hartle and S. W. Hawking: Phys. Rev. **D28** (1983) 2960.
2) A. Vilenkin: Phys. Rev. **D33** (1986) 3560.
3) For recent reviews see, *e.g.*, J. J. Halliwell: *Quantum Cosmology: An Introductory Review*, ITP prepring NSF-ITP-88-131 (1988); J. B. Hartle: in *Highlights in Gravitation and Cosmology*, eds. B. R. Iyer, A. Kembhavi, J. V. Narlikar, C. V. Vishveshwara (Cambridge University Press, Cambridge, 1989); and J. B. Hartle: in *Proceedings of the 12th International Conference on General Relativity and Gravitation* (Cambridge University

Press, Cambridge, 1990). For a bibliography of papers on quantum cosmology see, J. J. Halliwell: ITP preprint NSF-ITP-88-132.
4) As, for example, in recent discussions of the value of the cosmological constant see, *e.g.*, S. W. Hawking: Phys. Lett. **B195** (1983) 337; S. Coleman: Nucl. Phys. **B310** (1988) 643; and S. Giddings and A. Strominger: Nucl. Phys. **B307** (1988) 854.
5) D. Finkelstein: Trans. N.Y. Acad. Sci. **25** (1963) 621; J. B. Hartle: Am. J. Phys. **36** (1968) 704; R. N. Graham: in *The Many Worlds Interpretation of Quantum Mechanics*, eds. B. DeWitt and R. N. Graham (Princeton University Press, Princeton, 1973), and E. Farhi, J. Goldstone, and S. Gutmann: to be published.
6) See the essays "The Unity of Knowledge" and "Atoms and Human Knowledge" reprinted in N. Bohr: *Atomic Physics and Human Knowledge* (John Wiley, New York, 1958).
7) L. Landau and E. Lifshitz: *Quantum mechanics* (Pergamon, London, 1958).
8) For clear statements of this point of view see F. London and E. Bauer: *La théorie de l'observation en mécanique quantique* (Hermann, Paris, 1939) and R. B. Peierls: in *Symposium on the Foundations of Modern Physics*, eds. P. Lahti and P. Mittelstaedt (World Scientific, Singapore, 1985).
9) H. Everett: Rev. Mod. Phys. **29** (1957) 454.
10) J. A. Wheeler: Rev. Mod. Phys. **29** (1957) 463.
11) B. DeWitt: Physics Today **23** no. 9 (1970).
12) R. Geroch: *Noûs* **18** (1984) 617.
13) V. F. Mukhanov: in *Proceedings of the Third Seminar on Quantum Gravity*, eds. M. A. Markov, V. A. Berezin and V. P. Frolov (World Scientific, Singapore, 1985).
14) M. Gell-Mann: unpublished (1963).
15) L. Cooper and D. Van Vechten: Am. J. Phys. **37** (1969) 1212.
16) B. DeWitt and R. N. Graham: *The Many Worlds Interpretation of Quantum Mechanics* (Princeton University Press, Princeton, 1973).
17) Some elements of the program have been reported earlier. See, M. Gell-Mann: in *Proceedings of the Nobel Symposium No. 67*: "The Unification of the Fundamental Interactions", Physica Scripta **T15** (1987) 202.
18) H. Zeh: Found. Phys. **1** (1971) 69.
19) W. Żurek: Phys. Rev. **D24** (1981) 1516; Phys. Rev. **D26** (1982) 1862.
20) E. Joos and H. D. Zeh: Zeit. Phys. **B59** (1985) 223.
21) R. Griffiths: J. Stat. Phys. **36** (1984) 219.
22) R. Omnès: J. Stat. Phys. **53** (1988) 893; *ibid.* **53** (1988) 933; *ibid.* **53** (1988) 957.
23) See, *e.g.*, J. B. Hartle: in *Proceedings of the Osgood Hill Conference on the Conceptual Problems of Quantum Gravity*, eds. A. Ashtekar and J. Stachel (Birkhauser, Boston, 1990); in *Proceedings of the 5th Marcel Grossmann Meeting on Recent Developments in General Relativity*, eds. D. Blair and M. Buckingham (World Scientific, Singapore, 1990); Phys. Rev. **D37** (1988) 2818; Phys. Rev. **D38** (1988) 2985; and in *Quantum Cosmology and Baby Universes* (Proceedings of the 1989 Jerusalem Winter School on Theoretical Physics), eds. S. Coleman, J. B. Hartle and T. Piran (World Scientific, Singapore, 1990). For a concise discussion see M. Gell-Mann: Physics Today, February (1989) p. 50.
24) The utility of this Heisenberg picture formulation of quantum mechanics has been stressed by many authors, among them E. Wigner: Am. J. Phys. **31** (1963) 6; Y. Aharonov, P. Bergmann and J. Lebovitz: Phys. Rev. **B134** (1964) 1410; W. Unruh: in *New Techniques and Ideas in Quantum Measurement Theory*, ed. D. M. Greenberger (Ann. N.Y. Acad. Sci. **480**) (New York Academy of Science, New York, 1986); M. Gell-Mann: *Physica Scripta* **T15** (1987) 202.
25) A. O. Caldeira and A. J. Leggett: Physica **121A** (1983) 587.
26) W. Żurek: in *Non-Equilibrium Quantum Statistical Physics*, eds. G. Moore and M. Scully, (Plenum Press, New York, 1984).
27) R. P. Feynman and J. R. Vernon: Ann. Phys. (N.Y.) **24** (1963) 118.
28) See, *e.g.*, the papers reprinted in *E. T. Jaynes: Papers on Probability, Statistics, and Statistical Physics*, ed. R. D. Rosenkrantz (D. Reidel, Dordrecht, 1983) or A. Hobson: *Concepts in Statistical Mechanics* (Gordon and Breach, New York, 1970).
29) W. H. Żurek: Phys. Rev. **A40** (1989) 4731.
30) S. Lloyd: private communication.
31) See, *e.g.*, E. Joos: Phys. Lett. **A116** (1986) 6; H. Zeh: Phys. Lett. **A116** (1986) 9; C. Kiefer: Class. Quant. Grav. **4** (1987) 1369; T. Fukuyama and M. Morikawa: Phys. Rev. **D39** (1989) 462; J. Halliwell: Phys. Rev. **D39** (1989) 2912; T. Padmanabhan: Phys. Rev. **D39** (1989) 2924.
32) For discussion of how such hydrodynamic variables are distinguished in non-equilibrium statistical mechanics in not unrelated ways see, *e.g.*, L. Kadanoff and P. Martin: Ann. Phys. (N.Y.) **24** (1963) 419; D. Forster: *Hydrodynamic Fluctuations, Broken Symmetry, and Correlation Functions* (Benjamin, Reading, Mass., 1975); and J. Lebovitz: Physica **140A** (1986) 232.
33) For an interesting effort at precision see A. Daneri, A. Loinger and G. M. Prosperi: Nucl. Phys. **33** (1962) 297.
34) M. Renninger: Zeit. Phys. **158** (1960) 417.
35) R. H. Dicke: Am. J. Phys. **49** (1981) 925.

* * *

M. Namiki: I have two questions: (1) I think, the notion of "history" must be ascribed to a local system with a finite space-time interval. Different local systems should have their own histories different from each other. This idea is closely related to your coarse-graining procedure of defining the "history". What about this?

(2) You defined the "history", choosing a set of observables. Do we have the position dependent history and the momentum-dependent history? They must be generally different from each other. (Your definition of the "history" may not be unique.) But this question may be related to the problem on how to make the "quasiclassical world". What about this?

J. B. Hartle: (1) In principle, given an initial condition for the universe and the hamiltonian we should be able to compute the maximal sets of coarse grained, alternative, decohering histories with high classicity that define a quasiclassical world. The arguments of §5 and §9 support the idea that the results of such a calculation would include the histories of local systems. I agree that this must come out to be consistent with observations, but in a given theory we can calculate and see.

(2) We do not "choose" the histories of a quasiclassical world; we compute them as described above. A history in which, say the position of an electron is measured and one in which the momentum is measured, are two decohering alternatives because of the decoherence of the alternative configuration of the apparatus or memory of an observer.

(3) We do not know if there is essentially one quasiclassical world or many, but in the context of quantum cosmology it is a decidable question.

J. Anandan: I am interested in the reason for your choice of quasiclassical variables. I believe that the world looks classical because the interactions, which we believe to be gravitational and gauge fields, are functions of space-time and therefore diagonal in basis of position eigenstates. But is your reason for choosing the quasiclassical variables that these variables decohere?

J. B. Hartle: We do not "choose" quasiclassical variables. We could, in principle, *compute* them from the initial condition, the hamiltonian, and the information about our specific history. All three forms of information are needed to determine them, not just how they enter into interactions. However, in §9 arguments are offered that the result of such a computation will include the local densities of conserved or approximately conserved quantities as well as the variables describing space-time above the Planck scale.

W. H. Zurek: In the context of the two previous questions it is perhaps useful to emphasize that the reason why the position may be selected as a preferred classical observable! The interactions usually *depend on the position* (i.e., potentials are $V(x)$). Therefore they *commute* with the position observable. Hence, it is the position observable which gets "monitored" by the environment. Consequently, it will inevitably decohere first. What I am suggesting is that IGUS'es have found it convenient to describe objects in the universe in terms of this observable which decoheres first, and began singling it out or position only "with the benefit of hindsight": They would have ended up calling "position" *whatever* observable happened to commute with the typical interaction hamiltonians.

P. Mittelstaedt: Comment: on the beginning of your lecture you mentioned that Copenhagen interprectation can not be applied to quantum cosmology. This statement should be specified since there are many faces of the Copenhagen interpretation. Let me consider the Neumann programm: the quantum mechanical measuring process should be described in terms of quantum mechanics as an interaction between the object system and the measuring instrument. Many parts of this process are now very well understood. However there is an important missing link: The "objectification", i.e., the transition from a superposition to mixture of objectively decided outcomes has not yet been properly understood. In order to overcome this problem, Everett formulated the "many world interpretation". But even today we don't know any mechanism which could explain the objectification. For many practical situations the Copenhagen interprectation is sufficient, but the creation of the universe is not such a case. Hence the problems mentioned become relevant here.

Question: I have no doubt that the formalism which was presented here is very worthwhile and useful for the discription of the early universe. However, I wonder whether your considerations can help to solve the open problems of the quantum theory of the measuring process. Can you explain by means of your results—at least partly—the transition from a superposition to a mixture which appears in any quantum mechanical measuring process?

J. B. Hartle: I believe that it is useful to distinguish two kinds of "problems" in quantum mechanics. First, there are the problems concerned with giving a coherent, consistent and precise formulation of the theory without special external assumptions. The questions of the definition and origin of a quasicalssical world that I have discussed here are the "problems" in this sense. Then there are problems concerned with the adequacy of quantum mechanics relative to standards for physical theory individual physicists may have. These I have not discussed. I think that your problem concerning "objectification" (which I hope I understand correctly as what is sometimes called the "reduction of the wave packet") falls into this latter category. For us, this reassessment of probabilities when new information is required is no more mysterious and no more quantum mechanical than the reassessment that occurs of the joint probability for the winners of eight horse races after the first five races have been run. We do not see it as a "problem" for quantum mechanics.

J.-P. Vigier: The problem with this way of presenting the problems of the history of the Universe since its origin is that we do not ever know (to quote Voltaire) that the Universe was created or not, i.e., it might have an infinite history. The various questions discussed in present day cosmology, the redshift controversy, the physical composition and distribution of the matter and of its dominant forms (plasma?) and the existence or not of a "big bang", are not introduced on this type of models. Unless those it is difficult to distinguish scientific speculation from science fiction.

J. B. Hartle: Thank you for your comment.

M. D. Levenson: This picture demotes the "observer" from the role of creator of classical reality through measurement to that of an "IGUS" that evolved by accident and is at best a spectator or profiter of the quasiclassical "reality" produced by de-correlation due to the initial conditions. The claim seems to be that an "IGUS" must focus on classical aspects in order to make profitable predictions and modify its behavior accordingly. This may not be true. In particular this conference of professional quantum mechanics profits from the non-classical aspects of the quasiclassical universe. Are we a quantum-fluctuation or an evolutionary dead end? Might there be even less classical "IGUSes" than us? (Could we tell if they did exist?) Can any of this be simulated, for example, by cellular automata in a "quantum" artificial universe? Might it not *still* be the case that the possible "initial conditions on the universe" in this picture are entirely constrained by our own present apparant existence? Are some IGUSes not now learning how to use non-classical (*e.g.*, EPR) correlations to modify the universe, at least to the extent of translating 300 IGUSes to Kokubunji?

J. B. Hartle: You have asked several different and interesting questions. Let me respond to them separately:

(1) It is our position that IGUSes evolve to exploit a quasiclassical world presented by this particular universe. However, some would estimate the probability of the evolution of IGUSes as high, even ones sophisticated enough to use quantum mechanics. Their occurance in the universe may thus not be much of an accident.

(2) In discussing the use by an IGUS of specifically quantum mechanical phenomena, it is important to remember that, with the definition of quasiclassical world given here, measured quantum phenomena are *included* in it. Quantum measurements are one reason why we discuss a quasiclassical world and not a classical one. A quasiclassical world consists of many alternative histories. IGUSes carrying out quantum measurements and reporting on them are thus using a quasiclassical world.

(3) I think it would be very interesting to construct models of the decoherence of histories and we are studying this.

(4) I believe that it is very unlikely that the initial conditions of the universe are determined by our existence. We need so little to exist. One can ask oneself: "From our existence, what can we predict for new observations beyond the reach of present telescopes?" I think the answer is "Almost nothing".

M. Peshkin: You never mentioned chaos. Is chaos central to the overall strategy in your program, or does it merely play a technical role in some of the calculation?

J. B. Hartle: Chaos does not play a fundamental role in the framework of quantum mechanics sketched here. However, chaos is important in the universe, because it is a mechanism for the amplification of small, probabilistically distributed, fluctuations to events of significance in the quasiclassical world. Thus, certain chaotic phenomena may be ultimately traceable to quantum fluctuations.

DICK FEYNMAN—THE GUY IN THE OFFICE DOWN THE HALL

A brilliant, vital and amusing neighbor, Feynman was a stimulating (if sometimes exasperating) partner in discussions of profound issues. His sum-over-paths method may turn out to be not just useful, but fundamental.

Murray Gell-Mann

I hope someday to write a lengthy piece about Richard Feynman as I knew him (for nearly 40 years, 33 of them as his colleague at Caltech), about our conversations on the fundamental laws of physics, and about the significance of the part of his work that bears on those laws. In this brief note, I restrict myself to a few remarks and I hardly touch on the content of our conversations.

When I think of Richard, I often recall a chilly afternoon in Altadena shortly after his marriage to the charming Gweneth. My late wife, Margaret, and I had returned in September 1960 from a year in Paris, London and East Africa; Richard had greeted me with the news that he was "catching up with me"—he too was to have an English wife and a small brown dog. The wedding soon took place, and it was a delightful occasion. We also met the dog (called Venus, I believe) and found that Richard was going overboard teaching her tricks (leading his mother, Lucille, with her dry wit, to wonder aloud what would become of a child if one came along). The Feynmans and we both bought houses in Altadena, and on the afternoon in question Margaret and I were visiting their place.

Richard started to make a fire, crumpling up pages of a newspaper and tossing them into the fireplace for kindling. Anyone else would have done the same, but the way he made a game out of it and the enthusiasm that he poured into that game were special and magical. Meanwhile, he had the dog racing around the house, up and down the stairs, and he was calling happily to Gweneth. He was a picture of energy, vitality and playfulness. That was Richard at his best.

He often worked on theoretical physics in the same way, with zest and humor. When we were together discussing physics, we would exchange ideas and silly jokes in between bouts of mathematical calculation—we struck sparks off each other, and it was exhilarating.

What I always liked about Richard's style was the lack of pomposity in his presentation. I was tired of theorists who dressed up their work in fancy mathematical language or invented pretentious frameworks for their sometimes rather modest contributions. Richard's ideas, often powerful, ingenious and original, were presented in a straightforward manner that I found refreshing.

I was less impressed with another well-known aspect of Richard's style. He surrounded himself with a cloud of myth, and he spent a great deal of time and energy generating anecdotes about himself.

Sometimes it did not require a great deal of effort. For example, during my first decade at Caltech there was a rule at our faculty club, the Athenaeum, that men had to wear jackets and ties at lunch. Richard usually came to work quite conventionally dressed (for those days) and hung his jacket and tie in his office. He rarely ate lunch at the Athenaeum, but when he did, he would often make a point of walking over in his shirt sleeves, tieless, and then putting on one of the ragged sport coats and one of the loud ties that the Athenaeum provided in the cloakroom for men who arrived unsuitably attired.

Many of the anecdotes arose, of course, through the stories Richard told, of which he was generally the hero, and in which he had to come out, if possible, looking smarter than anyone else. I must confess that as the years went by I became uncomfortable with the feeling of being a

Murray Gell-Mann is Robert A. Millikan Professor of Theoretical Physics at the California Institute of Technology.

PRINCETON UNIVERSITY

Portrait of Richard Feynman by Jirayr Zorthian, one of Feynman's art teachers. The painting hangs in Jadwin Hall at Princeton University.

rival whom he wanted to surpass; and I found working with him less congenial because he seemed to be thinking more in terms of "you" and "me" than "us." Probably it was difficult for him to get used to collaborating with someone who was not just a foil for his own ideas (especially someone like me, since I thought of Richard as a splendid person to bounce my ideas off!).

At first, none of that was much of a problem, and we had many fine discussions in those days. In the course of those talks not only did we "twist the tail of the cosmos," but we also exchanged a good many lively reminiscences about our experiences in research.

Summing over histories

He told me, of course, of his graduate student days at Princeton and his adventures with his adviser, John Wheeler. Wheeler judged their work on the "absorber theory of radiation" to be too much of a collaboration to qualify as a dissertation for the PhD, and so Richard pursued his interest in Paul Dirac's work on the role of the action S in quantum mechanics. In his book on quantum mechanics, and even more in his article in the *Physikalische Zeitschrift der Sowjetunion* in 1932, Dirac had carried the idea quite far. He had effectively shown how a quantum mechanical amplitude for the transition from a set of values of the coordinates at one time to another set of values at a later time could be represented as a multiple integral, over the values of the coordinates at closely spaced intermediate times, of $\exp(iS/\hbar)$, where S is the value of the classical action along each sequence of intermediate coordinate values. What Dirac had not done was to state the result in so many words, to point out that this method could be used as the starting point for all quantum mechanics, and to suggest it as a practical way of doing quantum mechanical calculations.

Richard did just those things, I understand, in his 1942 dissertation, and then used the "path integral" or "sum over paths" approach in a great deal of his subsequent research. It was the basis, for example, of his way of arriving at the now standard covariant method of calculation in quantum field theory (which Ernst Stueckelberg reached in a different manner). That method is, of course, always presented in terms of "Feynman diagrams" such as the ones Dick later had painted on his van.

The sum-over-paths formulation is particularly convenient for integrating out one set of coordinates to concentrate on the remaining set. Thus the photon propagator in quantum electrodynamics is obtained[1] by "integrating out" the photon variables, leaving electrons and positrons, both real and virtual, to interact by means of the covariant function $\delta(x^2) + (\pi i x^2)^{-1}$.

In 1963 Feynman and his former student F. L. Vernon Jr, carrying further some research Ugo Fano had earlier done in a different way, showed how in a wide variety of problems of concern to laser physicists, condensed matter physicists and others of a practical bent, one can integrate out variables that are not of interest to throw light on the behavior of the ones that are kept. If initially the density matrix factors into one part depending on the interesting variables and another part depending on the rest, then the subsequent time development of the reduced density matrix for the interesting variables can be expressed in terms of a double path integral in which the coefficient of the initial reduced density matrix is $\exp[i(S - S' + W)/\hbar]$, where S is the action along the path referring to the left-hand side of the density matrix, S' is the action along the path referring to the right-hand side of density matrix, and W is the "influence functional," depending on both paths, that comes from integrating out all the uninteresting variables. Feynman and Vernon worked out a number of cases in detail, and subsequent research by A. O. Caldeira and Anthony Leggett, among others, further clarified some of the issues involved.

Shedding light on quantum mechanics

More recently, in the work of H. Dieter Zeh, of Erich Joos and of Wojciech Żurek and others, this line of research has thrown important light on how quantum mechanics produces decoherence, one of the conditions for the nearly classical behavior of familiar objects. For a planet, or even a dust grain, undergoing collisions with, for example, the photons of the 3-K radiation, the imaginary part of the functional W resulting from the integration over those quanta can yield, in $\exp(iW/\hbar)$, a factor that decreases exponentially with some measure of the separation between the coordinate trajectory on the left side of the density matrix and that on the right. The density matrix can thus be constrained to remain nearly diagonal in the coordinates of the particle, giving rise to decoherence. If in addition the dust grain's inertia is large enough that the grain resists, for the most part, the disturbances of its trajectory caused by the quantum and thermal fluctuations of the background, and also large enough that the quantum spreading of the coordinate is slow, then the behavior of the grain's position operator will be nearly classical.

When an operator comes into correspondence with a nearly classical operator, then the first operator can be measured or observed. Thus work such as that of Feynman and Vernon has led not only to practical applications but also to a better understanding of how quantum mechanics produces the world with which we are familiar.

The path integral approach has proved in numerous situations to be a useful alternative to the conventional formulation of quantum mechanics in terms of operators in Hilbert space. It has many advantages besides the ease of integrating out, under suitable conditions, some of the variables. The path integral method, making use as it does of the action, can usually display in an elegant manner the invariances of the theory and can point the way toward exhibiting those invariances in a perturbation expansion. It is obviously a good approach for deriving the classical limit, and it can also be very helpful in semiclassical approximations, for example, in the description of tunneling. For certain effects, such as tunneling via instantons, it permits calculations that are highly nonperturbative in the usual sense. It is also particularly good for the global study of field configurations in quantum field theory, as it permits a straightforward discussion of topological effects.

Of course the conventional approach is superior for certain purposes, such as exhibiting the unitarity of the S matrix and the fact that probabilities are not negative. Richard would never have contemplated, as he did around 1948, the consistent omission of all closed loops in quantum electrodynamics if he had been thinking in terms of a Hamiltonian formulation, where unitarity, which rules out such an omission, is automatic. (The impossible theory without closed loops could, by the way, realize the remarkable vision of Wheeler, which Richard said Wheeler once awakened him to explain: Not only are positrons electrons going backward in time, but all electrons and positrons represent the same electron going backward and forward, thus explaining why they all have the same absolute value of the electric charge!)

In any case, the path integral formulation remained merely a reformulation of quantum mechanics, equivalent to the usual formulation. I say "merely" because Richard, with his great talent for working out, sometimes in dramatically new ways, the consequences of known laws, was unnecessarily sensitive on the subject of discovering new ones. He wrote, in connection with the discovery of the universal vector and axial vector weak interaction in 1957: "It was the first time, and the only time, in my career that I knew a law of nature that nobody else knew. (Of course, it wasn't true, but finding out later that at least Murray Gell-Mann—and also [E. C. George] Sudarshan and [Robert] Marshak—had worked out the same theory didn't spoil my fun.) . . . It's the only time I ever discovered a new law."[2]

Thus it would have pleased Richard to know (and perhaps he did know, without my being aware of it) that there are now some indications that his PhD dissertation may have involved a really basic advance in physical theory and not just a formal development. The path integral formulation of quantum mechanics may be more fundamental than the conventional one, in that there is a crucial domain where it may apply and the conventional formulation may fail. That domain is quantum cosmology.

Seeking rules for quantum gravity

Of all the fields in fundamental physical theory, the gravitational field is picked out as controlling, in Einsteinian fashion, the structure of space–time. This is true even in a unified description of all the fields and all the particles of nature. Today, in superstring theory, we have the first

JOE MUNROE/CALIFORNIA INSTITUTE OF TECHNOLOGY

In the 'Court of the Oak' at Caltech in 1959, Feynman and the author enjoy a relaxed moment.

respectable candidate for such a theory, apparently finite in perturbation theory and describing, roughly speaking, an infinite set of local fields, one of which is the gravitational field linked to the metric of space–time. If all the other fields are dropped, the theory becomes an Einsteinian theory of gravitation.

Now the failure of the conventional formulation of quantum mechanics, if it occurs, is connected with the quantum mechanical smearing of space–time that is inevitable in any quantum field theory that includes Einsteinian gravitation.

If there is a dominant background metric for space–time, especially a Minkowskian metric, and one is treating the behavior of small quantum fluctuations about the background (for example, the scattering of gravitons by gravitons), then the deep questions about space–time in quantum mechanics do not come to the fore.

Dick played a major part in working out the rules of quantum gravity in that approximation. It so happened that I was peripherally involved in the story of that research. We first discussed it when I visited Caltech during the Christmas vacation of 1954–55 and he was my host. (I was offered a job within a few days—such things would take longer now.) I had been interested in a similar approach, sidestepping the difficult cosmological issues, and when I found that he had made considerable progress I encouraged him to continue, to calculate one-loop effects and to find out whether quantum gravity was really a divergent theory to that order. He was always very suspicious of unrenormalizability as a criterion for rejecting theories, but he did pursue the research on and off. In 1960 he complained to me that he was having trouble. His covariant diagram method was giving results incompatible with unitarity. The imaginary part of the amplitude for a fourth-order process should be related directly to the product of a second-order amplitude and the complex conjugate of a second-order amplitude. That relation was failing.

I suggested that he try the analogous problem in Yang–Mills theory, a much simpler nonlinear gauge theory than Einsteinian gravitation. Richard asked what Yang–Mills theory was. (He must have forgotten, because in 1957 we worked out the coupling of the photon to the charged intermediate boson for the weak interaction and noticed that it was the right coupling for a Yang–Mills theory of those quanta.) Anyway, it didn't take long to teach him the rudiments of Yang–Mills theory, and he threw himself with renewed energy into resolving the contradiction. He found, eventually, that in the Lorentz-covariant formulation of either theory it was necessary to introduce some weird supplementary fields called "ghosts," and they have been used ever since, acquiring more and more importance. He described them at a meeting in Poland (in 1963, I think). Usually they are called "Faddeev–Popov ghosts" after L. D. Faddeev and V. N. Popov, who also studied them.

Thus Feynman was able to report in the 1960s that Einsteinian gravitation was terribly divergent when interacting with electrons, photons or other particles. (The divergences in *pure* quantum gravitation theory turned out to be serious too, but that was shown much later, in the two-loop approximation, by two Caltech graduate students, Marc Goroff and Augusto Sagnotti.)

Those problems may be rectified by unification of all the particles and interactions, as they are in superstring theory. But we must still face up to the issues raised by the fact that the metric is up for quantum mechanical grabs and cannot in general be treated as a simple classical background plus small quantum fluctuations.

Quantum cosmology

Recently there has been great progress in thinking about the cosmological aspects of quantized Einsteinian gravitation. The work of Stephen Hawking and James Hartle, as well as Claudio Teitelboim, Alexander Vilenkin, Jonathan Halliwell and several others, has shown how the path integral method can probably deal with the situation and how it may be possible to generalize the method so as to describe *not only the dynamics of the universe but also its initial boundary condition* in terms of the classical action S. Furthermore, there are now, as I mentioned above, some indications that the conventional formulation of quantum mechanics may not be justifiable except to the extent that a background space–time emerges with small quantum fluctuations. Hartle in particular has emphasized such a possibility.

One crude way to see the argument is to express the wavefunction of the universe (which we assume to be in a pure state) as a path integral over all the fields in nature (for example, the infinity of local fields represented, roughly speaking, by the superstring), reserving the integral over the metric $g_{\mu\nu}$ for last. The total action S can be represented as the Einstein action S_G for pure gravitation plus the actions S_M for all the other, "matter" fields, including their coupling to gravitation. We have, then, crudely,

Amplitude =

$$\int \mathscr{D} g_{\mu\nu} \exp \frac{iS_G}{\hbar} \times \int \mathscr{D} (\text{everything else}) \exp \frac{iS_M}{\hbar}$$

For the moment, suppose only $g_{\mu\nu}$ configurations corresponding to a simple topology for space–time are allowed.

Before the integration over $g_{\mu\nu}$ is performed, there is a definite space–time, with the possibility of constructing well-defined space-like surfaces in a definite succession described by a time-like variable. There is an equivalent Hilbert-space formalism; we have unitarity (conservation of positive probability); and we can have conventional causality (it corresponds in the Hilbert-space formulation to the requirement of time ordering of operators in the formula for probabilities).

Now, when the integral over $g_{\mu\nu}$ is done, it is no longer clear that any of that machinery remains, since we are integrating over the structure of space–time and once the integral is performed it is hard to point to space-like surfaces or a succession described by a time-like variable. Of course it may be possible to construct a Hilbert-space formulation, with unitarity and causality, in some new way, perhaps employing a new, external time variable of some kind (what Feynman liked to call a fifth wheel), but it is by no means certain that such a program can be carried out.

At this stage, we may admit the possibility of summing over all topologies of space–time (or of the corresponding space–time with a Euclidean metric). If that is the correct thing to do, then we are immediately transported into the realm of baby universes and wormholes, so beloved of Stephen Hawking and now so fashionable, in which it seems to be demonstrable that the cosmological constant vanishes. In that realm the path integral method appears able to cope, and it remains to be seen to what extent the conventional formulation of quantum mechanics can keep up.

For Richard's sake (and Dirac's too), I would rather like it to turn out that the path integral method is the real foundation of quantum mechanics and thus of physical theory. This is true despite the fact that, having an algebraic turn of mind, I have always personally preferred the operator approach, and despite the added difficulty, in the absence of a Hilbert-space formalism, of interpreting the wavefunction or density matrix of the universe (already a bit difficult to explain in any case, as anyone attending my classes will attest). If notions of transformation theory, unitarity and causality really emerge from the mist only after a fairly clear background metric appears (that metric itself being the result of a quantum mechanical probabilistic process), then we may have a little more explaining to do. Here Dick Feynman's talents and clarity of thought would have been a help.

Turning things around

Richard, as is well known, liked to look at each problem, important or unimportant, in a new way—"turning it around," as he would say. He told how his father, who died when he was young, taught him to do that. This approach went along with Richard's extraordinary efforts to be different, especially from his friends and colleagues.

Of course any of us engaged in creative work, and in fact anyone having a creative idea even in everyday life, has to shake up the usual patterns in some way in order to get out of the rut (or the basin of attraction!) of conventional thinking, dispense with certain accepted but wrong notions, and find a new and better way to formulate some problem. But with Dick, "turning things around" and being different became a passion.

The result was that on certain occasions, in scientific work or in ordinary living, when an imaginative new way of looking at things was needed, he could come up with a remarkably useful innovation. But on many other occasions, when the usual way of doing business had its virtues, he was not the ideal person to consult. Remember his television appearance in which he made fun of the daily habit of brushing one's teeth? (And he didn't even suggest flossing!) Or take his occasional excursions into far-out political choices in the 1950s, during his second marriage. Those certainly set him off from most of his friends. But one day during that time, he called me and sheepishly admitted having voted for a particularly outrageous candidate for statewide office—and then asked me if in the future I would check over such names beforehand and tell him when he was really going off the deep end!

None of the aberrations mentioned here changes the fact that Dick Feynman was a most inspiring person. I have referred to his originality and straightforwardness and to his energy, playfulness and vitality. All of those characteristics showed up in his work and also in the other facets of his life. Indeed, that vitality may be related to the kind of biological (and probably psychological) vitality that enabled him to resist so remarkably and for so long the illness to which he finally succumbed.

When I think of him now, it is usually as he was during that first decade that we were colleagues, when we were both young and everything seemed possible. We phoned each other with good ideas and crazy ones, with serious messages and farcical gags. We yelled at each other in front of the blackboard. We taught stewardesses to say "quark–quark scattering" and "quark–antiquark scattering." We delivered a peacock to the bedroom of our friend Jirayr Zorthian on his birthday, while our wives distracted him. We argued about everything under the Sun.

Later on, we drifted apart to a considerable extent, but I was aware, all the time we were colleagues, that if a really profound question in science came up, there would be fun and profit in discussing it with Dick. Even though on many occasions during the last 20 years, I passed up the opportunity to talk with him in such a case, I knew that I *could* do so, and that made a great difference.

Besides, I did not always pass it up. For example, during the last few months and even weeks of his life, we kept up a running discussion of one of the most basic subjects, the role of "classical objects" in the interpretation of quantum mechanics. We thus resumed a series of conversations on that topic that we had begun a quarter of a century earlier. In between 1963–64 and 1987 those talks about quantum mechanics were rare, but there was at least one remarkable occasion during the last few of those years. Richard sat in on one of my classes on the meaning of quantum mechanics, interrupting from time to time. He did not, however, object to what I was saying; rather, he reinforced the points I was making. The students must have been delighted as they heard the same arguments made by both of us in a kind of counterpoint.

It is hard for me to get used to the fact that now, when I have a deep issue in physics to discuss with someone, Dick Feynman is no longer around.

• • •

I should like to thank James B. Hartle for many instructive conversations about quantum mechanics and the path integral method in quantum cosmology.

References

1. R. P. Feynman, Phys. Rev. 76, 749, 769 (1949).
2. R. P. Feynman, *"Surely You're Joking, Mr. Feynman!" Adventures of a Curious Character*, Bantam, New York (1986), p. 229. See also M. Gell-Mann, in *Proc. Int. Mtg. on the History of Scientific Ideas*, M. G. Doncel *et al.*, eds., Bellaterra, Barcelona (1987), p. 474.

22

Time Symmetry and Asymmetry in Quantum Mechanics and Quantum Cosmology

Murray Gell-Mann
Lauritsen Laboratory
California Institute of Technology
Pasadena, CA 91125, USA

James B. Hartle
Department of Physics
University of California
Santa Barbara, CA 93106-9530, USA

22.1 Introduction

The disparity between the time symmetry of the fundamental laws of physics and the time asymmetries of the observed universe has been a subject of fascination for physicists since the late 19th century.† The following general time asymmetries are observed in this universe:

- The thermodynamic arrow of time – the fact that approximately isolated systems are now almost all evolving towards equilibrium in the same direction of time.
- The psychological arrow of time – we remember the past, we predict the future.
- The arrow of time of retarded electromagnetic radiation.
- The arrow of time supplied by the CP non-invariance of the weak interactions and the CPT invariance of field theory.
- The arrow of time of the approximately uniform expansion of the universe.
- The arrow of time supplied by the growth of inhomogeneity in the expanding universe.

All of these time asymmetries could arise from time-symmetric dynamical laws solved with time-asymmetric boundary conditions. The thermodynamic arrow of time, for example, is implied by an initial condition in which the progenitors of today's approximately isolated systems were all far from equilibrium at an initial time. The CP arrow of time could arise as a spontaneously broken symmetry of the Hamiltonian. The approximate uniform expansion of the universe and the growth of inhomogeneity follow from an initial "big bang" of sufficient spatial

† For clear reviews, see [1], [2], [3].

Reprinted from *Physical Origins of Time Asymmetry*, eds. J. J. Halliwell et al., (© Cambridge Univ. Press, 1994), pp. 311–337, reproduced with permission.

homogeneity and isotropy, given the attractive nature of gravity. Characteristically such arrows of time can be reversed temporarily, locally, in isolated subsystems, although typically at an expense so great that the experiment can be carried out only in our imaginations. If we could, in the classical example of Loschmidt [4], reverse the momenta of all particles and fields of an isolated subsystem, it would "run backwards" with thermodynamic and electromagnetic arrows of time reversed.

Quantum cosmology is that part of physics concerned with the theory of the boundary conditions of our universe. It is, therefore, the natural and most general context in which to investigate the origin of observed time asymmetries. In the context of contemporary quantum cosmology, several such investigations have been carried out [2, 5–11], starting with those of Penrose [2] on classical time-asymmetric initial and final conditions and those of Page [5] and Hawking [6] on the emergence of the thermodynamic arrow of time from the "no-boundary" theory of the initial condition of the universe. It is not our purpose to review these results or the status of our understanding of the time asymmetries mentioned above. Rather, we shall discuss in this essay, from the perspective of quantum cosmology, a time asymmetry not specifically mentioned above. That is the arrow of time of familiar quantum mechanics.

Conventional formulations of quantum mechanics incorporate a fundamental distinction between the future and the past, as we shall review in Section 2. This quantum-mechanical arrow of time has, in a way, a distinct status in the theory from the time asymmetries discussed above. It does not arise, as they do, from a time-asymmetric choice of boundary conditions for time-neutral dynamical laws. Rather, it can be regarded as a time asymmetry of the laws themselves. However, the quantum mechanics of cosmology does not have to be formulated in this time-asymmetric way. In Section 3, extending discussions of Aharonov, Bergman, and Lebowitz [12] and of Griffiths [13], we consider a generalized quantum mechanics for cosmology that utilizes both initial and final conditions to give a time-neutral, two-boundary formulation that does not necessarily have an arrow of time [14]. In such a formulation all time asymmetries arise from properties of the initial and final conditions, in particular differences between them, or, at particular epochs, from nearness to the beginning or end. A theory of both initial and final conditions would be the objective of quantum cosmology.

In the context of a time-neutral formulation, the usual quantum mechanics results from utilizing a special initial condition, together with what amounts to a final condition representing complete indifference with respect to the future states, thus yielding the quantum-mechanical arrow of time, which is sufficient to explain the observed time asymmetries of this universe. However, a time-neutral formulation of quantum mechanics allows us to investigate to what extent the familiar final condition of indifference with respect to future states is mandated by our observations. In particular, it allows us to investigate whether quantum cosmologies with less blatantly asymmetric initial and final conditions might also be consistent with the observed general time asymmetries. As a step in this direction we discuss a quantum cosmology that would be, in a sense, the opposite extreme – a cosmology with a time-symmetric pair of initial and final conditions leading to a universe that is statistically symmetric about a moment of time. Such boundary conditions imply deviations from the thermodynamic arrow of time and the arrow of time supplied by the CP non-invariance of the weak interactions. We investigate such deviations to see if they are inconsistent with observations. The classical statistical models reviewed in Section 4 and the models of CP symmetry breaking discussed in Section 5 suggest that the predicted deviations may be insufficient to exclude time-symmetric boundary conditions if the interval between initial and final conditions is much longer than our distance in time from the initial condition. Next, we review and augment the arguments of Davies and Twamley that electromagnetic radiation may supply a probe of the final condition that *is* sufficiently accurate to rule out time-symmetric boundary conditions.

We should emphasize that we are not advocating a time-symmetric cosmology but only using it as a foil to test the extent to which observation now requires the usual asymmetric boundary conditions and to search for more refined experimental tests. The important result of this paper is a quantum framework for examining cosmologies with less asymmetric boundary conditions than the usual ones, so that the quantum-mechanical arrow of time (with its consequent time asymmetries) can be treated, or derived, as one possibility out of many, to be confronted with observation, rather than as an axiom of theory.

Relations between the initial and final conditions of a quantum-mechanical universe *sufficient* for both CPT-symmetric cosmologies and time-symmetric cosmologies are discussed in Section 5. Ways in which the T-violation exhibited by the weak interaction could arise in such universes are described there as well. In Section 6 we discuss the limitations on time-symmetric quantum boundary conditions following from the requirements of decoherence and classicality. Specifically, we show that for a set of alternative histories to have the negligible interference between its individual members that is necessary for them to be assigned probabilities at all, there must be some impurity in the initial or final density matrices or both, except in the highly unorthodox case in which there are only one or two coarse-grained histories with non-negligible probability.

We should make clear that our discussion of time-symmetric cosmologies, based on speculative generalizations of quantum mechanics and causality, with separate initial and final density matrices that are related by time symmetry, is essentially different from the conjecture that has sometimes been made that *ordinary* causal quantum or classical mechanics, with just a single boundary condition or a single prescribed wave function, CPT-invariant about some time in the distant future, might lead to a T-symmetric or CPT-symmetric cosmology with a contracting phase in which the arrows of time are reversed. [15–17, 6, 18] It is the latter notion, by the way, that Hawking refers to as his "greatest mistake"[19]. We shall return to this topic in Section 5.

22.2 The Arrow of Time in Quantum Mechanics

As usually formulated, the laws of quantum mechanics are not time-neutral but incorporate an arrow of time. This can be seen clearly from the expression for the probabilities of histories consisting of alternatives at definite moments of time $t_1 < t_2 < \cdots < t_n$. Let $\{\alpha_k\}$ be an exhaustive set of alternatives at time t_k represented by $\{P^k_{\alpha_k}(t_k)\}$, a set of projection operators in the Heisenberg picture. For example, the alternatives $\{\alpha_k\}$ might be defined by an exhaustive set of ranges for the center-of-mass position of a massive body. A particular history corresponds to a specific sequence of alternatives $(\alpha_1, \cdots, \alpha_n)$. The probability for a particular history in the exhaustive set of histories is

$$p\,(\alpha_n, \cdots, \alpha_1) = Tr\Big[P^n_{\alpha_n}(t_n) \cdots P^1_{\alpha_1}(t_1) \rho P^1_{\alpha_1}(t_1) \cdots P^n_{\alpha_n}(t_n)\Big], \tag{22.1}$$

where ρ is the density matrix describing the initial state of the system and the projection operators are time-ordered from the density matrix to the trace.†

The expression for the probabilities (22.1) is not time-neutral. This is not because of the time ordering of the projection operators. Field theory is invariant under CPT and the ordering of the operators could be reversed by a CPT transformation of the projection operators and density matrix, leaving the probabilities unchanged. (See e.g. [20] or [14]). Either time ordering may therefore be used; it is by convention that we usually use the one with the condition represented by the density matrix ρ in the past.

Rather, (22.1) is not time-neutral because there is a density matrix on one end of the chain of projections representing a history while at the other end there is the trace [12, 13, 14]. Whatever conventions are used for time ordering, there is thus an asymmetry between future and past exhibited in the formula for probabilities (22.1). That asymmetry is the arrow of time in quantum mechanics.

The asymmetry between past and future exhibited by quantum mechanics implies the familiar notion of causality. From an effective density matrix describing the present *alone* it is possible to predict the probabilities for the future. More precisely, given that alternatives $\alpha_1, \cdots, \alpha_k$ have "happened" at times $t_1 < \cdots < t_k$ before time t, the conditional probability for alternatives $\alpha_{k+1}, \cdots, \alpha_n$ to occur in the future at times $t_{k+1}, \cdots, t_n$ may be determined from an effective density matrix $\rho_{\rm eff}(t)$ at time t. Specifically, the conditional probabilities for future prediction are

$$p(\alpha_n, \cdots, \alpha_{k+1} | \alpha_k, \cdots, \alpha_1) = \frac{p(\alpha_n, \cdots, \alpha_1)}{p(\alpha_k, \cdots, \alpha_1)}\,. \tag{22.2}$$

† This compact expression of the probabilities of ordinary quantum mechanics has been noted by many authors. For more details of this and other aspects of the quantum-mechanical formalism we shall employ the reader is referred to [20] and [14] where references to earlier literature may be found.

These can be expressed as

$$p(\alpha_n,\cdots,\alpha_{k+1}|\alpha_k,\cdots,\alpha_1) = Tr\left[P^n_{\alpha_n}(t_n)\cdots P^{k+1}_{\alpha_{k+1}}(t_{k+1})\rho_{\rm eff}(t_k)P^{k+1}_{\alpha_{k+1}}(t_{k+1})\cdots P^n_{\alpha_n}(t_n)\right], \tag{22.3}$$

where the effective density matrix $\rho_{\rm eff}$ is

$$\rho_{\rm eff}(t_k) = \frac{P^k_{\alpha_k}(t_k)\cdots P^1_{\alpha_1}(t_1)\rho P^1_{\alpha_1}(t_1)\cdots P^k_{\alpha_k}(t_k)}{Tr\left[P^k_{\alpha_k}(t_k)\cdots P^1_{\alpha_1}(t_1)\rho P^1_{\alpha_1}(t_1)\cdots P^k_{\alpha_k}(t_k)\right]}\,. \tag{22.4}$$

The density matrix $\rho_{\rm eff}(t_k)$ can be said to define the effective state of the universe at time t_k, given the history $(\alpha_1,\cdots,\alpha_k)$.

What is the physical origin of the time asymmetry in the basic laws of quantum mechanics and what is its connection with the other observed time asymmetries of our universe? The rest of this Section addresses that question.

The reader may be most familiar with the expression for probabilities (22.1) in the context of the approximate "Copenhagen" quantum mechanics of measured subsystems. In that case operators, the density matrix, etc. all refer to the Hilbert space of the subsystem. The sets of projection operators $\{P^k_{\alpha_k}(t_k)\}$ describe alternative outcomes of measurements of the subsystem.

Formula (22.1) for the probabilities of a sequence of measured outcomes is then a unified expression of the "two forms of evolution" usually discussed in the quantum mechanics of subsystems — unitary evolution in between measurements and the "reduction of the state vector" on measurement. The time asymmetry of (22.1) does not arise from the unitary evolution of the projection operators representing the measured quantities in the Heisenberg picture; that is time-reversible. Rather, it can be said to arise from the successive reductions represented by the projections in (22.4) that occur on measurement. The common explanation for the origin of the arrow of time in the quantum mechanics of measured subsystems is that measurement is an irreversible process and that quantum mechanics inherits its arrow of time from the arrow of time of thermodynamics.† If that is the case, then the origin of the quantum-mechanical arrow of time must ultimately be cosmological, for the straightforward explanation of the thermodynamic arrow of time is a special initial condition for the universe implying that its constituents were far from equilibrium across a spacelike surface. Let us, therefore, investigate more fundamentally the quantum-mechanical arrow of time, not in an approximate quantum mechanics of

† This connection between the thermodynamic arrow of time and the quantum-mechanical arrow of time can be ambiguous. Suppose, for example, a measuring apparatus is constructed in which the local approach to equilibrium is in the opposite direction of time from that generally prevailing in the larger universe. If that apparatus interacts with a subsystem (perhaps previously measured by other apparatus adhering to the general thermodynamic arrow of time) should the operators representing those measurements be ordered according to the thermodynamic arrow of the apparatus or of the larger universe with respect to which it is running backwards? Such puzzles are resolvable in the more general quantum mechanics of closed systems to be discussed below, where "measurements", the "thermodynamic arrow of time", and any connection between the two are all approximate notions holding in only special situations.

measured subsystems, but in the quantum mechanics of a closed system — most realistically and generally the universe as a whole.

The formula (22.1) for the probabilities of histories also holds in the quantum mechanics of a closed system such as the universe as a whole, at least in an approximation in which gross fluctuations in the geometry of spacetime are neglected. The sets of projection operators describe alternatives for the whole system, say the universe, and the density matrix can be thought of as describing its initial condition.† Not every set of histories that may be described can be assigned probabilities according to (22.1). In the quantum mechanics of closed systems consistent probabilities given by (22.1) are predicted only for those sets of histories for which there is negligible interference between the individual members of the set [14] as a consequence of the particular initial ρ. Such sets of histories are said to "decohere". We shall defer until Section 5 a discussion of the precise measure of the coherence between histories and the implications of decoherence for time symmetry in quantum mechanics. We concentrate now on the theoretical status of the arrow of time exhibited by (22.1) in the quantum mechanics of cosmology.

An arrow of time built into a basic quantum mechanics of cosmology may not (as in the approximate "Copenhagen" quantum mechanics of measured subsystems) be attributed to the thermodynamic arrow of an external measuring apparatus or larger universe. In general, these external objects are not there. An arrow of time in the quantum mechanics of cosmology would be a fundamental time asymmetry in the basic laws of physics. Indeed, given that, as we mentioned in the Introduction, the other observed time asymmetries could all arise from time-symmetric dynamical laws solved with time-asymmetric boundary conditions, a fundamental arrow of time in the laws of quantum mechanics could be the only fundamental source of time asymmetry in all of physics.

There is no inconsistency between known data and a fundamental arrow of time in quantum mechanics. General time asymmetries *are* exhibited by our universe and there is no evidence suggesting any violation of causality. The observed time asymmetries such as the thermodynamic arrow of time, the arrow of retarded electromagnetic radiation, the absence of white holes, etc. *could* all be seen to follow from a fundamental quantum-mechanical distinction between the past and future. That is, they could all be seen to arise from a special initial ρ in a quantum-mechanical framework based on (22.1).

But might it not be instructive to generalize quantum mechanics so that it does not so blatantly distinguish past from future? One could then investigate a more general class of quantum cosmologies and identify those that are compatible with the observed time asymmetries. Even if it is highly unlikely that ordinary quantum mechanics needs to be replaced by such a generalization, the generalization can still provide an instructive way of viewing the origin of time asymmetry in the

† For a more detailed exposition of this quantum mechanics of cosmology, the reader is referred to our previous work [20], [21], and [14], where references to the earlier literature may also be found.

universe and provide a framework for discussing tests of the usual assumptions. We shall discuss in the next section a quantum mechanics that employs two boundary conditions, one for the past and one for the future, to give a time-neutral formulation. Each condition is represented by a density matrix and the usual theory is recovered when the future density matrix is proportional to the unit matrix while the one for the past is much more special.

22.3 A Time-Neutral Formulation of Quantum Mechanics for Cosmology

Nearly thirty years ago, Aharonov, Bergmann, and Lebowitz [12] showed how to cast the quantum mechanics of measured subsystems into time-neutral form by considering final conditions as well as initial ones. The same type of framework for the quantum mechanics of closed systems has been discussed by Griffiths [13] and ourselves [14]. In this formulation the probabilities for the individual members of a set of alternative histories is given by

$$p(\alpha_n, \cdots, \alpha_1) = N Tr\left[\rho_f P^n_{\alpha_n}(t_n) \cdots P^1_{\alpha_1}(t_1)\ \rho_i\ P^1_{\alpha_1}(t_1) \cdots P^n_{\alpha_n}(t_n)\right], \qquad (22.5)$$

where

$$N^{-1} = Tr\left(\rho_f \rho_i\right) . \qquad (22.6)$$

Here, ρ_i and ρ_f are Hermitian, positive operators that we may conventionally call Heisenberg operators representing initial and final conditions for the universe respectively. They need not be normalized as density matrices with $Tr(\rho) = 1$ because (22.5) is invariant under changes of normalization.

The expression (22.5) for the probabilities of histories is time-neutral. There is a density matrix at both ends of each history. Initial and final conditions may be interchanged by making use of the cyclic property of the trace. Therefore, the quantum mechanics of closed systems based on (22.5) need not have a fundamental arrow of time.

Different quantum-mechanical theories of cosmology are specified by different choices for the initial and final conditions ρ_i and ρ_f. For those cases with $\rho_f \propto I$, where I is the unit matrix, this formulation reduces to that discussed in the previous Section because then (22.5) coincides with (22.1).

Of course, the condition for decoherence must also be extended to incorporate initial and final conditions. That extension, however, is straightforward [13, 14] and will be reviewed briefly in Section 5. The result is a generalized quantum mechanics in the sense of Refs. [14] and [21].

Lost in this generalization is a built-in notion of causality in quantum mechanics. Lost also, when ρ_f is not proportional to I, is any notion of a unitarily evolving "state of the system at a moment of time". There is generally no effective density matrix like $\rho_{\text{eff}}(t)$ in (22.4) from which *alone* probabilities for either the future or past could be computed. What is gained is a quantum mechanics without a fundamental arrow of time in which all time asymmetries may arise in particular cosmologies because of differences between ρ_i and ρ_f or at particular epochs from their being near to the beginning or the end. That generalized quantum mechanics embraces a richer variety of possible universes, allowing for the possibility of violations of causality and advanced as well as retarded effects. These, therefore, become testable features of the universe rather than axioms of the fundamental quantum framework.

From the perspective of this generalized quantum mechanics the task of quantum cosmology is to find a theory of *both* the initial and final conditions that is theoretically compelling and fits our existing data as well as possible. Certainly a final condition of indifference, $\rho_f \propto I$, and a special initial condition, ρ_i, seem to fit our data well, and there is no known reason for modifying them. But how accurately is $\rho_f \propto I$ mandated by the data? What would be the observable consequences of a completely time-symmetric boundary condition that is, in a sense, the opposite extreme?

Our ability to detect the presence of a final condition differing from $\rho_f \propto I$ depends on our experimental access to systems whose behavior today predicted with $\rho_f \not\propto I$ would be measurably different from the predictions of that behavior with $\rho_f \propto I$. Loosely speaking, it depends on our finding physical systems which can "see" the final condition of the universe today. In the following we examine several candidates for such systems, beginning with simple classical analyses in Section 4 and proceeding to more quantum-mechanical ones in Section 5.

22.4 Classical Two-Time Boundary Problems

22.4.1 A Simple Statistical Model

The simplest explanation of the observed thermodynamic arrow of time is the asymmetry between a special, low-entropy,† initial condition and a maximal-entropy final condition describable as indifference with respect to final state (or no condition at all!). Studying deviations of the entropy increase predicted by statistical mechanics with these boundary conditions from that predicted with time-symmetric boundary conditions is a natural way to try to discriminate between the two. Such studies were carried out in classical statistical models by by Cocke [22], Schulman [23], Wheeler [24], and others in the late '60s and early '70s. Schulman, in particular, has written extensively on these problems both in classical and quantum mechanics [25]. We briefly review such statistical models here.

Relaxation to equilibrium is a time-symmetric process in a universe with an underlying dynamics that is time reversal invariant. Without boundary conditions, a system out of equilibrium is just as likely to have evolved from a state of higher entropy as it is to evolve to a state of higher entropy. The characteristic relaxation

† For quantitative estimates of how low the initial entropy is, see [2].

time for a system to evolve to equilibrium depends on the size of the system and the strength of the interactions that equilibrate it. Other factors being equal, the larger the system the longer the relaxation time.

There is no simpler instructive model to illustrate the approach to equilibrium than the Ehrenfest urn model [26]. For this reason, it and related models have been much studied in connection with statistical two-time boundary problems [22], [23]. The model consists of two boxes and a numbered set of n balls that are distributed between them. The system evolves in time according to the following dynamical rule: At each time step a random number between 1 and n is produced, and the ball with that number is moved from the box containing it to the other box. This dynamical rule is time-symmetric.

The fine-grained description of this system specifies which ball is in which box (a "microstate"). An interesting coarse-grained description involves following just the total number of balls in each box (a "macrostate") irrespective of *which* balls are in which box. Let us use this coarse graining to consider an initial condition in which all the balls are in one box and follow the approach to equilibrium as a function of the number of time steps, with no further conditions. Figure 22.1 shows a numerical calculation of how the entropy averaged over many realizations of this evolution grows with time to approach its maximum, equilibrium value. The relaxation time, obtained either analytically or numerically, is approximately the total number of balls, $t_{relax} \sim n$. If there are no further constraints, the system tends to relax to equilibrium and remain there.

Consider evolution in the Ehrenfest urn model when a final condition identical to the initial one is imposed at a later time T. Specifically, construct an ensemble of evolutions consisteht with these boundary conditions by evolving forward from an initial condition where all the balls are in one box but accepting only those evolutions where all the balls are back in this box at time T. Figure 22.1 shows the results of two such calculations, one for a system with a small number of balls (where the relaxation time is significantly smaller than T) and the other for a system with a larger number of balls (where it is significantly larger than T.)

For both systems the time-symmetric boundary conditions imply a behavior of the average entropy that is time-symmetric about the midpoint, $T/2$. For the system with a relaxation time short compared to the time at which the final condition is imposed, the initial approach to equilibrium is nearly indistinguishable from that in the case with no final condition. That is because, in equilibrium, the system's coarse-grained dynamics is essentially independent of its initial *or* final condition. It, in effect, "forgets" both from whence it started and whither it is going.

By contrast, if the relaxation time is comparable to or greater than the time interval between initial and final condition, then there will be significant deviations from the unconstrained approach to equilibrium. Such systems typically do not reach equilibrium before the effect of the final condition forces their entropy to decrease.

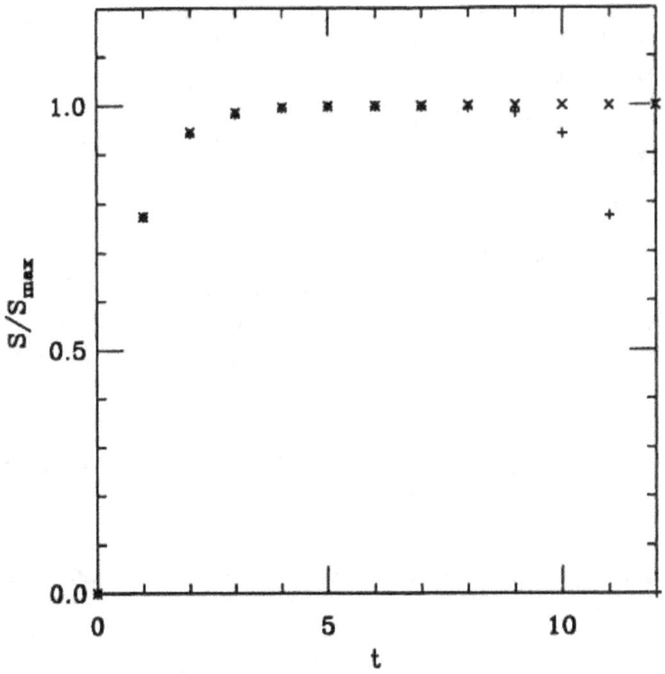

Fig. 22.1. The entropy of a coarse-grained state, in which only the total number of balls in each box is followed, is the logarithm of the number of different ways of distributing the balls between the boxes consistent with a given total number in each. This figure shows this entropy averaged over a large number of different evolutions of the system for several situations. These simulations were carried out by the authors but are no different in spirit from those discussed by Cocke [22].

The left figure shows the evolution of a system of four balls. In each case the system starts with all balls in one box – a configuration of zero entropy as far from equilibrium as it is possible to get. The ×'s show how the average entropy of 12,556 cases approaches equilibrium when there are no further constraints. The entropy approaches its equilibrium value in a relaxation time given approximately by its size, $t_{relax} \sim 4$, and remains there. The curve of +'s shows the evolution when a time-symmetric final condition is imposed at $T = 12$ that all balls have returned to the one box from whence they started at $t = 0$. A total of 100,000 evolutions were tried. (Figure 22.1 continued on the next page).

The evolution of the entropy in the presence of time-symmetric initial and final conditions must itself be time-symmetric when averaged over many evolutions, as the simulations in Figure 22.1 show. However, in a statistical theory with time-symmetric boundary conditions the individual histories need not be time-symmetric. Figure 22.2 shows an example of a single history from an urn model calculation for which the average behavior of the entropy is shown in Figure 22.1. The ensemble of histories is time-symmetric by construction; the individual histories need not be. Since, by definition, we experience only one history of the universe, this leaves open the possibility that the time-asymmetries that we see could be the result of

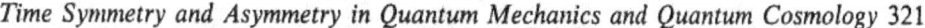

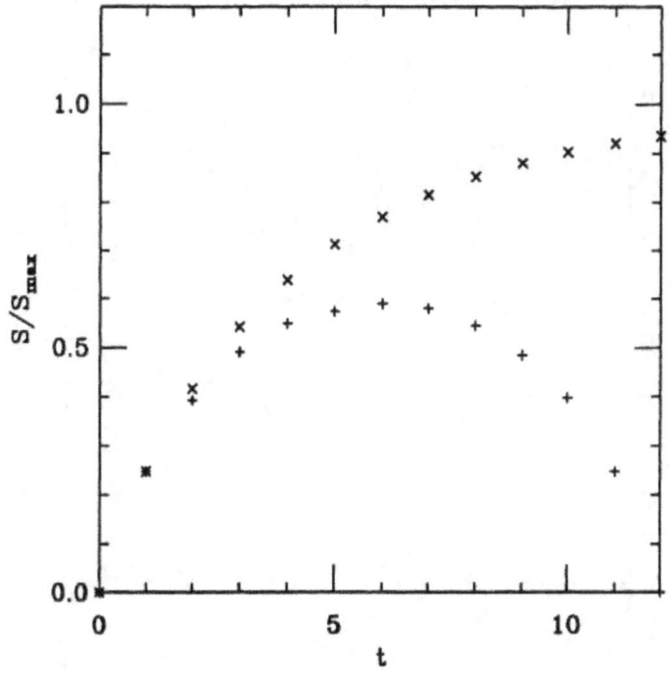

Fig. 22.1. *Continued* – The average entropy of the 12,556 cases that met the final condition is shown. It is time symmetric about the midpoint, $T/2 = 6$. The initial approach to equilibrium is virtually indistinguishable from the approach without a final condition because $t_{relax} \sim 4$ is significantly less than the time $T = 12$ at which the final condition is imposed. Only within one relaxation time of the final condition is the deviation of the evolution from the unconstrained case apparent.

The right figure shows the same two cases for a larger system of twenty balls. The unconstrained approach to equilibrium shown by the $\times$'s was calculated from the average of 1010 evolutions and exhibits a relaxation time, $t_{relax} \sim 20$. The average entropy when time-symmetric boundary conditions are imposed at $T = 12$ is shown in the curve of $+$'s. 10,000,000 evolutions were tried of which 1010 met the time-symmetric final condition. (This demonstrates vividly that it is very improbable for the entropy of even a modest size system to fluctuate far from equilibrium as measured by the CPU time needed to find such fluctuations.) The deviation from the unconstrained approach to equilibrium caused by the imposition of a time-symmetric final condition is significant from an early time of about $t = 3$ as the differences between the $+$'s and the $\times$'s show. These models suggest that to detect the effects of a time-symmetric final condition for the universe we must have access to systems for which the relaxation time is at least comparable to the time difference between initial and final conditions.

a statistical fluctuation in a universe with time-symmetric boundary conditions. In quantum cosmology, we would not count such a theory of the initial and final conditions as successful if the fluctuations required were very improbable. However, in some examples the magnitude of the fluctuation need not be so very large. For instance, consider a classical statistical theory in which the boundary conditions

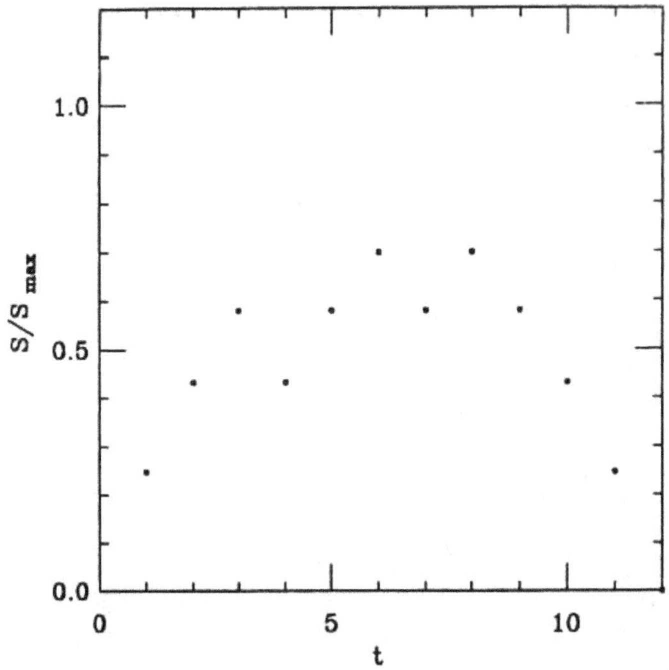

Fig. 22.2. An individual evolution in the urn model with time-symmetric initial and final conditions. The history of the entropy averaged over many evolutions must clearly be time-symmetric in a statistical theory with time-symmetric initial and final conditions as Figure 22.1 shows. However, the individual evolutions need not be separately time-symmetric. This figure shows the entropy for the case of twenty balls in the first evolution among the 10,000,000 described in Figure 22.1 that met the time-symmetric final condition. It is not time-symmetric. For systems such as this with relaxation time, t_{relax}, larger than the time between initial and final conditions, significant deviations from exact time symmetry may be expected.

allow an ensemble of classical histories each one of which displays an arrow of time and also, with equal probability, the time-reverse of that history displaying the opposite arrow of time. The boundary conditions and the resulting ensemble are time-symmetric, but predict an observed time-asymmetry with probability one. Of course, such a theory might be indistinguishable from one that posited boundary conditions allowing just one arrow of time. However, other theoretical considerations may make it reasonable to consider such proposals, for example, the "no boundary" initial condition which is believed to have this character [5]. In the subsequent discussion of time-symmetric cosmological boundary conditions we shall assume that they predict with high probability *some* observable differences from the usual special initial conditions and final condition of indifference and investigate what these are.

22.4.2 Classical Dynamical Systems with Two-Time Statistical Boundary Conditions

The analysis of such simple classical statistical models with two-time boundary conditions suggests the behavior of a classical cosmology with time-symmetric initial and final conditions. A classical dynamical system is described by a phase space and a Hamiltonian, H. We write $z = (q, p)$ for a set of canonical phase-space coördinates. The histories of the classical dynamical system are the phase-space curves $z(t)$ that satisfy Hamilton's equations of motion.

A statistical classical dynamical system is described by a distribution function on phase space $\rho^{cl}(z)$ that gives the probability of finding it in a phase-space volume. For analogy with quantum mechanics it is simplest to use a "Heisenberg picture" description in which the distribution function does not depend explicitly on time but time-dependent coördinates on phase space are used to describe the dynamics. That is, if z_0 is a set of canonical coördinates at time $t = 0$, a set z_t appropriate to time t may be defined by $z_t = z_0(t)$ where $z_0(t)$ is the classical evolution of z_0 generated in a time t by the Hamiltonian H. The statistical system at time t is then distributed according to the function ρ^{cl} expressed in terms of the coördinates z_t, viz. $\rho^{cl}(z_t)$. The distributions $\rho^{cl}(z_t)$ and $\rho^{cl}(z_{t'})$ will therefore typically have *different* functional forms.

An ensemble of histories distributed according to the probabilities of a statistical classical dynamical system with boundary conditions at two times t_i and t_f might be constructed as follows: Evolve a large number of systems distributed according to the initial distribution function $\rho_i^{cl}(z_i)$ forward from time t_i to time t_f. If a particular system arrives at time t_f in the phase space volume Δv centered about point z_f, select it for inclusion in the ensemble with probability $\rho_f^{cl}(z_f)\Delta v$ where ρ_f^{cl} is the distribution function representing the final boundary condition. Thus, if ρ_i^{cl} and ρ_f^{cl} are referred to a common set of phase-space coördinates, say z_t, the time-symmetric ensemble of systems will be distributed according to the function

$$\bar{\rho}^{cl}(z_t) = N\rho_f^{cl}(z_t)\rho_i^{cl}(z_t) , \tag{22.7}$$

where

$$N^{-1} = \int dz_t \; \rho_f^{cl}(z_t)\rho_i^{cl}(z_t) . \tag{22.8}$$

Referred to the initial time, (22.7) has a simple interpretation: Since classical evolution is unique and deterministic the selection at the final time could equally well be carried out at the initial time with ρ_f^{cl} evolved back to the initial time. The distribution $\bar{\rho}^{cl}$ is the result.

We now discuss the relation between ρ_i^{cl} and ρ_f^{cl} that is necessary for the probabilities of this classical cosmology to be symmetric about a moment of time. Take this time to be $t = 0$ and introduce the operation, $\mathcal{T}$, of time-reversal about it.

$$\mathcal{T}\rho^{cl}(q_0, p_0) \equiv \rho^{cl}(q_0, -p_0) . \tag{22.9}$$

If we assume that the Hamiltonian is itself time-reversal invariant

$$H(q_0, p_0, t) = H(q_0, -p_0, -t) \,, \tag{22.10}$$

this implies

$$\mathcal{T}\rho^{cl}(q_t, p_t) = \rho^{cl}(q_{-t}, -p_{-t}) \,. \tag{22.11}$$

The distribution function (22.7) may then be conveniently rewritten

$$\bar{\rho}^{cl}(q_t, p_t) = N\rho_i^{cl}(q_t, p_t)\mathcal{T}\rho_f^{cl}(q_{-t}, -p_{-t}) \,. \tag{22.12}$$

A relation between ρ_i^{cl} and ρ_f^{cl} sufficient to imply the time-symmetry of the distribution $\bar{\rho}^{cl}$ is now evident, namely

$$\rho_f^{cl}(q_t, p_t) = \mathcal{T}^{-1}\rho_i^{cl}(q_t, p_t) \,. \tag{22.13}$$

The final condition is just the time-reversed version of the initial one.

The imposition of time-symmetric statistical boundary conditions on a classical cosmology means in particular that the entropy must behave time-symmetrically if it is computed utilizing a coarse graining that is itself time-symmetric. The entropy of the final distribution must be the same as the initial one. The thermodynamic arrow of time will run backwards on one side of the moment of time symmetry as compared to the other side. This does not mean, of course, that the histories of the ensemble need be individually time-symmetric, as the example in Figure 22.2 shows. In particular, subsystems with relaxation times long compared to the interval between initial and final conditions might have non-negligible probabilities for fluctuations from exactly time-symmetric behavior. There would appear to be no principle, for example, forbidding us to live on into the recontracting phase of the universe and see it as recontracting. It is just that as time progressed events judged to be unexpected on the basis of just an initial condition would happen with increasing frequency. It is by the frequency of such unexpected events that we could detect the existence of a final condition.

Could we infer the existence of a time-symmetric final condition for the universe from the deviations that it would imply for the approach to equilibrium that would be expected were there no such final condition? The statistical models reviewed in above suggest that to do so we would need to study systems with relaxation times comparable to or longer than the lifetime of the universe between the "big bang" and the "big crunch". If the lifetime of the universe is comparable to the present age of the universe from the "big bang", then we certainly know such systems. Systems of stars such as galaxies and clusters provide ready examples. Any single star, with the ambient radiation, provides another as long as the star's temperature is above that of the cosmic background radiation. Black holes with lifetime to decay by the Hawking radiation longer than the Hubble time are further examples. Indeed, from the point of view of global cosmological geometry, the singularities contained within black holes can be considered to be parts of the final singularity of the universe, where a final condition would naturally be imposed [27]. The singularities of detectable black holes may be the parts of this final singularity closest to us. On smaller scales, samples of radioactive material with very long half-lives may be other such examples and Wheeler [24] has discussed experiments utilizing them to search for a time-symmetric final condition. We may hope, as mentioned above, that the evolving collective complex adaptive system of which we are a part could be such a long-lasting phenomenon!

However, if the lifetime of the universe is much *longer* than the present age from the "big bang", then it might be much more difficult to find systems that remain out of equilibrium long enough for their initial approach to equilibrium to be significantly affected by a time-symmetric final condition. That could be the case with the Ω-near-one universe that would result from a rapid initial inflation. If its lifetime were long enough, we might never be able to detect the existence of a time-symmetric final condition for the universe.

The lifetime of our classical universe obeying the Einstein equation is, of course, in principle determinable from present observations (for example, of the Hubble constant, mean mass density, and deceleration parameter). Unfortunately we do not have enough data to distinguish observationally between a lifetime of about twenty-five billion years and an infinite lifetime. Very long lifetimes are not only consistent with observations, but also, as we now describe, are suggested theoretically by quantum cosmology as a consequence of inflation.

The quasiclassical cosmological evolution that we observe should be, on a more fundamental level, a likely prediction of a quantum theory of the universe and its boundary conditions. We shall discuss time symmetry in the context of quantum cosmology in later sections, but for the present discussion we note that, in quantum cosmology, the probabilities for different lifetimes of the universe are predictable from a theory of its initial and final conditions. That is because, in a quantum theory that includes gravitation, the geometry of spacetime, including such features as the time between the "big bang" and a "big crunch", if any, is quantum-probabilistic.

A quantum state that predicts quasiclassical behavior does not typically predict a unique classical history but rather an ensemble of possible classical histories with different probabilities. This is familiar from wave functions of WKB form, which do not predict single classical trajectories but only the classical connection between momentum and classical action. Similarly, in the quantum mechanics of closed cosmologies, we expect a theory of quantum boundary conditions to determine an ensemble of different classical cosmological geometries with different probabilities.† The geometries in the ensemble will have *different times* between the "big bang" and the "big crunch" because in quantum gravity that time is a dynamical variable and not a matter of our choice. In this way, the probability distribution of lifetimes of the universe becomes predictable in quantum cosmology.

† For further discussion see e.g. [14].

Cosmological theories that predict inflation lead to very large expected values for the lifetime of the universe; and inflation seems to be implied by some currently interesting theories of the boundary conditions of the universe. The question has been analyzed only for theories with a special initial condition, such as the "no-boundary proposal" and the "tunneling-from-nothing proposal". Analyses by Hawking and Page [10], Grishchuk and Rozhansky [28], and Barvinsky and Kamenshchik [29] suggest expected lifetimes for the "no-boundary proposal" that are at least large enough to give sufficient inflation to explain the present size of the universe. Analyses by Vilenkin [30] and by Mijić, Morris, and Suen [31] also suggest very large expected lifetimes for the "tunneling-from-nothing" case.

22.4.3 *Electromagnetic Radiation*

The above discussion suggests that in order to probe the nature of a non-trivial final condition, one should study processes today that are sensitive to that final condition no matter how far in the future it is imposed. At the conference, P.C.W. Davies suggested that electromagnetic radiation might provide such a mechanism for "seeing" a final condition in the arbitrarily far future in realistic cosmologies. In an approximately static and homogeneous cosmology, radiation must travel through ever more material the longer the time separation between initial and final conditions. For sufficiently large separations, the universe becomes opaque to the electromagnetic radiation necessary to probe the details of the final condition directly. However, in an expanding universe the dilution of matter caused by the expansion competes with the longer path length as the separation between initial big bang and final big crunch becomes longer and longer. Davies and Twamley [32] show that, under reasonable conditions, the expansion wins and that the future light cone is transparent to photons all the way to a distance from the final singularity comparable to ours from the big bang.

Transparency of the forward light cone raises the possibility of constraining the final condition by present observations of electromagnetic radiation and perhaps ruling out time-symmetric boundary conditions. Partridge [33] has actually carried out experiments which could be interpreted in this way and Davies and Twamley discuss others. The following is an example of a further argument of a very direct kind.

Suppose the universe to have initial and final classical distributions that are time-symmetric in the sense of (22.13). Suppose further that these boundary conditions imply with high probability an initial epoch with stars in galaxies distributed approximately homogeneously and a similar final epoch of stars in galaxies at the symmetric time. Consider the radiation emitted from a particular star in the present epoch. If the universe is transparent, it is likely to reach the final epoch without being absorbed or scattered. There it may either be absorbed in the stars or proceed past them towards the final singularity. If a significant fraction of the radiation proceeds past, then by time-symmetry we should expect a corresponding amount of radiation to have been emitted from the big bang. Observations of the brightness of the night sky could therefore constrain the possibility of a final boundary condition time-symmetrically related to the initial one. The alternative that the radiation is completely absorbed in future stars implies constraints on present emission that are probably inconsistent with observation because the total cross section of future stars is only a small fraction of the whole sky, as it is today.†

By such arguments, made quantitative, and extended to neutrinos, gravitational and other forms of radiation, we may hope to constrain the final condition of the universe no matter how long the separation between the big bang and the big crunch.

22.5 Hypothetical Quantum Cosmologies with Time Symmetries

22.5.1 CPT- and T-Symmetric Boundary Conditions

The time-neutral generalized quantum mechanics with initial and final conditions developed in Section 3 permits the construction of model quantum cosmologies that exhibit symmetries with respect to reflection about a moment of time. By this we mean that the probabilities given by (22.5) for a set of alternative histories are identical to those of the symmetrically related set. This section explores the relations between ρ_f and ρ_i and the conditions on the Hamiltonian under which such symmetries exist.

CPT-symmetric universes are the most straightforward to implement because local field theory in flat spacetime is invariant under CPT. We expect CPT invariance as well for field theories in curved cosmological spacetimes such as closed Friedmann universes that are symmetric under a space inversion and symmetric about a moment of time.

To construct a CPT-invariant quantum cosmology, choose the origin of time so that the time reflection symmetry is about $t = 0$. Let Θ denote the antiunitary CPT transformation and for simplicity consider alternatives $\{P^k_{\alpha_k}(t_k)\}$ such that their CPT transforms, $\{\widetilde{P}^k_{\alpha_k}(-t_k)\}$, are given by

$$\widetilde{P}^k_{\alpha_k}(-t_k) = \Theta^{-1} P^k_{\alpha_k}(t_k)\Theta \,. \tag{22.14}$$

A CPT-symmetric universe would be one in which the probabilities of histories of alternatives at times $t_1 < t_2 < \cdots < t_n$ would be identical to the probabilities of the CPT-transformed histories of alternatives at times $-t_n < \cdots < -t_2 < -t_1$. Denote by C_α the string of projection operators representing one history

$$C_\alpha = P^n_{\alpha_n}(t_n) \cdots P^1_{\alpha_1}(t_1) \,, \tag{22.15}$$

and by $\widetilde{C}_\alpha$ the corresponding string of CPT-transformed alternatives written in

† Thanks are due to D. Craig for discussions of this example.

standard time order with the earliest alternatives to the right

$$\widetilde{C}_\alpha \equiv \widetilde{P}^1_{\alpha_1}(-t_1)\cdots\widetilde{P}^n_{\alpha_n}(-t_n)\,. \tag{22.16}$$

Thus,

$$\widetilde{C}_\alpha = \Theta^{-1}C^\dagger_\alpha\Theta\,. \tag{22.17}$$

The requirement of CPT symmetry is then, from (22.5),

$$Tr\left(\rho_f C_\alpha \rho_i C^\dagger_\alpha\right) = Tr\left(\rho_f \widetilde{C}_\alpha \rho_i \widetilde{C}^\dagger_\alpha\right)\,. \tag{22.18}$$

Using (22.14), (22.16), the cyclic property of the trace, and the identity $Tr[A\Theta^{-1}B\Theta] = Tr[B^\dagger\Theta A^\dagger\Theta^{-1}]$, following from the antiunitarity of Θ, the right hand side of (22.18) may be rewritten to yield the following form of the requirement of CPT symmetry

$$Tr\left(\rho_f C_\alpha \rho_i C^\dagger_\alpha\right) = Tr\left(\Theta\rho_i\Theta^{-1}C_\alpha\Theta\rho_f\Theta^{-1}C^\dagger_\alpha\right)\,. \tag{22.19}$$

Evidently a sufficient condition for a CPT-symmetric universe is that the initial and final conditions be CPT transforms of each other:

$$\rho_f = \Theta\rho_i\Theta^{-1} \tag{22.20}$$

because, acting on Bose operators, Θ^2 is effectively unity, and as a consequence of (22.20), $\rho_i = \Theta\rho_f\Theta^{-1}$.

As stressed by Page [34], a CPT-symmetric universe can also be realized with within the usual formulation of quantum mechanics with an initial ρ_i and a final $\rho_f = I$, provided the ρ_i representing the condition at the initial instant is CPT-*invariant* about some time in the future. Thus, initial and final conditions that are not related by (22.20) do not *necessarily* imply differing probabilities for sets of histories connected by CPT. Further, as discussed in the previous section, both ways of realizing a CPT-symmetric universe can, with appropriate kinds of initial and final conditions and coarse-graining, lead to sets of histories in which each individual member is CPT-asymmetric about the moment of symmetry. Thus, neither are CPT-symmetric boundary conditions *necessarily* inconsistent with arrows of time that extend consistently over the whole of the universe's evolution.

A universe is time-symmetric about a moment of time if the probabilities of any set of alternative histories are identical to those of the time-inverted set. The relation between initial and final conditions necessary for a purely time-symmetric universe is analogous to that for a CPT-symmetric one and derived in the same way. However, we cannot expect boundary conditions to impose time symmetry if the Hamiltonian itself distinguishes past from future. We must assume that the Hamiltonian is symmetric under time inversion, $\mathscr{T}$,

$$\mathscr{T}^{-1}H(t)\mathscr{T} = H(-t)\,. \tag{22.21}$$

Given (22.21), a time-symmetric universe will result if the initial and final conditions are related by time inversion:

$$\rho_f = \mathscr{T} \rho_i \mathscr{T}^{-1} . \tag{22.22}$$

For realistic quantum cosmologies, the time-neutral quantum mechanics of universes in a box, described in Section 3, must be generalized to allow for significant quantum fluctuations in spacetime geometry, and notions of space and time inversion must be similarly generalized. A sketch of a generalized quantum mechanics for spacetime can be found in [14,42] and discussions of time inversion in the quantum mechanics of cosmology in [34], [5], and [6].

22.5.2 T Violation in the Weak Interactions

The effective Hamiltonian describing the weak interaction on accessible energy scales is not CP-invariant. As a consequence of the CPT invariance of field theory it is also not T-invariant. T violation of this kind is a small effect in laboratory experiments but is thought to be of central importance in the evolution of the matter content of the universe. It is believed to be responsible, together with the non-conservation of baryons, for the emergence of a matter-dominated universe from an initial equality of matter and antimatter, as originally pointed out by Sakharov.† Can the symmetric universes just discussed be consistent with this effective T violation in the weak interaction?

The violation of time-inversion symmetry that we observe in the effective weak interaction Hamiltonian could arise in three ways: First, it could be the result of T violation in the fundamental Hamiltonian. Second, it could arise throughout the universe, even if the fundamental Hamiltonian were time-inversion-symmetric, from asymmetries in the cosmological boundary conditions of the universe. Third, it could be an asymmetry of our particular epoch and spatial location arising dynamically in extended domains from a time-inversion symmetric Hamiltonian and boundary conditions. We shall now offer a few comments on each of these possibilities.

If the fundamental Hamiltonian is time-inversion asymmetric, then we cannot expect a time-symmetric universe, as we have already discussed. One could investigate whether such a fundamental time asymmetry is the source of the other observed time asymmetries. So far such an approach has neither been much studied nor shown much promise.

Even though a T-symmetric universe is inconsistent with a T-asymmetric fundamental Hamiltonian, a CPT-symmetric universe could be realized if the initial and final density matrices were related by (22.20). That is because a field-theoretic Hamiltonian is always CPT-symmetric even if it is not T-symmetric. But CPT symmetry needs to be reconciled with the observed matter-antimatter asymmetry over large

† Ref [36]. For an accessible recent review of these ideas see [37].

domains‡ of the universe and the classical behavior of their matter content. If the universe is *homogeneously* matter-dominated now, then CPT symmetry would imply that it will be homogeneously antimatter-dominated at the time-inverted epoch in the future. What evolution of the present universe could lead to such an inversion? One possibility is a universe that lasts much longer than the proton lifetime.§

There is no evidence for CP violation in the basic dynamics of superstring theory. If it is the correct theory, the effective CP violation in the weak interaction in four dimensions has to arise in the course of compactification or from some other form of spontaneous symmetry breaking. From the four-dimensional point of view, which we are taking for convenience in this article, this would correspond to having a non-zero expected value of a CP-odd quantity. Then, as discussed above, it is possible to investigate time-symmetric universes with initial and final conditions related by (22.22). An effective CP violation could arise from CP asymmetries of the initial or final states or both. Typical theories of these boundary conditions relate them to the Hamiltonian or an equivalent action. Each density matrix, ρ_i or ρ_f, may either inherit the symmetries of the fundamental Hamiltonian or be an asymmetrical member of a symmetrical family of density matrices determined by it. This is the case, for example, with "spontaneous symmetry breaking" of familiar field theory where there are degenerate candidates for the ground state not individually symmetrical under the symmetries of the Hamiltonian. Before discussing the possibility of effective CP violation in time-symmetric universes, let us review how an effective CP violation can arise in familiar field theory and in usual quantum cosmology with just an initial condition.

Effective CP violation can arise in field theory even when the fundamental Hamiltonian is CP-invariant, provided there is a non-vanishing vacuum expected value of a CP-odd field $\phi(\vec{x},t)$ [39], i.e. one such that

$$\phi(-\vec{x},t) = -(\mathscr{C}\mathscr{P})^{-1}\phi(\vec{x},t)(\mathscr{C}\mathscr{P}) . \tag{22.23}$$

Usually the vacuum state $|\Psi_0\rangle$ inherits the symmetry of the Hamiltonian that determines it and the vacuum expected value of a CP-odd field would vanish if the Hamiltonian is CP-invariant. However, if there is a symmetrical family of degenerate candidates for the ground state that are individually not CP-invariant, then the expected value

$$\langle\phi(\vec{x},t)\rangle = Tr\left[\phi(\vec{x},t)|\Psi_0\rangle\langle\Psi_0|\right] \tag{22.24}$$

may be non-zero for the physical vacuum.

Similarly, in usual quantum cosmology with just an initial condition ρ_i, a non-zero value of

$$\langle\phi(\vec{x},t)\rangle = Tr\left[\phi(\vec{x},t)\rho_i\right] \tag{22.25}$$

‡ For a classic review of the observational evidence that there is a matter-antimatter asymmetry over a domain at least the size of the local group of galaxies see [38].

§ We owe this suggestion to W. Unruh.

can lead to effective CP violation. The "no-boundary" wave function of the universe [34] is the generalization of the flat space notion of ground state, i.e. vacuum, to the quantum mechanics of closed cosmological spacetimes. The "no-boundary" prescription with matter theories that would lead to spontaneous CP violation in flat space thereby becomes an interesting topic for investigation. In such situations, we expect the "no-boundary" construction to yield a CP-symmetric set of possible wave functions for the universe that are individually CP-asymmetric.

We now turn to effective CP violation in time-symmetric universes with initial and final states related by (22.22). An expected value for a field is defined when probabilities are assignable to its alternative values — that is, when there is decoherence among the alternatives. The requirements of decoherence will be discussed in the next Section. They are automatically satisfied for alternatives at a single moment of time when $\rho_f \propto I$ but they are non-trivial when ρ_f is non-trivial. We have not analyzed the circumstances in which the values of the field decohere but we assume those circumstances to obtain here so that the expectation value of the field may be defined.

A consequence of decoherence and the probability formula (22.5) is the validity of two equivalent expressions for the expected value of the field that are analogous to (22.24) and (22.25):

$$\langle\phi(\vec{x},t)\rangle = NTr\left[\rho_f\phi(\vec{x},t)\rho_i\right] = NTr\left[\rho_i\phi(\vec{x},t)\rho_f\right] . \tag{22.26}$$

These are demonstrated in the Appendix. The symmetry between the initial and final conditions in (22.26) can be understood from the fact that it is not probabilities at one moment of time that distinguish the future from the past. We shall now show that for a CP-odd field this expected value is odd under time inversion for a time-symmetric universe.

We carry over from flat space field theory the assumption that we are dealing with a CPT-even field $\phi(\vec{x},t)$. In flat space that is necessary if the field is to have a non-vanishing vacuum expected value. The CPT invariance of field theory then means that it is possible to choose a (real) representation of $\phi(\vec{x},t)$ such that

$$\phi(-\vec{x},-t) = (\mathscr{C}\mathscr{P}\mathscr{T})^{-1}\phi(\vec{x},t)(\mathscr{C}\mathscr{P}\mathscr{T}) . \tag{22.27}$$

Therefore, since $\phi(\vec{x},t)$ is CP-odd it must be T-odd and then

$$\langle\phi(\vec{x},-t)\rangle = -Tr\left[\rho_f\mathscr{T}^{-1}\phi(\vec{x},t)\mathscr{T}\rho_i\right] . \tag{22.28}$$

But if ρ_i and ρ_f are related by (22.22) this relation may be written

$$\begin{aligned}\langle\phi(\vec{x},-t)\rangle &= -Tr\left[\mathscr{T}^{-1}\rho_i\mathscr{T}\mathscr{T}^{-1}\phi(\vec{x},t)\mathscr{T}\mathscr{T}^{-1}\rho_f\mathscr{T}\right] & (22.29)\\ &= -Tr\left[\rho_i\phi(\vec{x},t)\rho_f\right] = -\langle\phi(\vec{x},t)\rangle . & (22.30)\end{aligned}$$

The conclusion is that it is possible to choose initial and final conditions so that a universe is time-symmetric and has a non-vanishing expected value of a CP-odd field. That expected value is odd in time (the correct time-*symmetric* behavior for a T-odd field.) As a consequence the sign of CP violation would be opposite on opposite sides of the moment of time symmetry and the magnitude of CP violation would decrease on cosmological time scales as the moment of time symmetry is approached. The CP violation in the early universe might well be larger than generally supposed and Sakharov's mechanism for the generation of the baryons more effective. However, if the moment of time symmetry is far in our future, then such variation in the strength of CP violation would be small and it would be difficult to distinguish this time-symmetric situation from the kind of CP violation that arises from just an initial condition as discussed above.

In the class of time-symmetric universes just discussed, CP violation arises from initial and final conditions that are not CP-symmetric. However, an effective CP violation could also exist in our epoch, in local spatial domains, even if both Hamiltonian and initial and final states were CP-symmetric:

$$H = (\mathscr{C}\mathscr{P})^{-1}H(\mathscr{C}\mathscr{P})\,,\ \rho_i = (\mathscr{C}\mathscr{P})^{-1}\rho_i(\mathscr{C}\mathscr{P})\,,\ \rho_f = (\mathscr{C}\mathscr{P})^{-1}\rho_f(\mathscr{C}\mathscr{P})\,. \qquad (22.31)$$

Dynamical mechanisms would need to exist that make likely the existence of large spacetime domains in which CP is effectively broken, say by the expected value of a CP-odd field that grows to be homogeneous over such a domain. In such a picture the set of histories of the universe would be overall CP-symmetric and T-symmetric, as follows from (22.31). Individual histories would display effective CP violation in domains with sizes and durations that are quantum-probabilistic. If very large sizes and durations were probable it would be difficult to distinguish this kind of mechanism from any of those discussed above.

Overall matter-antimatter symmetry would be expected for such universes with matter or anti-matter predominant only in local domains. Their size must therefore be larger than the known scales on which matter is dominant [38]. The calculation of the probabilities for these sizes and durations thus becomes an important question in such pictures. An extreme example occurs in the proposal of Linde [40], in which such domains are far larger than the present Hubble radius.

22.6 The Limitations of Decoherence and Classicality

As we mentioned in Section 2, the quantum mechanics of a closed system such as the universe as a whole predicts probabilities only for those sets of alternative histories for which there is negligible interference between the individual members in the set. Sets of histories that exhibit such negligible interference as a consequence of the Hamiltonian and boundary conditions are said to decohere. A minimal requirement on any theory of the boundary conditions for cosmology is that the universe exhibit a decoherent set of histories that corresponds to the quasiclassical domain of everyday experience. This requirement places significant restrictions on the relation between ρ_i and ρ_f in the generalized quantum mechanics for cosmology, as we shall now show.

22.6.1 Decoherence

Coherence between individual histories in an exhaustive set of coarse-grained histories is measured by the decoherence functional [20]. This is a complex-valued functional on each pair of histories in the set. If the cosmos is replaced by a box, so that possible complications from quantum gravity disappear, then individual coarse-grained histories are specified by sequences of alternatives $\alpha = (\alpha_1, \cdots, \alpha_n)$ at discrete moments of time, $t_1, \cdots, t_n$. The decoherence functional for the case of two-time boundary conditions is given by [14]

$$D(\alpha', \alpha) = N Tr\left[\rho_f C_{\alpha'} \rho_i C_\alpha^\dagger\right] \ . \tag{22.32}$$

A set of histories decoheres when the real parts of the "off-diagonal" elements of the decoherence functional — those between two histories with any $\alpha_k \neq \alpha'_k$ — vanish to sufficient accuracy. As first shown by Griffiths [13], this is the necessary and sufficient condition that the probabilities (22.5), which are the "diagonal" elements of D, satisfy the sum rules defining probability theory.

The possibility of decoherence is limited by the choice of initial and final density matrices ρ_i and ρ_f. To see an example of this, consider the case in which both are pure, $\rho_i = |\Psi_i><\Psi_i|$ and $\rho_f = |\Psi_f><\Psi_f|$. The decoherence functional would then factor:

$$D(\alpha', \alpha) = N <\Psi_f|C_{\alpha'}|\Psi_i><\Psi_i|C_\alpha^\dagger|\Psi_f> , \tag{22.33}$$

where N now is $| <\Psi_i|\Psi_f> |^{-2}$. In this circumstance the requirement that the real part of D vanish for $\alpha' \neq \alpha$ can be satisfied only if there are at most two non-vanishing quantities $<\Psi_i|C_\alpha|\Psi_f>$, with phases differing by 90°, giving at most two histories with non-vanishing probabilities! Thus initial and final states that are both pure, such as those corresponding to a "wave function of the universe", leads to a highly unorthodox quantum mechanics in which there are only one or two coarse-grained histories. All the apparent accidents of quantum mechanics would be determined† by the boundary conditions ρ_i and ρ_f. The usual idea of a simple ρ_i (or ρ_i and ρ_f), with the algorithmic complexity of the universe contained almost entirely in the throws of the quantum dice, would here be replaced by a picture in which the algorithmic complexity is transferred to the state vectors $|\Psi_i\rangle$ and $|\Psi_f\rangle$. Presumably these would be described by a simple set of rules plus a huge amount of specific information, unknowable except by experiment and described in practice by a huge set of parameters with random values.

This bizarre situation refers to the use of a pure ρ_i and a pure ρ_f, whether or not there is any kind of time symmetry relating them.

† This situation is closely related to the one described by L. Schulman [41].

22.6.2 *Impossibility of a Universe with $\rho_f = \rho_i$.*

We shall now give a very special example of a relation between ρ_i and ρ_f, stronger than time symmetry, that is inconsistent with the existence of a quasiclassical domain. More precisely, we shall show that in the extreme case

$$\rho_f = \rho_i \equiv \rho \tag{22.34}$$

only sets of histories exhibiting trivial dynamics can exactly decohere. This condition means that ρ_f has the same form when expressed in terms of the initial fields $\phi(\vec{x}, t_0)$ as ρ_i does. Such a situation could arise if, in addition to time symmetry, we had ρ_i and ρ_f separately, individually time-symmetric and with effectively no time difference between the initial and final conditions. We know of no theoretical reason to expect such a situation, but it does supply an example that leads to a contradiction with experience.

Given the artificial condition, (22.34), we can write the decoherence condition as

$$(Tr\rho^2)^{-1} Re\, Tr(\rho C_{\alpha'} \rho C_\alpha^\dagger) = \delta_{\alpha'\alpha} p(\alpha)\ , \tag{22.35}$$

where $p(\alpha)$ is the probability of the history α. Summing over all the $\{\alpha_n\}$ and $\{\alpha'_n\}$ except α_k and α'_k, we have

$$(Tr\rho^2)^{-1} Re\, Tr[\rho P^k_{\alpha'_k}(t_k) \rho P^k_{\alpha_k}(t_k)] = \delta_{\alpha'_k \alpha_k} p(\alpha_k)\ . \tag{22.36}$$

We note that $P^k_{\alpha_k}(t_k)$ and $P^k_{\alpha'_k}(t_k)$ are just projection operators and thus of the form

$$\sum_n |n><n| \quad \text{and} \quad \sum_{n'} |n'><n'|$$

respectively, where the $|n>$ and $|n'>$ are mutually orthogonal for $\alpha_k \neq \alpha'_k$. Eq. (22.36) then tells us that

$$(Tr\rho^2)^{-1} \sum_{n,n'} |<n|\rho|n'>|^2 = 0, \quad \text{for } \alpha_k \neq \alpha'_k\ . \tag{22.37}$$

Thus ρ has no matrix elements between any $|n>$ and any $|n'>$ for $\alpha_k \neq \alpha'_k$. In other words, ρ commutes with all the P's and therefore with all the chains C_α of P's:

$$[C_\alpha, \rho] = 0 \quad \text{for all } \alpha\ . \tag{22.38}$$

This consequence of perfect decoherence for the special case (22.34) has some important implications. For one thing, the decoherence formula can now be written

$$(Tr\rho^2)^{-1} Tr(C_{\alpha'} \rho^2 C_\alpha^\dagger) = \delta_{\alpha'\alpha} p(\alpha)\ , \tag{22.39}$$

so that we are back to ordinary quantum mechanics with only an initial density matrix $\bar{\rho} \equiv (Tr\rho^2)^{-1}\rho^2$ [cf. (22.11) in the classical case] but with the very restrictive condition

$$[C_\alpha, \bar{\rho}] = 0 \quad \text{for all } \alpha\ . \tag{22.40}$$

The cosmology with the symmetry (22.34) was supposed to be in contrast to the usual one with only an initial density matrix, and yet it turns out to be only a special case of the usual one with the stringent set of conditions (22.40) imposed in addition. The resolution of this apparent paradox is that Eq. (22.40) permits essentially no dynamics and thus achieves symmetry between ρ_i and ρ_f in a rather trivial way. That is not surprising in view of the nature of this condition discussed above.

We have seen that any $P^k_{\alpha_k}(t_k)$ has to commute with $\bar{\rho}$ if it is to be permitted in a chain of P's constituting a member of a decohering set of alternative coarse-grained histories. Now it is unreasonable that for a given projection operator P there should be only a discrete set of times at which it is permissible to use it in a history (e.g., for a measurement). Thus we would expect that there should be a continuous range of such times, which means that $\dot{P} = -i[P, H]$ must commute with $\bar{\rho}$. But $\bar{\rho}$ and P, since they commute, are simultaneously diagonalizable, with eigenvalues π_i and q_i respectively. The time derivative of the probability $Tr(\bar{\rho}P)$ is

$$Tr(\bar{\rho}\dot{P}) = -iTr(\bar{\rho}[P,H]) = -i\sum_i \pi_i(q_i - q_i)H_{ii} = 0 \, . \tag{22.41}$$

The probabilities of the different projections P remain constant in time, so that there is essentially no dynamics and certainly no second law of thermodynamics.

22.6.3 Classicality

A theory of the boundary conditions of the universe must imply the quasiclassical domain of familiar experience. A set of histories describing a quasiclassical domain must, of course, decohere. That is the prerequisite for assigning probabilities in quantum mechanics. But further, the probabilities must be high that these histories are approximately correlated by classical dynamical laws, except for the intervention of occasional amplified quantum fluctuations.

There are, of course, limitations on classical two-time boundary conditions. We cannot, for example, specify both coördinates and their conjugate momenta at *both* an initial and a final time. There would, in general, be no corresponding solutions of the classical equations of motion. Even if initial and final conditions in quantum cosmology allow for decoherence as discussed above, they could still be too restrictive to allow for classical correlations. One would expect this to be the case, for example, if they required a narrow distribution of both coördinates and momenta both initially and finally. Quantum cosmologies with two boundary conditions are therefore limited by both decoherence and classicality.

22.7 Conclusions

Time-symmetric quantum cosmologies can be constructed utilizing a time-neutral generalized quantum mechanics of closed systems with initial and final conditions related by time-inversion symmetry. From the point of view of familiar quantum mechanics such time-symmetric cosmologies are highly unusual. If we think of Hilbert space as finite-dimensional, we could introduce a normalization $Tr(\rho_f)\ Tr(\rho_i) = Tr(I)$, which would agree with the usual case $Tr(\rho_i) = 1$, $Tr(\rho_f) = Tr(I)$. (Note that both $N = Tr(\rho_f \rho_i)$ and the quantity $Tr(\rho_f)\ Tr(\rho_i)$ are invariant under multiplication of ρ_i by a factor and ρ_f by the inverse factor.) With this normalization we may think of $N^{-1} = Tr(\rho_f \rho_i)$ as a measure of the likelihood of the final condition given the initial one. The similarly defined quantity N^{-1} in the analogous classical time-symmetric cosmologies is just that. It is the fraction of trajectories meeting the initial condition that also meet the final one [cf. (22.7)]. The measure N^{-1} is unity for the usual cases where $\rho_f = I$. It can be expected to be very small for large systems with time-symmetric boundary conditions, as the simple model described in Figure 22.1 suggests. The measure N^{-1} is likely to be *extraordinarily* small in the case of the universe itself. Were it exactly zero the initial and final boundary condition construction would become doubtful. We are unsure how much of that doubt survives if it is merely extraordinarily small.

As a prerequisite for a time-symmetric quantum cosmology, the fundamental Hamiltonian must be time-inversion symmetric to give a meaningful notion of time-symmetry and this restricts the mechanisms by which the effective CP violation in the weak interactions can arise. There must be some impurity in the initial or final density matrices or in both for any non-trivial probabilities to be predicted at all. If we wish to exclude the highly unorthodox quantum mechanics in which $|\Psi_i\rangle$ and $|\Psi_f\rangle$ determine all the throws of the quantum dice, then we could not have, for example, a time-symmetric quantum cosmology with both the initial and final conditions resembling something like the "no-boundary" proposal. These results have been obtained by assuming unrealistic exact decoherence and by neglecting gross quantum variations in the structure of spacetime, which may be important in the early universe. It would be desirable to extend the discussion to remove these special restrictions.

Even if these purely theoretical requirements for time-symmetry were met, observations might rule out such boundary conditions. Deviations from the usual thermodynamic or CP arrows of time may be undetectably small if the time between initial and final conditions is long enough. But, as suggested by Davies and Twamley, an expanding and contracting time-symmetric cosmology may be transparent enough to electromagnetic and other forms of radiation that the effects of certain time-symmetric initial and final conditions would be inconsistent with observations today. In the absence of some compelling theoretical principle mandating time symmetry, the simplest possibility seems to be the usually postulated universe where there is a fundamental distinction between past and future — a universe with a special initial state and a final condition of indifference with respect to state. Nevertheless, the notion of complete T symmetry or CPT symmetry remains sufficiently intriguing to warrant further investigation of how such a symmetry could occur or what observations could rule it out. In this paper we have provided a quantum-mechanical framework for such investigations.

Acknowledgments

An earlier version of this paper appeared in the *Proceedings of the 1st International Sakharov Conference on Physics*, Moscow, USSR, May 27–31, 1991 as a tribute to the memory of A.D. Sakharov.

Part of this research was carried out at the Aspen Center for Physics. The work of MG-M was supported by DOE contract DE-AC-03-81ER40050 and by the Alfred P. Sloan Foundation. That of JBH was supported by NSF grant PHY90-08502.

References

[1] P.C.W. Davies (1976) *The Physics of Time Asymmetry*, University of California Press, Berkeley.

[2] R. Penrose (1979) in *General Relativity: An Einstein Centenary Survey* ed. by S.W. Hawking and W. Israel, Cambridge University Press, Cambridge.

[3] H.D. Zeh (1989) *The Physical Basis of the Direction of Time*, Springer, Berlin.

[4] J. J. Loschmidt (1876) *Wiener Ber.*, **73**, 128; *ibid.* (1877) **75**, 67.

[5] D. Page (1985) *Phys. Rev.*, **D32**, 2496.

[6] S.W. Hawking (1985) *Phys. Rev.*, **D32**, 2989.

[7] J. Halliwell and S.W. Hawking (1985) *Phys. Rev.*, **D31**, 1777.

[8] R. Laflamme (Unpublished) "Wave Function of an $S^1 \times S^2$ Universe".

[9] S.W. Hawking (1987) *New Scientist*, **115**, 46.

[10] S.W. Hawking and D. Page (1988) *Nucl. Phys.*, **B298**, 789.

[11] S. Wada (1990) in *Proceedings of the 3rd International Symposium on the Foundations of Quantum Mechanics in the Light of New Technology*, ed. by S. Kobayashi, H. Ezawa, Y. Murayama, and S. Nomura, Physical Society of Japan, Tokyo.

[12] Y. Aharonov, P. Bergmann, and J. Lebovitz (1964) *Phys. Rev.*, **B134**, 1410.

[13] R. B. Griffiths (1984) *J. Stat. Phys.*, **36**, 219.

[14] J.B. Hartle (1991) in *Quantum Cosmology and Baby Universes, Proceedings of the 1989 Jerusalem Winter School*, ed. by S. Coleman, J. Hartle, T. Piran, and S. Weinberg, World Scientific, Singapore.

[15] T. Gold (1958) in *La structure et l'evolution de l'universe, Proceedings of the 11th Solvay Congress*, Editions Stoops, Brussels; (1962) *Amer. J. Phys.*, **30**, 403.

[16] M. Gell-Mann (unpublished) comments at the 1967 Temple University Panel on Elementary Particles and Relativistic Astrophysics.

[17] Y. Ne'eman (1970) *Int. J. Theor. Phys.*, **3**, 1.

[18] H.-D. Zeh (1994) This volume.

[19] S.W. Hawking (1994) This volume.

[20] M. Gell-Mann and J.B. Hartle (1990) in *Complexity, Entropy, and the Physics of Information, SFI Studies in the Sciences of Complexity*, Vol. VIII, ed. by W. Zurek, Addison Wesley, Reading or in *Proceedings of the 3rd International Symposium on the Foundations of Quantum Mechanics in the Light of New Technology* ed. by S. Kobayashi, H. Ezawa, Y. Murayama, and S. Nomura, Physical Society of Japan, Tokyo.

[21] M. Gell-Mann and J.B. Hartle (1990) in the *Proceedings of the 25th International Conference on High Energy Physics, Singapore, August, 2-8, 1990*, ed. by K.K. Phua and Y. Yamaguchi, South East Asia Theoretical Physics Association and Physical Society of Japan, distributed by World Scientific, Singapore.

[22] W.J. Cocke (1967) *Phys. Rev.*, **160**, 1165.

[23] L.S. Schulman (1973) *Phys. Rev.*, **D7**, 2868; L.S. Schulman (1977) *J. Stat. Phys.*, **16**, 217; L.S. Schulman and R. Shtokhamer (1977) *Int. J. Theor. Phys.*, **16**, 287.

[24] J.A. Wheeler (1979) in *Problemi dei fondamenti della fisica*, Scuola internazionale di fisica "Enrico Fermi", Corso 52, ed. by G. Toraldo di Francia, North-Holland, Amsterdam.

[25] L.S. Schulman (1991) *Physica A*, **177**, 373.

[26] M. Kac (1959) *Probability and Related Topics in Physical Sciences*, Interscience, New York.

[27] R. Penrose (1978) in *Confrontation of Cosmological Theories with Observational Data* (IAU Symp. 63) ed. by M. Longair, Reidel, Boston (1974) and in *Theoretical Principles in Astrophysics and Relativity*, ed. by N.R. Lebovitz, W.H. Reid, and P.O. Vandervoort, University of Chicago Press, Chicago.

[28] L. Grishchuk and L.V. Rozhansky (1988) *Phys. Lett.*, **B208**, 369; (1990) *ibid.*, **B234**, 9.
[29] A. Barvinsky and A. Kamenshchik (1990) *Class. Quant. Grav.*, **7**, L181.
[30] A. Vilenkin (1988) *Phys. Rev.*, **D37**, 888.
[31] M. Mijić, M. Morris and W.-M. Suen (1989) *Phys. Rev.*, **D39**, 1486.
[32] P.C.W. Davies and J. Twamley (1993) *Class. Quant. Grav.*, **10**, 931.
[33] R.B. Partridge (1973) *Nature*, **244**, 263.
[34] D. Page (1993), *No Time Asymmetry from Quantum Mechanics*, University of Alberta preprint.
[35] J.B. Hartle and S.W. Hawking (1983) *Phys. Rev.*, **D28**, 2960.
[36] A.D. Sakharov (1967) *ZhETF Pis'ma*, **5**, 32; [(1967) *Sov. Phys. JETP Lett.*, **5**, 24]; (1979) *ZhETF*, **76**, 1172, 1979; [(1979) *Sov. Phys. JETP*, **49**, 594].
[37] E.W. Kolb and M.S. Turner (1990) *The Early Universe*, Addison-Wesley, Redwood City, Ca.
[38] G. Steigman (1976) *Ann. Rev. Astron. Astrophys.*, **14**, 339.
[39] T.D. Lee (1974), *Physics Reports*, **9C**, 144.
[40] A. Linde (1986) *Mod. Phys. Lett. A*, **1**, 81; (1987) *Physica Scripta*, **T15**, 169.
[41] L. Schulman (1986) *J. Stat. Phys.*, **42**, 689.
[42] J.P. Paz and W.H. Zurek (1993) *Phys. Rev.* **D48**, 2728.
[43] M. Gell-Mann and J.B. Hartle (1993) *Phys. Rev.* **D47**, 3345.
[44] J. Finkelstein (1993) *Phys. Rev.* **D47**, 5430.
[45] M. Gell-Mann and J.B. Hartle (to be published).

47 Progress in elementary particle theory, 1950–1964

MURRAY GELL-MANN

Born 1929, New York City; Ph.D., Massachusetts Institute of Technology, 1951; theoretical physics; Nobel Prize, 1969, for his contributions and discoveries concerning the classification of elementary particles and their interactions; California Institute of Technology

I should like to begin by expressing regret for not having been able to be present for the last Fermilab meeting, at which I had been asked to give the final talk about what it was like to be a student of theoretical physics in the late forties, the end of the period covered by that meeting. I still hope to present that material somewhere. Between that conference and this one there were two others, one in Paris in the summer of 1982, where I gave a talk about my experiences with strangeness,[1] and one in 1983, in Sant Feliu de Guíxols in Catalunya, where I spoke on the subject "Particle Theory from *S*-Matrix to Quarks."[2] That second conference was called a "trobada" in the Catalan language, a word that reminds us of the troubadours of the Middle Ages who flourished in Catalunya. It reminds us also that we have become much like those medieval minstrels. We spend a great deal of time now traveling from one orgy of reminiscence to another, each one held in the capital of some princely state. Here I am helping to close another "trobada."

Let me repeat something I said in Catalunya: I shall once again commit the sin of hindsight in the eyes of some of those historians of science who come from the historical tradition. Like most scientists who try to provide material for the history of science, I have, of course, no objection to situating the discoveries in the intellectual context and even in the social, economic, and political contexts of the times in which they were made, but I, along with many other physicists, think it is important to inquire also how close we came to what we now perceive are the right ways of looking at things, to list what we missed, to speculate about why we missed it, and to note things we got right and why we got them right.

Work supported in part by the U.S. Department of Energy under contract DE-AC03-81-ER40050 and by the Alfred P. Sloan Foundation.

In Catalunya, I talked about the mistakes and hesitations and confusions that some theorists went through, particularly the ones that I experienced myself during the fifties and early sixties. The mood of the presentation was rather sad; I mentioned that I was like a person who could not concentrate on the doughnut but kept looking at the hole. In this chapter, I shall deal instead with the doughnut and try to emphasize what I would call the envelope of understanding that some theorists were able to achieve during the years that we are discussing. [Actually, I was assigned to discuss a period (1950–64) different from that discussed by everybody else, and I have taken that mission seriously.] What I mean by treating the envelope of understanding is that instead of talking about the hesitations and the backtracking, I shall pretend that when one of us got a correct idea for a good reason, the idea stuck. I shall include each such idea, and not stress that we doubted it or that we forgot it for a while. Of course, I shall try to point out when we failed to understand something, or when we needed another crucial idea in order to make things work.

If we add together the ideas and the omissions that I list, then we get pretty much the standard model of 1972–3. That model, as I pointed out at Shelter Island a couple of years ago,[3] solves virtually all the problems that we considered to be such when I was a graduate student. But, of course, there are other problems, just as serious, that are left. Few of us believe today that the standard model is anything but the low-energy phenomenology of a much better theory, and we hope that much better theory can be found. Ideally, it will be a completely unified, parameter-free, finite theory of all the interactions, including gravity, a theory that, for all we know, may have been finally written down during the past year. I shall return to that point at the end of this chapter. Let me first review briefly what we learned up to 1964.

I am discussing a close partnership of experiment and theory in a period during which there were sensational experiments. Just to mention a few, there were experiments on strange particles and their decays, antinucleons, hadron scattering cross sections, nucleon form factors, explicit neutrino discovery, neutrino collisions, hadron resonances, parity violation, K^0 and $\bar{K}^0$, *CP* violation, and the Ω^-. Hearing about these experiments, in some cases anticipating them, interacting with the people who were carrying them out, was a fantastic adventure. However, I propose to address the subject here only from the point of view of the theoretician, in fact from the point of view of a very small number of theoreticians who were in the forefront of speculation about the fundamentals of particle physics. (It may be difficult for some young people today to realize what a small set of theoreticians it was.) In most cases I shall adopt my own point of view, even in describing advances with which I did not have a great deal to do.

At the beginning of the period we are discussing, a world picture was developing according to which all physical phenomena could be ascribed to ν and $\bar{\nu}$, $e^{\pm}$ (along with the puzzling $\mu^{\pm}$), the neutron and proton and their presumed antiparticles, the pions $\pi^{\pm}$ and π^0, the photon, possibly some kind of intermediate bosons for the universal weak interaction, and presumably the graviton (although it was not polite to talk about that).

The pion was strongly emphasized. It was hoped that the success of the perturbative renormalization program in QED would carry over to a field theory of nucleon and pion treated as elementary particles. As the pseudoscalar and isovector character of the pion became clear from experiment, speculations about the nuclear force centered on the renormalizable meson theory with the coupling

$$ig\bar{N}\gamma_5\tau N \cdot \pi \tag{47.1}$$

Here, I use Enrico Fermi's notation in which the field is denoted by the particle to which it corresponds. The nonrelativistic approximation is

$$\frac{g}{2M_N}\bar{N}\sigma\tau N \cdot \nabla\pi \tag{47.2}$$

which had already been used in 1942 by Wolfgang Pauli and Sidney M. Dancoff. What a heavy responsibility that poor pion had to bear during those years, and how some theorists laughed at mixed meson theories in which one combined the pion with vector mesons, for example, in order to correct singularities in the nuclear force at short distances.

From that somewhat naïve and disconnected picture of elementary particle phenomena, we progressed by 1964 to a notion of fundamental leptons and quarks, coupled to the vector fields of a generalized Yang–Mills theory of the strong, electromagnetic, and weak interactions, as in the standard model of today, but lacking several crucial ideas, especially color (which removes the clash between the Yang–Mills theories of the strong and the electroweak interactions by making the two kinds of charge variables independent of each other, and which also can lead to quark confinement), asymptotic freedom (which involves effective charge decreasing with decreasing distance, as in confinement, and which allows quarks to be nearly free in the deep interior of the nucleon), and the role of charm in canceling the neutral strangeness-changing current.

I have listed in the following sections a number of advances in understanding during the period in question, advances that helped us to progress from the earlier picture to the later one. By using square brackets, I have indicated from time to time ingredients that were still lacking or comparisons with the present level of understanding. Following some of the items in the list of advances, I have inserted comments and reminiscences about those advances and particularly about how they appeared to me at the time. I have divided my remarks into two sections: quantum field theory, and interactions and elementary particles.

Quantum field theory

A: P, C, *and* T *are empirical symmetries, although local field theory requires* PCT *invariance*

Around 1949, it was not clear to everyone that *P*, *C*, and *T* separately are merely possible symmetries of particular field theories and may be conserved or not depending on the form of the interaction. My experience with *P* may be illuminating. When I was a graduate student at MIT in 1949, I took a course taught by Herman Feshbach, who assigned as a problem proving parity conservation by reflection of coordinates without specifying the theory. I worked on the problem over the weekend and was unable to solve it; on Monday, I turned in a statement that conservation of parity does not follow from transformation of coordinates, but is an empirical law that depends on the transformation properties of the Hamiltonian. I do not remember what impression that made, or what kind of grade I received in the course, but I retained from then on the idea that parity is an empirical symmetry, and charge conjugation as well. Neither Fermi nor P. A. M. Dirac ever believed in parity conservation as a fundamental principle, but the impression was surprisingly widespread that such symmetries were fundamental rules that could not be violated.

B: *Program for extracting exact results from field theories*

1. *"Dispersion-theory program," just quantum field theory formulated on the mass shell (renamed "S-matrix program" and misunderstood by some as a departure from quantum field theory)*
   - (a) *Crossing relations*
   - (b) *Dispersion relations, forward light scattering, generalized to spin-flip, nonzero mass, nonforward, and so forth*
   - (c) *Unitarity relations generalized with amplitudes extended to unphysical values of momentum, but still on the mass shell*
   - (d) *Boundary conditions as various momentum variables approach infinity help specify which field theory is meant*

In Catalunya, I discussed the program that Marvin Goldberger and I worked on, starting around 1952, for extracting exact results from field theories. We abstracted the crossing relations from diagrams to all orders. The dispersion relations were known, but only for forward elastic light scattering, and we generalized them to the spin-flip case. After I left Chicago, Goldberger and his colleagues generalized them to the case of nonzero mass, and then many of us extended them to non-forward dispersion relations. Next we added unitarity, and generalized unitarity so as to carry the amplitudes out of the physical region to imaginary momenta while always keeping them on the mass shell. Finally we said that there needed to be boundary conditions as various momentum variables go to infinity. This was the dispersion-theory program that I outlined at the Rochester meeting in 1956, pointing out that it was a way of formulating field theory on the mass shell. At Rochester, I showed how one could in fact obtain the whole set of scattering amplitudes on the shell iteratively starting with the particle poles, but admitted that I did not know what the boundary conditions would have to be for a particular theory. I casually mentioned Werner Heisenberg's *S*-matrix program, almost as a joke, but later on, after we had convinced Geoffrey Chew of the value of the program, he renamed it "*S*-matrix theory." As we know, L. D. Landau (around 1959) and Chew (around 1961) insisted that it was somehow distinct from field theory.

2. *Renormalization group [but possibility of asymptotic freedom overlooked]*

In 1953, I went on with extracting exact results from field theory, working with Francis Low on the renormalization group. But here, in brackets, we find the first of the crucial gaps in our understanding. Nobody noticed until the early 1970s that the renormalization group function has the sign reversed in a Yang–Mills theory, leading to asymptotic freedom.

C: *Yang–Mills theory*

1. *Group*
   (a) *SU(2) trivially generalized to products of SU(2) and U(1) factors*
   (b) *Generalized to products of arbitrary simple Lie groups and U(1) factors*
2. *Properties*
   (a) *Renormalizable if exact*
   (b) *Perfect gauge invariance*
   (c) *Massless vector bosons*
   (d) *Other multiplets degenerate with respect to the group*
3. *Soft-mass mechanism called for*
   that would break symmetries (b), (c), and even (d)
   renormalizably
   [but not found until later]

The Yang–Mills theory was developed in 1954, using SU(2), and it was immediately clear that one could generalize gauge theory to any product of SU(2) and U(1) factors, with U(1) behaving as in quantum electrodynamics. Later on, at the end of the decade, Sheldon Glashow and I showed that the SU(2) factors could be replaced by arbitrary simple Lie groups. Around the same time, it was shown by Abdus Salam and K. Kumar that Yang–Mills theory is renormalizable if the gauge invariance is exact, but that any masses for the vector mesons, or any other known kind of violation of symmetry, ruined the renormalizability. That was a very serious matter, because we did not know how to make use of perfectly unbroken non-Abelian gauge symmetry. I advertised for a soft-mass mechanism, something that would allow the renormalizability of the pure Yang–Mills theory to persist in the presence of vector-meson masses, but it took a few years for the relevance of the so-called Higgs mechanism to the soft-mass problem to become clear.

D: *Spinless bosons and symmetry breaking*

1. *Exact continuous symmetries,*
   when not producing degeneracy,
   can have massless spinless "Nambu–Goldstone bosons" instead
2. *Approximate symmetry can correspond*
   to low-mass spinless bosons
3. *"Higgs–Kibble. . .Anderson mechanism":*
   way of breaking continuous symmetry
   without degeneracy or massless spinless bosons
   if these bosons are eaten by vector gauge bosons,
   which then acquire mass [not yet recognized as a solution to the soft-mass problem]

In 1959–60, while Yoichiro Nambu and Jeffrey Goldstone were pursuing their separate investigations of how exact continuous symmetry could fail to produce degeneracy, but lead instead to the presence of massless spin-0 bosons, I was studying the problem in Paris with Maurice Lévy, Jeremy Bernstein, Sergio Fubini, and Walter Thirring. We considered the isovector axial vector current in the weak interaction of hadrons and its relation to the pion. Inspired by the success of the Goldberger–Treiman relation and the lack of a convincing derivation of it, we concentrated on the idea of a "partially conserved" axial vector current with a divergence dominated by the low-mass pion pole. In the limit of exact conservation, the pion would become massless, and thus we found the Nambu–Goldstone boson independently, and we saw how approximate conservation would lead to a low-mass boson. The models that Lévy and I used as examples included the nonlinear σ model, later to become a favorite subject of research (even more important today) and the ordinary σ model, which was useful later on in the theoretical discovery of "Higgs bosons."

Those spinless bosons represented still another way that exact continuous symmetry could manifest itself, without either degeneracy or massless spinless bosons. The work of Philip Anderson, Peter W. Higgs, T. W. B. Kibble, and many others showed how the new spinless bosons could be eaten by massless vector bosons, which would then acquire mass. In Chapter 44, Nambu describes how he and his colleague were actually thinking of this mechanism as early as 1959–60.

In any case, the "Higgs" mechanism was available by the end of the period we are discussing, but it was not yet understood that it was the soft mass that we needed for Yang–Mills theory, to supply symmetry breaking and gauge boson masses without destroying renormalizability. Steven Weinberg, in his letter about electroweak theory in 1967, suggested that such might be the case, and Gerard 't Hooft finally proved it around 1971.

E: *"Reggeism"*

1. *Composite particles tend to lie on "Regge trajectories" in a plot of J versus M^2*

During the 1960s we learned a great deal about singularities in the complex angular-momentum plane and their variation with mass. Here I consider them in their relation to quantum field theory (especially in dispersion theory or "*S*-matrix" theory form), and later I mention them again in connection with hadron physics and then once more at the end of this chapter.

These singularities are just as important as ever in spite of the fact that they are a less fashionable research topic than they once were. I remember how excited many of us were when we studied, around 1961 and 1962, the relationship between hadron trajectories in the domain of positive mass squared, where stable particles and resonances lie along them, and the high-energy behavior of scattering amplitudes in the cross channel with nontrivial quantum numbers exchanged. (That work was initiated by Goldberger and R. Blankenbecler, and developed and emphasized by Chew and S. C. Frautschi.)

2. *In simple problems*
   in nonrelativistic quantum mechanics,
   these correspond to "Regge" or "Sommerfeld" poles
   in the complex J plane
3. *In quantum field theory, there are still trajectories,*
   but the poles tend to become
   more complicated singularities (like poles accompanied by branch points)
   [details not clarified until much later]

As it became clear that in field theory the poles are accompanied by cuts that are only logarithmically weaker at high energies, the situation became somewhat less striking, but still not too bad.

It was after our assigned period that the fear was expressed, at the Berkeley conference of 1966, that the cuts would actually be dominant away from $t = 0$, but that turned out to be excessively pessimistic. The situation was cleared up much later by Fredrik Zachariasen and his collaborators, in work that has not received sufficient attention.

4. *In vector field theories,*
   elementary spinor fields give Regge trajectories,
   not fixed poles as in lowest order

To return to the early 1960s: Chew wanted to identify Regge poles as peculiar to composite particles, so that the Regge behavior of all hadrons would be evidence that they were all composed of one another in accordance with the bootstrap picture. I was delighted with the bootstrap idea, but not with the notion that the elementary particles of field theory were somehow doomed to inhabit fixed poles, while only composite particles were allowed the mobility of "Regge" singularities. Goldberger and I, and later several collaborators, pointed out that spinor elementary particles in a vector field theory would become "Reggeized" by radiative corrections. We had some difficulty in getting our work published, perhaps because it contradicted a popular dogma. Later on, of course, other "Reggeizations" were demonstrated.

5. *"Regge behavior" can provide boundary conditions*
   at high momenta for the dispersion-theory program
6. *High-energy behavior of scattering amplitudes*
   depends on leading "Regge" singularities in the cross channel
   as a function of momentum transfer
   (for a pole $s^{\alpha(t)}$ or $s^{\alpha(u)}$)
   [pole dominance over cuts at very high s
   not clarified until much later
   except at $t = 0$ or $u = 0$]

Interactions and elementary particles

A: *Division into strong, electromagnetic, weak, and gravitational interactions*

At the beginning of our period, it was clear to some of us that the known interactions should be divided into strong, electromagnetic, and weak (and, of course, gravitational), each with its characteristic symmetry properties. That notion was not at all universal, however, and in my first letter on strangeness in 1953, I had to describe the classification, labeling the first three interactions (i), (ii), and (iii), before going on to suggest how each behaved with respect to isotopic spin.

1. *All conserve baryon number,*
   but not understood why conservation exact
   [now thought to be probably approximate!]
2. (a) *Lepton number conserved*
   [now thought to be probably approximate]
   (b) *Electron and muon number conserved*
   [now thought to be probably approximate]

It was believed, of course, that all the interactions conserved baryon number, with the proton lifetime being not simply very long (as was known from observation) but infinite. In the absence of any known massless gauge boson coupled to baryon number (and the Eötvös experiment sets a rather stringent limit on such a coupling), exact baryon conservation was rather mysterious. Nowadays, of course, we solve that mystery by believing that the conservation is only approximate. Lepton number and also (with the knowledge that $\nu_e \neq \nu_\mu$) electron number and muon number, separately, are in the same situation. In the standard model, of course, all these quantities are conserved, and the hypothetical violations are always connected with a more general, at least partially unified theory.

B: *Electromagnetism*

1. *QED renormalizable for spin* $\frac{1}{2}$, *also for spin 0 if* $\lambda(\phi^+\phi)^2$ *renormalized*

By about 1950 it was known that QED is renormalizable for charged spinor particles (and also for charged scalar particles if an additional parameter is renormalized). The second-order renormalizability of the charge in QED had been established in 1934 by Dirac and by Heisenberg, and that of the mass by a number of authors in 1948. Of those, the first ones to complete correct relativistic calculations of the Lamb shift (Willis Lamb and Norman Kroll and J. Bruce French and Victor Weisskopf) actually used the clumsy old non-covariant method. The place where the new covariant methods played a crucial role, particularly those of E. C. G. Stueckelberg and Richard P. Feynman, which are still used today, was in permitting calculations to be done quickly, especially calculations to fourth and higher orders (which would have been impractical with the old methods), and in making possible the proof of renormalizability to all orders. The last was accomplished by Salam and P. T. Matthews in a very long paper and by John Ward in concise form, following a crude sketch by Freeman Dyson.

2. *Electromagnetic interaction minimal (Ampère's law)* $p_\mu \rightarrow p_\mu - eA_\mu$ *(always right, but sometimes ambiguous)*

I described at the Paris meeting how in fixing the isotopic spin properties of the electromagnetic interaction in the presence of strange particles (in 1952–3) it was necessary to exclude arbitrary new electromagnetic interactions by insisting on a "minimal" interaction obtained by the rule $p_\mu \rightarrow p_\mu - eA_\mu$, a modern version of Ampere's law. This prescription is not always unique, but it always includes the right answer.

3. *C and P conserved* *to the extent they are conserved elsewhere*

4. *QED of vector charged particles*
   thought to be related to broken Yang–Mills theory, including the photon
   [but not known how to do this renormalizably:
   same as problem of soft mass]

In 1957–8, when some of us were constructing theories of the weak interaction mediated by a spin-1 charged intermediate boson (which Feynman and I called $X^{\pm}$, and which I still call $X^{\pm}$, although some authors seem to use $W^{\pm}$), we were faced with the question of how to resolve the ambiguity in the electromagnetic coupling of that boson, and several of us adopted the prescription that comes from a Yang–Mills theory of $X^{\pm}$ and the photon. That prescription, which is now part of the standard model, leads to finite radiative corrections (even without the Higgs mechanism) in some situations where other prescription would yield infinities.

C: *Strong interaction and "strongly interacting particles" or "hadrons"*

1. (a) *Strong interaction conserves isotopic spin*

The conservation of isotopic spin by the strong interaction, leading to charge independence of the nuclear force, and probably involving an isovector meson, was an old idea that was confirmed at the beginning of our period by the discovery of $\pi^{\pm}$ and π^0 and then by the π–N scattering experiments, especially the ones at Chicago. (I remember Fermi searching in 1952 for a convenient way to describe charge-independent elastic scattering in terms of real quantities and fixing on phase shifts for $J = \frac{1}{2}$ and $J = \frac{3}{2}$ separately and for $I = \frac{1}{2}$ and $I = \frac{3}{2}$ separately.)

(b) *Isotopic spin multiplets can be displaced*
in center of charge by S/2;
then S conserved by strong interaction

(c) *Because electromagnetism is minimal and Q is linear in* I_z,
electromagnetism obeys $|\Delta I| = 0, 1; \Delta I_z = 0 \rightarrow \Delta S = 0$

(d) *Strong interaction conserves C and P*
[today, its conservation of CP to high accuracy looks remarkable,
and the mechanism is much discussed – Higglet (axion), etc.]

(e) *Particle–antiparticle symmetry really works*
for nucleons!

During the early 1950s, the failure to observe antinucleons in cosmic rays gave rise to occasional doubts about the applicability of C to the nucleon, but the experimental discovery of $\bar{p}$ and $\bar{n}$ at the Bevatron laid those doubts to rest.

2. (a) *"Nuclear democracy" or hadronic egalitarianism; no observable hadron is more fundamental than any other*

I have described the dispersion-theory program, renamed the "*S*-matrix" program, and how it requires boundary conditions as various momentum variables tend toward infinity. The condition of having amplitudes dominated in those limits by expressions corresponding to moving singularities in the J plane was suggested by Chew and Frautschi as the way to implement the bootstrap idea that all the hadrons are composed of one another in a consistent way, with no (observable) hadron being more fundamental than any other. (The insertion of the word "observable" is, of course, my modification, and it makes the bootstrap notion perfectly compatible with the existence of quarks and gluons that are fundamental but confined.) The success of QCD has brought research on the dispersion theory of hadrons pretty much to an end; so we do not know what kinds of conditions on that program would reproduce QCD – but presumably there are some that would do the job. It would be fascinating to know exactly how to formulate QCD on the mass shell.

3. *Rich spectrum of baryon and meson states classified by J^P, I, S, Q (or I_z), G (for mesons)*

The first known excited hadron state was, of course, the $J = (\frac{3}{2})^+$, $I = (\frac{3}{2})$ resonance in π–N scattering. It had been predicted theoretically from a strong or intermediate coupling of the foregoing form (47.2) long before it was discovered experimentally by Fermi and his collaborators. Curiously, Fermi did not accept the idea of a resonant state. Starting with the experimental cross sections, he found a solution for the phase shifts that had the $p_{3/2,3/2}$ phase shift increasing with energy to about 60° and then falling again. Later on, one of the graduate students, after a conversation with a visiting theorist (Hans Bethe, I believe), discovered that there was another perfectly good solution in which the $p_{3/2,3/2}$ phase shift increased through 90° as in a resonance.

(c) *Baryons and mesons lie on nearly straight Regge trajectories extending high in spin [still rather mysterious]*

(d) *Approximate exchange degeneracy [still rather mysterious]*

(e) *Hadron high-energy scattering amplitudes really dominated by appropriate Regge singularities in the cross channel*

(f) *Hadron high-energy elastic scattering dominated by "Pomeranchuk" singularity ($J = 1$ for $t = 0$)*

(g) *At $t = 0$, this could not be a simple pole;*
total cross sections could not be asymptotically constant
[situation not further clarified until much later]

3. (a) *Strong interaction has approximate "flavor" SU(3) invariance*
(b) *Baryons in 8 ($J^P = \frac{1}{2}^+$), 10 ($J^P = \frac{3}{2}^+$),*
and higher multiplets
(c) *Mesons in 8 and 1 ($J^P = 0^-$), 8 and 1 ($J^P = 1^-$),*
and higher multiplets
(d) *Violation of SU(3) by octet eighth component*
[and a little bit of third]
(e) *First order predominates, giving mass formula*
[still a little mysterious]

4. (a) *Strong interaction has*
quite good "chiral SU(2) × SU(2)" invariance
and fair "chiral SU(3) × SU(3)" invariance
(b) *Approximate symmetry under "vector current" charges*
gives approximate degeneracy, but
approximate symmetry under "axial vector current" charges
gives nearly massless modified Nambu–Goldstone pseudoscalar
bosons: the pseudoscalar mesons
(c) *So terms violating conservation of axial vector current*
are somehow soft – "PCAC": divergence of axial vector current
has matrix elements dominated by pseudoscalar meson pole;
they transform like $(3, \bar{3})$ and $(\bar{3}, 3)$ under SU(3) × SU(3)
(d) *But there is not a very light pseudoscalar boson corresponding to*
the total axial vector current charge
[explained much later – in QCD – using anomalous divergence of
the current]
(e) *Just as form factor*
for divergence of axial vector current
is dominated at low q^2 by pseudoscalar meson pole,
so form factor of vector current
is dominated at low q^2 by vector meson poles ("vector
dominance")
(f) *These are important, of course,*
because vector current occurs
in electromagnetic and weak couplings,
and axial vector current occurs in weak couplings

5. *Strong interaction thought to come from a vector theory,*
most likely Yang–Mills,
but how can that be arranged in "flavor" space
without clashing with weak and electromagnetic gauge theory?
[this difficulty was overcome when we found color]

D: *Weak interaction*

1. (a) *Charge-exchange weak interaction obeying "Puppi triangle"*

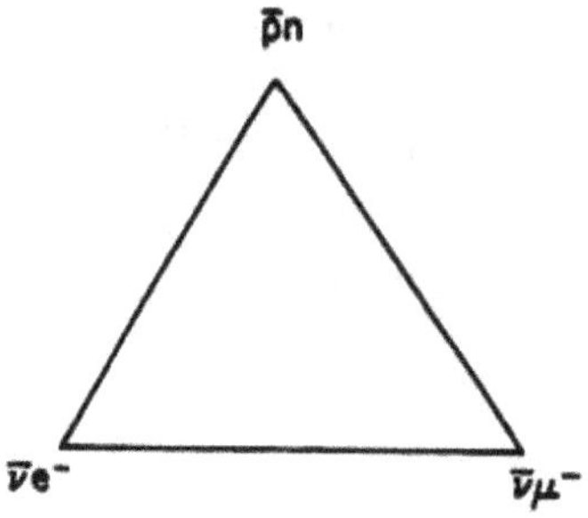

generalized to "tetrahedron"

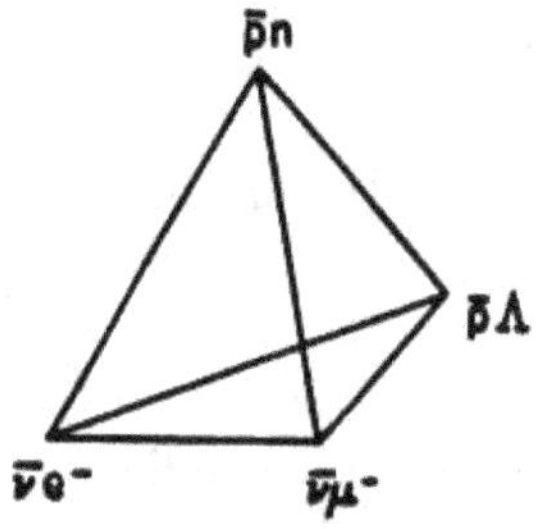

(b) *Leptonic interactions of "$\bar{p}n$," $|\Delta I_z| = 1$, $|\Delta I| = 1$, $\Delta S = 0$; leptonic interactions of "$\bar{p}\Lambda$," $|\Delta I_z| = \frac{1}{2}$, $|\Delta I| = \frac{1}{2}$, $(\Delta S/\Delta Q) = +1 \rightarrow$ nonleptonic interactions have no $|\Delta S| = 2$ and nonleptonic interactions with $|\Delta S| = 1$ have $|\Delta I| = \frac{1}{2}, \frac{3}{2}$ + electromagnetic corrections*

(c) *Approximate predominance of nonleptonic $|\Delta I| = \frac{1}{2}$ [still somewhat mysterious]*

The charge-exchange weak interaction was described around 1949 by the "Puppi triangle," with vertices $\bar{\nu}e^-$, $\bar{\nu}\mu^-$, and $\bar{p}n$, also discussed by Tsung Dao Lee, Marshall Rosenbluth, and Chen Ning Yang, by Jayme Tiomno and John A. Wheeler, and perhaps by others. Around 1954, N. Dallaporta and I, among others, generalized the triangle to a tetrahedron, with something like $\bar{p}\Lambda$ at the fourth vertex, to take account of the strangeness-changing weak interaction, both leptonic and nonleptonic.

Most of us assumed that some kinds of intermediate bosons were involved and that therefore the vertices would interact with themselves as well as with one another. The $\Delta S/\Delta Q = +1$ property of the weak strangeness-changing current would then prevent $|\Delta S| = 2$ nonleptonic interactions from occurring. (We were aware first of the rather mild limitation on $|\Delta S| = 2$ from the failure to observe $\Xi \rightarrow \pi \rightarrow + N$, and only later of the very stringent limitation connected with $K^0 \leftrightarrow \bar{K}^0$.) The rules $|\Delta\mathbf{I}| = 1$ and $|\Delta\mathbf{I}| = \frac{1}{2}$ at the hadronic vertices give rise to $|\Delta\mathbf{I}| = \frac{1}{2}, \frac{3}{2}$ for the nonleptonic strangeness-changing interaction (apart from electromagnetic corrections), as I understood as early as 1954, but I was unable to account in this way for the predominance of $|\Delta\mathbf{I}| = \frac{1}{2}$ over $|\Delta\mathbf{I}| = \frac{3}{2}$, which I described in an unpublished but widely distributed letter in the fall of 1953. It never occurred to me that more than thirty years later the reasons for that approximate $|\Delta\mathbf{I}| = \frac{1}{2}$ rule would still be a subject of research, while the rest of the issues would be understood in the framework of a dynamical theory like the standard model.

2. (a) *K^0 and $\bar{K}^0$ made and absorbed as such, but decay as K_1^0, K_2^0*
(defined by C at first)
   (b) *P and C maximally violated in lepton weak interactions with*
hadrons
and with other leptons
(only two components of neutrino utilized –
maybe only two components exist)
   (c) *PC apparently conserved (K_1^0, K_2^0 now defined by CP)*
   (d) *PC slightly violated;*
beautiful clean-cut result, but not well understood
at the time
[present favorite explanation requires three fermion families
(at least),
but whole CP question is complicated, as mentioned earlier];
now K_1^0, K_2^0 slightly mixed
3. (a) *Coupling is $\gamma_\alpha(1+\gamma_5)$ (V–A), universal*
   (b) *Most likely charged spin-1 intermediate boson*
   (c) *As its mass $\to \infty$, $(G/\sqrt{2})J_\alpha^+ J_\alpha$, with $J_\alpha = \bar{\nu}\gamma_\alpha(1+\gamma_5)e +$*
$\bar{\nu}\gamma_\alpha(1+\gamma_5)\mu$ + hadronic V–A current
with preceding properties
   (d) *Hadron $\Delta S = 0$ vector current*
is just component of isotopic spin current,
conserved and not changed by renormalization
   (e) *Two neutrinos different*
to prevent $\mu \to e + \gamma$ when intermediate boson exists,
$J_\alpha = \bar{\nu}_e\gamma_\alpha(1+\gamma_5)e + \bar{\nu}_\mu\gamma_\alpha(1+\gamma_5)\mu$
4. (a) *"$\bar{p}\Lambda$" term in hadronic current smaller than "$\bar{p}n$" term*
ratio $(\varepsilon/\sqrt{1+\varepsilon^2})$ to $(1/\sqrt{1+\varepsilon^2})$ or $\sin\theta$ to $\cos\theta$, $\theta \approx 15°$
   (b) *Whole hadronic current must be current of a generator of some*
sensible algebra.
Looks like

$$\bar{\nu}_e\gamma_\alpha(1+\gamma_5)e + \bar{\nu}_\mu\gamma_\alpha(1+\gamma_5)\mu + \text{``}\bar{p}\gamma_\alpha(1+\gamma_5)\left(\frac{n}{\sqrt{1+\varepsilon^2}} + \varepsilon\frac{\Lambda}{\sqrt{1+\varepsilon^2}}\right)\text{''}$$

(c) *In fact, total weak and electromagnetic currents probably correspond to generators of $SU(2) \times U(1)$ (as today)*

(d) *Most likely neutral weak interaction through intermediate boson Z^0 (as today),*
but if $SU(2) \times U(1)$, why no appreciable strangeness-changing neutral current?
[not understood until 1970]

(e) *Z^0 and γ result of mixture using weak angle (as today)*

(f) *Z^0 and $X^{\pm}$ masses $\sim$ 100 GeV (as today)*

(g) *$SU(2) \times U(1)$ thought to be broken Yang–Mills theory (as today) [but soft breaking that gives fermion and boson masses not understood]*
[and neutral-current problem $\rightarrow$ trouble]

(h) *Is $SU(2) \times U(1)$ embedded in a larger weak and electromagnetic gauge group including heavier boson?*
[still not really known]

E: *Quarks (1963–4)*

1.

	"flavors"		
	u	d	s
Q	$+\frac{2}{3}$	$-\frac{1}{3}$	$-\frac{1}{3}$

2. *Statistics peculiar, Bose-like for three quarks like "parafermions of rank 3," but where are the baryons with peculiar statistics that would result? [not understood until much later that paraquarks with parabaryons, etc., suppressed, equivalent to color with colored hadrons suppressed, and that quark confinement is related to color suppression]*

3. (a) *Confinement ("mathematical quarks") as opposed to emergence of single quarks ("real quarks")*

   (b) *"Mathematical quarks" defined first (1963) as limit as quark mass $\rightarrow \infty$, then (1965–6) as limit of infinite potential barrier (as today)*

   (c) *If quarks never emerge as "real" particles, then compatible with "nuclear democracy"*

4. *Easily see (with peculiar statistics) $8\,(\frac{1}{2}^+)$ and $10\,(\frac{3}{2}^+)$*

with p-states above, etc.,
for baryons, and
8 and 1 (0^- and 1^-)
with p-states above, etc., for mesons

5. (a) *Many properties abstracted*
from theory of quarks
coupled to single neutral vector gluon
(keep pheasant, throw away veal)
[unfortunately not Yang–Mills color octet
of neutral vector gluons, as today]

With the proposal of quarks in 1963, it became possible to sum up in a very few hypotheses a long list of properties of hadrons and their weak and strong interactions that we had put together painfully over the preceding fifteen years or so. Even before that, I had suggested that we could abstract many of these properties from a quantum field theory of three flavors of spin-$\frac{1}{2}$ fermions coupled to a single neutral vector gluon field. With the quark idea, it became possible to abstract even more correct properties from that theory. In discussing the use of that field theory for such a purpose and then throwing it away, I used a simile in which I borrowed from my friend Valentine Telegdi the notion of a French recipe for cooking a pheasant between two slices of veal and then throwing away the veal and eating the pheasant. If I had used the eight neutral vector gluons of an SU(3) Yang–Mills theory of color, the veal would have been QCD, and there would have been no need to discard it, but the single-gluon theory evidently was not completely correct.

(b) *Chiral $SU(3) \times SU(3)$ broken by mass terms with right behavior*
$(3, \bar{3})$ and $(\bar{3}, 3)$ [but not yet understood why soft]

(c) *Vector current charges generate $SU(3)$ of flavor,*
masses give octet-breaking, weak current now

$$\bar{\nu}_e \gamma_\alpha (1 + \gamma_5) e + \bar{\nu}_\mu \gamma_\alpha (1 + \gamma_5) \mu + \bar{u} \gamma_\alpha (1 + \gamma_5) d'$$
$$d' = d \cos\theta + s \sin\theta$$

electromagnetic current now, of course,

$$\tfrac{3}{2}\bar{u}\gamma_\alpha u - \tfrac{1}{3}\bar{d}\gamma_\alpha d - \tfrac{1}{3}\bar{s}\gamma_\alpha s$$

(d) *1964 (D. J. Bjorken and S. Glashow): charm suggested;*
add $+\frac{2}{3}\bar{c}\gamma_\alpha c$ to electromagnetic current; add $\bar{c}$
$\gamma_\alpha(1 + \gamma_5)s'$ to weak current;
$s' = s\cos\theta - d\sin\theta$; [fits beautifully with $SU(2) \times U(1)$
Yang–Mills theory
of weak and electromagnetic currents and solves problem of
neutral current
having no appreciable strangeness-changing term,
but those consequences not noticed until 1970]

In closing, let me say a word about what may be the most important result to emerge from the theoretical research done during the period with which we are concerned. It has been mentioned that we do not know what hypotheses should have been introduced into the mass-shell formulation of field theory in order to obtain QCD, but we do not know what did actually happen in the evolution of the bootstrap idea. I remember complaining during the middle 1960s that although the bootstrap idea was a good one, the implementation by the Berkeley group was mostly focused on one channel (two pions) and one intermediate state (the ϱ meson). I favored instead an infinite number of states, approximated by narrow resonances, for the intermediate states in all channels and for all the external legs. This rough idea was made much more concrete in the seminal "duality" paper by the three Caltech postdoctoral fellows R. Dolen, D. Horn, and C. Schmid. It was the search for a concrete realization of the dual bootstrap that led G. Veneziano to his famous model, with its remarkable properties.

Of course, the Veneziano model contained a particle with negative mass squared, included no fermions, and turned out to work only in twenty-six dimensions. In 1971, A. Neveu and J. H. Schwarz suggested a different model, with both bosons and fermions and a critical dimension of 10. After the proof by F. Gliozzi, J. Scherk, and D. I. Olive that a sector of that theory could be consistently omitted, it was shown that the remaining theory contained no particles of negative mass squared or negative probability. This "superstring theory" thus appeared to be a consistent bootstrap theory, but it had serious difficulties (such as predicting a massless spin-2 particle) as a description of hadrons. (Besides, QCD was then available and looked very promising.)

Scherk and Schwarz then suggested that the characteristic slope of Regge trajectories in the theory be modified by a factor of 10^{38} or so, and they showed that the theory then becomes a generalization of a quantum version of Einsteinian gravitation, in fact containing supergravity (which had just been proposed) as an approximation, provided the extra six dimensions spontaneously curl up into a tiny structure comparable in size to the Planck length. Moreover, the theory contains fields that might describe the elementary particles we know, such as quarks, leptons, and gauge bosons.

More recently, as a result of the work of Green and Schwarz, it has been found that there are at least five such theories, at least some of which seem to be finite in perturbation theory, not even requiring renormalization. At least one of them, proposed by the Princeton "string quartet" at the end of 1985, offers a serious hope of being the long-sought unified field theory of all the elementary particles and interactions of nature.

Notes

1 M. Gell-Mann, "Strangeness," in Colloque International sur l'Histoire de la Physique des Particules, *J. Phys.* (*Paris*) *43* (December 1982), C8-395–408.

2 M. Gell-Mann, "Particle theory from *S*-Matrix to Quarks," in *Symmetries in Physics (1600–1980), Proceedings of the First International Meeting on the History of Scientific Ideas*, edited by M. G. Doncel, A. Hermann, L. Michel, and A. Pais (Barcelona: Bellaterra. 1987), pp. 474–97.

3 Murray Gell-Mann, "From Renormalizability to Calculability?" in *Shelter Island II, Proceedings of the 1983 Shelter Island Conference on Quantum Field Theory and Fundamental Problems of Physics* (Cambridge, Mass.: MIT Press, 1985), pp. 3–23.

ESSAYS & COMMENTARIES

Nature Conformable to Herself

Some arguments for a unified theory of the universe

BY MURRAY GELL-MANN

Nobel laureate Murray Gell-Mann has many intellectual passions including natural history, linguistics, archeology, history, depth psychology, and creative thinking. As a professor and at the Santa Fe Institute and co-chairman of the Science Board, he works with many researchers in their search for connections between basic laws of physics and the complexity and diversity of the natural world. He is also affiliated with the Los Alamos National Laboratory and the University of New Mexico.

Why are elegance and simplicity suitable criteria to apply in seeking to describe nature, especially at the fundamental level? Science has made notable progress in elucidating the basic laws that govern the behavior of all matter everywhere in the universe—the laws of the elementary particles and their interactions, which are responsible for all the forces of nature. And it is well known that a theory in elementary particle physics is more likely to be successful in describing and predicting observations if it is simple and elegant. Why should that be so? And what exactly do simplicity and elegance really mean in this connection?

To answer those questions, we need to deal first with the widespread notion that all scientific theory is nothing but a set of constructs with which the human mind attempts to grasp reality, a notion associated with the German philosopher Immanuel Kant. Although I had heard of that belief many times, I first came into collision with it thirty-six years ago, in Paris.

At that time, I was a visiting professor at the Collège de France, founded by Francis I more than four hundred years earlier. (As far as I know, I was the first visiting professor in the history of that venerable institution.) My office was in the laboratory of experimental physics established by Francis Perrin, a well-known scientist who was a permanent professor at the Collège. On my visits to the offices of the junior experimentalists down the hall, I noticed that they spent a certain amount of time drawing little pictures in their notebooks, which I assumed at first must be diagrams of experimental apparatus. Many of the drawings turned out, however, to be sketches of a gallows for hanging the vice-director of the lab, whose rigid ideas drove them crazy.

I soon got to know the *sous-directeur,* and we conversed on various subjects, one of which was Project Ozma, an early attempt to detect possible signals from other technical civilizations on planets orbiting nearby stars. The corresponding project nowadays is called the Search for Extraterrestrial Intelligence. We discussed how communication might take place if alien intelligences broadcasting signals were close enough to the solar system, assuming that both interlocutors would have the patience to wait years for the signals to be transmitted back and forth. I suggested that we might try beep, beep-beep, beep-beep-beep, etc. to indicate the numbers 1, 2, 3, and so forth, and then perhaps 1, 2, 3....42, 44.....60, 62.......92 for the atomic numbers of the 90 chemical elements that are stable—1 to 92 except for 43 and 61. "Wait," said the *sous-directeur*, "that is absurd. Those numbers up to 92 would mean nothing to such aliens....Why, if they have 90 stable chemical elements as we do, then they must also have the Eiffel Tower and Brigitte Bardot."

That is how I became acquainted with the fact that French schools taught a kind of neo-Kantian philosophy, according to which the laws of nature are nothing but Kantian "categories" used by the *human mind* to describe reality.

That discussion with the *sous-directeur* is, in fact, a good starting place for thinking about this matter.

Are the mathematics and the mathematical descriptions of physical phenomena used by an alien technological civilization on another planet likely to resemble what human beings come up with on Earth, even if the notation is very different? At present, we can only speculate about the answer, but the question is deep and meaningful, and the speculation can be instructive.

I regard it as likely that there are many advanced civilizations on planets revolving around stars scattered through the universe. Just consider the enormous number of stars in a galaxy and the number of galaxies in the universe; the non-negligible probability of a star possessing a planet something like the Earth; the ease with which complex adaptive systems can then get started as life got started on Earth some four billion years ago; and the fact that the evolution of complex adaptive systems amounts to a search process that allows more and more sophisticated forms to emerge as time goes on. In my view, the main unknowns in the search for extraterrestrial intelligence are the actual density in space of the planets on which such intelligence evolves and the length of time that a technical civilization lasts before it destroys itself or ceases to utilize technology. These unknowns affect the likelihood that an alien technological civilization broadcasting signals is close enough for the signals to be detectable here, but such a civilization still may have evolved somewhere, even if not close by. Thus I believe that the question of what kinds of mathematical science such a civilization can possess is probably a question about the actual universe, even though we may or may not get to compare our answers with any facts.

Are the laws of science just constructs of the human mind so that alien intelligences seeking the laws of nature would most likely arrive at something very different? Or does nature determine to a great extent how an intelligent being would have to describe its laws? And what about the relation of mathematics to science? Need the description of the fundamental laws of nature make use of mathematics as we understand the term, or is there some totally different way of describing the same laws?

These are difficult questions, and we cannot give definitive answers to them, but it is instructive to examine the way in which the human scientific enterprise has penetrated deeper and deeper into the domain of the basic laws. As an example of how a new mathematical theory of fundamental physics is discovered, take the case of the Yang-Mills theory, first discussed by Chen Ning (Frank) Yang and Robert Mills about forty years ago. They developed it in the hope that it would contribute somehow to the search for the fundamental laws, but at first it was a purely abstract construct with no known application. It was a generalization of the marvelously successful theory of the electromagnetic field developed by James Clerk Maxwell in the middle of the nineteenth century. A number of us elementary particle theorists showed how to generalize the Yang-Mills idea further to include higher symmetries and also broken symmetries. We also suggested, over the years, ways in which such slightly generalized theories of the Yang-Mills type might actually account for the forces—the so-called strong and weak forces—that were known to exist in addition to electromagnetism and gravitation. Whereas the last two are long-range forces—that is, they die out slowly with increasing distance—the strong and weak forces are short-range and, in fact, negligible at distances much larger than the size of an atomic nucleus. Before going further into the uses of Yang-Mills theory, let us review the history of the long-range forces.

Gravitation is universal, in the sense that all matter possesses energy, and all energy is subject to gravitation. Since all matter attracts and is attracted to all other matter, gravitational attractions add up. As a result, gravity is very conspicuous.

Electromagnetism is almost universal. It is produced by—and acts on—electric charges, which are not in short supply: every electron in each atom is electrically charged, and so is every atomic nucleus. However, unlike the gravitational force, the electromagnetic force on a sample of matter does not simply increase with the weight of the sample. Since like electric charges repel and unlike charges attract, a great deal of cancellation takes place in bulk matter, which is nearly electrically neutral. That is why electromagnetic phenomena are somewhat less familiar in the everyday experience of prescientific human beings than gravitation, which pulls us all toward the center of the Earth.

It is not surprising that the first force to be described by an adequate theory was the gravitational force, which is both long-range and universal, and that the second one was the electromagnetic force, which is also long-range. Short-range forces were not discovered until the twentieth century, and it is only in the last few decades, with the development of the standard model, that we theoretical physicists have produced a reasonably good picture of those that are known from observation.

Gravitation was fairly well described some three hundred years before Maxwell wrote his equations for electromagnetism. The brilliant theorist who provided the first serious theory of gravitation was, of course, Isaac Newton, who guessed and then demonstrated that the same force with which we are familiar on Earth also governs the motion of the planets and moons. Historians of science still argue over whether it was really an apple falling from a tree on his mother's farm that originally inspired his magnificent insight into the universality of gravitation. But in looking back much later on his discovery of that law, what struck Newton most forcibly was the presence in nature of a kind of consistency. He put it this way:

> How the great bodies of the earth, Sun, moon, and Planets gravitate towards one another what are the laws and quantities of their gravitating forces at all distances from them and how all the motions of those bodies are regulated by those their gravities I shewed in my Mathematical Principles of Philosophy to the satisfaction of my readers: And if Nature be most simple and fully consonant to her self she observes the same method in regulating the motions of

> smaller bodies which she doth in regulating those of the greater. This principle of nature being very remote from the conceptions of Philosophers I forbore to describe it in that book least I should be accounted an extravagant freak and so prejudice my Readers against all those things which were the main designe of the Book.

Thus, when Newton reflected on the unity of ordinary terrestrial gravity with the force driving the heavenly bodies, he regarded it as an example of the consonance of Nature. We might treat in the same fashion the unity, revealed in Maxwell's equations, of the description of electrical and magnetic phenomena.

But let us not fail to note that the gravitational law of force discovered by Newton has the same form as the law of force for electrical attraction and repulsion found much later by Coulomb. In fact, Newton was thinking along such more general lines when he returned repeatedly in his writings to the idea that Nature is "consonant" or "conformable" to herself. As he writes in the *Opticks*,

> For Nature is very consonant and conformable to her self... For we must learn from the Phaenomena of Nature what Bodies attract one another, and what are the Laws and Properties of the Attraction, before we enquire the Cause by which the Attraction is perform'd. The Attractions of Gravity, Magnetism, and Electricity, reach to very sensible distances, and so have been observed by vulgar Eyes, and there may be others which reach to so small distances as hitherto escape Observation; and perhaps electrical Attraction may reach to such small distances, even without being excited by Friction.

What a wealth of wisdom is contained in these words! Newton suggests that at small distances electrical interactions may play an important role, going far beyond the attraction of bits of paper to an amber rod rubbed against a cat's fur. He anticipates the existence of a multitude of short-range forces such as we now know to exist. (Indeed, theoretical considerations now point to an infinite number of such forces.) He points out how empirical laws generally precede detailed dynamical explanations. *And he seems to think of the various "Phaenomena" as exhibiting conformability among themselves as well as within each one.* The last idea is the key to understanding why the criterion of simplicity should be helpful in the search for the fundamental laws of physics.

In discussing why that is so, let us make use of the familiar metaphor that relates the successive discoveries in fundamental physics at higher and higher energies (or, what is the same thing, at shorter and shorter distances) to the peeling of an onion. Removing the skins of the onion one by one, we encounter similarities between one layer and the next. We have remarked that over the centuries, in passing from the gravitational force to the electrical one (under familiar "nonrelativistic and classical" conditions), scientists observed a noticeable similarity between these forces: both fall off with the square of the distance. During the last few decades, in going from electromagnetic theory to the combined theory of the weak and electromagnetic forces and to the theory of the strong force, we have encountered profound similarities in proceeding from Maxwell's equations to slight generalizations of the equations of Yang and Mills, which are themselves an ingenious generalization of Maxwell's equations. This type of theory has not been drastically altered.

Are the laws of science just constructs of the human mind?

As we peel the skins of the onion, penetrating to deeper and deeper levels of the structure of the elementary particle system, mathematics with which we become familiar because of its utility at one level suggests new mathematics, some of which may be applicable at the next level down—or to another phenomenon at the same level. Sometimes even the old mathematics is sufficient.

A generalization may be performed by a theoretical physicist, or by a mathematician, or by both working in ignorance of each other's efforts. But the usefulness to science of a generalization is not just a function of these human activities. It depends on the fact that nature actually exhibits similarities between one level and the next, between one skin of the onion and the next. That is what Newton, with remarkable precocity, apparently noticed. The fundamental laws of nature are such that a rough self-similarity prevails in the set of effective theories that approximately describes the successive layers.

Now we can return to the question of why simplicity is a useful criterion to apply in the search for the fundamental laws of physics. Simplicity, of course, is the opposite of complexity. In discussing the simplicity of a theory, we are referring to the near-absence of what I have called effective complexity. The effective complexity of a thing means the length of a very concise description of its regularities. Of course, any such definition is somewhat context-dependent. For instance, the length of the short description depends on the language that is employed and on the knowledge and understanding of the world that is assumed. But that is just the point. Since the mathematics needed to describe one skin of the onion is similar to that already developed for the previous skins, the new theory can be very concisely written in notation already familiar from earlier work. One reason the Yang-Mills equations look simple and elegant is that they are easily described in mathematical language suitable for Maxwell's equations.

But where does this process of peeling the onion end? It seems likely today that it ends in the discovery of a unified theory of all the forces of nature. In fact, it is possible that we have already achieved that goal in the form of superstring theory. For the first time in history, we have a plausible candidate for the role of unified quantum field theory of all the elementary particles and interactions. That theory correctly predicts Einstein's sophisticated general-relativistic theory of gravitation (which replaced Newton's some eighty years ago), and it does so within the framework of quantum mechanics without producing the preposterous infinite corrections to calculations that plagued previous attempts to quantize Einsteinian gravitation.

Superstring theory predicts an infinite number of kinds of forces; all but a finite number, however, are too short-range to be detected by experiment in the foreseeable future. Many of the predicted short-range forces that are detectable, including the strong and weak forces already known, are describable in superstring theory by means of mathematics similar to that of the Maxwell or Yang-Mills equations. Moreover, Einstein's theory of gravitation shares with both Maxwell and Yang-Mills theory and with superstring theory itself the very important mathematical property of being a gauge theory

We can now turn the story around. We can start from the heart of the onion—the simple unified quantum field theory of all the elementary particles and their interactions, whether it is superstring theory or something else—and work outward. That fundamental unified theory is not only simple. It also has the property that its consequences exhibit similarities between skins of the onion, that is, between phenomena (forces, for example) that are conspicuous in one range of energies or distances and in another. Those similarities, which relate the mathematics useful in one context to the mathematics useful in another, can be regarded more as intrinsic properties of the underlying fundamental law than as flowing from the properties of the human mind (or of any other intelligence investigating the basic laws of nature).

In summary, then, we have this picture of the laws of physics: A simple unified theory (which may well be superstring theory) describes all the elementary particles and all the forces of nature. It is a property of nature, not of the human mind, although the way it is formulated by human beings may be peculiar to our species. It has the characteristic that its various manifestations in different ranges of energy or distance possess a great deal of similarity to one another and to the underlying theory itself. The mathematical structure of the unified theory is reflected in certain properties of the structures needed to describe those manifestations. Any group of intelligent beings attempting to describe the system is likely to be working in from large distances or up from low energies and will encounter these manifestations successively. Because of the similarity of these manifestations, those beings will find it natural to keep generalizing their mathematics, and the successive steps will appear fairly simple. If the beings persist, they are likely to discover the fundamental unified theory with its essential mathematical structure, which will explain the other structures encountered along the way and their similarities. That basic law is, then, what is responsible for the usefulness of certain kinds of mathematics in physics. Intelligent beings on another planet can arrive at the same law, even if each of them has seven tentacles, thirteen sense organs, and a brain shaped like a pretzel. Of course, their notation is very unlikely to resemble ours, but we already know from many examples that what is essentially the same mathematics can often be expressed in very different ways.

All three principles—the conformability of nature to herself, the applicability of the criterion of simplicity, and the utility of certain parts of mathematics in describing physical reality—are thus consequences of the underlying law of the elementary particles and their interactions. Those three principles need not be assumed as separate metaphysical postulates. Instead, they are emergent properties of the fundamental laws of physics.

In my opinion, a great deal of confusion can be avoided, in many different contexts, by making use of the notion of emergence. Some people may ask, "Doesn't life on Earth somehow involve more than physics and chemistry plus the results of chance events in the history of the planet and the course of biological evolution? Doesn't mind, including consciousness or self-awareness, somehow involve more than neurobiology and the accidents of primate evolution? Doesn't there have to be something more?" But they are not taking sufficiently into account the possibility of emergence. Life can perfectly well emerge from the laws of physics plus accidents, and mind, from neurobiology. It is not necessary to assume additional mechanisms or hidden causes. Once emergence is considered, a huge burden is lifted from the inquiring mind. We don't need something more in order to get something more.

Although the "reduction" of one level of organization to a previous one—plus specific circumstances arising from historical accidents—is possible in principle, it is not by itself an adequate strategy for understanding the world. At each level new laws emerge that should be studied for themselves; new phenomena appear that should be appreciated and valued at their own level.

It in no way diminishes the importance of the chemical bond to know that it arises from quantum mechanics, electromagnetism, and the prevalence of temperatures and pressures that allow atoms and molecules to exist. Similarly, it does not diminish the significance of life on Earth to know that it emerged from physics and chemistry and the special historical circumstances permitting the chemical reactions to proceed that produced the ancestral life form and thus initiated biological evolution. Finally, it does not detract from the achievements of the human race, including the triumphs of the human intellect and the glorious works of art that have been produced for tens of thousand of years, to know that our intelligence and self-awareness, greater than those of the other animals, have emerged from the laws of biology plus the specific accidents of hominid evolution.

When we human beings experience awe in the face of the splendors of nature, when we show love for one another, and when we care for our more distant relatives—the other organisms with which we share the biosphere—we are exhibiting aspects of the human condition that are no less wonderful for beirg emergent phenomena.

ACKNOWLEDGMENTS

A version of this article appeared *in The Bulletin of the Santa Fe Institute*, 7, 7-10, (1992).

35

Quarks, Color, and QCD

MURRAY GELL-MANN

Born New York City, 1929; Ph.D., 1951 (physics), Massachusetts Institute of Technology; Robert A. Millikan Professor of Physics Emeritus at the California Institute of Technology; Professor, Santa Fe Institute; Nobel Prize in Physics, 1969; elementary particle physics (theory).

The story of my early speculations about quarks begins in March 1963, when I was on leave from Caltech at MIT and playing with various schemes for elementary objects that could underlie the hadrons. On a visit to Columbia, I was asked by Bob Serber why I didn't postulate a triplet of what we would now call SU(3) of flavor, making use of my relation $\underset{\sim}{3} \times \underset{\sim}{3} \times \underset{\sim}{3} = \underset{\sim}{1} + \underset{\sim}{8} + \underset{\sim}{8} + \underset{\sim}{10}$ to explain baryon octets, decimets, and singlets. I explained to him that I had tried it. I showed him on a napkin (at the Columbia Faculty Club, I believe) that the electric charges would come out $+\frac{2}{3}, -\frac{1}{3}, -\frac{1}{3}$ for the fundamental objects. During my colloquium that afternoon, I mentioned the notion briefly, but meanwhile I was reflecting that if those objects could not emerge to be seen individually, then all observable hadrons could still have integral charge, and also the principle of "nuclear democracy" (better called "hadronic egalitarianism") could still be preserved unchanged *for observable hadrons.* With that proviso, the scheme appealed to me.

I didn't get a great deal of time to work on these "kworks," as I thought of them (not yet having spotted the word "quarks" in *Finnegans Wake*) until September, 1963, when I returned to Caltech.

While I was away, my student George Zweig had completed his final oral, with Richard Feynman replacing me, and had left during the summer for CERN, where he conceived his ideas about "aces." We did not overlap, so we had no discussions about quarks and were in ignorance of each other's work. I did talk with Viki Weisskopf, then Director General of CERN, in the early fall by telephone between Pasadena and Geneva, but when I started to tell him about quarks he said, "this is a transatlantic phone call and we shouldn't waste time on things like that." I presume, therefore, that there was no further discussion of my ideas at CERN that year.

Naturally, my first concern about the scheme was the matter of statistics. The quarks, viewed as *constituents* of the low-lying baryon states, were most simply interpreted as being in a symmetrical state of space, spin, and isotopic spin or SU(3) coordinates. But Fermi–Dirac statistics would predict the opposite. Thus, from the beginning I wanted some weird statistics for the quarks, and I associated that requirement ("quark statistics") with their not emerging to be viewed individually. But what kind of weird statistics was involved?

Back at Caltech, I tried various speculations about the statistics. I had heard, at MIT, about the idea of parafermions and wondered, naturally, whether quarks might be parafermions of rank 3. Yuval Ne'eman, who was visiting Caltech that academic year, recalls that he accompanied me to the library to look up formulae relating to such objects, to see whether certain combinations of three of them would behave like fermions in a symmetrical state of space, spin, and isotopic spin or SU(3) coordinates. He says that we consulted an article that contained a mistake or misprint and therefore missed the construction, although, in fact, three such parafermions would yield, among other things, a state behaving like a fermion, symmetrical in the other variables.

Thus I left the matter of quark statistics for a later time and wrote up the triplet idea during the fall of 1963 (using the spelling "quark" and a reference to Joyce) in a brief submission to *Physics Letters*, emphasizing the "current quark" aspects more than the "constituent quark" aspects. I employed the term "mathematical" for quarks that would not emerge singly and "real" for quarks that would. In the letter, to illustrate what I meant by "mathematical," I gave the example of the limit of infinite mass and infinite binding energy.

Later, in my introductory lecture at the 1966 International Conference on High Energy Physics in Berkeley, I improved the characterization of mathematical quarks by describing them in terms of the limit of an infinite potential, essentially the way confinement is regarded today. Thus what I meant by "mathematical" for quarks is what is now generally thought to be both true and predicted by QCD. Yet, up to the present, numerous authors keep stating or implying that when I wrote that quarks were likely to be "mathematical" and unlikely to be "real," I meant that they somehow weren't there. Of course I meant nothing of the kind.

During the 1960s and 1970s, numerous experiments were undertaken to search for real quarks. Peter Franken had the idea that quarks in sea water would be concentrated by oysters, and he would phone me at midnight to tell me of the progress of his experiments on those molluscs, which he ground up and checked for spectral lines that might be emitted by atoms containing real quarks. William Fairbank at Stanford thought at times that he had candidate events for real quarks. Others were looking for real quarks in the cosmic radiation. If they existed, then some fractionally charged particle would be stable, and that stable object could have practical applications, not only possible catalysis of nuclear fusion, but presumably others as well. I used to say that real quarks would lead to a quarkonics industry. But I have always believed instead in mathematical quarks, ones that do not emerge singly to be observed or utilized.

I did not want to call such quarks "real" because I wanted to avoid painful arguments with philosophers about the reality of permanently confined objects. In view of the widespread misunderstanding of my carefully explained notation, I should probably have ignored the philosopher problem and used different words.

The prescription that I frequently recommended for studying quarks was current algebra, for example as abstracted from a useful but obviously wrong field theory of quarks with a single neutral gluon. It was in that connection that I mentioned the recipe that Valentine Telegdi attributed to Escoffier, in which pheasant meat is cooked between two slices of veal and the veal is then thrown away. The veal was the incorrect single gluon theory, not field theory in general.

Later on, in the late 1960s, Bjorken gave some further, approximate generalizations of the quark current algebra, arriving at what was a kind of impulse approximation for suitable high energy, high momentum-transfer inclusive processes such as deeply inelastic collisions of electrons and nucleons. This was what was renamed by Feynman the "parton model." But acknowledgment that the "partons" were just quarks (and antiquarks and sometimes gluons) came slowly, and even when some authors began to refer to "quark partons," it was still often implied that they were somehow different from current quarks treated in a particular approximation. The approximation in question is, of course, one that is justified when the strength of the interaction between colored particles, such as quarks, becomes weaker and weaker at shorter and shorter distances.

Let us now return to the matter of quark statistics. After the appearance of my letter, Oscar Greenberg suggested that the quarks might be parafermions of rank 3. He did the mathematics correctly and showed that fermionic baryons would result. However, he did not, so far as I know, go on to require that the baryons with bizarre statistics be suppressed, leaving only fermions.

Likewise, when Moo-Young Han and Yōichirō Nambu suggested the existence of what we called "color" at Caltech, they did not consider the suppression of color nonsinglets. In fact, in a special twist, they assigned different electric charges to different colors, in such a way that all the quarks and antiquarks would have integral charges $+1, 0$, or -1. That way color nonsinglets would not only exist but would be excited by the electromagnetic interaction.

In both the Greenberg and Han–Nambu schemes, three quarks in a *symmetrical* configuration with respect to space, spin, and isospin or SU(3) coordinates would yield a *fermion* composite, but other kinds of baryons would also exist, in one case with outlandish statistics, and in the other case with non–singlet color.

Since I was always convinced that quarks would not emerge to be observed as single particles ("real quarks"), I never paid much attention to the Han–Nambu model, in which their emergence was supposed to be made plausible by giving them integral charges. However, it is a pity that I missed a follow-up article by Nambu. It appeared in the 1966 Festschrift volume devoted to my old teacher, Weisskopf, in which friends and admirers celebrated his sixtieth birthday. As one of his students, I should, of course, have contributed, but my habit of procrastination proved – as it had many times before – to be an obstacle, and I didn't produce an article for the volume in time for publication. I was so ashamed of that situation that I never opened the book.

In his 1966 paper Nambu pointed out how a color octet vector interaction between quarks would serve to lower the energy of the color singlet configuration of three quarks relative to octet and decimet representations, thus explaining the observation at modest energies of baryon configurations symmetrical in the other variables. However, sticking to the Han–Nambu point of view, he did not try to abolish the color non-singlet representations, and in his scheme they would appear at higher energies.

If I had seen Nambu's article, I might have concluded that his color octet vector interaction should be utilized without the Han–Nambu idea of integral charges and with confined ("mathematical") quarks. I might then have made progress in the direction of QCD.

Instead, that sort of insight was delayed until 1971. It was in that year, on the day of the earthquake that shook the Los Angeles area, that Harald Fritzsch arrived at Caltech. (In memory of that occasion, I left the pictures on the wall askew, until they were further disturbed by the 1987 earthquake.)

In the fall of 1971, both Fritzsch and I moved to the vicinity of Geneva to spend the year working at CERN. William Bardeen was there with us. We took up again the questions involved in putting the quarks in the known baryon states into symmetrical configurations of the usual variables. First, we reexamined parastatistics, getting the arithmetic right this time, and saw at once that there would occur, for three paraquarks of rank 3, one regular fermion pattern, in which the quarks would have a symmetrical wave function in the usual variables, as well as three patterns with unconventional statistics. If we somehow abolished those three extra ones, then baryons would all be fermions, and the configurations assigned to the known baryon states would be all right. Similarly, quark–antiquark states could be restricted to bosons.

Looking at color, we knew that the requirement of color singlets only would accomplish the same thing. Moreover, we discovered that each of the four patterns formed by three paraquarks of rank 3 could be identified with a single component of one of the representations $\underset{\sim}{1}$, $\underset{\sim}{8}$, $\underset{\sim}{8}$, and $\underset{\sim}{10}$ of color SU(3) arising from the product $\underset{\sim}{3} \times \underset{\sim}{3} \times \underset{\sim}{3}$. The fermion composed of three paraquarks of rank 3 is the color singlet, and the states of unconventional statistics are just single components of the color octet and decimet representations.

Thus it was borne in on us that the situation was very simple. Paraquarks with prohibition of unconventional statistics would yield baryons and mesons in agreement with observation and nothing crazy. Colored quarks with prohibition of nonsinglet color would yield a mathematically equivalent situation. We concluded that this must, in fact, be the right way to describe Nature.

That fall, everyone was talking about the demonstration, by means of anomalies, that the elementary perturbation theory formula for $\pi^\circ \to \gamma\gamma$ decay, more or less as given by Steinberger in 1950, when he was briefly a theorist at the Institute for Advanced Study in Princeton, must be correct to a good approximation (small pion–mass squared) provided the right elementary fermions are put into the diagram. With quarks, the decay rate comes out right if the factor 3 for color is inserted. In fact, one then gets the same result that Steinberger got by putting in the neutron and proton, because $1^2 - 0^2 = 3\left[(\frac{2}{3})^2 - (-\frac{1}{3})^2\right]$. We concluded that colored quarks (presumably with suppression of color nonsinglets) were correct.

Likewise the ratio R measured by the SLAC–LBL experiment on the cross section for e^+e^- annihilation is, to a good approximation, just the sum of the squares of the charges of the fundamental fermions, other than the electron, that are relevant up to the given energy. One gets a contribution $1^2 = 1$ from the muon, and, in the case of color, a contribution $3[(\frac{2}{3})^2 + (-\frac{1}{3})^2 + (-\frac{1}{3})^2] = 2$ from u, d, and s quarks. We noted that above the charm threshold one should add to the sum 3 an additional term $3[\frac{2}{3}]^2 = \frac{4}{3}$, giving $\frac{13}{3}$. We did not, of course, know about the contributions from the third family, starting with a term $1^2 = 1$ for the tau lepton.

Burton Richter of SLAC tried, in his Stockholm lecture, to bury this essentially correct prediction in a mess of irrelevant numbers, so as to make fun of theorists, but in fact *he* had been claiming that the ratio didn't level off at all, while we theorists subscribing to hidden color had the contribution per family correct.

In fact, color used in this way was not very well received in all circles. Some theorists at SLAC laughed at it, and, adding insult to injury, contrasted colored quarks with what they called "Gell-Mann and Zweig quarks"!

Fritzsch and I went on to construct, during the winter of 1971–72, a field theory of colored quarks and gluons. Although we were still ignorant of Nambu's 1966 work, we knew that it would be a good idea to have a Yang–Mills theory based on gauged SU(3) of color, with perfect conservation of color and perfect gauge invariance. Such a theory would be renormalizable and also compatible with the electroweak theory, with quark masses supplied in both cases by the same "soft-mass" mechanism. (I had advertised for such a mechanism, by the way, in my Caltech report of January 1961 on the Eightfold Way.) One of the most important virtues of describing the strong interaction by means of a Yang–Mills theory based on color, with electromagnetic and weak charges not color-dependent, is that the strong and electroweak Yang–Mills theories utilize separate internal variables and so do not clash with each other. Back in 1961, I had had to abandon my speculations about a flavor SU(3) Yang–Mills theory for the strong interaction, mainly because of such a clash.

Although enthusiastic about the beauty of this theory, we hesitated a bit in endorsing it in print, for three reasons:

1. We were worried about how to generate a nonzero trace of the stress–energy–momentum tensor in the limit of zero quark masses. We knew that such a nonzero trace was needed: the mass of the nucleon, unlike the masses of the lowest pseudoscalar mesons, had to be nonvanishing in that limit and scale invariance had to be broken. Somewhere there was a source of mass that would hold up as quark masses vanished.

Even without the explicit dimensional transmutation later demonstrated by Sidney Coleman and Erick Weinberg, it was easy to show that in such a theory the trace could be nonvanishing in the limit. John Ellis had been a visitor at Caltech during 1969–70, and he had lectured there on the possible generation of an anomalous trace, yielding what he called – appropriately for that era – POT, for partially zero trace. If I had remembered his work, I would not have troubled myself about the generation of mass from no mass.

2. We understood that some form of string theory (in terms of which the Veneziano model had just been reformulated) was the embodiment of the bootstrap, and in those days, of course, the bootstrap idea was thought to apply to hadrons alone rather than to all the elementary particles. Thus we thought at times that perhaps the Yang–Mills field theory of colored quarks and gluons ought to be replaced by some kind of related string theory. (Of course, it turns out that QCD structures behave like bags and, when they are elongated, somewhat like strings, but those are approximate and derived features of the theory, not fundamental ones.)

3. We didn't understand what was causing the suppression of color nonsinglets (i.e., the confinement of color or the "mathematical" character of quarks and gluons). We didn't know that it would follow from the color SU(3) Yang–Mills theory itself.

In the summer of 1972, I presented a paper on behalf of Fritzsch and myself at the International Conference on High Energy Physics, at a session chaired by David Gross. I discussed these ideas and, in the spoken version, emphasized the possibility of a Yang–Mills theory of SU(3) of color coupled to colored quarks, as well as the alternative possibility that fundamental colored strings might somehow be involved. In preparing the written version, unfortunately, we were troubled by the doubts just mentioned, and we retreated into technical matters.

Fritzsch and I continued to work on the Yang–Mills theory and its implications, and the next summer, together with H. Leutwyler, we composed in Aspen a letter on "The Advantages of the Color Octet Gluon Model."

We compared the Yang–Mills color octet gluons with the "throwaway" model theory with a single neutral gluon. We pointed out that the octet theory overcame, in the case of color, a fundamental difficulty noticed by Lev Okun in the singlet model, namely invariance of the interaction under SU($3n$), where n is the number of flavors.

We went on to discuss the asymptotic freedom that had recently been pointed out in the Yang–Mills theory by David Politzer and by Gross and Frank Wilczek (and also by Gerard 't Hooft, who may not have fully appreciated its significance). They used the "renormalization group" method that Francis Low and I had developed for QED. In connection with that method, he and I had shown that for a charged particle (the electron) in QED, the weight function of the propagator did not have to be positive, whereas for the neutral photon it did have to be, with the result that the strength of the force carried by the photon had to increase from the infrared (where it is the renormalized charge squared) toward the ultraviolet. In QCD, the gluon is itself charged, and the positivity of the weight function can therefore be violated, so that the strength of the coupling need not vary in the same way as for electromagnetism. In fact, the relevant parameter in perturbative Yang–Mills theory has the opposite sign to that of the corresponding parameter in QED, and so there is asymptotic freedom in the ultraviolet and the possibility of a confining potential in the infrared.

If such a confining potential does result, then that explains, as Gross and Wilczek remarked, why color nonsinglets are eliminated from the spectrum of particles that emerge singly; and that in turn explains why quarks are "mathematical."

Finally, we discussed the question of whether there is, in the limit of vanishing quark masses, conservation of the flavor-singlet axial vector current, which threatens to yield four light pseudoscalar mesons instead of three for SU(2) of flavor, and nine instead of eight for SU(3) of flavor, contrary to fact. That was an old preoccupation of mine:

$$\partial_\mu \mathcal{F}^5_{0\mu} \overset{?}{\to} 0 \text{ as masses} \to 0\,.$$

The theory seemed at first to have that difficulty;

BUT there is an anomaly proportional to $g^2 \epsilon_{\mu\nu\kappa\lambda} G^A_{\mu\nu} G^A_{\kappa\lambda}$, where G is the Yang–Mills field strength;

BUT the anomaly term itself is the divergence of a current $\mathcal{J}_\mu^5$, so

$$\partial_\mu(\mathcal{F}_{0\mu}^5 - \mathcal{J}_\mu^5) \to 0 \text{ as masses} \to 0\,,$$

which seems to revive the difficulty in another form;

BUT $\mathcal{J}_\mu^5$ is not gauge-invariant, so the difficulty does not seem serious.

Then there were two more BUTs (fortunately an even number!) that we did not cover in the letter:

BUT the *charge* $\int \mathcal{J}_0^5 d^3x$ does appear to be gauge-invariant, with

$$\frac{d}{dt}\Big(\int \mathcal{F}_{00}^5 d^3x - \int \mathcal{J}_0^5 d^3x\Big) \to 0 \text{ as masses} \to 0\,,$$

apparently gauge-invariant;

BUT, as was soon shown by Alexander Polyakov et al. and 't Hooft, in connection with instantons, this charge is locally but not globally gauge-invariant, so in fact there is no problem of a fourth or ninth light pseudoscalar boson.

Meanwhile, the asymptotic freedom of QCD not only suggested that there could be a corresponding "infrared slavery" that would yield confinement of colors, but at the same time it gave directly an explanation of the so-called parton model, which amounted to assuming that quarks (and antiquarks and gluons) had a weakened interaction at short distances or large momentum transfers. Again, Gross and Wilczek discussed this important point in their paper.

The theory had many virtues and no known vices. It was during a subsequent summer at Aspen that I invented the name quantum chromodynamics, or QCD, for the theory and urged it upon Heinz Pagels and others. Feynman continued to believe, for a while, that the "parton" picture was something other than an approximation to QCD, but finally, at the Irvine meeting in December of 1975, he admitted that it was nothing else but that.

The mathematical consequences of QCD have still not been properly extracted, and so, although most of us are persuaded that it is the correct theory of hadronic phenomena, a really convincing proof still requires more work. It may be that it would be helpful to have some more satisfactory method of truncating the theory, say by means of collective coordinates, than is provided by the brute-force lattice gauge theory approximation!

Effective Complexity

Murray Gell-Mann
Seth Lloyd

It would take a great many different concepts—or quantities—to capture all of our notions of what is meant by complexity (or its opposite, simplicity). However, the notion that corresponds most closely to what we mean by complexity in ordinary conversation and in most scientific discourse is "effective complexity." In nontechnical language, we can define the effective complexity (EC) of an entity as the length of a highly compressed description of its regularities [6, 7, 8].

For a more technical definition, we need a formal approach both to the notion of minimum description length and to the distinction between regularities and those features that are treated as random or incidental.

We can illustrate with a number of examples how EC corresponds to our intuitive notion of complexity. We may call a novel complex if it has a great many different characters, scenes, subplots, and so on, so that the regularities of the novel require a long description. The United States tax code is complex, since it is very long and each rule in it is a regularity. Neckties may be simple, like those with regimental stripes, or complex, like some of those designed by Jerry Garcia.

From time to time, an author presents a supposedly new measure of complexity (such as the "self-dissimilarity" of Wolpert and Macready [17]) without recognizing that when carefully defined it is just a special case of effective complexity.

Like some other concepts sometimes identified with complexity, the EC of an entity is context-dependent, even subjective to a considerable extent. It depends on the coarse graining (level of detail) at which the entity is described, the language used to describe it, the previous knowledge and understanding that are assumed, and, of course, the nature of the distinction made between regularity and randomness.

Like other proposed "measures of complexity," EC is most useful when comparing two entities, at least one of which has a large value of the quantity in question.

Now, how do we distinguish regular features of an entity from ones treated as random or incidental? There is, as we shall see, a way to make a nearly absolute distinction between the two kinds of features, but that approach is of limited usefulness because it always assigns very low values of EC, attributing almost all information content to the random category rather than the regular one.

In most practical cases, the distinction between regularity and randomness—or between regular and random information content—depends on some judgment of what is important and what is unimportant, even though the judge need not be human or even alive.

Take the case of neckties, as discussed above. We tacitly assumed that effective complexity would refer to the pattern of the tie, while wine stains, coffee stains, and so on, would be relegated to the domain of the random or incidental. But suppose we are dry cleaners. Then the characteristics of the stains might be the relevant regularities, while the pattern is treated as incidental.

Often, regularity and randomness are envisaged as corresponding to signal and noise, respectively, for example in the case of music and static on the radio. But, as is well known, an investigation of sources of radio static by Karl Jansky et al. (at Bell Telephone Laboratories in the 1930s) revealed that one of those sources lies in the direction of the center of our galaxy, thus preparing the way for radio astronomy. Part of what had been treated as random turned into a very important set of regularities.

It is useful to encode the description of the entity into a bit string, even though the choice of coding scheme introduces another element of context dependence. For such strings we can make use of the well-known concept of algorithmic information content (AIC), which is a kind of minimum description length.

The AIC of a bit string (and, hence, of the entity it describes) is the length of the shortest program that will cause a given universal computer U to print out the string and then halt [3, 4, 11]. Of course, the choice of U introduces yet another form of context dependence.

For strings of a particular length, the ones with the highest AIC are those with the fewest regularities. Ideally they have no regularities at all except the length. Such strings are sometimes called "random" strings, although the terminology does not agree precisely with the usual meaning of random (stochastic, especially with equal probabilities for all alternatives). Some authors call AIC "algorithmic complexity," but it is not properly a measure of complexity, since randomness is not what we usually mean when we speak of complexity. Another name for AIC, "algorithmic randomness," is somewhat more apt.

Now we can begin to construct a technical definition of effective complexity, using AIC (or something very like it) as a minimum description length. We split the AIC of the string representing the entity into two terms, one for regularities and the other for features treated as random or incidental. The first term is then the effective complexity, the minimum description length of the regularities of the entity [8].

It is not enough to define EC as the AIC of the regularities of an entity. We must still examine how the regularities are described and distinguished from features treated as random, using the judgment of what is important. One of the best ways to exhibit regularities is the method used in statistical mechanics, say, for a classical sample of a pure gas. The detailed description of the positions and momenta of all the molecules is obviously too much information to gather, store, retrieve, or interpret. Instead, certain regularities are picked out. The entity considered—the real sample of gas—is embedded conceptually in a set of comparable samples, where the others are all imagined rather than real. The members of the set are assigned probabilities, so that we have an ensemble. The entity itself must be a typical member of the ensemble (in other words, not one with abnormally low probability). The set and its probability distribution will then reflect the regularities.

For extensive systems, the statistical-mechanical methods of Boltzmann and Gibbs, when described in modern language, amount to using the principle of maximum ignorance, as emphasized by Jaynes [9]. The ignorance measure or Shannon information I is introduced. (With a multiplicative constant, I is the entropy.) Then the probabilities in the ensemble are varied and I is maximized subject to keeping fixed certain average quantities over the ensemble. For example, if the average energy is kept fixed—and nothing else—the Maxwell-Boltzmann distribution of probabilities results.

We have, of course,

$$I = -\sum_r P_r \log P_r \,, \tag{1}$$

where log means logarithm to the base 2 and the P's are the (coarse-grained) probabilities for the individual members r of the ensemble. The multiplicative constant that yields entropy is $k \ln 2$, where k is Boltzmann's constant.

In this situation, with one real member of the ensemble and the rest imagined, the fine-grained probabilities are all zero for the members of the ensemble other than e, the entity under consideration (or the bit string describing it). Of course, the fine-grained probability of e is unity. The typicality condition previously mentioned is just

$$-\log P_e \lesssim I\,. \tag{2}$$

Here the symbol "$\lesssim$" means "less than or equal" to within a few bits.

We can regard the quantities kept fixed (while I is maximized) as the things judged to be important. In most problems of statistical mechanics, these are, of course, the averages of familiar extensive quantities such as the energy. The choice of quantities controls the regularities expressed by the probability distribution.

In some problems, the quantities being averaged have to do with membership in a set. (For example, in Gibbs's microcanonical ensemble, we deal with the set of states having energies in a narrow interval.) In such a case, we would make use of the membership function, which is one for members of the set and zero otherwise. When the average of that function over all the members of the ensemble is one, every member with nonzero probability is in the set.

In discussing an ensemble E of bit strings used to represent the regularities of an entity, we shall apply a method that incorporates the maximizing of ignorance subject to constraints. We introduce the AIC of the ensemble and call it Y. We then have our technical definition of effective complexity: it is the value of Y for the ensemble that is finally employed. In general, then, Y is a kind of candidate for the role of effective complexity.

Besides $Y = K(E)$, the AIC of the ensemble E (for a given universal computer U), we can also consider $K(r|E)$, the contingent AIC of each member r given the ensemble. The weighted average, with probabilities P_r, of this contingent AIC can be related to I in the following way.

We note that Rüdiger Schack [15] has discussed converting any universal computer U into a corresponding U' that incorporates an efficient recoding scheme (Shannon-Fano coding). Such a scheme associates longer bit strings with less probable members of the ensemble and shorter ones with more probable members. Schack has then shown that if K is defined using U', then the average contingent AIC of the members lies between I and $I+1$. We shall adopt his procedure and thus have

$$\sum_r P_r K(r|E) \approx I\,, \tag{3}$$

where $\approx$ means equal to within a few bits (here actually one bit).

Let us define the total information Σ as the sum of Y and I. The first term is, of course, the AIC of the ensemble and we have seen that the second is, to within a bit, the average contingent AIC of the members given the ensemble.

To throw some light on the role of the total information, consider the situation of a theoretical scientist trying to construct a theory to account for a large body of data. Suppose the theory can be represented as a probability distribution over a set of bodies of data, one of which consists of the real data and the rest of which are imagined. Then Y corresponds to the complexity of the theory and I measures the extent to which the predictions of the theory are distributed widely over different possible bodies of data. Ideally, the theorist would like both quantities to be small, the first so as to make the theory simple and the second so as to make it focus narrowly on the real data. However, there may be trade-offs. By adding bells and whistles to the theory, along with a number of arbitrary parameters, one may be able to focus on the real data, but at the expense of complicating the theory. Similarly, by allowing appreciable probabilities for very many possible bodies of data, one may be able to get away with a simple theory. (Occasionally, of course, a theorist is fortunate enough to be able to make both Y and I small, as James Clerk Maxwell did in the case of the equations for electromagnetism.) In any case, the first desideratum is to minimize the sum of the two terms, the total information Σ. Then one can deal with the possible trade-offs.

We shall show that to within a few bits the smallest possible value of Σ is $K \equiv K(e)$, the AIC of the string representing the entity itself. Here we make use of the typicality condition (2) that the log of the (coarse-grained) probability for the entity is less than or equal to I to within a few bits. We also make use of certain abstract properties of the AIC:

$$K(A) \lesssim K(A,B) \tag{4}$$

and

$$K(A,B) \lesssim K(B) + K(A|B)\,, \tag{5}$$

where again the symbol $\lesssim$ means "less than or equal to" up to a few bits. A true information measure would, of course, obey the first relation without the caveat "up to a few bits" and would obey the second relation as an equality.

Because of efficient recoding, we have

$$K(e|E) \lesssim -\log P_e\,. \tag{6}$$

We can now prove that $K = K(e)$ is an approximate lower bound for the total information $\Sigma = K(E) + I$:

$$\begin{aligned} K &= K(e) \lesssim K(e,E)\,, && \text{(7a)}\\ K(e,E) &\lesssim K(E) + K(e|E)\,, && \text{(7b)}\\ K(e|E) &\lesssim -\log P_e\,, && \text{(7c)}\\ -\log P_e &\lesssim I\,. && \text{(7d)} \end{aligned}$$

We see, too, that when the approximate lower bound is achieved, all these approximate inequalities become approximate equalities:

$$\begin{aligned} K &\approx K(e,E)\,, && \text{(8a)}\\ K(e,E) &\approx Y + K(e|E)\,, && \text{(8b)}\\ K(e|E) &\approx -\log P_e\,, && \text{(8c)}\\ -\log P_e &\approx I\,. && \text{(8d)} \end{aligned}$$

The treatment of this in Gell-Mann and Lloyd [8] is slightly flawed. The approximate inequality (7b), although given correctly, was accidentally replaced later on by an approximate equality, so that condition (8b) came out as a truism. Thus (8b) was omitted from the list of new conditions that hold when the total information achieves its approximate lower bound. As a result, we gave only three conditions of approximate equality instead of the four quoted here in (8a)–(8d).

Also, in the discussion at the end of the paragraph preceding eq. (2) of Gell-Mann and Lloyd [8], we wrote $\log K_U(a)$ by mistake in place of $\log K_U(b) + 2 \log\log K_U(b)$, but that does not affect any of our results.

Clearly the total information Σ achieves its approximate minimum value K for the singleton distribution, which assigns probability one to the bit string representing our entity and zero probabilities to all other strings. For that distribution, Y is about equal to K and the measure of ignorance I equals zero.

There are many other distributions for which $\Sigma \approx K$. If we plot Y against I, the line along which $Y + I = K$ is a straight line with slope minus one, with the singleton at the top of the line. We are imposing on the ensemble—the one that we actually use to define the effective complexity—the condition that the total information approximately achieve its minimum. In other words, we want to stay on the straight line or within a few bits of it.

All ensembles of which e is a typical member lie, to within a few bits, above and to the right of a boundary. That boundary coincides with our straight line all the way from the top down to a certain point, where we run out of ensembles that have $Y + I \approx K$. Below that point the actual boundary for ensembles in the $Y - I$ plane no longer follows the straight line but veers off to the right.

Now, as we discussed, we maximize the measure of ignorance I subject to staying on that straight line. If we do that and impose no other conditions, we end up at the point where the boundary in the $I - Y$ plane departs from the straight line. As described in the paper of Gell-Mann and Lloyd (who are indebted to Charles H. Bennett for many useful discussions of this manner), that point always corresponds to an effective complexity Y that is very small. If we imposed no other conditions, every entity would come out simple! In certain circumstances, that is all right, but for most problems it is an absurd result. What went wrong? The answer is that, as in statistical mechanics, we must usually impose some more conditions, fixing the values of certain average quantities treated as important by a judge. If we maximize I subject to staying (approximately) on the straight line and to keeping those values fixed, we end up with a meaningful effective complexity, which can be large in appropriate circumstances.

The situation is made easier to discuss if we narrow the universe of possible ensembles in a drastic manner suggested by Kolmogorov, one of the inventors (or discoverers?) of AIC, in work reviewed in the books by Cover and Thomas [4] and by Li and Vitányi [11]. Instead of using arbitrary probability distributions over the space of all bit strings, one restricts the ensembles to those obeying two conditions. The set must contain only strings of the same length as the original bit string and all the nonzero probabilities must be equal. In this simplified situation, every allowable ensemble can be fully characterized as a subset of the set of all bit strings that have the same length as the original one. Here I is just the logarithm of the number of members of the subset. Also, being a typical member of the ensemble simply means belonging to the subset.

Vitányi and Li describe how, for this model problem, Kolmogorov suggested maximizing I subject only to staying on the straight line. In that case, as pointed out above, one is led immediately to the point in the $I - Y$ plane where the boundary departs from the straight line. Kolmogorov called the value of Y at that point the "minimum sufficient statistic." His student L. A. Levin (now a professor at Boston University) kept pointing out to him that this "statistic" was always small and therefore of limited utility, but the great man paid insufficient attention [10].

In the model problem, the boundary curve comes near the I axis at the point where I achieves its maximum, the string length l. At that point the subset is the entire set of strings of the same length as the one describing the entity e. Clearly, that set has a very short description and thus a very small value of Y.

What should be done, whether in this model problem or in the more general case that we discussed earlier, is to utilize the lowest point on the straight line such that the average quantities judged to be important still have their fixed values. Then Y no longer has to be tiny and the measure of ignorance I can be much less than it was for the case of no further constraints.

We have succeeded, then, in splitting K into two terms, the effective complexity and the measure of random information content, and they are equal to the values of Y and I, respectively, for the chosen ensemble. We can think of the separation of K into Y and I in terms of a distinction between a basic program (for printing out the string representing our entity) and data fed into that basic program.

We can also treat as a kind of coarse graining the passage from the original singlet distribution (in which the bit string representing the entity is the only member with nonzero probability) to an ensemble of which that bit string is a typical member. In fact, we have been labeling the probabilities in each ensemble as coarse-grained probabilities P_r. Now it often happens that one ensemble can be regarded as a coarse graining of another, as was discussed in Gell-Mann and Lloyd [8]. We can explore that situation here as it applies to ensembles that lie on or very close to the straight line $Y + I = K$.

We start from the approximate equalities (8a)–(8d) (accurate to within a few bits) that characterize an ensemble on or near the straight line. There the coarse graining acts on initial "singleton" probabilities that are just one for the original string and zero for all others. We want to generalize the above formulae to the case of an ensemble with any initial fine-grained probability distribution $p \equiv \{p_r\}$, which gets coarse grained to yield another ensemble with probability distribution $P \equiv \{P_r\}$ and approximately the same value of Σ. We propose the following formulae as the appropriate generalizations:

$$\begin{aligned} K(p) &\approx K(p,P)\,, && \text{(9a)}\\ K(p,P) &\approx K(P)+K(p|P)\,, && \text{(9b)}\\ K(p|P) &\approx -\Sigma_r p_r \log P_r + \Sigma_r p_r \log p_r\,, && \text{(9c)}\\ -\Sigma_r p_r \log P_r &\approx -\Sigma_r P_r \log P_r\,. && \text{(9d)} \end{aligned}$$

These equations reduce to (8a) through (8d) respectively for the case in which the fine-grained distribution is the "singleton" distribution. Also, it is easy to see that Σ is approximately conserved by these approximate equalities, as a result of our including the last term in eq. (9c).

Equation (9a) tells us that, to within a few bits, the coarse-grained probability distribution P contains only algorithmic information that is in the fine-grained distribution p. Equation (9b) tells us that the ordinary relation between joint and conditional mutual information holds here to within a few bits even though that relation does not always hold for joint and conditional *algorithmic* information.

We can compare this discussion of coarse graining to the treatment in Gell-Mann and Lloyd [8]. There we required three properties of a coarse-graining transformation from p to P: that the transformation actually yield a probability distribution, that if iterated it produce the same set of P's, and that it obey eq. (9d) above. We attained these objectives by maximizing the ignorance associated with the P's while keeping some averages involving the P's equal to the corresponding averages involving the p's (linear constraint conditions).

Here we emphasize that we are generalizing that work to the case where Y is introduced and the sum of Y and I is kept approximately fixed at its minimum value while we maximize I subject to some constraint conditions linear in the probabilities.

Say we start with the singleton ensemble in which only the original string has a nonzero probability and move down the straight line in a succession of coarse grainings until we reach the ensemble for which Y is the effective complexity. The above equations are then applied over and over again for the successive coarse grainings, and they apply also between the original (singleton) probability distribution and the final one.

Alternatively, we can, if we like, regard the transition from P to p as a fine graining, using the same formulae. We can start at the point where the boundary curve departs from the straight line and move up the line in a sequence of fine grainings. In fact, we can utilize the linear constraints successively. We apply first one of them, then that one and another, then those two and a third, and so forth, until all the constraints have been applied to the maximization of I subject to staying on the straight line. Each additional constraint yields a fine graining.

There are at least four issues that we feel require discussion at this point, even though many questions about them remain. Two of these issues relate to certain generalizations of the notion of algorithmic information content.

AIC as it stands is technically uncomputable, as shown long ago by Chaitin [3]. That is not so if we modify the definition by introducing a finite maximum execution time T within which the program must cause the modified universal computer U' to print out the bit string. Such a modification has another, more important advantage. We can vary T and, thus, explore certain situations where apparent complexity is large but effective complexity as defined above (for $T \to \infty$) is small.

Take the example [6] of energy levels of heavy nuclei. Fifty years ago, it seemed that any detailed explanation of the pattern involved would be extremely long and complicated. Today, however, we believe that an accurate calculation of the positions of all the levels is possible, in principle, using a simple theory: QCD, the quantum field theory of quarks and gluons, combined with QED, the quantum field theory of photons and electromagnetic interactions, including those of quarks. Thus, for T very large or infinite, the modified AIC of the levels is small—they are simple. But the computation time required is too long to permit the calculations to be performed using existing hardware and software. Thus, for moderate values of T the levels appear complex.

In such a case, the time around which the modified AIC declines from a large value to a small one (as T increases) is related to "logical depth" as defined by Charles H. Bennett [2]. Roughly, logical depth is the time (or number of steps) necessary for a program to cause U to print out the coded description of an entity and then halt, averaged over programs in such a way as to emphasize short ones.

There are cases where the modified AIC declines, as T increases, in a sequence of steps or plateaus. In that case we can say that certain kinds of regularities are buried more deeply than others.

While it is very instructive to vary T in connnection with generalizing K—the AIC of the bit string describing our entity—we encounter problems if we try to utilize a finite value of T in our whole discussion of breaking up K into effective complexity and random information. Not all the theorems that allow us to treat AIC as an approximate information measure apply to the generalization with variable T.

In addition to logical depth, we can utilize a quantity that is, in a sense, inverse to it, namely Bennett's "crypticity," [2] which is, in rough terms, the time necessary to go from the description of an entity to a short program that yields that description. As an example of a situation where crypticity is important, consider a discussion of pseudorandomness. These days, when random numbers are called for in a calculation, one often uses instead a random-looking sequence of numbers produced by a deterministic process. Such a pseudorandom sequence typically has a great deal of crypticity. A lengthy investigation of the sequence could reveal its deterministic nature and, if it is generated by a short program, could correctly assign to it a very low AIC. Given only a modest time, however, we could fail to identify the sequence as one generated by a simple deterministic process and mistake it for a truly random sequence with a high value of AIC.

The concept of crypticity can also be usefully applied to situations where a bit string of modest AIC appears to exhibit large AIC in the form of effective complexity rather than random information. We might call such a string "pseudocomplex." An example of a pseudocomplex string would be one recording an image, at a certain scale, of the Mandelbrot set. Another would be an apparently complex pattern generated by a simple cellular automaton from a simple initial condition. Note that a pseudorandom string, which has passed many of the usual statistical tests for randomness, is not appreciably compressed by conventional data compression algorithms, such as the one known as LZW [4]. By contrast, a pseudocomplex string typically possesses a large number of obvious statistical regularities and is, therefore, readily compressible to some extent by LZW, but not all the way to the very short program that actually generated the string.

We should mention that a number of authors have considered mutual information as a measure of complexity in the context of dynamical systems [1, 5, 12]. Without modification, that idea presents a conflict with our intuitive notion of complexity. Consider two identical very long bit strings consisting entirely of ones. The mutual information between them is very large, yet each is obviously very simple. Moreover, the statement that they are the same is also very simple. The pair of strings is not at all complex in any usual sense of the word.

Typically, the authors in question have recognized that a more acceptable quantity in a discussion of complexity is mutual algorithmic information, defined for two strings as the sum of their AIC values minus the AIC of the two taken together. If two strings are simple and identical, though very long, their mutual AIC is small.

Of course, identical long strings could be "random," in which case their very large mutual algorithmic information does not correspond to what we usually mean by complexity. EC is still the best measure of complexity.

We can easily generalize the definition of mutual information to the case of any number of strings (or entities described by them). For example, for three strings we have

$$K_{\text{mut}} = K(1) + K(2) + K(3) - K(1,2) - K(2,3) - K(1,3) + K(1,2,3)\,. \quad (10)$$

Under certain conditions we can see a connection between mutual algorithmic information and effective complexity. For example, suppose we are presented not with a single entity but with N entities that are selected at random from among the typical members of a particular ensemble. The mutual algorithmic information content among these entities is then a good estimate of the AIC of the ensemble from which they are selected, and that quantity is, under suitable conditions, equal to the effective complexity candidate Y attributed to each of the entities.

The way the calculation goes is roughly the following. On average the K value for m arguments is approximately $Y + mI$, and the sum in eq. (10) then comes out equal to Y. It is easily shown that such an equality yielding Y holds not just for three entities but for any number N, with the appropriate generalization of eq. (10). The elimination of the I term produces the connection of K_{mut} with the effective complexity candidate.

At last we arrive at the questions relevant to a nontraditional measure of ignorance. Suppose that for some reason we are dealing, in the definition of I, not with the usual measure given in eq. (1), but rather with the generalization discussed in this volume, namely

$$I_q = -\frac{[\Sigma_r(P_r)^q - 1]}{(q-1)}, \tag{11}$$

which reduces to eq. (1) in the limit where q approaches 1. Should we be maximizing this measure of ignorance—while keeping certain average quantities fixed—in order to arrive at a suitable ensemble? (Presumably we average using not the probabilities P_r but their qth powers normalized so as to sum to unity—the so-called Escort probabilities.) Do we, while maximizing I, keep a measure of total information at its minimum value? Is a nonlinear term added to $I + Y$? What happens to the lower bound on $I + Y$? Can we make appropriate changes in the definition of AIC that will preserve or suitably generalize the relations we discuss here? What happens to the approximate equality of I and the average contingent AIC (given the ensemble)? What becomes of the four conditions in eqs. (8a) to (8d)? What happens to the corresponding conditions (9a) to (9d) for the case where we are coarse graining one probability distribution and thus obtaining another one?

As is well known, a kind of entropy based on the generalized information or ignorance of eq. (11) has been suggested [16] as the basis for a full-blown alternative, valid for certain situations, to the "thermostatistics" (thermodynamics and statistical mechanics) of Boltzmann and Gibbs. (The latter is, of course, founded on eq. (1) as the formula for information or ignorance.) Such a basic interpretation of eq. (11) has been criticized by authors such as Luzzi et al. [13] and Nauenberg [14]. We do not address those criticisms here, but should they prove justified—in whole or in part—they need not rule out, at a practical level, the applicability of eq. (11) to a variety of cases, such as systems of particles attracted by $1/r^2$ forces or systems at the so-called "edge of chaos."

ACKNOWLEDGMENTS

This research was supported by the National Science Foundation under the Nanoscale Modeling and Simulation initiative. In addition, the work of Murray Gell-Mann was supported by the C.O.U.Q. Foundation and by Insight Venture Management. The generous help provided by these organizations is gratefully acknowledged.

REFERENCES

[1] Adami, C., C. Ofria, and T. C. Collier. "Evolution of Biological Complexity." *PNAS (USA)* **97** (2000): 4463–4468.

[2] Bennett, C. H. "Dissipation, Information, Computational Complexity and the Definition of Organization." In *Emerging Syntheses in Science*, edited by D. Pines, 215–234. Santa Fe Institute Studies in the Sciences of Complexity, Proc. Vol. I. Redwood City: Addison-Wesley, 1987.

[3] Chaitin, G. J. *Information, Randomness, and Incompleteness.* Singapore: World Scientific, 1987.

[4] Cover, T. M., and J. A. Thomas. *Elements of Information Theory.* New York: Wiley, 1991.

[5] Crutchfield, J. P., and K. Young. "Inferring Statistical Complexity." *Phys. Rev. Lett.* **63** (1989): 105–108.

[6] Gell-Mann, M. *The Quark and the Jaguar.* New York: W. H. Freeman, 1994.

[7] Gell-Mann, M. "What is Complexity?" *Complexity* **1/1** (1995): 16–19.

[8] Gell–Mann, M., and S. Lloyd. "Information Measures, Effective Complexity, and Total Information." *Complexity* **2/1** (1996): 44–52.

[9] Jaynes, E. T. *Papers on Probability, Statistics and Statistical Physics*, edited by R. D. Rosenkrantz. Reidel: Dordrecht, 1982.

[10] Levin, L. A. Personal communication, 2000.

[11] Li, M., and P. M. B. Vitanyi. *An Introduction to Kolmogorov Complexity and Its Applications.* New York: Springer-Verlag, 1993.

[12] Lloyd, S., and H. Pagels. "Complexity as Thermodynamic Depth." *Ann. Phys.* **188** (1988): 186–213.

[13] Luzzi, R., A. R. Vasconcellos, and J. G. Ramos. "On the Question of the So-Called Non-Extensive Thermodynamics." IFGW-UNICAMP Internal Report, Universidade Estadual de Campinas, Campinas, Sao Paulo, Brasil, 2002.

[14] Nauenberg, M. "A Critique of Nonextensive q-Entropy for Thermal Statistics. Dec. 2002. lanl.gov e-Print Archive, Quantum Physics, Cornell University. ⟨http://eprints.lanl.gov/abs/cond-mat/0210561⟩.

[15] Schack, R. "Algorithmic Information and Simplicity in Statistical Physics." *Intl. J. Theor. Phys.* **36** (1997): 209–226.

[16] Tsallis, C. "Possible Generalization of Boltzmann-Gibbs Statistics." *J. Stat. Phys.* **52** (1988): 479–487.

[17] Wolpert, D. H., and W. G. Macready. "Self-Dissimmilarity: An Empirically Observable Measure of Complexity." In *Unifying Themes in Complex Systems: Proceedings of the First NECSI International Conference*, edited by Y. Bar-Yam, 626–643. Cambridge, Perseus, 2002.

PNAS PNAS PNAS PNAS

Asymptotically scale-invariant occupancy of phase space makes the entropy S_q extensive

Constantino Tsallis*[†‡], Murray Gell-Mann*[‡], and Yuzuru Sato*

*Santa Fe Institute, 1399 Hyde Park Road, Santa Fe, NM 87501; and [†]Centro Brasileiro de Pesquisas Físicas, Rua Xavier Sigaud 150, 22290-180 Rio de Janeiro, Brazil

Contributed by Murray Gell-Mann, July 25, 2005

Phase space can be constructed for N equal and distinguishable subsystems that could be probabilistically either *weakly* correlated or *strongly* correlated. If they are locally correlated, we expect the Boltzmann–Gibbs entropy $S_{BG} \equiv -k \sum_i p_i \ln p_i$ to be *extensive*, i.e., $S_{BG}(N) \propto N$ for $N \to \infty$. In particular, if they are independent, S_{BG} is *strictly additive*, i.e., $S_{BG}(N) = NS_{BG}(1)$, $\forall N$. However, if the subsystems are globally correlated, we expect, for a vast class of systems, the entropy $S_q \equiv k[1 - \sum_i p_i^q]/(q-1)$ (with $S_1 = S_{BG}$) for some special value of $q \neq 1$ to be the one which is extensive [i.e., $S_q(N) \propto N$ for $N \to \infty$]. Another concept which is relevant is strict or asymptotic *scale-freedom* (or *scale-invariance*), defined as the situation for which all marginal probabilities of the N-system coincide or asymptotically approach (for $N \to \infty$) the joint probabilities of the $(N-1)$-system. If each subsystem is a binary one, scale-freedom is guaranteed by what we hereafter refer to as the *Leibnitz rule*, i.e., the sum of two successive joint probabilities of the N-system coincides or asymptotically approaches the corresponding joint probability of the $(N-1)$-system. The kinds of interplay of these various concepts are illustrated in several examples. One of them justifies the title of this paper. We conjecture that these mechanisms are deeply related to the very frequent emergence, in natural and artificial complex systems, of scale-free structures and to their connections with nonextensive statistical mechanics. Summarizing, we have shown that, for asymptotically scale-invariant systems, it is S_q with $q \neq 1$, and not S_{BG}, the entropy which matches standard, clausius-like, prescriptions of classical thermodynamics.

The entropy S_q (1) is defined through[§¶]

$$S_q \equiv k \frac{1 - \sum_{i=1}^{W} p_i^q}{q-1} \quad \left(q \in \mathcal{R};\ S_1 = S_{BG} \equiv -k \sum_{i=1}^{W} p_i \ln p_i\right). \tag{1}$$

where k is a positive constant ($k = 1$ from now on) and BG stands for Boltzmann–Gibbs. This expression is the basis of *nonextensive statistical mechanics* (16–18) (see http://tsallis.cat.cbpf.br/biblio.htm for a regularly updated bibliography), a current generalization of BG statistical mechanics. For $q \neq 1$, S_q is nonadditive (hence nonextensive) in the sense that for a system composed of (probabilistically) *independent* subsystems, the total entropy differs from the sum of the entropies of the subsystems. However, the system may have special probability correlations between the subsystems such that extensivity is valid, not for S_{BG}, but for S_q with a particular value of the index $q \neq 1$. In this paper, we address the case where the subsystems are all equal and distinguishable. Their correlations may exhibit a kind of scale-invariance. We may regard some of the situations of correlated probabilities as related to the remark (see refs. 19–23 and references therein) that S_q for $q \neq 1$ can be appropriate for nonlinear dynamical systems that have phase space unevenly occupied. We return to this point later.

We shall consider two types of models. The first one involves N binary variables ($N = 1, 2, 3, \ldots$), and the second one involves N continuous variables ($N = 1, 2, 3$). In both cases, certain correlations that are scale-invariant in a suitable limit can create an intrinsically inhomogeneous occupation of phase space. Such systems are strongly reminiscent of the so called scale-free networks (24, 25), with their hierarchically structured hubs and spokes and their nearly forbidden regions.

Discrete Models

Some Basic Concepts. The most general probabilistic sets for N equal and distinguishable binary subsystems are given in Fig. 1 with

$$\sum_{n=0}^{N} \frac{N!}{(N-n)!\,n!}\, \pi_{N,n} = 1$$

$$(\pi_{N,n} \in [0, 1];\ N = 1, 2, 3, \ldots;\ n = 0, 1, \ldots, N). \tag{2}$$

Let us from now on call *Leibnitz rule* the following recursive relation:

$$\pi_{N,n} + \pi_{N,n+1} = \pi_{N-1,n}\ (n = 0, 1, \ldots, N-1;\ N = 2, 3, \ldots). \tag{3}$$

This relation guarantees what we refer to as *scale-invariance* (or *scale-freedom*) in this article. Indeed, it guarantees that, for any value of N, the associated *joint probabilities* $\{\pi_{N,n}\}$ produce *marginal probabilities* which coincide with $\{\pi_{N-1,n}\}$. Assuming $\pi_{10} + \pi_{11} = 1$, and taking into account that the Nth row has one more element than the $(N-1)$th row, a particular model is characterized by giving *one* element for each row. We shall adopt the convention of specifying the set $\{\pi_{N,0} \in [0, 1], \forall N\}$. Everything follows from it. There are many sets $\{\pi_{N,0}\}$ that satisfy Eq. **3**. Let us illustrate with a few simple examples:

(*i*) $\pi_{N,0} = (2\pi_{10})^N/N + 1$ $(0 \leq \pi_{10} \leq 1/2;\ N = 1, 2, 3, \ldots)$. We have that all 2^N states have nonzero probability if $0 < \pi_{10} \leq 1/2$.

[‡]To whom correspondence may be addressed. E-mail: tsallis@santafe.edu or mgm@santafe.edu.

[§]In the field of cybernetics and control theory, the form $S_\alpha = 2^{\alpha-1}/2^{\alpha-1} - 1(1 - \sum_i p_i^\alpha)$ was introduced in ref. 2, and was further discussed in ref. 3. With a different prefactor, it was rediscovered in ref. 4, and further commented in ref. 5. More historical details can be found in refs. 6–8. This type of entropic form was rediscovered once again in 1988 (16–18) and it was postulated as the basis of a possible generalization of Boltzmann–Gibbs statistical mechanics, nowadays known as *nonextensive statistical mechanics*.

[¶]Many entropic forms are related with S_q. A special mention is deserved by the Renyi entropy $S_q^R \equiv (\ln \sum_i p_i^q)/(1-q) = \ln[1 + (1-q)S_q]/(1-q)$, and by the Landsberg–Vedral–Abe–Rajagopal entropy (or just *normalized S_q entropy*) $S_q^{LVAR} \equiv S_q/\sum_{i=1}^{W} p_i^q = [1 - (\sum_{i=1}^{W} p_i^q)^{-1}]/(1-q) = S_q/[1 + (1-q)S_q]$. The Renyi entropy was, according to ref. 9, first introduced in ref. 10, and then in ref. 11. The Landsberg–Vedral–Abe–Rajagopal entropy was independently introduced in ref. 12 and in ref. 13. Both S_q^R and S_q^{LVAR} are monotonic functions of S_q; consequently, under identical constraints, they are all optimized by the same probability distribution. A two-parameter entropic form was introduced in ref. 14 which reproduces both S_q and Renyi entropy as particular cases. This scheme has been recently enlarged elegantly in ref. 15. S_{BG} and S_q (as well as a few other entropic forms that we do not address here) are concave and Lesche-stable for *all* $q > 0$, and provide a *finite* entropy production per unit time; S_q^R, S_q^{LVAR}, the Sharma–Mittal, and the Masi entropic forms (as well as others that we do not address here) violate all of these properties.

$$
\begin{array}{lc}
(N=0) & 1 \\
(N=1) & \pi_{10}\quad \pi_{11} \\
(N=2) & \pi_{20}\quad \pi_{21}\quad \pi_{22} \\
(N=3) & \pi_{30}\quad \pi_{31}\quad \pi_{32}\quad \pi_{33} \\
(N=4) & \pi_{40}\quad \pi_{41}\quad \pi_{42}\quad \pi_{43}\quad \pi_{44}
\end{array}
$$

Fig. 1. Most general sets of joint probabilities for N equal and distinguishable binary subsystems.

$$
\begin{array}{lc}
(N=0) & (1,1) \\
(N=1) & (1,1/2)\ (1,1/2) \\
(N=2) & (1,1/3)\ (2,1/6)\ (1,1/3) \\
(N=3) & (1,1/4)\ (3,1/12)\ (3,1/12)\ (1,1/4) \\
(N=4) & (1,1/5)\ (4,1/20)\ (6,1/30)\ (4,1/20)\ (1,1/5)
\end{array}
$$

Fig. 2. The left numbers within the parentheses correspond to Pascal triangle. The right numbers correspond to the Leibnitz harmonic triangle ($d = N$).

The particular case $\pi_{10} = 1/2$ recovers the original Leibnitz triangle itself (26) (see Fig. 2).

(*ii*) $\pi_{N,0} = (\pi_{10})^{N^\alpha}$ ($\alpha \geq 0$; $N = 1, 2, 3 \ldots$). The $\alpha = 1$ instance corresponds to independent systems, i.e., $\pi_{N,n} = (\pi_{10})^{N-n}(1 - \pi_{10})^n$. If $0 < \pi_{10} < 1$, then all 2^N states have nonzero probability. The $\alpha = 0$ instance corresponds to $\pi_{N,0} = \pi_{10}$, $\pi_{N,n} = 0$ ($n = 1, 2, \ldots, N-1$) and $\pi_{N,N} = 1 - \pi_{10}$. If $0 < \pi_{10} < 1$, then only two among the 2^N states have nonzero probability, $\forall N$, namely the states associated with $\pi_{N,0}$ and $\pi_{N,N}$.

We may relax the Leibnitz rule to some extent by considering those cases where the rule is satisfied only asymptotically, i.e.,

$$\lim_{N\to\infty} \frac{\pi_{N,n} + \pi_{N,n+1}}{\pi_{N-1,n}} = 1 \quad (n = 0, 1, 2, \ldots). \tag{4}$$

Such cases will be said to be not strictly but *asymptotically scale-invariant* (or *asymptotically scale-free*). This is, for a variety of reasons, the situation in which we are primarily interested. The main reason is that what vast classes of natural and artificial systems typically exhibit is not precisely power-laws, but behaviors *which only asymptotically become power-laws* (once we have corrected, of course, for any finite size effects). This is consistent with the fact that within nonextensive statistical mechanics S_q is optimized by q-exponential functions (see ref. 1 and references therein and refs. 27 and 28), which only asymptotically yield power-laws. It is consistent also with a new central limit theorem that has been recently conjectured (29) for specially correlated random variables.[‖]

Let us now introduce a further concept, namely *q-describability*. A model constituted by N equal and distinguishable subsystems will be called *q-describable* if a value of q exists such as $S_q(N)$ is *extensive*, i.e., $\lim_{N\to\infty} S_q(N)/N < \infty$. If that special value of q equals unity, this corresponds to the usual BG universality class. If that value of q differs from unity, we will have nontrivial universality classes. If the subsystems $\{A_i\}$ are not necessarily equal, the system is q-describable if an entropic index q exists such that $\lim_{N\to\infty} [S_q(A_1 + A_2 + \ldots + A_N)/\sum_{i=1}^N S_q(A_i)] < \infty$. It should be clear that we could equally well demand the extensivity of say S_{2-q} [or even of $S_{Q(q)}$, where $Q(q)$ is some monotonically decreasing function of q satisfying $Q(1) = 1$] instead of that of S_q. This would of course have the effect of having nontrivial solutions for $q > 1$ whenever we had solutions for $q < 1$ if the extensivity that was imposed was that of S_q.

Finally, let us point out that we might consider the subsystems of a probabilistic system to be either *strongly* (or *globally*) *correlated* or *weakly* (or "*locally*") *correlated*. The trivial case of *independence*, i.e., when the subsystems are *uncorrelated*, is of course a particular case of weakly correlated. Let us make these notions more precise. A system is weakly correlated if for every generic (different from zero and from unity) joint probability $\pi_{i_1 i_2 \ldots i_N}^{A_1+A_2+\cdots+A_N}$ a set of individual probabilities $\{\pi_{i_r}^{A_r}\}$ exists such that $\lim_{N\to\infty} (\pi_{i_1 i_2 \ldots i_N}^{A_1+A_2+\cdots+A_N})/\prod_{r=1}^N \pi_{i_r}^{A_r} = 1$. Otherwise, the system is said to be strongly correlated. The particular case of independence corresponds to

$$\pi_{i_r}^{A_r} = \Sigma_{i_1 i_2 \ldots i_{r-1} i_{r+1} \ldots i_N} \pi_{i_1 i_2 \ldots i_N}^{A_1+A_2+\cdots+A_N} \quad (r = 1, 2, \ldots, N).$$

If the subsystems are equal and binary, this definition becomes as follows: a system is weakly correlated if, for generic $\pi_{N,n}$, a probability p_0 exists such that $\lim_{N\to\infty} \pi_{N,n}/p_0^{N-n}(1 - p_0)^n = 1$. Otherwise the system is said to be strongly correlated. The particular case of independence corresponds to $p_0 = \pi_{10}$. In the present sense, weakly correlated systems could also be thought and referred to as *asymptotically uncorrelated*. The interplay of scale-invariance, q-describability, and global correlation is schematized in Fig. 3.

We have verified that all systems illustrated in *i* and *ii* above belong to the $q = 1$ class (see examples in Fig. 4). We next address $q \neq 1$ systems.

A Discrete Model That Is Not Asymptotically Scale-Invariant. Let us consider the probabilistic structure indicated in Fig. 5, where, for given N, only the $d + 1$ first elements are different from zero, with $d = 0, 1, 2, \ldots, N$.

As we see, $\pi_{N,n}^{(d)} = 0$ for $N \geq d + 1$ and $n = d + 1, d + 2,$

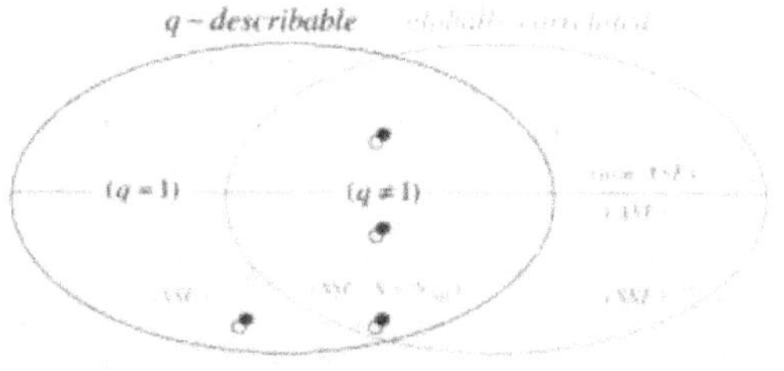

Fig. 3. Scheme representing the systems that are q-describable, globally correlated, asymptotically scale-free (*ASF*), and strictly scale-free (*SSF*). The $q = 1$ region corresponds to "locally" correlated systems. The Leibnitz rule is strictly satisfied for *SSF*, but only asymptotically satisfied for *ASF*. Below (above) the continuous red line we have the *ASF* (*non ASF*) systems. The *SSF* systems (below the dashed red line) constitute a subset of the *ASF* subset. The red spots correspond to the four families of discrete systems illustrated in the present paper: $q \neq 1$ *non ASF* (upper spot; Eqs. **12** and **14**); $q \neq 1$ *ASF* but *non SSF* (middle spot; Eqs. **17** and **24**); $q \neq 1$ *SSF* (right bottom spot; Eq. **8**); $q = 1$ *SSF* (left bottom spot; examples *i* and *ii* in the text).

‖On the basis of what we have called here the *Leibnitz rule*, L. G. Moyano, C.T., and M.G.-M. (44) obtained interesting preliminary numerical results based on the so called q-product (30, 31) and its relation to the possible q-generalization of the central limit theorem. More precisely, imposing the Leibnitz rule with $\pi_{N,0}^{-1} = p^{-1} \otimes_q p^{-1} \otimes_q \ldots \otimes_q p^{-1} = [N p^{q-1} - (N-1)]^{1/(1-q)}$ (with $\pi_{N,0} = p^N$ for $q = 1$), one verifies for $p = 1/2$ that, as N increases, the distribution probability appears to approach a q-generalized Gaussian $P(n, N)$. The centered and rescaled distribution $P(n, N)N/2$ gradually becomes (say for even N) proportional to $(1 - x^2)^{1/(1-q_{exp})}$, where $x = [n - (N/2)]/(N/2)$. Numerically, the exponent appears to satisfy $q_{exp} = 2 - (1/q)$. This relation is obtained by applying the $q \to (2 - q)$ transformation after the $q \to 1/q$ transformation (notice that this relation can be rewritten as $q = 1/(2 - q_{exp})$, which is the application of the same two transformations in the other possible order). The combinations of these two transformations define an interesting mathematical structure which might well be at the basis of the q-triplet conjectured in (32) and recently confirmed (33) with data received from the spacecraft Voyager 1 in the distant heliosphere. The q-triplet observed in the solar wind is given by $q_{sen} \simeq -0.6 \pm 0.2$, $q_{rel} = 3.8 \pm 0.3$, and $q_{stat} = 1.75 \pm 0.06$ (33). These values are consistent with $q_{rel} + (1/q_{sen}) = 2$ and $q_{stat} + (1/q_{rel}) = 2$, hence $1 - q_{sen} = [1 - q_{stat}]/[3 - 2q_{stat}]$. *Therefore, we expect only one q of the triplet to be independent.* The most precisely determined value in ref. 33 is $q_{stat} = 1.75 = 7/4$. It immediately follows that $q_{sen} = -1/2$ (neatly consistent with -0.6 ± 0.2) and $q_{rel} = 4$ (neatly consistent with 3.8 ± 0.3). There may be some difficulties with this approach, and efforts are being made to clear up the situation.

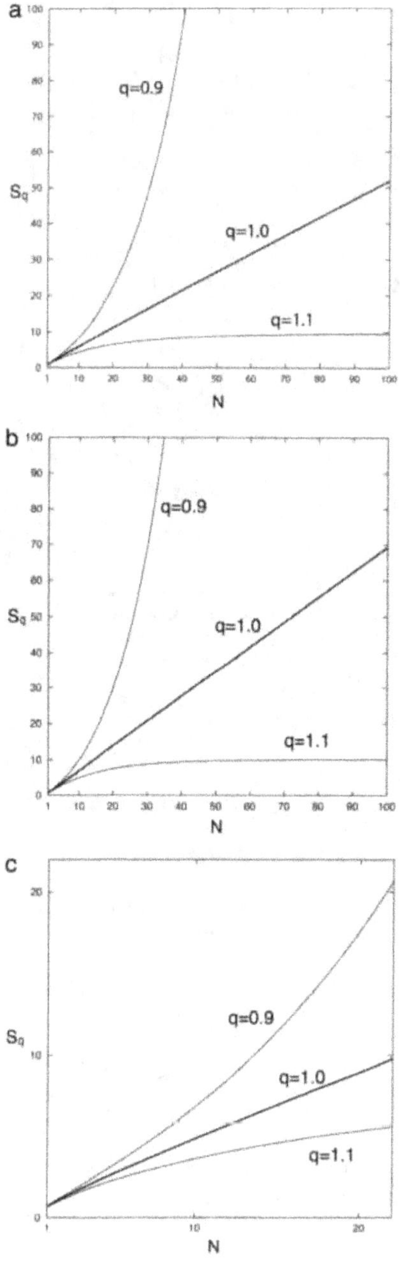

Fig. 4. $S_q(N)$ for the Leibnitz triangle [the explicit expression $\pi_{N,n} = 1/(N+1)(N-n)!n!/N!$ has been used to calculate $S_q(N)$] (*a*) $\alpha = 1$ (i.e., independent subsystems) with $\pi_{10} = 1/2$ [the explicit expression $\pi_{N,n} = (\pi_{10})^{N-n}(1-\pi_{10})^n$ has been used to calculate $S_q(N)$ (*b*) and $\alpha = 1/2$ with $\pi_{10} = 1/2$ [the recursive relation **3** has been used to calculated $S_q(N)$] (*c*). Only for $q = 1$ we have a *finite* value for $\lim_{N\to\infty} S_q(N)/N$; it *vanishes* (*diverges*) for $q > 1$ ($q < 1$).

$\ldots, N$. The total number of states is given by $W(N) = 2^N$ $(\forall d)$, but the number of states with nonzero probability is given by

$$W_{\text{eff}}(N, d) = \sum_{k=0}^{d} \frac{N!}{(N-k)!k!}, \quad [5]$$

where eff stands for effective. For example, $W_{\text{eff}}(N, 0) = 1$, $W_{\text{eff}}(N, 1) = N + 1$, $W_{\text{eff}}(N, 2) = \frac{1}{2}N(N+1) + 1$, $W_{\text{eff}}(N, 3) = \frac{1}{6}N(N^2+5) + 1$, and so on. For fixed d and $N \to \infty$ we have that

$$W_{\text{eff}}(N, d) \sim \frac{N^d}{d!}. \quad [6]$$

Let us now make a simple choice for the nonzero probabilities, namely *equal probabilities*. In other words,

$$\pi_{N,n}^{(d)} = 1/2^N \text{ (if } N \le d\text{)},$$

$$\pi_{N,n}^{(d)} = \frac{1}{W_{\text{eff}}(N, d)} \text{ (if } N > d \text{ and } n \le d\text{), and} \quad [7]$$

$$\pi_{N,n}^{(d)} = 0 \text{ (if } N > d \text{ and } n > d\text{)}.$$

See Fig. 6 for an illustration of this model.

The entropy for this model is given by

$$S_q(N) = \ln_q W_{\text{eff}}(N, d) \equiv \frac{[W_{\text{eff}}(N, d)]^{1-q} - 1}{1-q}$$

$$\sim \frac{N^{d(1-q)}}{(1-q)(d!)^{1-q}}, \quad [8]$$

where we have used now Eq. **6**. Consequently, S_q is *extensive* [i.e., $S_q(N) \propto N$ for $N \to \infty$] if and only if

$$q = 1 - \frac{1}{d}. \quad [9]$$

Hence, if $d = 1, 2, 3 \ldots,$ the entropic index monotonically approaches the BG limit from below. We can immediately verify in Fig. 6 (and using Eq. **7**) that this model violates the Leibnitz rule for all N, including asymptotically when $N \to \infty$. Consequently, it is neither strictly nor asymptotically scale-free. However, it is q-describable (see Fig. 3).

An Asymptotically Scale-Invariant Discrete Model. Starting with the Leibnitz harmonic triangle, we shall construct a heterogeneous distribution $\pi_{N,n}^{(d)}$. The Leibnitz triangle is given in Fig. 2 and satisfies

$$p_{N,n} = p_{N+1,n} + p_{N+1,n+1}, \quad [10]$$

$$p_{N,n} = \frac{1}{(N+1)} \frac{(N-n)!n!}{N!}. \quad [11]$$

We now define

$$\pi_{N,n}^{(d)} \equiv \begin{cases} p_{N,n} + l_{N,n}^{(d)} s_N^{(d)} & (n \le d) \\ 0 & (n > d) \end{cases} \quad [12]$$

where the *excess probability* $s_N^{(d)}$ and the *distribution ratio* $l_{N,n}^{(d)}$ (with $0 < \varepsilon < 1$) are defined through

$$s_N^{(d)} \equiv \sum_{k=d+1}^{N} p_{N,k} = \frac{N-d}{N+1} \quad [13]$$

PHYSICS

$$
\begin{array}{lcc}
(N=0) & (1,1) & (1,1) \\
(N=1) & (1,\pi^{(1)}_{10})\ (1,\pi^{(1)}_{11}) & (1,\pi^{(2)}_{10})\ (1,\pi^{(2)}_{11}) \\
(N=2) & (1,\pi^{(1)}_{20})\ (2,\pi^{(1)}_{21})\ (1,0) & (1,\pi^{(2)}_{20})\ (2,\pi^{(2)}_{21})\ (1,\pi^{(2)}_{22}) \\
(N=3) & (1,\pi^{(1)}_{30})\ (3,\pi^{(1)}_{31})\ (3,0)\ (1,0) & (1,\pi^{(2)}_{30})\ (3,\pi^{(2)}_{31})\ (3,\pi^{(2)}_{32})\ (1,0) \\
(N=4) & (1,\pi^{(1)}_{40})\ (4,\pi^{(1)}_{41})\ (6,0)\ (4,0)\ (1,0) & (1,\pi^{(2)}_{40})\ (4,\pi^{(2)}_{41})\ (6,\pi^{(2)}_{42})\ (4,0)\ (1,0)
\end{array}
$$

Fig. 5. Probabilistic models with $d = 1$ (*Left*) and $d = 2$ (*Right*).

$$
\begin{array}{lcc}
(N=0) & (1,1) & (1,1) \\
(N=1) & (1,1/2)\ (1,1/2) & (1,1/2)\ (1,1/2) \\
(N=2) & (1,1/3)\ (2,1/3)\ (1,0) & (1,1/4)\ (2,1/4)\ (1,1/4) \\
(N=3) & (1,1/4)\ (3,1/4)\ (3,0)\ (1,0) & (1,1/7)\ (3,1/7)\ (3,1/7)\ (1,0) \\
(N=4) & (1,1/5)\ (4,1/5)\ (6,0)\ (4,0)\ (1,0) & (1,1/11)\ (4,1/11)\ (6,1/11)\ (4,0)\ (1,0)
\end{array}
$$

Fig. 6. Uniform distribution model with $d = 1$ (*Left*) and $d = 2$ (*Right*).

$$
\begin{array}{lcc}
(N=0) & (1,1) & (1,1) \\
(N=1) & (1,1/2)\ (1,1/2) & (1,1/2)\ (1,1/2) \\
(N=2) & (1,1/2)\ (2,1/4)\ (1,0) & (1,1/3)\ (2,1/6)\ (1,1/3) \\
(N=3) & (1,1/2)\ (3,1/6)\ (3,0)\ (1,0) & (1,3/8)\ (3,5/48)\ (3,5/48)\ (1,0) \\
(N=4) & (1,1/2)\ (4,1/8)\ (6,0)\ (4,0)\ (1,0) & (1,2/5)\ (4,3/40)\ (6,1/20)\ (4,0)\ (1,0)
\end{array}
$$

Fig. 7. Leibnitz-triangle-based $\varepsilon = 0.5$ probability sets: $d = 1$ (*Left*), and $d = 2$ (*Right*).

$$
\pi^{(d)}_{N,n} \equiv \begin{cases} 1-\varepsilon & (n=0) \\ (1-\varepsilon)\varepsilon^{n}\dfrac{(N-n)!n!}{N!} & (0<n<d) \\ \varepsilon^{d}\dfrac{(N-d)!d!}{N!} & (n=d) \end{cases} \qquad [14]
$$

(see Fig. 7). We have verified for $d = 1, 2, 3, 4$ and $N \to \infty$ a result that we expect to be correct for all $d < N/2$, namely that $0 < \pi_{N,n+1} \ll \pi_{N,n} \sim \pi_{N-1,n} \ll 1$, hence

$$
\lim_{N\to\infty} \frac{\pi^{(d)}_{N-1,n}}{\pi^{(d)}_{N,n} + \pi^{(d)}_{N,n+1}} = 1, \qquad [15]
$$

$$
\lim_{N\to\infty} \frac{\pi^{(d)}_{N-1,d}}{\pi^{(d)}_{N,d} + 0} = 1. \qquad [16]
$$

In other words, the Leibnitz rule is asymptotically satisfied for the *entire* probability set $\{\pi_{N,n}\}$, i.e., this system has asymptotic scale invariance. Its entropy is given by

$$
S_q(N,d) = \frac{1 - \sum_{k=0}^{d} [N!/(N-k)!k!][\pi^{(d)}_{N,k}]^q}{q-1}, \qquad [17]
$$

and we verify that a value of q exists such that $\lim_{N\to\infty} S_q(N,d)/N$ is finite. Our numerical results suggest that, for $0 < \varepsilon < 1$ (see Fig. 8),

$$
q = 1 - \frac{1}{d}. \qquad [18]
$$

For a description of a strictly scale-invariant discrete model and a continuous model, see *Supporting Text* and Figs. 9–17, which are published as supporting information on the PNAS web site.

Final Remarks

Let us now critically re-examine the physical entropy, a concept which is intended to measure the nature and amount of our ignorance of the state of the system. As we shall see, extensivity may act as a guiding principle. Let us start with the simple case of an isolated classical system with *strongly* chaotic nonlinear dynamics, i.e., at least one *positive* Lyapunov exponent. For almost all possible initial conditions, the system quickly visits the various admissible parts of a *coarse-grained* phase space in a virtually homogeneous manner. Then, when the system achieves *thermodynamic equilibrium*, our knowledge is as meager as possible (*microcanonical ensemble*), i.e., just the Lebesgue measure W of the appropriate (hyper) volume in phase space (continuous degrees of freedom), or the number W of possible states (discrete degrees of freedom). The entropy is given by $S_{BG}(N) = k \ln W(N)$ [*Boltzmann principle* (34)].** If we consider independent equal subsystems, we have $W(N) = [W(1)]^N$, hence $S_{BG}(N) = NS_{BG}(1)$. If the N subsystems are only *locally* correlated, we expect $W(N) \sim \mu^N$ ($\mu \geq 1$), hence $\lim_{N\to\infty} S_{BG}(N)/N = \mu$, i.e., the entropy is *extensive* (i.e., *asymptotically additive*).

Consider now a strongly chaotic case for which we have more information, e.g., the set of probabilities $\{p_i\}$ ($i = 1, 2, \ldots, W$) of the states of the system. The form $S_{BG} \equiv -k \sum_{i=1}^{W} p_i \ln p_i$ yields $S_{BG}(A+B) = S_{BG}(A) + S_{BG}(B)$ in the case of independence ($p_{ij}^{A+B} = p_i^A p_j^B$). This form, although more general than $k \ln W$ (corresponding to equal probabilities), still satisfies additivity. It frequently happens, though, that we do not know the *entire set* $\{p_i\}$, but only some constraints on this set, besides the trivial one $\sum_{i=1}^{W} p_i = 1$. The typical case is Gibbs' canonical ensemble (Hamiltonian system in longstanding contact with a thermal

**A. Einstein: "Usually W is set equal to the number of ways (complexions) in which a state, which is incompletely defined in the sense of a molecular theory (i.e. coarse grained), can be realized. To compute W one needs a complete theory (something like a complete molecular-mechanical theory) of the system. For that reason it appears to be doubtful whether Boltzmann's principle alone, i.e. without a complete molecular-mechanical theory (Elementary theory) has any real meaning. The equation $S = k \log W + const.$ appears [therefore] without an Elementary theory—or however one wants to say it—devoid of any meaning from a phenomenological point of view." [translated by E. G. D. Cohen (34)]. A slightly different translation also is available: ["Usually W is put equal to the number of complexions. . . . In order to calculate W, one needs a *complete* (molecular-mechanical) theory of the system under consideration. Therefore it is dubious whether the Boltzmann principle has any meaning without a complete molecular-mechanical theory or some other theory which describes the elementary processes. $S = R/N \log W + const.$ seems without content, from a phenomenological point of view, without giving in addition such an *Elementartheorie*" (35)].

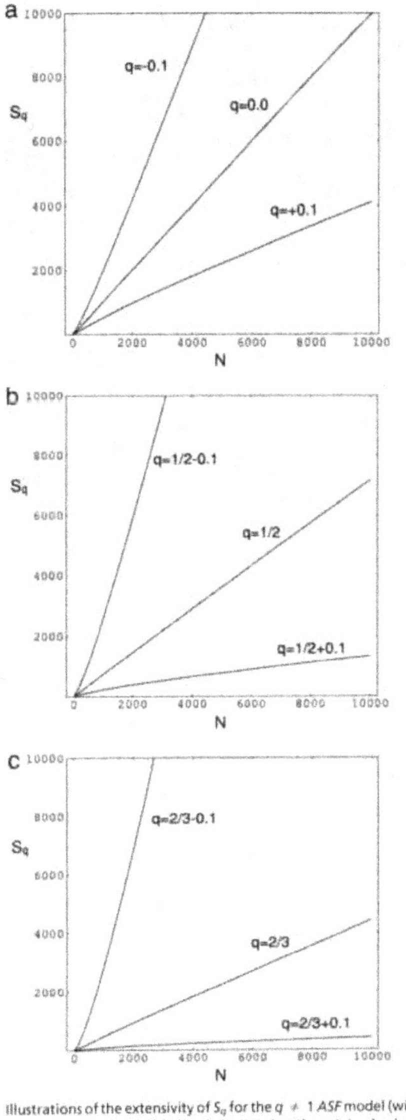

Fig. 8. Illustrations of the extensivity of S_q for the $q \neq 1$ *ASF* model (with $\varepsilon = 0.5$): (*a*) $d = 1$; (*b*) $d = 2$, and (*c*) $d = 3$. Notice that the minimal value of N equals $d - 1$. $\lim_{N\to\infty} S_q(N)/N$ *vanishes* (*diverges*) if $q > 1 - 1/d$ ($q < 1 - 1/d$), whereas it is *finite* for $q = 1 - 1/d$.

bath), in which case we know the mean value of the energy (*internal energy*). Extremization of S_{BG} yields, as well known, the celebrated BG weight, i.e., $p_i \propto e^{-\beta E_i}$, with $\beta \equiv 1/kT$ and $\{E_i\}$ being the set of possible energies. This distribution recovers the microcanonical case (equal probabilities) for $T \to \infty$.

Let us now address more subtle physical systems (still within the class associated with strong chaos), namely those in which the particles are indistinguishable (bosons, fermions). This new constraint leads to a substantial modification of the description of the states of the system, and the entropy form has to be consistently modified, as shown in any textbook. These expressions may be seen as further generalizations of S_{BG}, and the extremizing probabilities constitute, *at the level of the one-particle states*, generalizations of the just mentioned BG weight, recovered asymptotically at high temperatures. It is remarkable that, through these successive generalizations (and even more, since correlations due to local interactions might exist in addition to those connected with quantum statistics), *the entropy remains extensive*. Another subtle case is that of thermodynamic critical points, where correlations at all scales exist. There we can still refer to S_{BG}, but it exhibits singular behavior.[††]

Finally, we address the completely different class of systems for which the condition of independence is severely violated (typically because the system is only *weakly chaotic*, i.e., its sensitivity to the initial conditions grows slowly with time, say as a *power-law*, with the maximal Lyapunov exponent vanishing). In such systems, long range correlations typically exist that unavoidably point toward generalizing the entropic functional, essentially because the effective number of visited states grows with N as something like a power law instead of exponentially. We exhibited here such examples for which (either exact or asymptotic) *scale-invariant correlations* are present. There the entropy S_q for a special value of $q \neq 1$ is *extensive*, whereas S_{BG} is *not*.

Weak departures from independence make S_{BG} lose strict additivity, but not *extensivity*. Something quite analogous is expected to occur for scale-invariance in the case of S_q for $q \neq 1$. Amusingly enough, we have shown (see also refs. 29 and 38) that the "nonextensive" entropy S_q—indeed nonextensive for independent subsystems—*acquires extensivity in the presence of suitable asymptotically scale-invariant correlations*. Thus arguments presented in the literature that involve S_q (with $q \neq 1$) concomitantly with the assumption of independence should be revisited. In contrast, those arguments based on extremizing S_q, without reference to the composition of probabilities, remain unaffected. Although reference to "nonextensive statistical mechanics" still makes sense, say for long-range interactions, we see that the usual generic labeling of the entropy S_q for $q \neq 1$ as "nonextensive entropy" can be misleading.

The asymptotic scale invariance on which we focus appears to be connected with the asymptotically scale-free occupation of phase space that has been conjectured (1) to be dynamically generated by the complex systems addressed by nonextensive statistical mechanics (see also refs. 39 and 40). *Extensivity*—together with *concavity*, *Lesche-stability* (41–43), and *finiteness of the entropy production per unit time*—increases the suitability of the entropy S_q for linking, with no major changes, statistical mechanics to thermodynamics.

Last but not least, the probability structure of our discrete cases is, interestingly enough, intimately related to both the Pascal and the Leibnitz triangles.

[††]This is due, as well known, to the fractal structure of the correlation clusters existing at critical points. An instructive description in nonextensive terms of such special situations has been recently advanced in refs. 36 and 37.

We are grateful to R. Hersh for pointing out to us that the joint-probability structure of one of our discrete models is analogous to that of the Leibnitz triangle. We have also benefited from very fruitful remarks by J. Marsh and L. G. Moyano. Y.S. was supported by the Postdoctoral Fellowship at Santa Fe Institute. Support from SI International and AFRL is acknowledged as well. Finally, the work of one of us (M.G.M.) was supported by the C.O.U.Q. Foundation and by Insight Venture Management. The generous help provided by these organizations is gratefully acknowledged.

1. Gell-Mann, M. & Tsallis, C., eds. (2004) *Nonextensive Entropy: Interdisciplinary Applications* (Oxford Univ. Press, New York), pp. 1–422.
2. Harvda, J. & Charvat, F. (1967) *Kybernetica* **3,** 30–35.
3. Vajda, I. (1968) *Kybernetika* **4,** 105–110 (in Czech).
4. Daroczy, Z. (1970) *Inf. Control* **16,** 36–51.
5. Wehrl, A. (1978) *Rev. Mod. Phys.* **50,** 221–260.
6. Tsallis, C. (1995) *Chaos Solitons Fractals* **6,** 539–559.
7. Abe, S. & Okamoto, Y., eds. (2001) *Nonextensive Statistical Mechanics and Its Applications, Lecture Notes in Physics* (Springer, Heidelberg).
8. Tsallis, C. (2002) *Chaos Solitons Fractals* **13,** 371–391.
9. Csiszar, I. (1974) in *Transactions of the Seventh Prague Conference on Information Theory, Statistical Decision Functions, Random Processes, and the European Meeting of Statisticians, 1974* (Reidel, Dordrecht), pp. 73–86.
10. Schutzenberger, P.-M. (1954) *Publ. Inst. Statist. Univ. Paris* **3,** 3.
11. Renyi, A. (1961) *Proc. Fourth Berkeley Symp.* **1,** 547–561 (See also A. Renyi, A. (1970), *Probability Theory* (North-Holland, Amsterdam).)
12. Landsberg, P. T. & Vedral, V. (1998) *Phys. Lett. A* **247,** 211–217.
13. Rajagopal, A. K. & Abe, S. (1999) *Phys. Rev. Lett.* **83,** 1711–1714.
14. Sharma, B. D. & Mittal, D. P. (1975) *J. Math. Sci.* **10,** 28–40.
15. Masi, M. (2005) *Phys. Lett. A* **338,** 217–224.
16. Tsallis, C. (1988) *J. Stat. Phys.* **52,** 479–487.
17. Curado, E. M. F. & Tsallis, C. (1991) *J. Phys. A* **24,** L69–L72, and errata (1991) **24,** 3187 and (1992) **25,** 1019.
18. Tsallis, C., Mendes, R. S. & Plastino, A. R. (1998) *Physica A* **261,** 534–554.
19. Lyra, M. L. & Tsallis, C. (1998) *Phys. Rev. Lett.* **80,** 53–56.
20. Borges, E. P., Tsallis, C., Ananos, G. F. J. & Oliveira, P. M. C. (2002) *Phys. Rev. Lett.* **89,** 254103-1–254103-4.
21. Ananos, G. F. J. & Tsallis, C. (2004) *Phys. Rev. Lett.* **93,** 020601-1–20601-4.
22. Mayoral, E. & Robledo, A. (2005) *Phys. Rev. E* **72,** 026209-1–026209-7.
23. Casati, G., Tsallis, C. & Baldovin, F. (2005) *Europhys. Lett.*, in press.
24. Watts, D. J. & Strogatz, S. H. (1998) *Nature* **393,** 440–442.
25. Albert, R. & Barabasi, A.-L. (2002) *Rev. Mod. Phys.* **74,** 47–98.
26. Polya, G. (1962) *Mathematical Discovery* (Wiley, New York), Vol. 1, p. 88.
27. Plastino, A. R. & Plastino, A. (1995) *Physica A* **222,** 347–354.
28. Tsallis, C. & Bukman, D. J. (1996) *Phys. Rev. E* **54,** R2197–R2200.
29. Tsallis, C. (2005) *Milan J. Math.* **73,** in press.
30. Nivanen, L., Le Mehaute, A. & Wang, Q. A. (2003) *Rep. Math. Phys.* **52,** 437–444.
31. Borges, E. P. (2004) *Physica A* **340,** 95–101.
32. Tsallis, C. (2004) *Physica A* **340,** 1–10.
33. Burlaga, L. F. & Vinas, A. F. (2005) *Physica A* **356,** 375–384.
34. Cohen, E. G. D. (2005) *Pramana J. Phys.* **64,** 635–642.
35. Pais, A. (1982) *Subtle Is the Lord: The Science and the Life of Albert Einstein* (Oxford Univ. Press, New York).
36. Robledo, A. (2004) *Physica A* **344,** 631–636.
37. Robledo, A. (2005) *Mol. Phys.*, in press.
38. Tsallis, C. (2005) in *Complexity, Metastability and Nonextensivity*, eds. Beck, C., Benedek, G., Rapisarda, A. & Tsallis, C. (World Scientific, Singapore), pp. 13–32.
39. Soares, D. J. B., Tsallis, C., Mariz, A. M. & Silva, L. R. (2005) *Europhys. Lett.* **70,** 70–76.
40. Thurner, S. & Tsallis, C. (2005) *Europhys. Lett.* **72,** 197–203.
41. Lesche, B. (1982) *J. Stat. Phys.* **27,** 419–422.
42. Abe, S. (2002) *Phys. Rev. E* **66,** 046134-1–046134-6.
43. Lesche, B. (2004) *Phys. Rev. E* **70,** 017102-1–017102-4.
44. Moyano, L. G., Tsallis, C. & Gell-Mann, M. (2005) arXiv: cond-mat/0509229.

PHYSICAL REVIEW A **76**, 022104 (2007)

Quasiclassical coarse graining and thermodynamic entropy

Murray Gell-Mann[1,*] and James B. Hartle[1,2,†]
[1]*Santa Fe Institute, Santa Fe, New Mexico 87501*
[2]*Department of Physics, University of California, Santa Barbara, California 93106-9530*
(Received 24 December 2006; published 20 August 2007)

Our everyday descriptions of the universe are highly coarse grained, following only a tiny fraction of the variables necessary for a perfectly fine-grained description. Coarse graining in classical physics is made natural by our limited powers of observation and computation. But in the modern quantum mechanics of closed systems, some measure of coarse graining is inescapable because there are no nontrivial, probabilistic, fine-grained descriptions. This essay explores the consequences of that fact. Quantum theory allows for various coarse-grained descriptions, some of which are mutually incompatible. For most purposes, however, we are interested in the small subset of "quasiclassical descriptions" defined by ranges of values of averages over small volumes of densities of conserved quantities such as energy and momentum and approximately conserved quantities such as baryon number. The near-conservation of these quasiclassical quantities results in approximate decoherence, predictability, and local equilibrium, leading to closed sets of equations of motion. In any description, information is sacrificed through the coarse graining that yields decoherence and gives rise to probabilities for histories. In quasiclassical descriptions, further information is sacrificed in exhibiting the emergent regularities summarized by classical equations of motion. An appropriate entropy measures the loss of information. For a "quasiclassical realm" this is connected with the usual thermodynamic entropy as obtained from statistical mechanics. It was low for the initial state of our universe and has been increasing since.

DOI: 10.1103/PhysRevA.76.022104 PACS number(s): 03.65.Yz, 65.40.Gr, 98.80.Qc, 05.70.Ln

I. INTRODUCTION

Coarse graining is of the greatest importance in theoretical science, for example in connection with the meaning of regularity and randomness and of simplicity and complexity [1]. Of course it is also central to statistical mechanics, where a particular kind of coarse graining leads to the usual physicochemical or thermodynamic entropy. We discuss here the crucial role that coarse graining plays in quantum mechanics, making possible the decoherence of alternative histories and enabling probabilities of those decoherent histories to be defined. The "quasiclassical realms" that exhibit how classical physics applies approximately in this quantum universe correspond to coarse grainings for which the histories have probabilities that exhibit a great deal of approximate determinism. In this paper we connect these coarse grainings to the ones that lead, in statistical mechanics, to the usual entropy.

Our everyday descriptions of the world in terms of nearby physical objects like tables and clouds are examples of very *coarse-grained* quasiclassical realms. At any one time, these everyday descriptions follow only a tiny fraction of the variables needed for a fine-grained description of the whole universe; they integrate over the rest. Even those variables that are followed are not tracked continuously in time, but rather sampled only at a sequence of times.[1]

A classical gas of a large number of particles in a box provides other excellent examples of fine and coarse graining. The exact positions and momenta of all the particles as a function of time give the perfectly fine-grained description. Useful coarse-grained descriptions are provided by dividing the box into cells and specifying the volume averages of the energy, momentum, and particle number in each. Coarse graining can also be applied to ranges of values of the variables followed. This is an example in a classical context of what we will call a *quasiclassical coarse graining*. As we will describe in more detail below, under suitable initial conditions and with suitable choices for the volumes, this coarse graining leads to a deterministic, hydrodynamic description of the material in the box. We need not amplify on the utility of that.

In classical physics, coarse graining arises from practical considerations. It might be forced on us by our puny ability to collect, store, recall, and manipulate data. Particular coarse grainings may be distinguished by their utility. But there is always available, *in principle*, for every physical system, an exact fine-grained description that could be used to answer any question, whatever the limitations on using it in practice.

In the modern formulation of the quantum mechanics of a closed system—a formulation based on histories—the situa-

*mgm@santafe.edu
†hartle@physics.ucsb.edu

[1]This is not the most general kind of coarse graining. One can have instead a probability distribution characterized by certain parameters that correspond to the expected values of particular quantities, as in the case of absolute temperature and the energy in a Maxwell-Boltzmann distribution [1]

1050-2947/2007/76(2)/022104(16)

tion is different. The probability that either of two exclusive histories will occur has to be equal to the sum of the individual probabilities. This will be impossible in quantum mechanics unless the interference term between the two is made negligible by coarse graining. That absence of interference is called *decoherence*. There *is no (nontrivial) fine-grained, probabilistic description of the histories of a closed quantum-mechanical system*. Coarse graining is therefore inescapable in quantum mechanics if the theory is to yield any predictions at all.

Some of the ramifications of these ideas are the following:

(a) Quantum mechanics supplies probabilities for the members of various decoherent sets of coarse-grained alternative histories—different "realms" for short. Different realms are *compatible* if each one can be fine-grained to yield the same realm. (We have a special case of this when one of the realms is a coarse graining of the other.) Quantum mechanics also exhibits mutually *incompatible* realms for which there is no finer-grained decoherent set of which they are both coarse grainings.

(b) Quantum mechanics by itself does not favor one realm over another. However, we are interested for most purposes in the family of quasiclassical realms underlying everyday experience—a very small subset of the set of all realms. Roughly speaking, a quasiclassical realm corresponds to coarse graining which follows ranges of the values of variables that enter into the classical equations of motion.

(c) Having nontrivial probabilities for histories requires the sacrifice of information through the coarse graining necessary for decoherence.

(d) Coarse graining beyond that necessary for decoherence is needed to achieve the predictability represented by the approximate deterministic laws governing the sets of histories that constitute the quasiclassical realms.

(e) An appropriately defined entropy that is a function of time is a useful measure of the information lost through a coarse graining defining a quasiclassical realm.

(f) Such a coarse graining, necessary for having both decoherence and approximate classical predictability, is connected with the coarse graining that defines the familiar entropy of thermodynamics.

(g) The entropy defined by a coarse graining associated with a quasiclassical realm must be sufficiently low at early times in our universe (that is, low for the coarse graining in question) so that it tends to increase in time and exhibit the second law of thermodynamics. But for this the relaxation time for the increase of entropy must also be long compared with the age of the universe. From this point of view the universe is very young.

(h) The second law of thermodynamics, the family of quasiclassical realms, and even probabilities of any kind, are features of our universe that depend, not just on its quantum state and dynamics, but also crucially on the inescapable coarse graining necessary to define those features, since there is no exact, completely fine-grained description.

Many of these remarks are contained in our earlier work [2–10] or are implicit in it or are contained in the work of others, e.g., [11,12] and the references therein. They are developed more extensively here in order to emphasize the central and inescapable role played by coarse graining in quantum theory. What is new in this paper includes the detailed analysis of the triviality of decoherent sets of completely fine-grained histories, the role of narrative, and, most importantly, the central result that the kind of coarse graining defining the quasiclassical realms in quantum theory is closely related to the kind of coarse graining defining the usual entropy of chemistry and physics.

The reader who is familiar with our previous work and notation can immediately jump to Sec. III. But for those who are not we offer a very brief summary in the next section.

II. THE QUANTUM MECHANICS OF A CLOSED SYSTEM

This section gives a bare-bones account of some essential elements of the modern synthesis of ideas characterizing the quantum mechanics of closed systems [7,11,12].

To keep the discussion manageable, we consider a closed quantum system, most generally the universe, in the approximation that gross quantum fluctuations in the geometry of spacetime can be neglected. (For the generalizations that are needed for quantum spacetime see, e.g., [13,14].) The closed system can then be thought of as a large (say $\gtrsim$20 000 Mpc), perhaps expanding the box of particles and fields in a fixed background spacetime. Everything is contained within the box, in particular galaxies, planets, observers and observed, measured subsystems, and any apparatus that measures them. This is the most general physical context for prediction.

The fixed background spacetime means that the notions of time are fixed and that the usual apparatus of Hilbert space, states, and operators can be employed in a quantum description of the system. The essential theoretical inputs to the process of prediction are the Hamiltonian H governing evolution (which we assume for simplicity to be time-reversible[2]) and the initial quantum condition (which we assume to be a pure state[3] $|\Psi\rangle$). These are taken to be fixed and given.

The most general objective of quantum theory is the prediction of the probabilities of individual members of sets of coarse-grained alternative histories of the closed system. For instance, we might be interested in alternative histories of the center-of-mass of the earth in its progress around the sun, or in histories of the correlation between the registrations of a measuring apparatus and a property of the subsystem. Alternatives at one moment of time can always be reduced to a set of yes or no questions. For example, alternative positions of the earth's center of mass can be reduced to asking, "Is it in this region—yes or no?," "Is it in that region—yes or no?," etc. An exhaustive set of yes or no alternatives is represented in the Heisenberg picture by an exhaustive set of orthogonal projection operators $\{P_\alpha(t)\}$, $\alpha=1,2,3\ldots$. These satisfy

[2]For the issues arising from the time asymmetry of the effective low-energy theory of the elementary particles in our local neighborhood of the universe, see, e.g., [6].

[3]From the perspective of the time-neutral formulation of quantum theory (e.g., [6]), we are assuming also a final condition of ignorance.

$$\sum_{\alpha} P_\alpha(t) = I, \quad \text{and } P_\alpha(t)P_\beta(t) = \delta_{\alpha\beta}P_\alpha(t), \qquad (2.1)$$

showing that they represent an exhaustive set of exclusive alternatives. In the Heisenberg picture, the operators $P_\alpha(t)$ evolve with time according to

$$P_\alpha(t) = e^{+iHt/\hbar}P_\alpha(0)e^{-iHt/\hbar}. \qquad (2.2)$$

The state $|\Psi\rangle$ is unchanging in time.

Alternatives at one time can be described by expressing their projections in terms of fundamental coordinates, say a set of quantum fields and their conjugate momenta. In the Heisenberg picture these coordinates evolve in time. A given projection can be described in terms of the coordinates at any time. Thus at any time it represents some alternative [8].

An important kind of set of exclusive histories is specified by sets of alternatives at a sequence of times $t_1 < t_2 < \cdots < t_n$. An individual history α in such a set is a particular sequence of alternatives $\alpha \equiv (\alpha_1, \alpha_2, \ldots, \alpha_n)$. Such a set of histories has a branching structure in which a history up to any given time $t_m \leq t_n$ branches into further alternatives at later times. "Branch" is thus an evocative synonym for such a history.[4]

In such sets, we denote a projection at time t_k by

$$P^k_{\alpha_k}(t_k; \alpha_{k-1}, \ldots, \alpha_1). \qquad (2.3)$$

We now explain this notation.

Alternatives at distinct times can differ and are distinguished by the superscript k on the P's. For instance, projections on ranges of position at one time might be followed by projections on ranges of momentum at the next time, etc.

In realistic situations the alternatives will be *branch dependent*. In a branch-dependent set of histories the *set* of alternatives at one time depends on the particular branch. In Eq. (2.3) the string $\alpha_{k-1}, \ldots \alpha_1$ indicates this branch dependence. For example, in describing the evolution of the earth, starting with a protostellar cloud, a relatively coarse-grained description of the interstellar gas might be appropriate in the beginning, to be followed by finer and finer-grained descriptions at one location on the branch where a star (the sun) condensed, where a planet (the Earth) at 1 A.U. won the battle of accretion in the circumstellar disk, etc.—all events which happen only with some probability. Adaptive mesh refinement in numerical simulations of hydrodynamics provides a somewhat analogous situation.

An individual history α corresponding to a particular sequence of alternatives $\alpha \equiv (\alpha_1, \alpha_2, \ldots, \alpha_n)$ is represented by the corresponding chain of projections C_α, called a *class operator*. In the full glory of our notation this is

$$C_\alpha \equiv P^n_{\alpha_n}(t_n; \alpha_{n-1}, \ldots, \alpha_1) \cdots P^2_{\alpha_2}(t_2; \alpha_1)P^1_{\alpha_1}(t_1). \qquad (2.4)$$

To keep the notation manageable we will sometimes not indicate the branch dependence explicitly where confusion is unlikely.

[4]Sets of histories defined by sets of alternatives at a sequence of discrete times is the simplest case sufficient for our purposes. For more general continuous time histories see, e.g., [35,36].

Irrespective of branch dependence, a set of histories like the one specified by Eq. (2.4) is generally *coarse grained* because alternatives are specified at some times and not at every time and because the alternatives at a given time are typically projections on subspaces with dimension greater than 1 and not projections onto a complete set of states. Perfectly *fine-grained* sets of histories consist of one-dimensional projections at each and every time.

Operations of fine and coarse graining may be defined on sets of histories. A set of histories $\{\alpha\}$ may be *fine grained* by dividing up each class into an exhaustive set of exclusive subclasses $\{\alpha'\}$. Each subclass consists of some histories in a coarser-grained class, and every finer-grained subclass is in some class. *Coarse graining* is the operation of uniting subclasses of histories into bigger classes. Suppose, for example, that the position of the Earth's center of mass is specified by dividing space into cubical regions of a certain size. A coarser-grained description of position could consist of larger regions made up of unions of the smaller ones. Consider a set of histories with class operators $\{C_\alpha\}$ and a coarse graining with class operators $\{\bar{C}_{\bar{\alpha}}\}$. The operators $\{\bar{C}_{\bar{\alpha}}\}$ are then related to the operators $\{C_\alpha\}$ by summation, *viz.*

$$\bar{C}_{\bar{\alpha}} = \sum_{\alpha \in \bar{\alpha}} C_\alpha, \qquad (2.5)$$

where the sum is over the C_α for all finer-grained histories α contained within $\bar{\alpha}$.

For any individual history α, there is a *branch state vector* defined by

$$|\Psi_\alpha\rangle = C_\alpha|\Psi\rangle. \qquad (2.6)$$

When probabilities can be consistently assigned to the individual histories in a set, they are given by

$$p(\alpha) = \| |\Psi_\alpha\rangle \|^2 = \|C_\alpha|\Psi\rangle\|^2. \qquad (2.7)$$

However, because of quantum interference, probabilities cannot be consistently assigned to every set of alternative histories that may be described. The two-slit experiment provides an elementary example: An electron emitted by a source can pass through either of two slits on its way to detection at a farther screen. It would be inconsistent to assign probabilities to the two histories distinguished by which slit the electron goes through if no "measurement" process determines this. Because of interference, the probability for arrival at a point on the screen would not be the sum of the probabilities to arrive there by going through each of the slits. In quantum theory, probabilities are squares of amplitudes and the square of a sum is not generally the sum of the squares.

Negligible interference between the branches of a set

$$\langle\Psi_\alpha|\Psi_\beta\rangle \approx 0, \quad \alpha \neq \beta, \qquad (2.8)$$

is a sufficient condition for the probabilities (2.7) to be consistent with the rules of probability theory. The orthogonality of the branches is approximate in realistic situations. But we mean by Eq. (2.8) equality to an accuracy that defines probabilities well beyond the standard to which they can be checked or, indeed, the physical situation modeled [4].

Specifically, as a consequence of Eq. (2.8), the probabilities (2.7) obey the most general form of the probability sum rules

$$p(\bar{\alpha}) \approx \sum_{\alpha \in \bar{\alpha}} p(\alpha) \qquad (2.9)$$

for any coarse graining $\{\bar{\alpha}\}$ of the $\{\alpha\}$. Sets of histories obeying Eq. (2.8) are said to (medium) decohere. As Diósi has shown [15], medium decoherence is the weakest of known conditions that are consistent with elementary notions of the independence of isolated systems.[5] Medium-decoherent sets are thus the ones for which quantum mechanics consistently makes predictions of probabilities through Eq. (2.7). Weaker conditions, such as those defining merely "consistent" histories, are not appropriate.

The decoherent sets exhibited by our universe are determined through Eq. (2.8) by the Hamiltonian H and the quantum state $|\Psi\rangle$. We use the term *realm* as a synonym for a decoherent set of coarse-grained alternative histories.

A coarse graining of a decoherent set is again decoherent. A fine graining of a decoherent set risks losing decoherence.

It is more general to allow a density matrix ρ as the initial quantum state. Decoherence is then defined in terms of a *decoherence functional*.

$$D(\alpha,\beta) = \mathrm{Tr}(C_\alpha \rho C_\beta^\dagger). \qquad (2.10)$$

The rules for both decoherence and the existence of probabilities can then be expressed in a single formula

$$D(\alpha,\beta) \approx \delta_{\alpha\beta} p(\alpha). \qquad (2.11)$$

When the density matrix is pure, $\rho = |\Psi\rangle\langle\Psi|$, the condition (2.11) reduces to the decoherence condition (2.8) and the probabilities to Eq. (2.7).

An important mechanism of decoherence is the dissipation of phase coherence between branches into variables not followed by the coarse graining. Consider, by way of example, a dust grain in a superposition of two positions deep in interstellar space [16]. In our universe, about 10^{11} cosmic background photons scatter from the dust grain each second. The two positions of the grain become correlated with different, nearly orthogonal states of the photons. Coarse grainings that follow only the position of the dust grain at a few times therefore correspond to branch state vectors that are nearly orthogonal and satisfy Eq. (2.9).

Measurements and observers play no fundamental role in this general formulation of usual quantum theory. The probabilities of measured outcomes can be computed and are given to an excellent approximation by the usual story. But, in a set of histories where they decohere, probabilities can be assigned to the position of the moon when it is not receiving the attention of observers and to the values of density fluctuations in the early universe when there were neither measurements taking place nor observers to carry them out.

The probabilities of the histories in the various realms and the conditional probabilities constructed from them constitute the predictions of the quantum mechanics of a closed system given the Hamiltonian H and initial state $|\Psi\rangle$.

III. INESCAPABLE COARSE GRAINING

In this section and Appendix A, we show that there are no exactly decoherent, completely fine-grained sets of alternative histories describing a closed quantum system except trivial ones which essentially reduce to a description at only one time.[6] (In Appendix B, we also show that there is no certainty except that arising from the unitary evolution controlled by the Hamiltonian.)

A set of alternatives at one time is completely fine grained if its projections are onto an orthonormal basis for Hilbert space. Specifically,

$$P_i(t) = |i\rangle\langle i|, \qquad (3.1)$$

where $\{|i\rangle\}, i=1,2,\ldots$ are a set of orthonormal basis vectors. Any basis defines at any time some set of alternatives [8].

Probabilities are predicted for any fine-grained set of alternatives at one time. The branch state vectors are

$$|\Psi_i\rangle = P_i(t)|\Psi\rangle = |i\rangle\langle i|\Psi\rangle. \qquad (3.2)$$

These are mutually orthogonal and therefore decohere [cf. Eq. (2.8)]. The consistent probabilities are

$$p(i) = |\langle i|\Psi\rangle|^2. \qquad (3.3)$$

A completely fine-grained description does not consist merely of fine-grained alternatives at one time but rather of fine-grained alternatives at each and every available time. Specifically, a completely fine-grained set of histories is a set of alternative histories specified by sets of fine-grained alternatives like those in Eq. (3.1) at every available time.

To avoid the mathematical issues that arise in defining continuous infinite products of projections [cf. Eq. (2.4)], we will assume that the available times are restricted to a large discrete series $t_1,\ldots,t_n$ in a fixed interval $[0,T]$. Physically there can be no objection to this if the intervals between times is taken sufficiently small (if necessary of order of the Planck time). More importantly, the absence of decoherent sets with a finite number of times prohibits the existence of any finer-grained decoherent sets, however they are defined.

A completely fine-grained set of histories is then specified by a series of bases $\{|1,i_1\rangle\},\{|2,i_2\rangle\},\ldots,\{|n,i_n\rangle\}$ defining one-dimensional projections of the form (3.1). In $|k,i_k\rangle$ the first argument labels the basis, the second the particular vector in that basis. To get at the essentials of the argument, we will assume in this section the generic situations where there are no branch state vectors that vanish. More particularly, we assume that none of the matrix elements $\langle k+1,i_{k+1}|k,i_k\rangle$ or $\langle 1,i_1|\Psi\rangle$ vanishes. Put differently, we assume that none of the questions at each time is trivially related to a question at the next time or to the initial state. The general case where

[5]For a discussion of the linear positive, weak, medium, and strong decoherence conditions, see [3,10,37].

[6]The observation that coarse graining is necessary for nontrivial decoherence has been made since the start of decoherent histories quantum theory. See, e.g., [2,38–40] for some different takes on this.

these assumptions are relaxed is dealt with in Appendix A along with the related notion of trivial decoherence that arises.

The branch state vectors for the histories $(i_1,\ldots,i_n)$ in such a completely fine-grained set have the form

$$|n,i_n\rangle\langle n,i_n|n-1,i_{n-1}\rangle\cdots\langle 2,i_2|1,i_1\rangle\langle 1,i_1|\Psi\rangle. \tag{3.4}$$

The vectors $|i_n,n\rangle$ are orthogonal for different i_n. The condition for decoherence (2.8) then requires

$$\langle\Psi|1,i_1'\rangle\langle 1,i_1'|2,i_2'\rangle\cdots\langle n,i_n|n,i_n\rangle\cdots\langle 2,i_2|1,i_1\rangle\langle 1,i_1|\Psi\rangle=0 \tag{3.5}$$

whenever any $i_k'\neq i_k$ for $k=1,\ldots,n-1$. But, by assumption, none of these matrix elements vanishes so there are no exactly decoherent, completely fine-grained sets of this kind.

To get a different perspective, suppose that $\langle 1,i_1|\Psi\rangle$ is allowed to vanish for some i_1 but the restriction that $\langle k,i_k|k-1,i_{k-1}\rangle\neq 0$ is retained. Then Eq. (3.5) could be satisfied provided there is *only one* i_1 for which $\langle 1,i_1|\Psi\rangle\neq 0$. But since $\{|1,i_1\rangle\}$ is a basis, this means it must consist of $|\Psi\rangle$ and a complete set of orthogonal vectors. But that means that the first set of alternatives is trivial. It merely asks, "Is the state $|\Psi\rangle$ or not?" As explained in Appendix A, there are no exactly decoherent, completely fine-grained sets of alternative histories of a closed quantum system that are not trivial in this or similar senses. Coarse graining is therefore inescapable.

IV. COMMENTS ON REALMS

A perfectly fine-grained set of histories can be coarse grained in many different ways to yield a decoherent set whose probabilities can be employed in the process of prediction. Furthermore, there are many different completely fine-grained sets to start from, corresponding to the possible choices of the bases $\{|k,i_k\rangle\}$ at each time arising from different complete sets of commuting observables. Once these fine-grained sets are coarse grained enough to achieve decoherence, further coarse graining preserves decoherence. Some decoherent sets can be organized into compatible families all members of which are coarse grainings of a common finer-grained decoherent set. But there remain distinct families of decoherent sets which have no finer-grained decoherent sets of which they are all coarse grainings. As mentioned in the Introduction, these are called *incompatible* realms.

We may not draw inferences by combining probabilities from incompatible realms.[7] That would implicitly assume, contrary to fact, that probabilities of a finer-grained description are available. Incompatible decoherent sets provide different, incompatible descriptions of the universe. Quantum theory does not automatically prefer one of these realms over another without further criteria such as quasiclassicality.

Note that *incompatibility* is not *inconsistency* in the sense of predicting different probabilities for the same histories in different realms. The probability for a history α is given by Eq. (2.7) in any realm of which it is a member.

While quantum theory permits a great many incompatible realms for the description of a closed system, we as observers utilize mainly sets that are coarse grainings of one family of such realms—the quasiclassical realms underlying everyday experience.[8] We now turn to a characterization of those.

The histories of a quasiclassical realm constitute *narratives*. That is, they describe how certain features of the universe change over time. A narrative realm is a particular kind of set of histories in which the projections at successive times are related by a suitable rule. That way the histories are stories about what happens over time and not simply unrelated or redundant scraps of information from a sequence of times.

The simplest way of defining a narrative is to take the same quantities at each time; histories then describe how those quantities change over time. The corresponding rule connecting the projections at different times can be simply stated in the Schrödinger picture, where operators corresponding to a given quantity do not change in time. The set of Schrödinger picture projections $\{\hat{P}_\alpha\}$ is the same at each time. Equivalently, in the Heisenberg picture the projections at each time are given by

$$P^k_\alpha(t_k)=e^{iHt_k/\hbar}\hat{P}_\alpha e^{-iHt_k/\hbar}. \tag{4.1}$$

The model coarse graining using the same quasiclassical variables at each time step that will be presented in the next section obeys Eq. (4.1). But in a more realistic situation the histories will be more complicated, especially because they exhibit branch dependence.

There are trivial kinds of non-narrative realms with a high level of predictability but little or no utility. Examples are realms that mindlessly repeat the same set of Heisenberg picture projection operators at each and every time, and the trivial fine-grained realms described in Sec. III A and Appendix A. Suppose, for instance, the same Heisenberg picture projection operators are repeated at each time so that $P^k_{\alpha_k}(t_k)=Q_{\alpha_k}$ for some set of orthogonal projections $\{Q_\alpha\}$. This leads to the following form for the C_α:

$$C_\alpha=Q_{\alpha_n}\cdots Q_{\alpha_1}. \tag{4.2}$$

The only nonvanishing C's are just the projections Q_α. Effectively the histories are specified by alternatives at one time. Nothing happens! Restricting to narrative realms eliminates this kind of triviality.

It remains to specify carefully what is a suitable rule for relating the projection operators in each history at successive

[7]It has been shown, especially by Griffiths [11], that essentially all inconsistencies alleged against consistent histories quantum mechanics (or, for that matter, decoherent histories quantum mechanics) arise from violating this logical prohibition.

[8]Some authors, notably Dowker and Kent [38], have suggested that quantum mechanics is incomplete without a fundamental set selection principle that would essentially single out the quasiclassical realms from all others. The following discussion of the properties of quasiclassical realms could be viewed as steps in that direction. We prefer to keep the fundamental formulation clean and precise by basing it solely on the notion of decoherence.

times, thus providing a general definition of a narrative realm. We are searching for the best way to do that, using ideas about simplicity, complexity, and logical depth presented in [1].

V. THE QUASICLASSICAL REALM(S)

As discussed in Sec. II, coarse graining is necessary for probability. The families of decoherent sets of coarse-grained histories give rise to descriptions of the universe that are often incompatible with one another. As information gathering and utilizing systems (IGUS's), we use, both individually and collectively, only a very limited subset of these descriptions belonging to a compatible family of realms with histories and probabilities that manifest certain regularities of the universe associated with classical dynamical laws. These regularities include ones that are exploitable in our various pursuits, such as getting food, reproducing, avoiding becoming food, and achieving recognition. Such sets of histories are defined by alternatives that include ones which our perception is adapted to distinguish. As we will see, coarse graining very far beyond that necessary for mere decoherence is necessary to define the sets that exhibit the most useful of these regularities.

Specific systems like the planet Mars, Western culture, and asparagus exhibit various kinds of particular exploitable regularities. But the most widely applicable class of regularities consists of the correlations in time governed by the deterministic laws of motion of classical physics. In our quantum universe, classical laws are approximately applicable over a wide range of times, places, scales, and epochs. They seem to hold approximately over the whole of the visible universe from a time shortly after the beginning to the present. Indeed, we expect them to hold into the far future. We refer to the family of decoherent sets of coarse-grained histories that describe these regularities as the family of *quasiclassical realms*.

The characteristic properties of a quasiclassical realm all follow from the approximate conservation of the variables that define them, which we call *quasiclassical variables*. As we will illustrate with a specific model below, these include averages of densities of quantities such as energy and momentum that are exactly conserved.[9] But they also include densities of quantities such as baryon number that may be conserved only to an approximation that varies with the epoch.

Approximate conservation leads to predictability in the face of the noise that accompanies typical mechanisms of decoherence. Approximate conservation allows the local equilibrium that leads to closed sets of classical equations of motion summarizing that predictability. This local equilibrium is the basis for the definition of an entropy implementing the second law of thermodynamics.

More specifically, in this paper, by a *quasiclassical realm* we mean an exhaustive set of mutually exclusive coarse-grained alternative histories that obey medium decoherence with the following additional properties: The histories consist largely of related but branch-dependent projections onto ranges of quasiclassical variables at a succession of times. Each history with a non-negligible probability constitutes a narrative, with individual histories exhibiting patterns of correlation implied by closed sets of effective equations of motion interrupted by frequent small fluctuations and occasional major branchings (as in measurement situations). By a *family of quasiclassical realms* we mean a set of compatible ones that are all coarse grainings of a common one. Useful families span an enormous range of coarse graining. Quasiclassical realms describing everyday experience, for instance, may be so highly coarse-grained as to refer merely to the features of a local environment deemed worthy of an IGUS's attention. At the other end of the range are the quasiclassical realms that are as fine-grained as possible given decoherence and quasiclassicality. Those realms extend over the wide range of time, place, scale, and epoch mentioned above. These coarse grainings defining the maximally refined quasiclassical realms are not a matter of our choice. Rather, those maximal realms are a feature of our universe that we exploit—a feature that is emergent from the initial quantum state, the Hamiltonian, and the extremely long sequence of outcomes of chance events.[10]

A. Quasiclassical variables

To build a simple model with a coarse graining that defines a quasiclassical realm, we begin with a large box containing local quantum fields. The interactions of these fields are assumed to be local and, for simplicity, to result in short-range forces on bulk matter. Spacetime is assumed to be nearly flat inside the box so that large fluctuations in the geometry of spacetime are neglected. This is a reasonably general model for much of physics in the late universe.

Space inside the box is divided into equal volumes of size V labeled by a discrete index $\vec{y}$. (The branch dependence of volumes described in Sec. II is thus ignored.) The linear dimensions of the volumes are assumed to be large compared to the ranges of the forces. Quasiclassical variables are constructed as averages over these volumes of conserved or approximately conserved extensive quantities. These will include energy and linear momentum since spacetime is treated

[9]Their exact conservation reflects the local symmetries of the approximately fixed spacetime geometry that emerges from the quantum-gravitational fog near the beginning.

[10]In previous work we have taken the term "quasiclassical realm" to be defined in some respects more generally and in other respects more restrictively than we have here [2]. To investigate information-theoretic measures for quasiclassicality, we left open the possibility that there might be realms exhibiting deterministic correlations in time defined by variables different from the quasiclassical variables. The quasiclassical realms of this paper were called "usual quasiclassical realms." We also required that every quasiclassical realm be maximal in the sense discussed above so that it was an emergent feature of the universe and not our choice. For the limited objectives of this paper it seems best to employ the simpler terminology that we have used here rather than seek exact consistency. We leave open whether there are incompatible maximal realms that also exhibit high levels of predictability and utility.

as flat. There might be many conserved or approximately conserved species of particles. To keep the notation manageable we consider just one.

To be explicit, let $T^{\alpha\beta}(\vec{x},t)$ be the stress-energy-momentum operator for the quantum fields in the Heisenberg picture. The energy density $\epsilon(\vec{x},t)$ and momentum density $\pi^i(x,t)$ are $T^{tt}(\vec{x},t)$ and $T^{ti}(\vec{x},t)$, respectively. Let $\nu(\vec{x},t)$ denote the number density of the conserved or nearly conserved species. Then we define

$$\epsilon_V(\vec{y},t) \equiv \frac{1}{V}\int_{\vec{y}} d^3x\,\epsilon(\vec{x},t), \tag{5.1a}$$

$$\vec{\pi}_V(\vec{y},t) \equiv \frac{1}{V}\int_{\vec{y}} d^3x\,\vec{\pi}(\vec{x},t), \tag{5.1b}$$

$$\nu_V(\vec{y},t) \equiv \frac{1}{V}\int_{\vec{y}} d^3x\,\nu(\vec{x},t), \tag{5.1c}$$

where in each case the integral is over the volume labeled by $\vec{y}$. These are the quasiclassical variables for our model. We note that the densities in Eq. (5.1) are the variables for a classical hydrodynamic description of this system—for example, the variables of the Navier-Stokes equation. In more general situations, these might be augmented by further densities of conserved or nearly conserved species, field averages, and other variables, but we will restrict our attention to just these.

A set of alternative coarse-grained histories can be constructed by giving ranges of these quasiclassical variables at a sequence of times. We will call these a set of quasiclassical histories. To be a quasiclassical realm, such a set must decohere and the probabilities must be high for the correlations in time specified by a closed set of deterministic equations of motion. In Sec. V C we will review the construction of these sets of histories, their decoherence, and their probabilities, referring to the work of Halliwell [17–19]. But here we anticipate the qualitative reasons for these results.

Typical realistic mechanisms of decoherence involve the dissipation of phases (between the branch state vectors) into variables that are not followed by the coarse graining. That is the case, for instance, with the many models in which the position of one particle is followed while it is coupled to a bath of others whose positions are ignored for all time [5,16,20,21]. Quasiclassical coarse grainings do not posit one set of variables that are ignored (integrated over) for all time, constituting a fixed "environment." Rather, at each time, the projection operators of the coarse graining define the ignored interior configurations of the volumes whose overall energy, momentum, and number are followed.[11]

[11]As shown in [41], for any coarse graining of the form (2.5) it is possible to factor an infinite-dimensional Hilbert space into a part that is followed by the coarse graining and a part that is ignored (an environment). However, that factorization may change from one time to the next, making it difficult to use. The linear oscillator chain studied in [41] provides another example of decoherence brought about by a coupling of followed variables to internal ones.

The coupling between followed and ignored variables that is necessary for decoherence is inevitably a source of noise for the followed quantities, causing deviations from predictability. The approximate conservation of quasiclassical variables allows them to resist the noise that typical mechanisms of decoherence produce and to remain approximately predictable because typically the inertia of each relevant degree of freedom is large [5,9].

Indeed, consider the limit where there is only one volume occupying the whole box. Then the quasiclassical variables are the total energy, total momentum, and total number of particles in the box. These are exactly conserved and mutually commuting. Histories of these quantities are therefore precisely correlated with the final values and trivially decohere in the sense discussed in Sec. III and Appendix A. Exact decoherence and persistence in the limit of one volume suggest efficient approximate decoherence for smaller volumes.

Decoherence, however, is not the only requirement for a quasiclassical realm. A quasiclassical realm must also exhibit the correlations in time implied by a closed set of classical equations of motion. It is to this property that we now turn.

B. Classical equations for expected values

Isolated systems generally evolve toward equilibrium. That is a consequence of statistics. But conserved or approximately conserved quantities such as energy, momentum, and number approach equilibrium more slowly than others. That means that a situation of *local equilibrium* will generally be reached before complete equilibrium is established, if it ever is. This local equilibrium is characterized by the values of conserved quantities constrained in small volumes. Even for systems of modest size, time scales for small volumes to relax to local equilibrium can be very, very much shorter than the time scale for reaching complete equilibrium.[12] Once local equilibrium is established, the subsequent evolution of the approximately conserved quantities can be described by closed sets of effective classical equations of motion such as the Navier-Stokes equation. The local equilibrium determines the values of the phenomenological quantities such as pressure and viscosity that enter into these equations and the relations among them. This section reviews the standard derivation of these equations of motion for the expected values of the approximately conserved quantities for our model universe in a box, as can be found, for example, in [22,23]. The next section considers the histories of these quantities, their decoherence, and their probabilities.

Central to the description of local equilibrium is the idea of an *effective density matrix*. For large systems the compu-

[12]In realistic situations there will generally be a hierarchy of time scales in which different kinds of equilibrium are reached on different distance scales. Star clusters provide a simple example. The local equilibrium of the kind described here is reached much more quickly for the matter inside the stars than the metastable equilibrium governed by weak, long range gravitation that may eventually characterize the cluster as a whole. By restricting our model to short-range forces we have avoided such realistic complications.

tation of the decoherence functional directly from the quantum state $\rho \equiv |\Psi\rangle\langle\Psi|$ may be practically impossible. However, it may happen for *certain classes of coarse grainings* that the decoherence functional for a coarse graining is given to a good approximation by an effective density matrix $\tilde{\rho}$ that requires less information to specify than ρ does (cf. Sec. VI). That is,

$$D(\alpha,\beta) \equiv \mathrm{Tr}(C_\alpha \rho C_\beta^\dagger) \approx \mathrm{Tr}(C_\alpha \tilde{\rho} C_\beta^\dagger) \tag{5.2}$$

for all histories α and β in the exhaustive set of alternative coarse-grained histories.[13]

A familiar example of an effective density matrix is the one describing a system in a box in thermal equilibrium,

$$\tilde{\rho}_{\mathrm{eq}} = Z^{-1} \exp[-\beta(H - \vec{U}\cdot\vec{P} - \mu N)]. \tag{5.3}$$

Here, H, $\vec{P}$, and N are the operators for total energy, total momentum, and total conserved number inside the box—all extensive quantities. The c-number intensive quantities β, $\vec{U}$, and μ are, respectively, the inverse temperature (in units where Boltzmann's constant is 1), the velocity of the box, and the chemical potential. A normalizing factor Z ensures $\mathrm{Tr}(\tilde{\rho}_{\mathrm{eq}})=1$. In the next section, we will give a standard derivation of this effective density matrix as one that maximizes the missing information subject to expected value constraints.

Local equilibrium is achieved when the decoherence functional for sets of histories of quasiclassical variables $(\epsilon, \vec{\pi}, n)$ is given approximately by the *local* version of the equilibrium density matrix (5.3),

$$\tilde{\rho}_{\mathrm{leq}} = Z^{-1} \exp\Big(-\sum_{\vec{y}} \beta(\vec{y},t)[\epsilon_V(\vec{y},t) - \vec{u}(\vec{y},t)\cdot\vec{\pi}_V(\vec{y},t) - \mu(\vec{y},t)\nu_V(\vec{y},t)]\Big). \tag{5.4}$$

This local equilibrium density matrix is constructed to reproduce the expected values of the quasiclassical variables such as the energy density averaged over a volume at a moment of time $\epsilon_V(\vec{y},t)$, viz.

$$\langle \epsilon_V(\vec{y},t)\rangle \equiv \mathrm{Tr}[\epsilon_V(\vec{y},t)\rho] = \mathrm{Tr}[\epsilon_V(\vec{y},t)\tilde{\rho}_{\mathrm{leq}}] \tag{5.5}$$

and similarly for $\vec{\pi}_V(\vec{y},t)$ and $\nu_V(\vec{y},t)$. The expected values of quasiclassical quantities are thus functions of the intensive c-number quantities $\beta(\vec{y},t)$, $\vec{u}(\vec{y},t)$, and $\mu(\vec{y},t)$. These are the local inverse temperature, velocity, and chemical potential respectively. They now vary with time as the system evolves toward complete equilibrium. In the next section, we will give a standard derivation of the local equilibrium density matrix as one that maximizes the missing information subject to expected value constraints of local quantities as in Eq. (5.5).

When there is not too much danger of confusion we will often replace sums over $\epsilon_V(\vec{y},t)$ with integrals over $\epsilon(\vec{x},t)$ and similarly with the other quasiclassical quantities $\vec{\pi}$ and ν. In particular, the differential equations of motion that we discuss below are a familiar kind of approximation to a set of difference equations.

A closed set of deterministic equations of motion for the expected values of $\epsilon(\vec{x},t)$, $\vec{\pi}(x,t)$, and $\nu(\vec{x},t)$ follows from assuming that $\tilde{\rho}_{\mathrm{leq}}$ is an effective density matrix for computing them. To see this, begin with the Heisenberg equations for the conservation of the stress-energy-momentum operator $T^{\alpha\beta}(\vec{x},t)$ and the number current $j^\alpha(\vec{x},t)$.

$$\frac{\partial T^{\alpha\beta}}{\partial x^\beta} = 0, \quad \frac{\partial j^\alpha}{\partial x^\alpha} = 0. \tag{5.6}$$

Noting that $\epsilon(\vec{x},t)=T^{tt}(\vec{x},t)$ and $\pi^i(\vec{x},t)=T^{ti}(\vec{x},t)$, Eqs. (5.6) can be written out in a 3+1 form and their expected values taken. The result is the set of five equations

$$\frac{\partial\langle\pi^i\rangle}{\partial t} = -\frac{\partial\langle T^{ij}\rangle}{\partial x^j}, \tag{5.7a}$$

$$\frac{\partial\langle\epsilon\rangle}{\partial t} = -\vec{\nabla}\cdot\langle\vec{\pi}\rangle, \tag{5.7b}$$

$$\frac{\partial\langle\nu\rangle}{\partial t} = -\vec{\nabla}\cdot\langle\vec{j}\rangle. \tag{5.7c}$$

The expected values are all functions of $\vec{x}$ and t.

To see how these conservation laws lead to a closed system of equations of motion, consider the right hand sides, for instance that of Eq. (5.7a). From the form of $\tilde{\rho}_{\mathrm{leq}}$ in Eq. (5.4) we find that the stress tensor $\langle T^{ij}(\vec{x},t)\rangle$ is a *function* of $\vec{x}$ and t, and a *functional* of the intensive multiplier functions $\beta(\vec{\xi},\tau)$, $\vec{u}(\vec{\xi},\tau)$, and $\mu(\vec{\xi},\tau)$. We write this relation as

$$\langle T^{ij}(\vec{x},t)\rangle = \check{T}^{ij}[\beta(\vec{\xi},\tau), u(\vec{\xi},\tau), \mu(\vec{\xi},\tau); \vec{x}, t). \tag{5.8}$$

In the same way, the expected values of $\epsilon(\vec{x},t)$, $\vec{\pi}(x,t)$, and $\nu(\vec{x},t)$ become functionals of $\beta(\vec{\xi},\tau)$, $\vec{u}(\vec{\xi},\tau)$, and $\mu(\vec{\xi},\tau)$, e.g.,

$$\langle\epsilon(\vec{x},t)\rangle = \check{\epsilon}[\beta(\vec{\xi},\tau), \vec{u}(\vec{\xi},\tau), \mu(\vec{\xi},\tau); \vec{x}, t). \tag{5.9}$$

Inverting the five relations like Eq. (5.9) and substituting in the expressions for the right hand side like Eq. (5.8), we get

$$\langle T^{ij}(\vec{x},t)\rangle = \check{T}^{ij}[\langle\epsilon(\vec{\xi},\tau)\rangle, \langle\vec{u}(\vec{\xi},\tau)\rangle, \langle\nu(\vec{\xi},\tau)\rangle; \vec{x}, t). \tag{5.10}$$

Thus the set of equations (5.7) can be turned into a closed set of deterministic equations of motion for the expected values of the quasiclassical variables $\langle\epsilon(\vec{x},t)\rangle$, $\langle\vec{\pi}(\vec{x},t)\rangle$, and $\langle\nu(\vec{x},t)\rangle$.

The process of expression and inversion adumbrated above could be difficult to carry out in practice. The familiar classical equations of motion arise from further approximations, in particular from assuming that the gradients of all quantities are small. For example, for a nonrelativistic fluid of particles of mass m, the most general Galilean-invariant form of the stress tensor that is linear in the gradients of the fluid velocity $\vec{u}(x)$ has the approximate form [24]

[13]The $\approx$ in Eq. (5.2) means equality to the extent that there is local equilibrium. This is a different approximation from that in Eq. (2.11), where it is the off-diagonal elements of the decoherence functional that are negligible.

$$\check{T}^{ij} = p\delta^{ij} + m\nu u^i u^j - \eta\left[\frac{\partial u^i}{\partial x^j} + \frac{\partial u^j}{\partial x^i} - \frac{2}{3}\delta_{ij}(\vec{\nabla}\cdot\vec{u})\right] - \zeta\delta_{ij}(\vec{\nabla}\cdot\vec{u}). \tag{5.11}$$

The pressure p and coefficients of viscosity η and ζ are themselves functions, say, of β and ν determined by the construction leading to Eq. (5.10). This form of the stress tensor in Eq. (5.7a) leads to the Navier-Stokes equation.

C. Quasiclassical histories

A quantum system can be said to behave quasiclassically when, in a suitable realm, the probability is high for its quasiclassical variables to be correlated in time by approximate, deterministic classical laws of motion. For instance, according to quantum mechanics there is a probability for the Earth to move on any orbit around the Sun. The Earth moves classically when the probability is high that histories of suitably coarse-grained positions of the Earth's center of mass are correlated by Newton's laws of motion. The probabilities defining these correlations are probabilities of *time histories* of coarse-grained center-of-mass positions of the Earth. The behavior of the expected values of position as a function of time is not enough to evaluate these probabilities. Similarly the classical behavior of the expected values of quasiclassical hydrodynamic variables derived in the last subsection does not necessarily imply either the decoherence or the classicality of these variables except in very special situations.

One example of such a special situation for the motion of a nonrelativistic particle in one dimension starts from Ehrenfest's theorem:

$$m\frac{d^2\langle x\rangle}{dt^2} = -\left\langle\frac{dV(x)}{dx}\right\rangle. \tag{5.12}$$

If the initial state is a narrow wave packet that does not spread very much over the time of interest this becomes approximately

$$m\frac{d^2\langle x\rangle}{dt^2} \approx -\frac{dV(\langle x\rangle)}{d\langle x\rangle}. \tag{5.13}$$

This is a classical equation of motion for the expected value $\langle x\rangle$ not dissimilar in character from those for quasiclassical variables in the previous subsection.

Suppose we study the history of the particle using a set of histories coarse grained by ranges of x at a sequence of times with the ranges all large compared to the width of the initial wave packet. Then, at sufficiently early times, the only history with a non-negligible amplitude consists of the ranges traced out by the center of the wave packet along $\langle x(t)\rangle$. Decoherence of this set is immediate since only the one coarse-grained history tracking $\langle x(t)\rangle$ has any significant amplitude. And, since that history is correlated in time by Eq. (5.13), the probability for classical correlations in time is near unity.

In this example the initial state is very special. Also, the coarse graining is very coarse—too coarse, for instance, to exhibit any quantum corrections to classical behavior.

In several interesting papers [17–19] Halliwell has provided a demonstration of the classical behavior of quasiclassical hydrodynamic variables. This has something of the character of the Ehrenfest example although it is more technically complex and less special and correspondingly provides more insight into the problem of classicality.

A very brief summary of his assumptions and results are as follows:

(a) Consider a system of N nonrelativistic particles in a box with a Hamiltonian H specified by two-body potentials with a characteristic range L. Consider an initial state $|\Psi\rangle$ that is an approximate eigenstate of the quasiclassical variables $(\epsilon_V(\vec{y},t), \vec{\pi}_V(\vec{y},t), \nu_V(\vec{y},t))$. (This can be achieved approximately even though these variables do not commute.) The volume V is chosen so that (i) $\langle\nu_V(y)\rangle$ is large, and (ii) $V \gg L^3$.

(b) Define a set of quasiclassical histories of the type discussed in Sec. III A, using ranges of the $(\epsilon_V(\vec{y},t), \vec{\pi}_V(\vec{y},t), \nu_V(\vec{y},t))$ at a sequence of times. Under the above assumptions it is possible to show that the fluctuations in any of the quasiclassical quantities are small and remain small, for instance

$$\langle[\Delta\nu_V(\vec{y},t)]^2\rangle/\langle\nu_V(\vec{y},t)\rangle^2 \ll 1. \tag{5.14}$$

The approximate conservation underlying the quasiclassical quantities ensures that the relations like Eq. (5.14) hold over time.

(c) As a consequence Halliwell shows that histories of the quasiclassical variables are approximately decoherent and that their probabilities are peaked about the evolution given by the classical equations of motion for the expected values. The crucial reason is that the small fluctuation relations like Eq. (5.14) mean that there is essentially only one history in each set, somewhat as in the Ehrenfest example.

Halliwell's result is the best we have today. But in our opinion there is still much further work to be done in demonstrating the quasiclassical behavior of *histories* of quasiclassical variables. In particular, as mentioned earlier, quasiclassical realms that are maximally refined consistent with decoherence and classicality are of interest. The coarse graining in the above analysis is likely to be much coarser than that.

The contrast with the studies of classicality in the much simpler oscillator models (e.g., [5,19,41]) is instructive. (The reader not familiar with this work should skip this paragraph.) On the negative side these models do not deal with the quasiclassical variables under discussion here but rather with positions of particles or averages of these. Consequently they assume an arbitrary system environment split. However, on the positive side it is possible to study various levels of refinement of the coarse graining, to provide quantitative estimates for such quantities as the decoherence time, and to exhibit the effects of quantum noise on classical behavior. We express the hope that it will someday be possible to achieve similar levels of precision with the more realistic quasiclassical realms.

VI. INFORMATION AND ENTROPY

Information must be sacrificed through coarse graining to allow the existence of probabilities for nontrivial sets of histories. Further coarse graining is needed to achieve sets of histories that exhibit predictable regularities such as those of the quasiclassical realms. These are two messages from the previous sections. Quantitative measures of the missing information are supplied by various candidates for entropy that can be constructed in connection with realms.[14] This section focuses on one measure of missing information and its connection with the usual entropy of chemistry and physics.

We review the general prescription for constructing entropies from coarse grainings. When information is missing about the state of a quantum system, that system can be described by a density matrix ρ. The natural and usual measure of the missing information is the entropy of ρ defined as

$$S(\rho) = -\,\mathrm{Tr}(\rho \ln \rho). \tag{6.1}$$

This is zero when ρ is a pure state, $\rho = |\Psi\rangle\langle\Psi|$, showing that a pure state is a complete description of a quantum system. For a Hilbert space of finite dimension $\mathcal{N}$, complete ignorance is expressed by $\rho = I/\mathrm{Tr}(I)$. The maximum value of the missing information is $\ln \mathcal{N}$.

A coarse-grained description of a quantum system generally consists of specifying the expected values of certain operators A_m $(m=1,\ldots,M)$ in the "mixed" state ρ. That is, the system is described by specifying

$$\langle A_m\rangle \equiv \mathrm{Tr}(A_m\rho)(m=1,\ldots,M). \tag{6.2}$$

For example, the $\{A_m\}$ might be an exhaustive set of orthogonal projection operators $\{P_\alpha\}$ of the kind used in Sec. II to describe yes or no alternatives at one moment of time. These projections might be onto ranges of the center-of-mass position of the planet Mars. In a finer-grained quasiclassical description, of the kind discussed in Sec. V, they might be projections onto ranges of values of the densities of energy, momentum, and other conserved or nearly conserved quantities averaged over small volumes in the interior of Mars. In a coarse graining of that, the $\{A_m\}$ might be the operators $\epsilon_V(\vec{y},t)$, $\vec{\pi}_V(\vec{y},t)$, and $\nu_V(\vec{y},t)$ themselves, so that the description at one time is in terms of the expected values of variables rather than expected values of projections onto ranges of values of those variables. As these examples illustrate, the $\{A_m\}$ are not necessarily mutually commuting.

The measure of missing information in descriptions of the form (6.2) for a quantum system with density matrix ρ is the maximum entropy over the effective density matrices $\tilde{\rho}$ that are consistent with the coarse-grained description (e.g., [25–27]). Specifically,

$$S(\{A_m\},\rho) \equiv -\,\mathrm{Tr}(\tilde{\rho} \ln \tilde{\rho}), \tag{6.3}$$

where $\tilde{\rho}$ maximizes this quantity subject to the constraints

$$\mathrm{Tr}(A_m\tilde{\rho}) = \langle A_m\rangle \equiv \mathrm{Tr}(A_m\rho), \quad m=1,\ldots,M. \tag{6.4}$$

The solution to this maximum problem is straightforward to obtain by the method of Lagrange multipliers and is

$$\tilde{\rho} = Z^{-1}\exp\left(-\sum_{m=1}^{M}\lambda^m A_m\right). \tag{6.5}$$

Here, the $\{\lambda_m\}$ are c-number Lagrange multipliers determined in terms of ρ and A_m by the coarse graining constraints (6.4); Z ensures normalization. The missing information is then the entropy

$$S(\{A_m\},\rho) = -\,\mathrm{Tr}(\tilde{\rho} \ln \tilde{\rho}). \tag{6.6}$$

As mentioned earlier the A's need not be commuting for this construction. What is important is that the constraints (6.4) are linear in ρ.

Suppose that a coarser-grained description is characterized by operators $\{\bar{A}_{\bar{m}}\}$, $\bar{m}=1,\ldots,\bar{M}<M$ such that

$$\bar{A}_{\bar{m}} = \sum_{m\in\bar{m}} c_{\bar{m}m}A_m, \tag{6.7}$$

where the $c_{\bar{m}m}$ are coefficients relating the operators of the finer graining to those of the coarser graining. For example, suppose that the A_m are averages over volumes $V(\vec{y})$ of one of the quasiclassical variables. Averaging over bigger volumes $\bar{V}(\vec{z})$ that are unions of these would be a coarser-grained description. The coefficients $c_{\bar{m}m}$ would become $c(\vec{z},\vec{y}) \equiv V(\vec{y})/\bar{V}(\vec{z})$. Then, we obtain

$$S(\{\bar{A}_{\bar{m}}\},\rho) \geq S(\{A_m\},\rho), \tag{6.8}$$

since there are fewer constraints to apply in the maximum-missing-information construction. Entropy increases with further coarse graining.

Among the various possible operators $\{A_m\}$ that could be constructed from those defining histories of the quasiclassical realm, which should we choose to find a connection (if any), with the usual entropy of chemistry and physics? Familiar thermodynamics provides some clues. The first law of thermodynamics connects changes in the *expected values* of energy over time with changes in volume and changes in entropy in accord with conservation of energy. This suggests that for our present purpose we should consider an entropy defined at one time rather than for a history. It also suggests that we should consider the entropy defined by the expected values of quasiclassical quantities and not use the finer-grained description by ranges of values of the quantities themselves, which enter naturally into histories.

To see that this is on the right track, let us compute the missing information specifying the expected values of the total energy H, total momentum $\vec{P}$, and total conserved number N for our model system in a box. The density matrix maximizing the missing information is,[15] from Eq. (6.5),

[14]For discussion and comparison of some of these measures see [2,42].

[15]Note that in Eq. (6.9) H means the Hamiltonian, not the enthalpy, and $\langle H\rangle$ is the same as what is often denoted by U in thermodynamics.

$$\tilde{\rho}_{\max} = Z^{-1} \exp[-\beta(H - \vec{U} \cdot \vec{P} - \mu N)]. \tag{6.9}$$

This is the equilibrium density matrix (6.3). The entropy S defined by Eq. (6.6) is straightforwardly calculated from the (Helmholtz) free energy F defined by

$$\mathrm{Tr}\{\exp[-\beta(H - \vec{U} \cdot \vec{P} - \mu N)]\} \equiv \exp[-\beta(F - \vec{U} \cdot \langle \vec{P} \rangle - \mu \langle N \rangle)]. \tag{6.10}$$

We find

$$S = \beta(\langle H \rangle - F). \tag{6.11}$$

This standard thermodynamic relation shows that we have recovered the standard entropy of chemistry and physics.

In an analogous way, the missing information can be calculated for a coarse graining in which $\langle \epsilon(\vec{y},t) \rangle$, $\langle \vec{\pi}(\vec{y},t) \rangle$, and $\langle \nu(\vec{y},t) \rangle$ are specified at one moment of time. The density matrix that maximizes the missing information according to Eq. (6.3) is that for the assumed *local* equilibrium (5.4). The entropy is

$$S = \sum_{\vec{y}} \beta(\vec{y},t)[\langle \epsilon(\vec{y},t) \rangle - \langle \phi(\vec{y},t) \rangle], \tag{6.12}$$

where $\langle \phi(\vec{y},t) \rangle$ is the free energy density defined analogously to Eq. (6.10). The integrand of Eq. (6.12) is then naturally understood as defining the average over small volumes of an *entropy density* $\sigma(\vec{x},t)$.

What is being emphasized here is that, in these ways, the usual entropy of chemistry and physics arises naturally from the coarse graining that is inescapable for quantum mechanical decoherence and for quasiclassical predictability of histories, together with the assumption of local equilibrium.[16]

VII. THE SECOND LAW OF THERMODYNAMICS

We can now discuss the time evolution in our universe of the entropy described in the previous section by the coarse graining defining a quasiclassical realm. Many pieces of this discussion are standard (see, e.g., [23]) although not often described in the context of quantum cosmology.

The context of this discussion continues to be our model universe of fields in a box. For the sake of simplicity, we are not only treating an artificial universe in a box but also sidestepping vital issues such as the accelerated expansion of the universe, possible eternal inflation, the decay of the proton, the evaporation of black holes, long range forces, gravitational clumping, etc. (See, for example, [28–30].)

For this discussion we will distinguish two connected but different features of the universe:

(a) the tendency of the total entropy[17] of the universe to increase;

(b) the tendency of the entropy of each presently almost isolated system to increase in the same direction of time. This might be called the homogeneity of the thermodynamic arrow of time.

Evidently these features are connected. The first follows from the second, but only in the late universe when almost isolated systems are actually present. In the early universe we have only the first. Together they may be called the second law of thermodynamics.

More particularly, isolated systems described by quasiclassical coarse grainings based on approximately conserved quantities can be expected to evolve toward equilibrium characterized by the total values of these conserved quantities. In our model universe in a box, the probabilities of $\epsilon_V(\vec{y},t)$, $\vec{\pi}_V(\vec{y},t)$, $\nu_V(\vec{y},t)$, and their correlations are eventually given by the equilibrium density matrix (5.3). The conditions that determine the equilibrium density matrix are sums of the conditions that determine the local equilibrium density matrix in the Jaynes construction (6.4). The smaller number of conditions means that the equilibrium entropy will be larger than that for any local equilibrium [cf. Eq. (6.8)].

Two conditions are necessary for our universe to exhibit a general increase in total entropy defined by quasiclassical variables:

(a) The quantum state $|\Psi\rangle$ is such that the initial entropy is near the minimum it could have for the coarse graining defining it. It then has essentially nowhere to go but up.

(b) The relaxation time to equilibrium is long compared to the present age of the universe so that the general tendency of its entropy to increase will dominate its evolution.

In our simple model we have neglected gravitation for simplicity, but for the following discussion we restore it. Gravity is essential to realizing the first of these conditions because in a self-gravitating system gravitational clumping increases entropy. The early universe is approximately homogeneous, implying that the entropy has much more room to increase through the gravitational growth of fluctuations. In a loose sense, as far as gravity is concerned, the entropy of the early universe is low for the coarse graining we have been discussing. The entropy then increases. Note that a smaller effect in the opposite direction is connected with the thermodynamic equilibrium of the matter and radiation in the very early universe. See [31] for an entropy audit of the *present* universe.

Coarse graining by approximately conserved quasiclassical variables helps with the second of the two conditions above—relaxation time short compared to present age. Small volumes come to local equilibrium quickly. But the approximate conservation ensures that the whole system will approach equilibrium slowly, whether or not such equilibrium is actually attained. Again gravity is important because its

[16]In some circumstances (for example, when we take gravitation into account in metastable configurations of matter) there may be other kinds of entropy that are appropriate such as q entropy with $q \neq 1$. Such circumstances have been excluded from our model for simplicity, but for more on them see, e.g., [43].

[17]In the context of our model universe in a box the total entropy is defined. For the more realistic cases of a flat or open universe the total entropy may be infinite and we should refer to the entropy of a large comoving volume.

effects concentrate a significant fraction of the matter into almost isolated systems such as stars and galactic halos, which strongly interact with one other only infrequently.[18]

Early in the universe there were no almost isolated systems. They arose later from the condensation of initial fluctuations by the action of gravitational attraction.[19] They are mostly evolving toward putative equilibrium in the same direction of time. Evidently this homogeneity of the thermodynamic arrow of time cannot follow from the approximately time-reversible dynamics and statistics alone. Rather the explanation is that the progenitors of today's nearly isolated systems were all far from equilibrium a long time ago and have been running downhill ever since. This provides a stronger constraint on the initial state than merely having low total entropy. As Boltzmann put it over a century ago: "The second law of thermodynamics can be proved from the [time-reversible] mechanical theory, if one assumes that the present state of the universe...started to evolve from an improbable [i.e., special] state" [32].

The initial quantum state of our universe must be such that it leads to the decoherence of sets of quasiclassical histories that describe coarse-grained spacetime geometry and matter fields. Our observations require this now, and the successes of the classical history of the universe suggests that there was a quasiclassical realm at a very early time. In addition, the initial state must be such that the entropy of quasiclassical coarse graining is low in the beginning and also be such that the entropy of presently isolated systems was also low. Then the universe can exhibit both aspects of the second law of thermodynamics.

The quasiclassical coarse grainings are therefore distinguished from others, not only because they exhibit predictable regularities of the universe governed by approximate deterministic equations of motion, but also because they are characterized by a sufficiently low entropy in the beginning and a slow evolution towards equilibrium, which makes those regularities exploitable.

This confluence of features suggests the possibility of a connection between the dynamics of the universe and its initial condition. The "no-boundary" proposal for the initial quantum state [33] is an example of just such a connection. According to it, the initial state is computable from the Euclidean action, in a way similar to the way the ground state of a system in flat space can be calculated.

VIII. CONCLUSIONS AND COMMENTS

Many of the conclusions of this paper can be found among the items in the list in the Introduction. Rather than reiterating all of them here, we prefer to discuss in this section some broader issues. These concern the relation between our approach to quantum mechanics, based on coarse-grained decoherent histories of a closed system, and the approximate quantum mechanics of measured subsystems, as in the "Copenhagen interpretation." The latter formulation *postulates* (implicitly for most authors or explicitly in the case of Landau and Lifshitz [34]) a classical world and a quantum world, with a movable boundary between the two. Observers and their measuring apparatus make use of the classical world, so that the results of a "measurement" are ultimately expressed in one or more "c numbers."

We have emphasized that this widely taught interpretation, although successful, cannot be the fundamental one because it seems to require a physicist outside the system making measurements (often repeated ones) of it. That would seem to rule out any application to the universe, so that quantum cosmology would be excluded. Also billions of years went by with no physicist in the offing. Are we to believe that quantum mechanics did not apply to those times?

In this discussion, we will concentrate on how the Copenhagen approach fits in with ours as a set of special cases and how the "classical world" can be replaced by a quasiclassical realm. Such a realm is not *postulated* but rather is *explained* as an emergent feature of the universe characterized by H, $|\Psi\rangle$, and the enormously long sequences of accidents (outcomes of chance events) that constitute the coarse-grained decoherent histories. The material in the preceding sections can be regarded as a discussion of how quasiclassical realms emerge.

We say that a "measurement situation" exists if some variables (including such quantum-mechanical variables as electron spin) come into high correlation with a quasiclassical realm. In this connection we have often referred to fission tracks in mica. Fissionable impurities can undergo radioactive decay and produce fission tracks with randomly distributed definite directions. The tracks are there irrespective of the presence of an "observer." It makes no difference if a physicist or other human or a chinchilla or a cockroach looks at the tracks. Decoherence of the alternative tracks induced by interaction with the other variables in the universe is what allows the tracks to exist independent of "observation" by an observer. All those other variables are effectively doing the observing. The same is true of the successive positions of the moon in its orbit not depending on the presence of observers and for density fluctuations in the early universe existing when there were no observers around to measure them.

The idea of "collapse of the wave function" corresponds to the notion of variables coming into high correlation with a quasiclassical realm, with its decoherent histories that give true probabilities. The relevant histories are defined only through the projections that occur in the expressions for these probabilities [cf. Eq. (2.7)]. Without projections, there are no questions and no probabilities. In many cases conditional probabilities are of interest. The collapse of the probabilities that occurs in their construction is no different from the collapse that occurs at a horse race when a particular horse wins and future probabilities for further races conditioned on that event become relevant.

The so-called "second law of evolution," in which a state is "reduced" by the action of a projection, and the probabilities renormalized to give ones conditioned on that projection, is thus not some mysterious feature of the measurement pro-

[18]For thoughts on what happens in the very long term in an expanding universe, see [44,45].

[19]Much later certain IGUS's create isolated systems in the laboratory which typically inherit the thermodynamic arrow of the IGUS and apparatus that prepared them.

cess. Rather it is a natural consequence of the quantum mechanics of decoherent histories, dealing with alternatives much more general than mere measurement outcomes.

There is thus no actual conflict between the Copenhagen formulation of quantum theory and the more general quantum mechanics of closed systems. Copenhagen quantum theory is an approximation to the more general theory that is appropriate for the special case of measurement situations. Decoherent histories quantum mechanics is rather a *generalization* of the usual approximate quantum mechanics of measured subsystems.

In our opinion decoherent histories quantum theory advances our understanding in the following ways among many others:

(a) Decoherent histories quantum mechanics extends the domain of applicability of quantum theory to histories of features of the universe irrespective of whether they are receiving attention of observers and in particular to histories describing the evolution of the universe in cosmology.

(b) The place of classical physics in a quantum universe is correctly understood as a property of a particular class of sets of decoherent coarse-grained alternative histories—the quasiclassical realms [5,9]. In particular, the *limits* of a quasiclassical description can be explored. Dechoherence may fail if the graining is too fine. Predictability is limited by quantum noise and by the major branchings that arise from the amplification of quantum phenomena as in a measurement situation. Finally, we cannot expect a quasiclassical description of the universe in its earliest moments where the very geometry of spacetime may be undergoing large quantum fluctuations.

(c) Decoherent histories quantum mechanics provides new connections such as the relation (which has been the subject of this paper) between the coarse graining characterizing quasiclassical realms and the coarse graining characterizing the usual thermodynamic entropy of chemistry and physics.

(d) Decoherent histories quantum theory helps with understanding the Copenhagen approximation. For example, measurement was characterized as an "irreversible act of amplification," "the creation of a record," or as "a connection with macroscopic variables." But these were inevitably imprecise ideas. How much did the entropy have to increase, how long did the record have to last, what exactly was meant by "macroscopic?" Making these ideas precise was a central problem for a theory in which measurement is fundamental. But it is less central in a theory where measurements are just special, approximate situations among many others. Then characterizations such as those above are not false, but true in an approximation that need not be exactly defined.

(e) Irreversibility clearly plays an important role in science as illustrated here by the two famous applications to quantum-mechanical measurement situations and to thermodynamics. It is not an absolute concept but context dependent like so much else in quantum mechanics and statistical mechanics. It is highly dependent on coarse graining, as in the case of the document shredding [7]. This was typically carried out in one dimension until the seizure by Iranian "students" of the U.S. Embassy in Tehran in 1979, when classified documents were put together and published. Very soon, in many parts of the world, there was a switch to two-dimensional shredding, which still appears to be secure today. It would now be labeled as irreversible just as the one-dimensional one was previously. The shredding and mixing of shreds clearly increased the entropy of the documents, in both cases by an amount dependent on the coarse grainings involved. Irreversibility is evidently not absolute but dependent on the effort or cost involved in reversal.

The founders of quantum mechanics were right in pointing out that something external to the framework of wave function and Schrödinger equation *is* needed to interpret the theory. But it is not a postulated classical world to which quantum mechanics does not apply. Rather it is the initial condition of the universe that, together with the action function of the elementary particles and the throws of quantum dice since the beginning, explains the origin of quasiclassical realm(s) within quantum theory itself.

ACKNOWLEDGMENTS

We thank J. Halliwell and S. Lloyd for useful recent discussions. We thank the Aspen Center for Physics for hospitality over several summers while this work was in progress. J.B.H. thanks the Santa Fe Institute for supporting several visits there. The work of J.B.H. was supported in part by the National Science Foundation under Grant No. PHY02-44764 and Grant No. PHY05-55669. The work of M.G.-M. was supported by the C.O.U.Q. Foundation, by Insight Venture Management, and by the KITP in Santa Barbara. The generous help provided by these organizations is gratefully acknowledged.

APPENDIX A: TRIVIAL DECOHERENCE OF PERFECTLY FINE-GRAINED SETS OF HISTORIES

A set of perfectly fine-grained histories trivially decoheres when there is a different final projection for each history with a nonzero branch state vector. Equivalently, we could say that a set of completely fine-grained histories is trivial if each final projection has only one possible prior history. The orthogonality of the final alternatives then automatically guarantees the decoherence of the realm. Such sets are trivial in the sense that what happens at the last moment is uniquely correlated with the alternatives at all previous moments. This appendix completes the demonstration, adumbrated in Sec. III A, that completely fine-grained realms are trivial.

The domain of discussion was described in Sec. III A. We consider sets of histories composed of completely fine-grained alternatives at any one time specified by a complete basis for Hilbert space. It is sufficient to consider a sequence of times t_k $(k=1,2,\ldots,n)$ because if such decoherent sets are trivial all finer-grained sets will also be trivial. In Sec. III A we denoted the bases by $\{|k,i_k\rangle\}$. Here, in the hope of keeping the notation manageable, we will drop the first label and just write $\{|i_k\rangle\}$. The condition for decoherence (2.8) is then

$$\langle\Psi_{i'_n,\ldots,i'_1}|\Psi_{i_n,\ldots,i_1}\rangle = p(i_n,\ldots,i_1)\delta_{i'_n i_n}\cdots\delta_{i'_1 i_1}, \quad \text{(A1)}$$

where the branch state vectors $|\Psi_{i_n,\ldots,i_1}\rangle$ are defined by Eq. (2.6) with alternatives at each time of the form (3.1) and where equality has replaced $\approx$ because we are insisting on exact decoherence.

Mathematically, the decoherence condition (A1) can be satisfied in several ways. Some of the histories could be represented by branch state vectors which vanish (zero histories for short). These have vanishing inner products with all branch state vectors including themselves. They therefore do not affect decoherence and their probabilities are equal to zero. Further discussion can therefore be restricted to the nonvanishing branches.

Decoherence in the final alternatives i_n is automatic because the projections $P_{i'_n}$ and P_{i_n} are orthogonal if different. Decoherence is also automatic if each nonzero history has its own final alternative or, equivalently, if each final alternative i_n has a unique prior history. Then decoherence of the final alternatives guarantees the decoherence of the set.

Two simple and trivial examples may help to make this discussion concrete. Consider the set of completely fine-grained histories defined by taking the same basis $\{|i\rangle\}$ at each time. The resulting orthogonality of the projections between times ensures that the only nonzero histories have $i_n = i_{n-1} = \cdots = i_2 = i_1$ in an appropriate notation. Evidently each i_n corresponds to exactly one chain of past alternatives and the decoherence condition (A1) is satisfied.

A related trivial example is obtained choosing the $\{|i_k\rangle\}$ to be the state $|\Psi\rangle$ and some set of orthogonal states at each of the times $t_1, \ldots, t_{n-1}$. That is, we are mindlessly asking the question, "Is the system in the state $|\Psi\rangle$ or not?" over and over again until the last alternative. The only non-zero branches are of the form

$$|i_n\rangle\langle i_n|\Psi\rangle\langle\Psi|\Psi\rangle\cdots\langle\Psi|\Psi\rangle = |i_n\rangle\langle i_n|\Psi\rangle. \tag{A2}$$

In this case, each final alternative is correlated with the same set of previous alternatives. The set decoheres because the only nontrivial branching is at the last time as the equality in Eq. (A2) shows. But then each history has a unique final end alternative and the set is thus trivial.[20]

Trivial sets of completely fine-grained histories have many zero histories as in the above examples. To see this more quantitatively, imagine for a moment that the dimension of Hilbert space is a very large but finite number $\mathcal{N}$. A generic, completely fine-grained set with n times would consist of $\mathcal{N}^n$ histories. But there can be at most $\mathcal{N}$ orthogonal branches—no more than are be supplied just by alternatives at one time. Indeed the trivially decohering completely fine-grained sets described above have at most $\mathcal{N}$ branches. Most of the possible histories must therefore be zero. Further, assuming that at least one nonzero branch is added at each time, the number of times n is limited to $\mathcal{N}$.

Were a final alternative in a completely fine-grained set to have more than one possible previous nonzero history, the set would not decohere. To see this, suppose some particular final alternative k_n had two possible, nonzero, past histories. Let the earliest (and possibly the only) moment the histories differ be time t_j. The condition for decoherence is [cf. Eq. (A1)]

$$\langle\Psi|i'_1\rangle\langle i'_1|i'_2\rangle\cdots\langle i'_{n-1}|k_n\rangle\langle k_n|i_{n-1}\rangle\cdots\langle i_2|i_1\rangle\langle i_1|\Psi\rangle = p(k_n,\ldots,i_1)\delta_{i'_{n-1}i_{n-1}}\cdots\delta_{i'_1 i_1}. \tag{A3}$$

Sum this over all the $i_1,\ldots,i_{j-1}$ and $i'_1,\ldots,i'_{j-1}$ to find

$$\langle\Psi|i'_j\rangle M(i'_j,\ldots,i'_{n-1},k_n,i_{n-1},\ldots,i_j)\langle i_j|\Psi\rangle = p(k_n,\ldots,i_j)\delta_{i'_{n-1}i_{n-1}}\cdots\delta_{i'_j i_j}, \tag{A4}$$

where we have employed an abbreviated notation for the product of all the remaining matrix elements. By hypothesis there are at least two chains $(i_j,\ldots,k_n)$ differing in the value of i_j for which M does not vanish and the history is not zero. But decoherence at time t_j then requires that $\langle i_j|\Psi\rangle$ vanish for all but one i_j, contradicting the assumption that there are two nonzero histories.

The only decohering, perfectly fine-grained sets are thus trivial.

We now discuss the connection between this notion of a trivial fine-grained realm and the idea of generalized records introduced in [3,10]. Generalized records of a realm are a set of orthogonal projections $\{R_\alpha\}$ that are correlated to a good approximation with the histories of the realm $\{C_\alpha\}$ so that

$$R_\alpha C_\beta|\Psi\rangle \approx \delta_{\alpha\beta}C_\alpha|\Psi\rangle. \tag{A5}$$

Generalized records of a particular kind characterize important physical mechanisms of decoherence. For instance, in the classic example of Joos and Zeh [16] a dust grain is imagined to be initially in a superposition of two places deep in intergalactic space. Alternative histories of coarse-grained positions of the grain are made to decohere by the interactions with the 10^{11} photons of the 3° cosmic background radiation that scatter every second. Through the interaction, records of the positions are created in the photon degrees of freedom, which are variables *not* followed by the coarse graining.

But the trivial generalized records that characterize fine-grained realms are not of this kind. They are in variables that *are* followed by the coarse graining. They could not be in other variables because the histories are fine-grained and follow everything. They are trivial records for a trivial kind of realm.[21]

APPENDIX B: NO NONTRIVIAL CERTAINTY ARISING FROM THE DYNAMICS PLUS THE INITIAL CONDITION

The Schrödinger equation is deterministic and the evolution of the state vector is certain. In this appendix we show

[20]Coarse-grained examples of both kinds of triviality can be given. Repeating the same Heisenberg picture sets of projections at all times is an example of the first. Histories defined by Heisenberg picture sets of projections onto a subspace containing $|\Psi\rangle$ and other orthogonal subspaces are examples of the second.

[21]In the general case of medium decoherence, we can construct projection operators onto orthogonal spaces each of which includes one of the states $C_\alpha|\Psi\rangle$ [10], but again these operators are not what we are really after in the way of generalized records *if they deal mainly with what is followed.*

that this kind of determinism is the only source of histories that are certain (probability equal to 1), coarse-grained or not, for a closed quantum mechanical system.

Suppose we have a set of exactly decoherent histories of the form (2.4), one member of which has probability 1, i.e., the history is certain. Denote the class operator for this particular history by C_1 and index its particular alternatives at each time so they are $\alpha_1 = \alpha_2 = \cdots \alpha_n = 1$. Thus we have

$$p(1) = \|C_1|\Psi\rangle\|^2 = 1, \quad p(\alpha) = \|C_\alpha|\Psi\rangle\|^2 = 0, \quad \alpha \neq 1. \tag{B1}$$

The second of these implies $C_\alpha|\Psi\rangle = 0$ for $\alpha \neq 1$. Given that $\Sigma_\alpha C_\alpha = I$, the first implies that $C_1|\Psi\rangle = |\Psi\rangle$. In summary we can write

$$C_\alpha|\Psi\rangle = P^1_{\alpha_n}(t_n)\cdots P^1_{\alpha_1}(t_1)|\Psi\rangle = \delta_{\alpha_n,1}\delta_{\alpha_{n-1},1}\cdots\delta_{\alpha_1,1}|\Psi\rangle. \tag{B2}$$

Summing both sides over this over all α's except those at time t_k, we get

$$P^k_{\alpha_k}(t_k)|\Psi\rangle = \delta_{\alpha_k,1}|\Psi\rangle. \tag{B3}$$

This means that all projections in a set of histories of which one is certain are either onto subspaces which contain $|\Psi\rangle$ or onto subspaces orthogonal to $|\Psi\rangle$. Effectively they correspond to questions that ask, "Is the state still $|\Psi\rangle$ or not?" That is how the certainty of unitary evolution is represented in the Heisenberg picture, where $|\Psi\rangle$ is independent of time.

Note that this argument does not exclude histories which are essentially certain over big stretches of time as in histories describing alternative values of a conserved quantity at many times. Then there would be, in general, initial probabilities for the value of the conserved quantity which then do not change through subsequent history.

[1] M. Gell-Mann and S. Lloyd, *Effective Complexity in Non-Extensive Entropy—Interdisciplinary Applications*, edited by M. Gell-Mann and C. Tsallis (Oxford University Press, New York, 2004).
[2] M. Gell-Mann and J. B. Hartle, in *Complexity, Entropy, and the Physics of Information*, SFI Studies in the Sciences of Complexity Vol. VIII, edited by W. Zurek (Addison-Wesley, Reading, MA, 1990).
[3] M. Gell-Mann and J. B. Hartle, in *Proceedings of the 25th International Conference on High Energy Physics*, Singapore, August, 2–8, 1990, edited by K. K. Phua and Y. Yamaguchi (South East Asia Theoretical Physics Association and Physical Society of Japan, distributed by World Scientific, Singapore, 1990).
[4] J. B. Hartle, in *Quantum Cosmology and Baby Universes: Proceedings of the 1989 Jerusalem Winter School for Theoretical Physics*, edited by S. Coleman, J. B. Hartle, T. Piran, and S. Weinberg (World Scientific, Singapore, 1991), pp. 65–157.
[5] M. Gell-Mann and J. B. Hartle, Phys. Rev. D **47**, 3345 (1993).
[6] M. Gell-Mann and J. B. Hartle, in *The Physical Origins of Time Asymmetry*, edited by J. Halliwell, J. Pérez-Mercader, and W. Zurek (Cambridge University Press, Cambridge, England, 1994).
[7] M. Gell-Mann, *The Quark and the Jaguar* (W. H. Freeman, New York, 1994).
[8] M. Gell-Mann and J. B. Hartle, e-print arXiv:gr-qc/9404013.
[9] J. B. Hartle, in *Proceedings of the Cornelius Lanczos International Centenary Conference*, edited by J. D. Brown, M. T. Chu, D. C. Ellison, and R. J. Plemmons (SIAM, Philadelphia, 1994).
[10] M. Gell-Mann and J. B. Hartle, in *Proceedings of the 4th Drexel Symposium on Quantum Non-Integrability—The Quantum-Classical Correspondence*, Drexel University, September 8–11, 1994, edited by D.-H. Feng and B.-L. Hu (International Press, Boston/Hong-Kong, 1995).
[11] R. B. Griffiths, *Consistent Quantum Theory* (Cambridge University Press, Cambridge, UK, 2002).
[12] R. Omnès, *Interpretation of Quantum Mechanics* (Princeton University Press, Princeton, NJ, 1994).
[13] J. B. Hartle, in *Proceedings of the 1992 Les Houches Summer School*, edited by B. Julia and J. Zinn-Justin, Les Houches Summer School Proceedings, Vol. LVII (North Holland, Amsterdam, 1995). A précis of these lectures is given in *Quantum Mechanics at the Planck Scale*, talk given at the *Workshop on Physics at the Planck Scale*, Puri, India, December 1994, eprint arXiv:gr-qc/9508023 (unpublished).
[14] J. B. Hartle, *Generalizing Quantum Mechanics for Quantum Spacetime, The Quantum Structure of Space and Time*, Proceedings of the 23rd Solvay Conference, e-print arXiv:gr-qc/0602013.
[15] L. Diósi, Phys. Rev. Lett. **92**, 170401 (2004).
[16] E. Joos and H. D. Zeh, Z. Phys. B: Condens. Matter **59**, 223 (1985).
[17] J. Halliwell, Phys. Rev. D **58**, 105015 (1998).
[18] J. J. Halliwell, Phys. Rev. Lett. **83**, 2481 (1999).
[19] J. J. Halliwell, Phys. Rev. D **68**, 025018 (2003).
[20] R. P. Feynman and J. R. Vernon, Ann. Phys. (N.Y.) **24**, 118 (1963).
[21] A. Caldeira and A. Leggett, Physica A **121**, 587 (1983).
[22] D. Forster, *Hydrodynamic Fluctuations, Broken Symmetry, and Correlation Functions* (Addison-Wesley, Redwood City, CA, 1975).
[23] D. N. Zubarev, *Nonequilibrium Statistical Thermodynamics*, edited by P. Gray and P. J. Shepherd (Consultants Bureau, New York, 1974).
[24] L. Landau and E. Lifshitz, *Fluid Mechanics* (Pergamon, London, 1959).
[25] E. T. Jaynes, *Papers on Probability Statistics and Statistical Mechanics*, edited by R. D. Rosenkrantz (D. Reidel, Dordrecht, 1983).
[26] A. Katz, *Principles of Statistical Mechanics: The Information Theory Approach* (W. H. Freeman, San Francisco, 1967).

[27] E. T. Jaynes, *Probability Theory* (Cambridge University Press, Cambridge, UK, 1993).

[28] A. Vilenkin, Nucl. Phys. B **226**, 527 (1983).

[29] A. Linde, Mod. Phys. Lett. A **1**, 81 (1986); Phys. Lett. B **175**, 395 (1986).

[30] A. Albrecht, in *Science and Ultimate Reality: From Quantum to Cosmos, Honoring John Wheeler's 90th Birthday*, edited by J. D. Barrow, P. C. W. Davies, and C. L. Harper (Cambridge University Press, Cambridge, England, 2003).

[31] B. Basu and D. Lynden-Bell, Q. J. R. Astron. Soc. **31**, 359 (1990).

[32] L. Boltzmann, Ann. Phys. **60**, 392 (1897).

[33] J. B. Hartle and S. W. Hawking, Phys. Rev. D **28**, 2960 (1983).

[34] L. Landau and E. Lifshitz, *Quantum Mechanics* (Pergamon, London, 1958).

[35] J. B. Hartle, Phys. Rev. D **44**, 3173 (1991).

[36] C. J. Isham, J. Math. Phys. **35**, 2157 (1994); C. J. Isham and N. Linden, *ibid.* **35**, 5452 (1994).

[37] J. B. Hartle, Phys. Rev. A **70**, 022104 (2004).

[38] H. F. Dowker and A. Kent, J. Stat. Phys. **82**, 1574 (1996).

[39] L. Diosi, Phys. Lett. B **203**, 267 (1995).

[40] D. Craig, e-print arXiv:gr-qc/9704031.

[41] T. Brun and J. B. Hartle, Phys. Rev. D **60**, 123503 (1999).

[42] T. Brun and J. B. Hartle, Phys. Rev. E **59**, 6370 (1999).

[43] C. Tsallis, J. Stat. Phys. **52**, 479 (1988); C. Tsallis, M. Gell-Mann, and Y. Sato, Proc. Natl. Acad. Sci. U.S.A. **153**, 15382 (2005); S. Umarov, C. Tsallis, and S. Steinberg (unpublished); S. Umarov, C. Tsallis, M. Gell-Mann, and S. Steinberg (unpublished).

[44] F. Dyson, Rev. Mod. Phys. **51**, 447 (1979).

[45] F. C. Adams and G. Laughlin, Rev. Mod. Phys. **69**, 337 (1997).

Progress in Elementary Particle Theory, 1946-1973
Outline in Technical Language

Murray Gell-Mann

California Institute of Technology, Pasadena, CA 91125

PROGRESS TO 1951

I) QUANTUM ELECTRODYNAMICS:

A) NON-RELATIVISTIC CALCULATION OF LAMB SHIFT
IN ROUGH AGREEMENT WITH OBSERVATION.

B) VERIFICATION THAT MASS RENORMALIZATION
ELIMINATES ALL INFINITIES REMAINING
IN SECOND ORDER
AFTER CHARGE RENORMALIZATION.

C) RELATIVISTIC CALCULATION OF LAMB SHIFT
GIVING FINITE RESULT
IN ACCURATE AGREEMENT WITH OBSERVATION

D) DEVELOPMENT OF ELEGANT COVARIANT METHODS
OF CALCULATION IN QUANTUM FIELD THEORY,
MAKING POSSIBLE
CALCULATION OF HIGHER ORDER CORRECTIONS
AND DEMONSTRATION THAT RENORMALIZATION WORKS
TO ALL ORDERS

II) MUONS AND PIONS

A) RECTIFICATION OF ERRONEOUS IDENTIFICATION
OF MUON WITH YUKAWA'S CHARGED MESON:
"TWO-MESON HYPOTHESIS,"
FOLLOWED BY EXPERIMENTAL DISCOVERY
OF CHARGED PIONS, THEN NEUTRAL PION,
ALL PSEUDOSCALAR.

B) UNDERSTANDING OF MUON DECAY AS

$$\mu^{\pm} \longrightarrow e^{\pm} + \nu + \bar{\nu}$$

C) PROPOSAL OF SOME SORT OF
"UNIVERSAL FERMI INTERACTION"
AMONG $\bar{\nu}e^-$, $\bar{\nu}\mu^-$, and $\bar{p}n$.

D) REVIVAL OF CHARGE INDEPENDENCE OF NUCLEAR FORCES
BASED ON ISOTOPIC SPIN CONSERVATION
AND ISOVECTOR PION.

E) CRUDE UNDERSTANDING OF
NUCLEAR FORCE AND $\pi - N$ INTERACTIONS
IN TERMS OF RENORMALIZABLE π, N FIELD THEORY
WITH COUPLING $ig\bar{N}\gamma_5 \underset{\sim}{\tau} N \cdot \underset{\sim}{\pi}$
LEADING TO NON-RELATIVISTIC APPROXIMATION
$\frac{ig}{2M}\bar{N}\underset{\sim}{\sigma}\underset{\sim}{\tau}N \cdot \underset{\sim}{\nabla}\underset{\sim}{\pi}$,
ALREADY INVESTIGATED BEFORE THE WAR.

F) PREDICTION THAT EVEN FOR INTERMEDIATE COUPLING
THE NON-RELATIVISTIC LAGRANGIAN

WOULD YIELD A $J = \frac{3}{2}^{+}$, $I = \frac{3}{2}$ EXCITED STATE
OF THE NUCLEON,
SOON CONFIRMED BY EXPERIMENT

G) SUSPICION THAT OTHER MESONS,
SUCH AS VECTOR MESONS,
MIGHT BE NEEDED AS WELL
TO MEDIATE NUCLEAR FORCE.

PROGRESS 1952-1962

I) QUANTUM FIELD THEORY

A) P, C, AND T ARE EMPIRICAL SYMMETRIES,
ALTHOUGH LOCAL FIELD THEORY REQUIRES PCT INVARIANCE

B) PROGRAM FOR EXTRACTING EXACT RESULTS FROM FIELD THEORIES:

1) "DISPERSION THEORY PROGRAM," JUST QUANTUM FIELD THEORY
FORMULATED ON MASS SHELL,
(RENAMED "S-MATRIX PROGRAM" AND MISUNDERSTOOD BY SOME
AS DEPARTURE FROM QUANTUM FIELD THEORY)

a) CROSSING RELATIONS

b) DISPERSION RELATIONS
FORWARD LIGHT SCATTERING GENERALIZED TO
SPIN-FLIP, NON-ZERO MASS, NON-FORWARD, ETC.

c) UNITARITY RELATIONS GENERALIZED
WITH AMPLITUDES EXTENDED
TO UNPHYSICAL VALUES OF MOMENTUM
BUT STILL ON MASS-SHELL

d) BOUNDARY CONDITIONS AS VARIOUS MOMENTUM VARIABLES $\to \infty$
HELP SPECIFY WHICH FIELD THEORY IS MEANT

2) RENORMALIZATION GROUP

a) FOR QUANTUM ELECTRODYNAMICS
AND OTHER QUANTUM FIELD THEORIES
STUDIED DURING THIS PERIOD
BY THE RENORMALIZATION GROUP METHOD
THE EFFECTIVE COUPLING STRENGTHS
INCREASE WITH DECREASING DISTANCE;
POSSIBILITY OF ASYMPTOTIC FREEDOM OVERLOOKED.

b) SINCE 1954 THERE HAS BEEN SOME DOUBT
THAT THEORIES BEHAVING LIKE QED
REALLY ARE CONSISTENT;
THIS QUESTION IS STILL NOT SETTLED.

C) YANG-MILLS THEORY

1) GROUP

a) SU_2 TRIVIALLY GENERALIZED TO PRODUCTS OF SU_2
AND U_1 FACTORS

b) GENERALIZED TO PRODUCTS OF ARBITRARY SIMPLE LIE GROUPS AND U_1 FACTORS

2) PROPERTIES

a) RENORMALIZABLE IF EXACT

b) PERFECT GAUGE INVARIANCE

c) MASSLESS VECTOR BOSONS

d) OTHER MULTIPLETS DEGENERATE WITH RESPECT TO THE GROUP

3) SOFT MASS MECHANISM CALLED FOR
(THAT WOULD BREAK SYMMETRIES b), c), AND EVEN d)
RENORMALIZABLY)
[BUT NOT FOUND UNTIL LATER]

D) SPINLESS BOSONS AND SYMMETRY BREAKING

1) EXACT CONTINUOUS SYMMETRIES,
WHEN NOT PRODUCING DEGENERACY,
CAN HAVE MASSLESS SPINLESS "NAMBU-GOLDSTONE BOSONS" INSTEAD

2) APPROXIMATE SYMMETRY CAN CORRESPOND
TO LOW-MASS SPINLESS BOSONS

3) "HIGGS-KIBBLE - - - - - ANDERSON MECHANISM:"
WAY OF BREAKING CONTINUOUS SYMMETRY
WITHOUT DEGENERACY OR MASSLESS SPINLESS BOSONS
IF THESE BOSONS ARE EATEN BY VECTOR GAUGE BOSONS,
WHICH THEN ACQUIRE MASS
[SUGGESTED ORIGINALLY IN 1962,
BUT NOT YET TAKEN SERIOUSLY
BY ELEMENTARY PARTICLE THEORISTS,
NOT YET RECOGNIZED AS A SOLUTION TO THE SOFT-MASS PROBLEM]

E) "REGGEISM":

1) COMPOSITE PARTICLES TEND TO LIE ON "REGGE TRAJECTORIES"
IN PLOT OF J vs M^2

2) IN SIMPLE PROBLEMS
IN NON-RELATIVISTIC QUANTUM MECHANICS
THESE CORRESPOND TO "REGGE" OR "SOMMERFELD" POLES
IN COMPLEX J PLANE

3) IN QUANTUM FIELD THEORY, THERE ARE STILL TRAJECTORIES
BUT THE POLES TEND TO BECOME
MORE COMPLICATED SINGULARITIES
(LIKE POLES ACCOMPANIED BY BRANCH POINTS)
[DETAILS NOT CLARIFIED UNTIL MUCH LATER]

4) IN VECTOR FIELD THEORIES,
ELEMENTARY SPINOR FIELDS GIVE REGGE TRAJECTORIES,
NOT FIXED POLES AS IN LOWEST ORDER

5) "REGGE BEHAVIOR" CAN PROVIDE BOUNDARY CONDITIONS
AT HIGH MOMENTA FOR THE DISPERSION THEORY PROGRAM

6) HIGH ENERGY BEHAVIOR OF SCATTERING AMPLITUDES
DEPENDS ON LEADING "REGGE" SINGULARITIES IN THE CROSS-CHANNEL
AS A FUNCTION OF MOMENTUM TRANSFER
(FOR A POLE, $s^{\alpha(t)}$ OR $s^{\alpha(u)}$)
[POLE DOMINANCE OVER CUTS AT VERY HIGH s
NOT CLARIFIED UNTIL MUCH LATER
EXCEPT AT $t = 0$ OR $u = 0$]

II) INTERACTIONS AND ELEMENTARY PARTICLES

A) DIVISION INTO STRONG, ELECTROMAGNETIC, WEAK, AND GRAVITATIONAL INTERACTIONS

1) ALL CONSERVE BARYON NUMBER
BUT NOT UNDERSTOOD WHY CONSERVATION EXACT
[NOW THOUGHT TO BE PROBABLY APPROXIMATE!]

2) a) LEPTON NUMBER CONSERVED
[NOW THOUGHT TO BE PROBABLY APPROXIMATE]

b) ELECTRON AND MUON NUMBER CONSERVED
[NOW THOUGHT TO BE TO BE PROBABLY APPROXIMATE]

B) ELECTROMAGNETISM

1) Q.E.D. RENORMALIZABLE FOR SPIN $\frac{1}{2}$,
ALSO FOR SPIN 0 IF $\lambda(\phi^+\phi)^2$ RENORMALIZED

2) ELECTROMAGNETIC INTERACTION MINIMAL (AMPÈRE'S LAW)
$p_\mu \to p_\mu - eA_\mu$ (ALWAYS RIGHT, BUT SOMETIMES AMBIGUOUS)

3) C AND P CONSERVED
TO THE EXTENT THEY ARE CONSERVED ELSEWHERE

4) QED OF VECTOR CHARGED PARTICLES
THOUGHT TO BE RELATED TO BROKEN YANG-MILLS THEORY
INCLUDING THE PHOTON
[BUT NOT KNOWN HOW TO DO THIS RENORMALIZABLY:
SAME AS PROBLEM OF SOFT MASS]

C) STRONG INTERACTION AND "STRONGLY INTERACTING PARTICLES" OR "HADRONS"

1) a) STRONG INTERACTION CONSERVES ISOTOPIC SPIN

b) ISOTOPIC SPIN MULTIPLETS CAN BE DISPLACED
IN CENTER OF CHARGE BY $S/2$;
THEN S CONSERVED BY STRONG INTERACTION
AND "STRANGE PARTICLES" EXPLAINED

c) SINCE ELECTROMAGNETISM IS MINIMAL AND Q IS LINEAR IN I_z,
ELECTROMAGNETISM OBEYS $|\Delta \underline{I}| = 0, 1$; $\Delta I_z = 0 \Rightarrow \Delta S = 0$

d) STRONG INTERACTION CONSERVES C AND P

[TODAY, ITS CONSERVATION OF CP TO HIGH ACCURACY
LOOKS REMARKABLE
AND THE MECHANISM IS MUCH DISCUSSED – HIGGLET (AXION), ETC.]

e) PARTICLE–ANTIPARTICLE SYMMETRY REALLY WORKS
FOR NUCLEONS!

2) a) "NUCLEAR DEMOCRACY" OR HADRONIC EGALITARIANISM
NO OBSERVABLE HADRON IS MORE FUNDAMENTAL THAN ANY OTHER

b) RICH SPECTRUM OF BARYON AND MESON STATES
CLASSIFIED BY J^P, I, S, Q (OR I_z), G (FOR MESONS)

c) BARYONS AND MESONS LIE ON
NEARLY STRAIGHT REGGE TRAJECTORIES
EXTENDING HIGH IN SPIN.
[STILL RATHER MYSTERIOUS]

d) APPROXIMATE EXCHANGE DEGENERACY
[STILL RATHER MYSTERIOUS]

e) HADRON HIGH-ENERGY SCATTERING AMPLITUDES
REALLY DOMINATED
BY APPROPRIATE REGGE SINGULARITIES IN THE CROSS CHANNEL

f) HADRON HIGH-ENERGY ELASTIC SCATTERING
DOMINATED BY "POMERANCHUK" SINGULARITY
($J = 1$ FOR $t = 0$)

g) AT $t = 0$ THIS COULD NOT BE A SIMPLE POLE
TOTAL CROSS SECTIONS COULD NOT BE ASYMPTOTICALLY CONSTANT
[SITUATION NOT FURTHER CLARIFIED UNTIL MUCH LATER]

3) a) STRONG INTERACTION HAS APPROXIMATE "FLAVOR" SU_3 INVARIANCE

b) BARYONS IN $\underset{\sim}{8}(J^P = \frac{1}{2}^+)$, $\underset{\sim}{10}$ $(J^P = 3/2^+)$
AND HIGHER MULTIPLETS

c) MESONS IN $\underset{\sim}{8}$ AND $\underset{\sim}{1}$ $(J^P = 0^-)$, $\underset{\sim}{8}$ and $\underset{\sim}{1}(J^P = 1^-)$
AND HIGHER MULTIPLETS

d) VIOLATION OF SU_3 BY OCTET 8TH COMPONENT
[AND A LITTLE BIT OF THIRD]

e) FIRST ORDER PREDOMINATES, $\Rightarrow$ MASS FORMULA
[STILL A LITTLE MYSTERIOUS]

4) a) STRONG INTERACTION HAS
QUITE GOOD "CHIRAL $SU_2 \times SU_2$" INVARIANCE
AND FAIR "CHIRAL $SU_3 \times SU_3$" INVARIANCE

b) APPROXIMATE SYMMETRY UNDER "VECTOR CURRENT" CHARGES
GIVES APPROXIMATE DEGENERACY, BUT
APPROXIMATE SYMMETRY UNDER "AXIAL VECTOR CURRENT" CHARGES
GIVES NEARLY MASSLESS
MODIFIED NAMBU-GOLDSTONE PSEUDOSCALAR BOSONS:

THE PSEUDOSCALAR MESONS

c) SO TERMS VIOLATING CONSERVATION OF AXIAL VECTOR CURRENTS
ARE SOMEHOW SOFT – "PCAC"
DIVERGENCE OF AXIAL CURRENT HAS MATRIX ELEMENTS
DOMINATED BY PSEUDOSCALAR MESON POLE;
THEY TRANSFORM LIKE $(\underset{\sim}{3},\bar{\underset{\sim}{3}})$ AND $(\bar{\underset{\sim}{3}},\underset{\sim}{3})$ UNDER $SU_3 \times SU_3$

d) BUT THERE IS NOT A VERY LIGHT PSEUDOSCALAR BOSON
CORRESPONDING TO THE TOTAL AXIAL VECTOR CURRENT CHARGE
[EXPLAINED MUCH LATER - IN QCD - USING ANOMALOUS DIVERGENCE OF THE CURRENT]

e) JUST AS FORM FACTOR
FOR DIVERGENCE OF AXIAL VECTOR CURRENT
DOMINATED AT LOW q^2 BY PSEUDOSCALAR MESON POLE
SO FORM FACTOR OF VECTOR CURRENT
DOMINATED AT LOW q^2 BY VECTOR MESON POLES
"VECTOR DOMINANCE"

f) THESE ARE IMPORTANT, OF COURSE,
BECAUSE VECTOR CURRENT OCCURS
IN ELECTROMAGNETIC AND WEAK COUPLINGS
AND AXIAL VECTOR CURRENT OCCURS IN WEAK COUPLINGS

g) FIRST INDICATION OF
ρ AND ω VECTOR MESONS
CAME FROM ELECTROMAGNETIC FORM FACTORS
FOUND IN $e - N$

ELASTIC SCATTERING EXPERIMENTS.

5) THERE IS NOT ONLY AN ALGEBRA $SU_3 \times SU_3$
OF VECTOR AND AXIAL VECTOR CHARGES
BUT ALSO A "CURRENT ALGEBRA"
OF VECTOR AND AXIAL VECTOR
CHARGE DENSITIES.

6) STRONG INTERACTION THOUGHT TO COME FROM A VECTOR THEORY,
MOST LIKELY YANG-MILLS,
BUT HOW CAN THAT BE ARRANGED IN "FLAVOR" SPACE
WITHOUT CLASHING
WITH WEAK AND ELECTROMAGNETIC GAUGE THEORY?
[THIS DIFFICULTY WAS OVERCOME WHEN WE FOUND COLOR]

D <u>WEAK INTERACTION</u>

1) a) CHARGE EXCHANGE WEAK INTERACTION OBEYING "PUPPI TRIANGLE"

GENERALIZED TO "TETRAHEDRON"

b) LEPTONIC INTERACTIONS OF "$\bar{p}n$", $|\Delta I_z| = 1, |\Delta \underset{\sim}{I}| = 1, \Delta S = 0$

LEPTONIC INTERACTIONS OF "$\bar{p}\Lambda$", $|\Delta I_z| = 1/2, |\Delta \underset{\sim}{I}| = 1/2, \frac{\Delta S}{\Delta Q} = +1$

$\Rightarrow$ NON-LEPTONIC INTERACTIONS HAVE NO $|\Delta S| = 2$

AND NON-LEPTONIC INTERACTIONS WITH $|\Delta S| = 1$

HAVE $|\Delta \underset{\sim}{I}| = \frac{1}{2}, \frac{3}{2} +$ ELECTROMAGNETIC CORRECTIONS

c) APPROXIMATE PREDOMINANCE OF NONLEPTONIC $|\Delta \underset{\sim}{I}| = 1/2$

[STILL SOMEWHAT MYSTERIOUS]

2) a) K^0 AND $\bar{K}^0$ MADE AND ABSORBED AS SUCH BUT DECAY AS K_1^0, K_2^0

(DEFINED BY C AT FIRST)

b) P AND C MAXIMALLY VIOLATED

IN LEPTON WEAK INTERACTIONS WITH HADRONS

AND WITH OTHER LEPTONS

(ONLY TWO COMPONENTS OF NEUTRINO UTILIZED –

<u>MAYBE</u> ONLY TWO COMPONENTS EXIST)

c) CP APPARENTLY CONSERVED (K_1^0, K_2^0 NOW DEFINED BY CP)

3) a) COUPLING IS $\gamma_\alpha(1 + \gamma_5)$ (V-A), UNIVERSAL

b) MOST LIKELY CHARGED SPIN 1 INTERMEDIATE BOSON

c) AS ITS MASS$\to\infty$, $\frac{G}{\sqrt{2}}J_\alpha^+ J_\alpha$

WITH $J_\alpha = \bar{\nu}\gamma_\alpha(1+\gamma_5)e + \bar{\nu}\gamma_\alpha(1+\gamma_5)\mu$ + HADRONIC V-A CURRENT

WITH PROPERTIES ABOVE

d) HADRON $\Delta S = 0$ VECTOR CURRENT

IS JUST COMPONENT OF ISOTOPIC SPIN CURRENT,

CONSERVED AND NOT CHANGED BY RENORMALIZATION

e) TWO NEUTRINOS DIFFERENT

TO PREVENT $\mu \to e + \gamma$ WHEN INTERMEDIATE BOSON EXISTS

$J_\alpha = \bar{\nu}_e\gamma_\alpha(1+\gamma_5)e + \bar{\nu}_\mu\gamma_\alpha(1+\gamma_5)\mu$

4) a) "$\bar{p}\Lambda$" TERM IN HADRONIC CURRENT SMALLER THAN "$\bar{p}n$" TERM

RATIO $\frac{\varepsilon}{\sqrt{1+\varepsilon^2}}$ TO $\frac{1}{\sqrt{1+\varepsilon^2}}$ OR $\sin\vartheta$ TO $\cos\vartheta$, $\vartheta \approx 15°$

b) WHOLE HADRONIC CURRENT MUST BE CURRENT OF A GENERATOR OF

SOME SENSIBLE ALGEBRA.

LOOKS LIKE

$\bar{\nu}_e\gamma_\alpha(1+\gamma_5)e + \bar{\nu}_\mu\gamma_\alpha(1+\gamma_5)\mu +$ "$\bar{p}\gamma_\alpha(1+\gamma_5)(\frac{n}{\sqrt{1+\varepsilon^2}} + \varepsilon\frac{\Lambda}{\sqrt{1+\varepsilon^2}})$"

c) IN FACT, TOTAL WEAK AND E-M CURRENTS

PROBABLY CORRESPOND TO GENERATORS OF $SU_2 \times U_1$,

(AS TODAY)

d) MOST LIKELY NEUTRAL WEAK INTERACTION

THROUGH INTERMEDIATE BOSON Z^0

(AS TODAY)

BUT IF $SU_2 \times U_1$ WHY NO APPRECIABLE

STRANGENESS-CHANGING NEUTRAL CURRENT?

[NOT UNDERSTOOD UNTIL 1970]

e) Z^0 AND γ RESULT OF MIXTURE USING WEAK ANGLE (AS TODAY)

f) Z^0 AND $X^{\pm}$ MASSES $\sim$ 100 GeV. (AS TODAY)

g) $SU_2 \times U_1$ THOUGHT TO BE BROKEN YANG-MILLS THEORY
(AS TODAY)
[BUT SOFT BREAKING THAT GIVES FERMION AND BOSON MASSES
NOT UNDERSTOOD]
[AND NEUTRAL CURRENT PROBLEM $\Rightarrow$ TROUBLE]

h) IS $SU_2 \times U_1$ EMBEDDED IN A LARGER
WEAK AND ELECTROMAGNETIC GAUGE GROUP
INCLUDING HEAVIER BOSONS?
[STILL NOT REALLY KNOWN]

PROGRESS 1963-1967

A) QUARKS AND GLUONS

1) QUARKS SUGGESTED:

"FLAVORS"	u	d	s
Q	$+\frac{2}{3}$	$-\frac{1}{3}$	$-\frac{1}{3}$

2) a) STATISTICS PECULIAR, BOSE-LIKE FOR 3 QUARKS
LIKE "PARAFERMIONS OF RANK 3"
BUT WHERE ARE THE BARYONS WITH PECULIAR STATISTICS

THAT WOULD RESULT?

b) UNDERSTOOD MUCH LATER (1971)
THAT PARAQUARKS WITH PARABARYONS, ETC. SUPPRESSED
EQUIVALENT TO COLOR WITH COLORED HADRONS SUPPRESSED
AND THAT QUARK CONFINEMENT IS RELATED
TO COLOR SUPPRESSION

3) a) CONFINEMENT ("MATHEMATICAL QUARKS")
AS OPPOSED TO EMERGENCE OF SINGLE QUARKS ("REAL QUARKS")

b) "MATHEMATICAL QUARKS" DEFINED FIRST (1963) AS LIMIT
AS QUARK MASS $\rightarrow \infty$
THEN (1965-66) AS LIMIT OF INFINITE POTENTIAL BARRIER
(AS TODAY)

c) IF QUARKS NEVER EMERGE AS "REAL" PARTICLES,
THEN COMPATIBLE WITH "NUCLEAR DEMOCRACY"

4) EASILY UNDERSTOOD (WITH PECULIAR STATISTICS):
$\underset{\sim}{8}(\frac{1}{2}^+)$ AND $\underset{\sim}{10}$ $(\frac{3}{2}^+)$
WITH P-STATES ABOVE, ETC.
FOR BARYONS, AND
$\underset{\sim}{8}$ AND $\underset{\sim}{1}$ (0^- AND 1^-)
WITH P-STATES ABOVE, ETC.
FOR MESONS

5) a) MANY PROPERTIES ABSTRACTED INITIALLY

FROM THEORY OF QUARKS

COUPLED TO SINGLE NEUTRAL VECTOR GLUON

(KEEP PHEASANT, THROW AWAY VEAL)

[NOT YET YANG-MILLS COLOR OCTET

OF NEUTRAL VECTOR GLUONS]

b) VECTOR CURRENT CHARGES GENERATE SU_3 OF FLAVOR;

MASSES GIVE OCTET BREAKING

c) CHIRAL $SU_3 \times SU_3$

OF VECTOR AND AXIAL VECTOR CHARGES BROKEN BY MASS TERMS

WITH RIGHT BEHAVIOR $(\underset{\sim}{3},\underset{\sim}{\bar{3}})$ AND $(\underset{\sim}{\bar{3}}, \underset{\sim}{3})$

[BUT NOT YET UNDERSTOOD WHY SOFT]

d) WEAK CURRENT NOW

$\bar{\nu}_e\gamma_\alpha(1+\gamma_5)e + \bar{\nu}_\mu\gamma_\alpha(1+\gamma_5)\mu + \bar{u}\gamma_\alpha(1+\gamma_5)d'$

$d' = d\cos\vartheta + s\sin\vartheta$

ELECTROMAGNETIC CURRENT NOW OF COURSE

$\frac{2}{3}\bar{u}\gamma_\alpha u - \frac{1}{3}\bar{d}\gamma_\alpha d - \frac{1}{3}\bar{s}\gamma_\alpha s$

e) CURRENT ALGEBRA FOR

VECTOR AND AXIAL VECTOR CHARGE DENSITIES

AT EQUAL TIMES.

f) CURRENT ALGEBRA

ON LIGHT PLANE

OR AT INFINITE MOMENTUM

ESPECIALLY USEFUL

g) AS MOMENTUM $P_z \to \infty$

"GOOD COMPONENTS" OF CURRENTS ARE $O(1)$,

"BAD COMPONENTS" ARE $O(P_z^{-1})$;

"GOOD-GOOD" AND

"GOOD-BAD" RELATIONS

ARE USEFUL ONES.

h) IF TENSOR CURRENTS INTRODUCED AS WELL,

"GOOD COMPONENTS" FORM ALGEBRA $[SU_6]_W$.

i) CURRENT ALGEBRA SUM RULES

TOGETHER WITH PCAC

GIVE FORMULAE

VERIFIED BY EXPERIMENT

6) 1964 (B.J. BJORKEN & S. GLASHOW): CHARM SUGGESTED;

ADD $+\frac{2}{3}\bar{c}\gamma_\alpha c$ TO ELECTROMAGNETIC CURRENT

ADD $\bar{c}\gamma_\alpha(1+\gamma_5)s'$ TO WEAK CURRENT

$s' = s\cos\vartheta - d\sin\vartheta$

[FITS BEAUTIFULLY WITH $SU_2 \times U_1$ YANG-MILLS THEORY

OF WEAK AND ELECTROMAGNETIC CURRENTS

AND SOLVES PROBLEM OF NEUTRAL CURRENT

HAVING NO APPRECIABLE STRANGENESS-CHANGING TERM,

BUT THOSE CONSEQUENCES NOT NOTICED UNTIL 1970]

7) SU_3 COLOR OCTET SUGGESTED FOR GLUONS,

GIVES RIGHT KIND OF FORCE

FOR LIGHT $q\bar{q}$ MESONS

TO BE COLOR SINGLETS.

B) WEAK INTERACTION:

1) IN $SU_2 \times U_1$ YANG-MILLS THEORY
OF WEAK AND ELECTROMAGNETIC INTERACTIONS
SYMMETRY BREAKING BY "HIGGS" BOSONS SUGGESTED
[CONJECTURED THAT THIS WOULD GIVE RENORMALIZABILITY]

2) AT LEAST FOR LEPTONS,
MASSES WOULD ARISE
FROM COUPLING TO "HIGGS" BOSONS.
[SAME IDEA WOULD WORK FOR QUARK BARE MASSES.
BUT NOT YET UNDERSTOOD
HOW CHARM WOULD CURE THE PROBLEM
OF STRANGENESS-CHANGING NEUTRAL CURRENT,
SO IDEA NOT YET APPLIED TO QUARKS.]

C) EXPERIMENTAL DISCOVERY
OF CP VIOLATION
MEANS K_1^0 AND K_2^0 SLIGHTLY MIXED
[BEST DESCRIPTION TODAY
INVOLVES THIRD FERMION FAMILY,
NOT FOUND UNTIL MUCH LATER.]

PROGRESS 1968-1973

I) SCALING BEHAVIOR, ESPECIALLY IN
DEEP INELASTIC $e - N$ SCATTERING

A) CURRENT ALGEBRA COMMUTATION RULES

FOR VECTOR AND AXIAL VECTOR CURRENTS
AT EQUAL TIMES
OR ON LIGHT PLANE
GENERALIZED TO NEW APPROXIMATE RULES
THAT TURN OUT LATER
TO BE LIGHT-CONE COMMUTATION RULES
EXACTLY VALID ONLY FOR FREE QUARKS.

B) THESE LEAD TO PREDICTION
OF SCALING BEHAVIOR
IN DEEP INELASTIC SCATTERING
OF ELECTRONS
ON PROTONS
AND ON NEUTRONS IN NUCLEI
IN AGREEMENT WITH EXPERIMENT

C) THE RULES ARE INTERPRETED
IN THE APPROXIMATE "QUARK-PARTON" MODEL
AS MEANING THAT QUARKS ACT NEARLY FREE
IN THE DEEP INTERIOR OF THE NUCLEON.

D) NEW APPLICATIONS OF THIS APPROXIMATION
ARE FOUND IN OTHER REACTIONS
THAT PROBE THE BEHAVIOR OF QUARKS
UNDER THESE CONDITIONS
INCLUDING MUON PAIR PRODUCTION
AND "INCLUSIVE" HADRON REACTIONS.

E) IN EXPERIMENTS THAT STUDY

THE INTERACTION OF QUARKS
WHEN THEY ARE CLOSE TOGETHER
THEIR INTERACTION IS STRONG

II) ELECTRO-WEAK THEORY

A) CHARM SEEN TO SOLVE THE PUZZLES
OF RECONCILING THE OLD
ELECTRO-WEAK YANG-MILLS THEORY
BASED ON $SU_2 \times U_1$
WITH THE ABSENCE OF A
STRANGENESS-CHANGING NEUTRAL CURRENT
AND OF A
$|\Delta S| = 2$ NON-LEPTONIC WEAK INTERACTION.

B) THUS THE $SU_2 \times U_1$ THEORY
CAN BE APPLIED TO QUARKS
AS WELL AS LEPTONS
AND ITS NEUTRAL CURRENT PREDICTIONS
CAN BE TAKEN SERIOUSLY.

C) THE "HIGGS" MECHANISM
THAT GIVES MASSES
TO THE INTERMEDIATE BOSONS
AND THE LEPTONS
CAN NOW BE APPLIED
TO THE BARE QUARK MASSES AS WELL.

D) THE ELECTRO-WEAK THEORY

REMAINS, HOWEVER,
A THEORY THAT MERELY MIXES
ELECTROMAGNETISM AND THE WEAK INTERACTION;
IT HAS TWO SEPARATE COUPLING CONSTANTS
AND DOES NOT UNIFY THE INTERACTIONS.

E) IT REMAINS TO PROVE
THAT THE "HIGGS" MECHANISM
GIVES SOFT ENOUGH MASSES
TO MAKE THE ELECTRO-WEAK THEORY
RENORMALIZABLE.
THIS IS DONE IN EUROPE AROUND 1971.

III) QUANTUM CHROMODYNAMICS(QCD)

A) PARASTATISTICS FOR QUARKS
WITH PARAHADRONS FORBIDDEN
SHOWN TO BE EQUIVALENT TO
COLOR FOR QUARKS
WITH COLORED HADRONS FORBIDDEN

B) COLOR CONFINEMENT THUS SUGGESTED:
COLORED QUARKS AND GLUONS
FORM ONLY COLOR SINGLET HADRONS.

C) SU_3 COLOR YANG-MILLS THEORY SUGGESTED
FOR GLUON COLOR OCTET
WITH EXACT SU_3 GAUGE SYMMETRY.
RESULTING "QCD" IS RENORMALIZABLE

D) ASYMPTOTIC FREEDOM POINTED OUT
FOR EXACT SU_3 YANG-MILLS THEORY

E) APPROXIMATE "QUARK PARTON" BEHAVIOR
OF HADRONS
THUS EXPLAINED.

F) INCREASE OF EFFECTIVE COUPLING CONSTANT
WITH INCREASING DISTANCE
CONJECTURED TO CONTINUE TO ∞
SO AS TO GIVE DYNAMICAL CONFINEMENT
OF QUARKS AND GLUONS

G) WORRY ABOUT EXACT SCALE INVARIANCE
OF QUARK-GLUON THEORY
IN LIMIT OF VANISHING QUARK MASSES
SETTLED BY UNDERSTANDING
OF ANOMALOUS VIOLATION
OF SCALING SYMMETRY
AND DIMENSIONAL TRANSMUTATION.

H) THUS COUPLING CONSTANT OF QCD
EFFECTIVELY REPLACED
BY A CHARACTERISTIC MASS
OF THE ORDER OF 100–200 MeV
THAT DETERMINES THE ENERGY SCALE
OF THE THEORY
IN THE LIMIT

OF VANISHING BARE QUARK MASSES.

I) IT IS NOT YET UNDERSTOOD
WHY THOSE BARE MASSES
ARE OF THE SAME GENERAL SIZE
AS THE CHARACTERISTIC MASS OF QCD
(FOR EXAMPLE, WHY THE K-MESON MASS
IS SIMILAR TO THE PROTON MASS).

IV) STANDARD MODEL

A) A RENORMALIZABLE YANG-MILLS THEORY
BASED ON $SU_3 \times SU_2 \times U_1$
NOW AVAILABLE
FOR THE STRONG, ELECTROMAGNETIC, AND WEAK INTERACTIONS
WITH MANY ARBITRARY PARAMETERS
ESPECIALLY FOR THE "HIGGS" BOSONS
AND THEIR COUPLINGS.

B) THE THEORY DOES NOT UNIFY
ANY OF THESE INTERACTIONS,
MERELY MIXING SU_2 AND U_1,
AND IT FAILS TO EXPLAIN
MANY ORDER-OF-MAGNITUDE AGREEMENTS
OF PARAMETERS.

C) IT DOES, HOWEVER, AGREE
WITH ALL WELL-ESTABLISHED EXPERIMENTAL FACTS
TO THIS DAY

AND PREDICTED THE INTERMEDIATE BOSONS
AND THEIR PROPERTIES
AS WELL AS DETAILS OF
DEEP INELASTIC SCATTERING EXPERIMENTS
INVOLVING HADRONS
(INCLUDING QUARK AND GLUON JETS, ETC.)

D) THE SPIN $\frac{1}{2}$ FERMION "FAMILIES"
ν_e, e^-, u, d AND ν_μ, μ^-, c, s
CONTRIBUTE CANCELLING ANOMALIES
SO THAT THE THEORY
REMAINS CONSISTENT AND RENORMALIZABLE.

E) THE SAME FAMILIES HAVE
THE SUM OF THE ELECTRICAL CHARGES
EQUAL TO ZERO,
SO THAT THE ELECTRIC CHARGE
COULD BE A GENERATOR
OF A SIMPLE GROUP
IN A FUTURE
UNIFIED YANG-MILLS THEORY.

F) THE STANDARD MODEL
OBEYS MANY SELECTION RULES
INCLUDING BARYON CONSERVATION
THAT COULD BE VIOLATED
IN A UNIFIED YANG-MILLS THEORY.

www.ingramcontent.com/pod-product-compliance
Lightning Source LLC
Chambersburg PA
CBHW082103220526
45472CB00009B/2021
* 9 7 8 9 8 1 4 2 6 1 6 2 3 *